Physiological Responses to Abiotic and Biotic Stress in Forest Trees

Physiological Responses to Abiotic and Biotic Stress in Forest Trees

Special Issue Editors

Heinz Rennenberg
Andrea Polle

MDPI • Basel • Beijing • Wuhan • Barcelona • Belgrade

Special Issue Editors

Heinz Rennenberg
Institut für Forstwissenschaften
Professur für Baumphysiologie
Georges-Köhler Allee
Germany

Andrea Polle
Forstbotanik und Baumphysiologie,
Büsgen-Institut, Georg-August
Universität Göttingen
Germany

Editorial Office
MDPI
St. Alban-Anlage 66
4052 Basel, Switzerland

This is a reprint of articles from the Special Issue published online in the open access journal *Forests* (ISSN 1999-4907) from 2018 to 2019 (available at: https://www.mdpi.com/journal/forests/special_issues/Abiotic_and_Biotic_Stress_in_Forest_Trees)

For citation purposes, cite each article independently as indicated on the article page online and as indicated below:

LastName, A.A.; LastName, B.B.; LastName, C.C. Article Title. *Journal Name* **Year**, *Article Number*, Page Range.

ISBN 978-3-03921-514-0 (Pbk)
ISBN 978-3-03921-515-7 (PDF)

Cover image courtesy of Chen Shaoliang.

Contents

About the Special Issue Editors

Heinz Rennenberg is a German professor whose research is in tree physiology (PhD, University of Cologne, 1977). He has research stations in Cologne, the DOE Plant Research Laboratory at Michigan State University in East Lansing (USA), Groningen (Netherlands), Fraunhofer Institute for Institute of Atmospheric Environmental Research, and the University of Freiburg, where he headed the Department for Tree Physiology from 1992 to 2017. He has been a member of the Deutsche Akademie der Naturforscher Leopoldina since 2004.

Andrea Polle is a German professor whose research is in forest botany and tree physiology. After studying biology and biophysics at the universities of Cologne and Osnabrück (PhD, 1986), she took a position as a researcher with the Fraunhofer Institute for Atmospheric Environment and the University of Freiburg. Since 1996, she has been the Head of the Department of Forest Botany and Tree Physiology and the director of the Forest Botanical Garden at the University of Göttingen. She has been a member of the Academy of Sciences (Göttingen) since 2006.

Editorial

Physiological Responses to Abiotic and Biotic Stress in Forest Trees

Andrea Polle [1],* and Heinz Rennenberg [2],*

[1]   Forstbotanik und Baumphysiologie, Büsgen-Institut, Georg-August Universität Göttingen, Büsgenweg 2, 37077 Göttingen, Germany

[2]   University of Freiburg, Institute of Forest Sciences, Chair of Tree Physiology, Georges-Köhler Allee, Geb. 53/54, 79085 Freiburg, Germany

*   Correspondence: apolle@gwdg.de (A.P.); heinz.rennenberg@ctp.uni-freiburg.de (H.R.)

Received: 15 August 2019; Accepted: 19 August 2019; Published: 21 August 2019

Abstract: Forests fulfill important ecological functions by sustaining nutrient cycles and providing habitats for a multitude of organisms. They further deliver ecosystem services such as carbon storage, protection from erosion, and wood as an important commodity. Trees have to cope in their environment with a multitude of natural and anthropogenic forms of stress. Resilience and resistance mechanisms to biotic and abiotic stresses are of special importance for long-lived tree species. Since trees exist for many decades or even centuries on the same spot, they have to acclimate their growth and reproduction to constantly changing atmospheric and pedospheric conditions. In this special issue, we invited contributions addressing the physiological responses of forest trees to a wide array of different stress factors. Among the eighteen papers published, seventeen covered drought or salt stress as major environmental cues, highlighting the relevance of this topic in times of climate change. Only one paper studied cold stress [1]. The dominance of drought and salt stress studies underpins the need to understand tree responses to these environmental threats from the molecular to the ecophysiological level. The papers contributing to this Special Issue cover these scientific aspects in different areas of the globe and encompass conifers as well as broadleaf tree species. In addition, two studies deal with bamboo (*Phyllostachys* sp., [1,2]). Bamboo, although botanically belonging to grasses, was included because its ecological functions and applications are similar to those of trees.

---

1. Drought Has Multifaceted Consequences That Impede Tree Performance

Several studies in this issue addressed traits of conifers experiencing large variations in environmental growth conditions. Fotelli et al. [3] assessed the physiological plasticity of Aleppo pine (*Pinus halepensis*), an important species in the Mediterranean area, to cope with seasonal changes in environmental conditions, especially summer drought. They found that Aleppo pine is an isohydric species, displaying a drought avoidance strategy. However, Lin et al. [4], using *Pinus massoniana*, showed significant acclimatory changes in needle carbohydrates after long-term exclusion of precipitation. This finding supports an induction of tolerance mechanisms (instead of avoidance) to prevent excessive water loss. Fan et al. [5] studied the physiology of *Pinus koraiensis*, a keystone species of temperate mountain forests, along an altitudinal gradient in China, where precipitation was among the strongest drivers for sapling growth. Interestingly, the saplings showed better growth in mixed than in pure *Pinus koraiensis* forests, suggesting facilitation effects, which obviously were not overruled by other fluctuating environmental constraints. Along similar lines, Magh et al. [6] investigated nitrogen nutrition of European beech (*Fagus sylvatica*) in pure and mixed stands with silver fir (*Abies alba*). Fir needles contained less N than beech leaves, which may dampen the competition for nitrogen in mixed stands. When soil N was low, beech benefited from the interaction with fir but not at high soil N. Together, these studies highlight the importance of distinguishing different evolutionary strategies coping with divergent nutrient and water availability and underpin that the responses of trees are context dependent.

A major issue of growing concern is the spread of pathogens in stressed forests. Therefore, the paper of Terhonen et al. [7] is particularly timely. They reported that water-deficient spruce (*Picea abies*) shows stronger disease symptoms and stronger growth suppression when exposed to *Heterobasidion* species than well-watered plants. It is, thus, urgently required to develop protective measures against forest pests.

Sudden tree death is a significant threat in many areas of the world [8]. The most likely reasons for sudden tree death are carbohydrate depletion or hydraulic failure. Here, Eckert et al. [9] provide an overview on how hardwoods acclimate their hydraulic system to cope with environmental stresses. They explain the production of basic wood structures, focusing mainly on molecular regulation, and compile information on how these structures are influenced to maintain water flow under environmental constraints. Their review is of interest for researchers who wish to obtain a glimpse into the complex regulation of wood production. Sun et al. [10] add an ecological perspective to this important topic. They studied ecophysiological markers such as ring width, $\delta^{13}C$ isotope ratios for water availability, and intrinsic water use efficiency in the wood of poplar (*Populus simonii*) to find early bio-indicators for trees that would succumb in the future. In a fragile ecosystem of the northern Chinese shelterbelt, they classified trees as dead, dying, and not affected. They report that already almost two decades before death, tree rings become smaller and $\delta^{13}C$ higher, suggesting that these traits can be used as earlier warning symptoms. Based on such traits, it is hoped that future genetic studies can develop markers that may allow the selection of resistant trees for reforestation programs in areas with strong tree decline.

2. Salinity and Combined Stresses: From Soil Amendment to Redox Balance

Salinity causes osmotic stress and, in this regard, some similarities with drought stress exist [11]. However, salinity also acts via ionic stress and, therefore, drought and salt responses are only partly overlapping [12]. Salt stressed plants often exhibit enhanced concentrations of reactive oxygen species (ROS) [13]. An overabundance of ROS, which cannot be compensated by antioxidative systems, leads to damage symptoms such as chlorophyll degradation, membrane leakage, and eventually necrosis [13]. Some examples for salt damage symptoms at the tissue and organelle levels are also demonstrated in this Special Issue [14,15]. Since soil degradation with enhanced salinization is an increasing problem, identification of salt-tolerant plant species and measures to enhance the resistance of crops and horticultural plants are urgently needed. In this Special Issue, several papers are devoted to improving salt tolerance.

Geng et al. [16] present a study on *Osmanthus fragans* (sweet olive), an ornamental plant, which has important commercial applications (e.g., for perfume production). They showed that moderate doses of γ-radiation of seeds have a long-lasting effect on the salt tolerance of seedlings that went along with a generally enhanced level of antioxidative enzyme activities and reduced superoxide accumulation. Under approximately 80 mM salt, the injury index of non-treated seedlings was twice that of γ-radiated ones, suggesting that the vitality of this horticultural species can be enhanced to cope with moderate salt stress. Further studies are required to elucidate how γ-radiation leads to this interesting effect.

Hornbeam (*Carpinus turczaninowii*) is known for its beautiful autumn colors and fine-textured wood. Zhou et al. [15,17] systematically tested the performance of this tree species as well as its European counterpart *Carpinus betulus* under moderate salt stress (up to 85 mM). They found that antioxidant systems of the European species collapsed after about 1 month and that of the Asian species after about 2 months in the presence of low salinity levels (30 to 50 mM). The plants showed distinct damage symptoms under these conditions, supporting the conclusion that both species are relatively salt sensitive. This finding calls for further research because *C. turczaninowii* is relatively drought tolerant, suggesting that the genetic basis for drought and salt tolerance diverged in this species, making it an interesting model to dissect drought and salt adaptation.

Soil amendments with hydrogels have a great potential to enhance plant performance under osmotic and ionic stress [18]. In this Special Issue, Li et al. [19] tested synthetic hydrogels and glucomannan-based biopolymer additions to salinized or dehydrated soils, in which *Metasequoia glyptostroboides* (dawn redwood) was grown. While the effects of hydrogels on growth rescue were moderate for single stress factors, all tested compounds had positive effects when drought and salinity occurred together, indicating that the performance of redwood, which is an endangered species, can be improved by retention of salt ions and better water provision.

Chemical treatments can also improve salt tolerance. Here, Rao et al. [20] showed that pre-exposure of *Populus euphratica* to salicylic acid (0.4 mM) enhanced the performance of plants under subsequent massive salt stress (300 mM). The beneficial effect was concentration dependent and disappeared in plants exposed to 1 mM salicylic acid. In another study, *Robinia pseudoacacia* seeds or seedlings were treated with 24-epibrassinolide (a brassinosteroid) and subsequently exhibited enhanced antioxidative protection and enhanced salt tolerance [14]. To obtain deeper insights into the signal transduction pathway activated by stress, the *TCP15* transcription factor (*TEOSINTE BRANCHED1, CYCLOIDEA, and PROLIFERATION CELL FACTOR*) was isolated from *Fraxinus mandshurica* (Mandchurian ash) [21]. TCP15 transient overexpression activated many downstream responses, for example, transcripts encoding antioxidative proteins and hormonal signals [21]. Along similar lines, Yuan et al. [22] tested the response of glutaredoxin *SRGRX1* of rubber (*Hevea brasiliensis*) to a large array of different hormones and stresses. Glutaredoxins are important modulators of the cellular redox state. The rubber SRGRX1 was quickly activated by ROS as well as by abscisic acid (ABA) and salicylic acid, indicating the need for redox regulation under stress [22]. These results underpin the importance of plant hormones for activating defense systems and the need for a better understanding of signaling pathways. However, for practical applications, the trees from current laboratory studies have to be transferred to the field and tested for the stability of the modified traits under natural conditions. A recent example for ABA-related transgenic trees shows that this may lead to unexpected results and novel insights into tree acclimation to their fluctuating environment [23].

3. Conclusions and Outlook

Overall, this issue covers an impressive range of trees species and their response to salinity or drought at different scales from ecophysiology to molecular mechanisms. We hope that efforts such as the current Special Issue can be used to identify similarities and divergences of stress responses, because in-depth knowledge on basal stress pathways can be exploited to develop protection strategies for trees on salt- or drought-affected soils. Furthermore, the distinctive responses of evolutionary stress-adapted tree species hold great promise for implementing specific protection measures in important crop trees.

Conflicts of Interest: The authors declare no conflict of interest.

References

1. Qian, Z.Z.; Zhuang, S.Y.; Li, Q.; Gui, R.Y. Soil Silicon Amendment Increases *Phyllostachys praecox* Cold Tolerance in a Pot Experiment. *Forests* **2019**, *10*, 405. [CrossRef]
2. Jing, X.; Cai, C.; Fan, S.; Wang, L.; Zeng, X. Spatial and Temporal Calcium Signaling and Its Physiological Effects in Moso Bamboo under Drought Stress. *Forests* **2019**, *10*, 224. [CrossRef]
3. Fotelli, M.N.; Korakaki, E.; Paparrizos, S.A.; Radoglou, K.; Awada, T.; Matzarakis, A. Environmental Controls on the Seasonal Variation in Gas Exchange and Water Balance in a Near-Coastal Mediterranean *Pinus halepensis* Forest. *Forests* **2019**, *10*, 313. [CrossRef]
4. Lin, T.; Zheng, H.; Huang, Z.; Wang, J.; Zhu, J. Non-Structural Carbohydrate Dynamics in Leaves and Branches of *Pinus massoniana* (Lamb.) Following 3-Year Rainfall Exclusion. *Forests* **2018**, *9*, 315. [CrossRef]
5. Fan, Y.; Moser, W.K.; Cheng, Y. Growth and Needle Properties of Young *Pinus koraiensis* Sieb. et Zucc. Trees across an Elevational Gradient. *Forests* **2019**, *10*, 54. [CrossRef]

6.   Magh, R.-K.; Yang, F.; Rehschuh, S.; Burger, M.; Dannenmann, M.; Pena, R.; Burzlaff, T.; Ivanković, M.; Rennenberg, H. Nitrogen Nutrition of European Beech Is Maintained at Sufficient Water Supply in Mixed Beech-Fir Stands. *Forests* **2018**, *9*, 733. [CrossRef]

7.   Terhonen, E.; Langer, G.J.; Bußkamp, J.; Răscuţoi, D.R.; Blumenstein, K. Low Water Availability Increases Necrosis in *Picea abies* after Artificial Inoculation with Fungal Root Rot *Pathogens Heterobasidion parviporum* and *Heterobasidion annosum*. *Forests* **2019**, *10*, 55. [CrossRef]

8.   Adams, H.D.; Zeppel, M.J.B.; Anderegg, W.R.L.; Hartmann, H.; Landhäusser, S.M.; Tissue, D.T.; Huxman, T.E.; Hudson, P.J.; Franz, T.E.; Allen, C.D.; et al. A multi-species synthesis of physiological mechanisms in drought-induced tree mortality. *Nat. Ecol. Evol.* **2017**, *1*, 1285–1291. [CrossRef]

9.   Eckert, C.; Sharmin, S.; Kogel, A.; Yu, D.; Kins, L.; Strijkstra, G.; Polle, A. What Makes the Wood? Exploring the Molecular Mechanisms of Xylem Acclimation in Hardwoods to an Ever-Changing Environment. *Forests* **2019**, *10*, 358. [CrossRef]

10.   Sun, S.; Qiu, L.; He, C.; Li, C.; Zhang, J.; Meng, P. Drought-Affected *Populus simonii* Carr. Show Lower Growth and Long-Term Increases in Intrinsic Water-Use Efficiency Prior to Tree Mortality. *Forests* **2018**, *9*, 564. [CrossRef]

11.   Polle, A.; Chen, S.; Eckert, C.; Harfouche, A. Engineering Drought Resistance in Forest Trees. *Front. Plant Sci.* **2018**, *9*, 1875. [CrossRef]

12.   Rizhsky, L.; Liang, H.; Shuman, J.; Shulaev, V.; Davletova, S.; Mittler, R. When Defense Pathways Collide. The Response of Arabidopsis to a Combination of Drought and Heat Stress. *Plant Physiol.* **2004**, *134*, 1683–1696. [CrossRef] [PubMed]

13.   Polle, A.; Chen, S. On the salty side of life: molecular, physiological and anatomical adaptation and acclimation of trees to extreme habitats. *Plant Cell Environ.* **2015**, *38*, 1794–1816. [CrossRef] [PubMed]

14.   Yue, J.; Fu, Z.; Zhang, L.; Zhang, Z.; Zhang, J. The Positive Effect of Different 24-epiBL Pretreatments on Salinity Tolerance in *Robinia pseudoacacia* L. Seedlings. *Forests* **2019**, *10*, 4. [CrossRef]

15.   Zhou, Q.; Zhu, Z.; Shi, M.; Cheng, L. Growth and Physicochemical Changes of *Carpinus betulus* L. Influenced by Salinity Treatments. *Forests* **2018**, *9*, 354. [CrossRef]

16.   Geng, X.; Zhang, Y.; Wang, L.; Yang, X. Pretreatment with High-Dose Gamma Irradiation on Seeds Enhances the Tolerance of Sweet Osmanthus Seedlings to Salinity Stress. *Forests* **2019**, *10*, 406. [CrossRef]

17.   Zhou, Q.; Shi, M.; Zhu, Z.; Cheng, L. Ecophysiological Responses of *Carpinus turczaninowii* L. to Various Salinity Treatments. *Forests* **2019**, *10*, 96. [CrossRef]

18.   Chen, S.; Hawighorst, P.; Sun, J.; Polle, A. Salt tolerance in Populus: Significance of stress signaling networks, mycorrhization, and soil amendments for cellular and whole-plant nutrition. *Environ. Exp. Bot.* **2014**, *107*, 113–124. [CrossRef]

19.   Li, J.; Ma, X.; Sa, G.; Zhou, D.; Zheng, X.; Zhou, X.; Lu, C.; Lin, S.; Zhao, R.; Chen, S. Natural and Synthetic Hydrophilic Polymers Enhance Salt and Drought Tolerance of *Metasequoia glyptostroboides* Hu and W.C.Cheng Seedlings. *Forests* **2018**, *9*, 643. [CrossRef]

20.   Rao, S.; Du, C.; Li, A.; Xia, X.; Yin, W.; Chen, J. Salicylic Acid Alleviated Salt Damage of *Populus euphratica*: A Physiological and Transcriptomic Analysis. *Forests* **2019**, *10*, 423. [CrossRef]

21.   Liang, N.; Zhan, Y.; Yu, L.; Wang, Z.; Zeng, F. Characteristics and Expression Analysis of FmTCP15 under Abiotic Stresses and Hormones and Interact with DELLA Protein *in Fraxinus mandshurica* Rupr. *Forests* **2019**, *10*, 343. [CrossRef]

22.   Yuan, K.; Guo, X.; Feng, C.; Hu, Y.; Liu, J.; Wang, Z. Identification and Analysis of a CPYC-Type Glutaredoxin Associated with Stress Response in Rubber Trees. *Forests* **2019**, *10*, 158. [CrossRef]

23.   Yu, D.; Wildhagen, H.; Tylewicz, S.; Miskolczi, P.C.; Bhalerao, R.P.; Polle, A. Abscisic acid signalling mediates biomass trade-off and allocation in poplar. *New Phytol.* **2019**, *223*, 1192–1203. [CrossRef] [PubMed]

 *forests*

Review

What Makes the Wood? Exploring the Molecular Mechanisms of Xylem Acclimation in Hardwoods to an Ever-Changing Environment

Christian Eckert *, Shayla Sharmin, Aileen Kogel, Dade Yu, Lisa Kins, Gerrit-Jan Strijkstra and Andrea Polle

Forstbotanik und Baumphysiologie, Georg-August Universität Göttingen, Büsgenweg 2, 37077 Göttingen, Germany; ssharmi@gwdg.de (S.S.); aglusch@gwdg.de (A.K.); dyu@gwdg.de (D.Y.); lkins@gwdg.de (L.K.); gstrijk@gwdg.de (G.-J.S.); apolle@gwdg.de (A.P.)
* Correspondence: eckert5@gwdg.de; Tel.: +49-551-39-33485

Received: 28 February 2019; Accepted: 23 April 2019; Published: 25 April 2019

Abstract: Wood, also designated as secondary xylem, is the major structure that gives trees and other woody plants stability for upright growth and maintains the water supply from the roots to all other plant tissues. Over recent decades, our understanding of the cellular processes of wood formation (xylogenesis) has substantially increased. Plants as sessile organisms face a multitude of abiotic stresses, e.g., heat, drought, salinity and limiting nutrient availability that require them to adjust their wood structure to maintain stability and water conductivity. Because of global climate change, more drastic and sudden changes in temperature and longer periods without precipitation are expected to impact tree productivity in the near future. Thus, it is essential to understand the process of wood formation in trees under stress. Many traits, such as vessel frequency and size, fiber thickness and density change in response to different environmental stimuli. Here, we provide an overview of our current understanding of how abiotic stress factors affect wood formation on the molecular level focussing on the genes that have been identified in these processes.

Keywords: wood formation; abiotic stress; nutrition; gene regulation; tree

---

1. Introduction: The Xylem Keeps the Stream of Life Flowing

The main function of the xylem besides granting plant stability is to ensure long-distance water transport driven by water transpiration from the leaves to the atmosphere [1,2]. Therefore, xylem vessels are connected to each other in order to form a large conduit system throughout the plant. During vessel formation, the cell wall between two adjacent vessel cells is degraded to build a continuum. A measure for the water transport capacity within the xylem is xylem hydraulic conductance, which is, among other factors, determined by the size of the xylem vessels. Vessels with higher diameter facilitate faster water transport, thereby resulting in a higher hydraulic conductance. The volumetric flow rate is proportional to the fourth power of the vessel radius according to the Hagen-Poiseuille law [3,4]. Thus, a small reduction in vessel lumen already results in a large reduction of hydraulic conductivity. A severe danger for plant viability is a disruption of this water flow, called embolism, which primarily occurs at the bordered pits [5,6]. When the conduits are filled with air or vapor instead of xylem sap, cavitation stops further water flux. This process is considered as one of the most life-threatening phenomena for plants [3,7]. Cavitation results in non-functional water conduits, which leads to a reduction in hydraulic conductivity and stomatal closure, thereby reduced photosynthetic activity, which finally can lead to the plant′s death [5,8]. It is therefore crucial for plants to develop mechanisms to prevent cavitation by modulating their xylem structure. The major adjustment is to form vessels with smaller diameter to reduce the risk of cavitation. An additional mechanism is to increase vessel stability, achieved by increasing vessel cell wall thickness [9]. As a result, vessel lumina decrease [9]. To counteract the reduction in hydraulic conductivity per vessel plants can increase the number of vessels in their xylem [9–11]. In this review, we will summarize the current knowledge on the molecular regulation of xylem acclimation in angiosperms with the focus on *Populus* species and identify knowledge gaps.

2. Anatomy and Molecular Biology of Wood Formation

2.1. What Makes the Stem?

Growth depends on the presence of cells, which are able to undergo division and differentiation processes. These cells are designated as stem cells [12]. In plants, four layers of stem cell harboring tissues are present: the root apical meristem, the shoot apical meristem, the vascular meristem (cambium) and the cork cambium. While the apical meristems are primarily responsible for longitudinal growth, which is not within the scope of this review, they also contribute to radial growth to a certain extent during primary growth. During primary growth of the stem, the stem cell layer deriving from the shoot apical meristem is designated as pro-cambium, which gives rise to the proto-xylem [13]. The procambial cells divide asymmetrically either into proto-phloem or proto-xylem precursor cells. Proto-phloem precursor cells further differentiate into sieve elements, phloem fibers, phloem parenchyma cells and companion cells, while proto-xylem precursor cells further diversify into xylem vessels, xylem fibers and ray parenchyma cells. The wood generated by the activity of the pro-cambium is designated as proto-xylem. An excellent overview on wood formation during primary growth has been published by Furuta and colleagues [14].

Wood is produced from the activity of vascular cambium that is composed of meristematic initials, which either differentiate to either phloem or xylem precursor cells. Xylem precursor cells further diversify into vessels, fibers and parenchyma cells in order to build up the xylem in angiosperms (Figure 1).

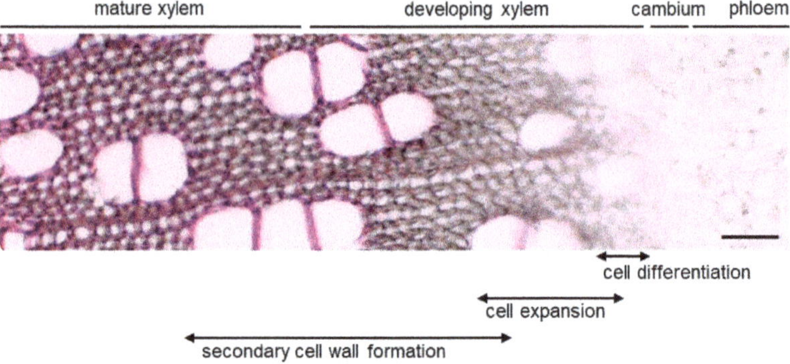

Figure 1. Cross-section of a 3 month old poplar stem illustrating the different steps of wood formation. Cross-section has been stained with Phloroglucinol/HCl. Purple color indicates lignification. Bar = 50 μm.

Secondary growth is what makes up the girth of the stem by generating the majority of the wood. The correct botanical denomination of this secondary wood is meta-xylem. It is also comprised of the three major components, vessels, fibers and parenchyma cells formed by the activity of the cambium. It should be noted that vessels are often surrounded by small living cells, the paratracheal parenchyma, which has important functions in the transport of compounds for example in *Populus tremula* x *tremuloides*, *Fraxinus excelsior* and *Acer pseudoplatanus* [15].

A major difference between the proto- and meta-xylem is the deposition of the secondary cell wall (SCW) compounds leading to structural differences: the SCW of proto-xylem vessels is build up in a helical shape, which allows for continuous growth as the SCW can be elongated during longitudinal growth [16]. The SCW of meta-xylem is deposited directly onto the primary cell wall resulting in a rigid structure around the vessel that is no longer able to expand.

The SCW is a complex structure, which is primarily composed of cellulose, hemicelluloses and lignin with addition of pectin and cell wall proteins. The composition of the SCW varies within the plant kingdom, e.g., in angiosperms SCWs contain 40%–50% cellulose, 20%–30% hemicelluloses and 25%–30% lignin [17]. Cellulose is a linear homopolymer of β-1,4-linked D-anhydroglucose molecules forming cellulose chains of varying length. These cellulose chains are then connected via Van-der-Waals interactions and hydrogen-bonds to form cellulose micro fibrils [18]. In contrast to cellulose, hemicellulose is chemically complex, consisting of polysaccharide heteropolymers of hexose and pentose sugars with a β-1,4-linked backbone of xyloglucans, xylans, mannans and glucomannans with a vast variety of side chains [19]. The main function of hemicellulose is the interconnection of cellulose micro fibrils and lignin [20]. Lignin provides rigidness and recalcitrance to the cell wall, with strong impact on wood properties by adding extra strength and water impermeability [21]. Lignin is built up by three phenylpropanoid compounds, also designated as monolignols: coniferyl alcohol (G), sinapyl alcohol (S) and p-coumaryl alcohol (H) [22] with G and S subunits being the main building blocks for lignin in angiosperms [23,24]. The combination of these monolignols via radical coupling constitutes a complex polymer, the composition of which varies with developmental and environmental stimuli. Interestingly, the lignin composition is also specific for different cell types, e.g., the lignin in fibers differs significantly from those of vessel elements as it shows a higher content of S-subunits [25].

Wood formation is, thus, characterized by a succession of four major steps, including cell division, cell expansion (elongation and radial enlargement), cell wall thickening (involving biosynthesis and deposition of cellulose, hemicelluloses, lignin and cell wall proteins), and finally programmed cell death [26].

2.2. Molecular Mechanism of Wood Formation

Our understanding of the molecular mechanisms of xylogenesis has massively increased in the last two decades. Several detailed reviews about the molecular mechanisms of wood formation have been published [13,16,26,27]. Here, we will shortly summarize the main aspects important for understanding the impact of environmental stimuli that alter wood formation (Figure 2). Wood formation is tightly controlled by two classes of transcription factors (TFs), namely the NAC family and the MYB family. Using *Arabidopsis thaliana* as a model system, several members of these two TF families have been shown to work in a hierarchical manner to regulate the transcription of genes necessary for wood formation [26]. The first level master switches all belong to the class of NAC TFs and their activation is crucial for the cell identity of wood cells. For example, SND1, NST1 and NST2 are responsible for fiber development, while VASCULAR-RELATED NAC DOMAIN 7 (VND7) controls differentiation of the proto-xylem and VASCULAR-RELATED NAC DOMAIN 6 (VND6) of the meta-xylem [28,29]. Several orthologs of these *A. thaliana* genes have been identified in *Populus trichocarpa* (Table 1). It is somewhat confusing that these master switch TFs are not termed VNDs, as in Arabidopsis but were named WOOD-related NAC DOMAIN (WND) transcription factors, yet they also belong to the family of NAC TFs and are closely related to the *A. thaliana* genes [30]. The expression of these master switches is modulated by fine-tuning factors like the HD-Zip transcription factors PtrAtHB.11 and PtrATHB.12 or PtrE2FC.1 [31,32]. The master switch WNDs further activate a group of MYB master switch TFs (PtrMYB2, PtrMYB3, PtrMYB20, PtrMYB21, [33,34]), which either directly regulate the biosynthesis of cellulose, hemicellulose, and lignin or lead to the activation of down-stream MYC and NAC TFs that promote or decrease the expression of cell wall biosynthesis genes (Table 1, Figure 2).

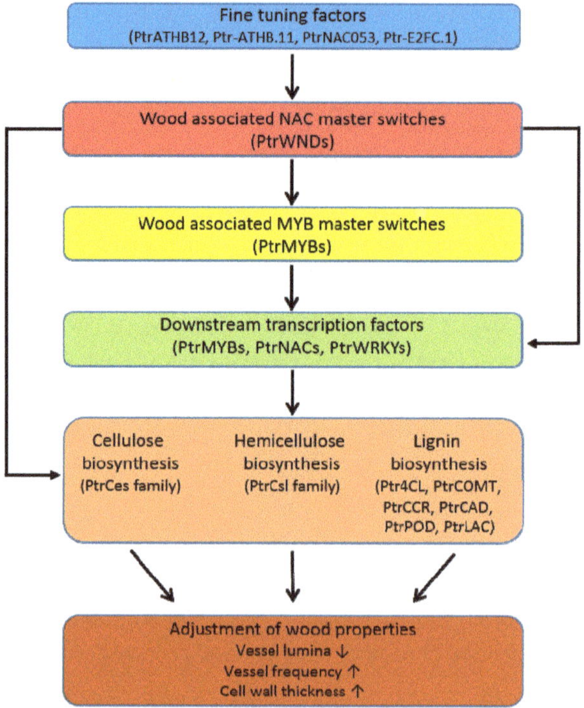

Figure 2. Schematic representation of the molecular mechanisms and gene families leading to xylem formation and acclimation under various stresses.

Cell wall biosynthesis requires the production of cellulose strands by cellulose synthases (CesA), which are homologs of prokaryotic *celA* genes [35] and were firstly characterized in the plant kingdom in *A. thaliana* [35]. The CesA proteins are localized across the plasma membrane using intracellular UDP-glucose to generate cellular strands towards the outside. Meanwhile, a whole *CesA* gene family comprised of 18 genes, of which four show xylem specific expression patterns, was discovered in *P. trichocarpa* [36,37]. A recent study revealed that during xylem formation, two different types of CesA complexes are necessary for cell expansion and cell wall thickening in Arabidopsis, indicating that CesA complex composition is dynamically changing [38]. Since these processes also occur in trees, it is likely that a similar mechanism is also present in woody plants, although there is no experimental evidence up to now.

The cell wall of the meta-xylem in trees is heavily lignified. Lignin biosynthesis is a very complex process starting with aromatic amino acids, mainly phenylalanine [39]. The main classes of dedicated enzymes required to produce lignin building units are 4-coumarate:CoA ligases (4CL), caffeic acid *O*-methyltransferases (COMT), cinnamoyl-CoA reductases (CCR), and cinnamyl alcohol dehydrogenases (CAD) [22]. Monolignols are transported from the cytosol into the cell wall [40], where they are polymerized by peroxidases (POD) and laccases (LAC), yielding lignin. However, only very few monolignol transporters have been identified by now, such as the Arabidopsis p-coumaryl alcohol transporter AtABCG29 [41,42].

As outlined above hemicelluloses are also very complex molecules. Therefore, it is not surprising that a variety of enzymes is necessary for their synthesis. These enzymes are summarized as Cellulose Synthase-Like (CSL) proteins [43], of which 30 have been identified in poplar [37]. Four CSL genes (*PtrCSLA1, PtrCSLA2, PtrCSLA5, PtrCSLD6*) are predominantly expressed in the xylem [37].

Table 1. List of orthologs of *Arabidopsis thaliana* wood-formation associated transcription factors present in *Populus trichocarpa.*

Gene Function	Potri ID	Populus Gene Name	AGI ID	Arabidopsis Gene Name
Fine-tuning factors	Potri.001G188800	*Ptr-ATHB.12*	At1G52150	*AtHB15/AtCNA/AtICU4*
	Potri.003G050100	*Ptr-ATHB.11*	At1G52150	*AtHB15/AtCNA/AtICU4*
	Potri.001G197000		AT3G13890	*AtMYB26*
	no annotated ortholog		At3G32090	*AtWRKY12*
	Potri.002G023400	*Ptr-E2FC.1/E2Fc*	At1G47870	*AtE2Fc*
	Potri.001G061200	*PtrNAC053*	AT5G13180	*AtVNI2/AtANAC083*
Master regulators (NAC)	no annotated ortholog		At1G32770	*AtNST3/AtSND1/AtANAC012*
	Potri.001G448400	*PtrWND1B/NAC063/PtVNS11*	At2G46770	*AtANAC043/AtNST1*
	Potri.002G178700	*PtrWND2B/NAC061/PtVNS10*	AT2G46770	*AtANAC043/AtNST1*
	Potri.011G153300	*PtrWND1A/NAC068/PtVNS12*	AT2G46770	*AtANAC043/AtNST1*
	Potri.014G104800	*PtrWND2A/NAC065/PtVNS09*	AT2G46770	*AtANAC043/AtNST1*
	Potri.015G127400	*PtrWND3A/NAC050/PtVNS05*	AT1G12260	*AtANAC007/AtNAC007/ATVND4*
	Potri.012G126500	*PtrWND3B/NAC037/PtVNS06*	AT1G12260	*AtANAC007/AtNAC007/ATVND4*
	Potri.001G120000	*PtrWND4A/NAC038/PtVNS03*	AT1G12260	*AtANAC007/AtNAC007/ATVND4*
	Potri.003G113000	*PtrWND4B/NAC046/PtVNS04*	AT1G12260	*AtANAC007/AtNAC007/ATVND4*
	Potri.007G014400	*PtrWND5A/NAC025/PtVNS01*	AT2G18060	*AtANAC037/AtVND1*
	Potri.005G116800	*PtrWND5B/NAC039/PtVNS02*	AT2G18060	*AtANAC037/AtVND1*
	Potri.013G113100	*PtrWND6A/NAC055/PtVNS07*	AT1G71930	*AtANAC030/AtVND7*
	Potri.019G083600	*PtrWND6B/NAC060/PtVNS08*	AT1G71930	*AtANAC030/AtVND7*
	Potri.015G002900	*PtrNAC147*	AT1G71930	*AtANAC030/AtVND7*
	no annotated ortholog		At3G61910	*AtNST2*
	Potri.004G107200		At5G62380	*AtVND6*
	Potri.004G107400		At5G62380	*AtVND6*
	Potri.005G082700	*PtrNAC144*	At5G62380	*AtVND6*
	Potri.006G231300		At5G62380	*AtVND6*
	Potri.014G163600		At5G62380	*AtVND6*
	no annotated ortholog		At4G36160	*AtVND2/AtANAC076*
	no annotated ortholog		At5G66300	*AtVND3/AtANAC105*
	no annotated ortholog		At1G62700	*AtVND5*
Second Level Regulators (MYB)	Potri.001G258700	*PtrMYB2*	At5G12870	*AtMYB46*
	Potri.009G053900	*PtrMYB21*	At5G12870	*AtMYB46*
	Potri.001G267300		At3G08500	*AtMYB83*
	Potri.009G061500		At3G08500	*AtMYB83*

Table 1. *Cont.*

Gene Function	Potri ID	Populus Gene Name	AGI ID	Arabidopsis Gene Name
Third Level Regulators	Potri.004G049300		AT4G28500	*AtSND2/AtANAC073*
	Potri.007G135300		AT4G28500	*AtSND2/AtANAC073*
	Potri.011G058400		AT4G28500	*AtSND2/AtANAC073*
	Potri.017G016700		AT4G28500	*AtSND2/AtANAC073*
	no annotated ortholog		AT1G28470	*AtSND3*
	no annotated ortholog		At1G63910	*AtMYB103*
	Potri.002G073500		At1G17950	*AtMYB52*
	Potri.005G186400	*PtrMYB158/PtrMYB.50*	At1G17950	*AtMYB52*
	Potri.007G134500	*PtrMYB161/PtrMYB.43*	At1G17950	*AtMYB52*
	Potri.012G039400	*PtrMYB167/PtrMYB.41*	At1G17950	*AtMYB52*
	Potri.015G033600	*PtrMYB090/PtrMYB.38*	At1G17950	*AtMYB52*
	no annotated ortholog		At3G48920	*AtMYB45*
	no annotated ortholog		At1G16490	*AtMYB58*
	Potri.005G096600		At1G79180	*AtMYB63*
	Potri.007G067600		At1G79180	*AtMYB63*
	Potri.019G118900		At1G79180	*AtMYB63*
	Potri.015G129100		At4G22680	*AtMYB85*
	Potri.001G112200	*PtrKNAT7.1*	At1G62990	*AtKNAT7*

This rough overview on the regulatory mechanisms and main processes that are required for the main cell wall compounds provides only a glimpse into the complexity of the underlying processes. The composition of wood is important for its further use, e.g., as construction material, where durable material is necessary and for paper making, where fibers of a certain length are favorable, or for secondary biofuel production where easily accessible cellulose is needed. Therefore, it is not only important to know how wood is being produced, but also to understand how wood properties are changed in response to environmental cues.

3. Abiotic Stresses Affecting Wood Formation

3.1. Drought Severely Changes Xylem Anatomy

Drought has a major impact on wood and wood formation processes. Since the xylem is the tissue that enables water transport throughout the whole tree, it is pertinent to keep the xylem architecture intact and to acclimate it to changing water supply to prevent embolism [44,45]. The effects of drought on xylogenesis have been studied on several angiosperm species over the last decade. Poplar, in particular, was intensely studied [46,47]. Cambial cell layers are reduced under drought compared to well-watered plants [10,48]. When growth is still possible under water-limited conditions, poplars, regardless of whether they originate from dry or moist habitats, show reduced vessel lumina and an increased number of vessels compared to non-stressed plants (*Populus* x *canescens*, [49]; *Populus euphratica*, [11]; different *Populus nigra* genotypes originating from dry and moist areas, [10]), Figure 3). Moreover, Schreiber and colleagues reported a strong correlation between vessel diameter and cavitation resistance in five hybrid poplar clones in Alberta, Canada [50]. However, these acclimatory anatomical changes are not only confined to water-spending trees species like poplar but also to more drought tolerant species such as oak depending on the level of acclimation (*Quercus pubescens* > *Quercus robur* > *Quercus petraea*, [51]). The importance of this safety strategy is further corroborated by a study on dead trees performed in Italian forests [52]. Here, the authors compared the wood anatomy of dead trees to that of surviving trees of the same age. They found that dead trees had formed wide early vessels and no vessels with reduced lumen during dry periods in contrast to trees that survived this period [52].

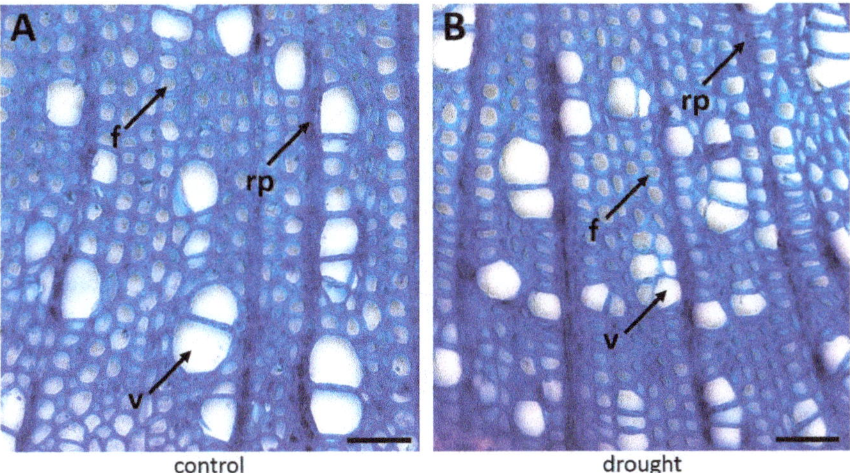

Figure 3. Comparison of non-stressed (**A**) and drought-stressed (**B**) xylem tissue of *Populus tremula* x *tremuloides*. B shows vessels with reduced size but increased vessel frequency. v: vessel cells, f: fiber cells, rp: ray parenchyma. Bar = 50 μm.

Taken together, these studies suggest that angiosperms have developed a common mechanism to acclimate their xylem to increasing drought. The xylem, as stated earlier, is formed by vessels and fibers, with the vessels forming the pipe system that facilitates water transport [26]. Under drought the plant has to fortify the vessels by thickening the secondary cell wall to prevent cavitation, which results in a reduced water transportation rate due to diminished vessel lumina [11]. To compensate the loss in vessel lumen, more vessels are formed to restore the water transportation rate that is necessary for tree growth (Figure 3). Improving the resistance against drought-induced xylem cavitation is a crucial acclimation mechanism of plants to dry environments [8,53]. It is notable that these alterations resemble those of seasonal acclimation of wood to shorter day lengths and cold, which lead to the formation of tree rings. Tree rings are a result of changes in the activity of xylem building processes [54]. Interestingly, Arend and Fromm observed in 2007 that trees responded to drought with smaller vessel lumina only during the main growth season but not in fall, when late wood with small lumina was formed [48]. Still, controlled experiments to disentangle temperature and day lengths effects on wood anatomy and the impact of drought under these conditions are lacking.

3.2. Drought Leads to Major Transcriptional Remodelling

The molecular mechanism behind the reduction of vessel volume and cell wall strengthening has not yet been unravelled. One important tool to tackle this question is the analysis of wood transcriptomes to identify key genes in this acclimation processes. Wildhagen and colleagues (2018) found an altered expression of genes involved in cell wall polysaccharide biosynthesis (e.g., xyloglucan endotransglucosylase/hydrolases, UDP-XYL synthase 6, expansin A) in three poplar genotypes. This finding suggests that denser wood under drought is achieved by regulation of cellulose and hemicellulose biosynthesis; notably the lignin content was unaltered [10]. These changes unexpectedly resulted in a higher saccharification potential of wood formed under drought compared to wood of well-irrigated poplars [10]. If these changes were also responsible for the reduction in vessel volume has not been clarified yet. Vessel expansion depends on the activity of potassium (K^+) uptake transporters (e.g., PtrKUP1) and outward rectifying K^+ channels (e.g., PTORK) [55]. In agreement with a high requirement for K^+ to regulate turgor pressure, the cambium contains the highest K^+ concentrations [56]. Furthermore, KUPs show seasonal regulation with high expression levels in wood [57]. Since drought limits K^+ acquisition from the soil [58], it is possible that changes in the osmotic control influence vessel size. However, this suggestion is currently speculative as an impact of drought on K^+ channel transcription levels has only been reported for an ortholog of the Arabidopsis *KUP2* which is positively correlated with fiber lumen under drought in *P. nigra* [10]. In addition, the concentration of soluble non-structural carbohydrates, sucrose in particular, has a major impact on vessel expansion. It has been shown that sucrose accumulation in the cambium correlates with cambial activity and cell expansion phase during xylogenesis in *Populus* x *canadensis* Mönch 'I-214' [59]. Interestingly, transgenic *Populus tremula* x *alba* suppressing the expression of the vacuolar sucrose efflux transporter PtaSUT4, are more susceptible to moderate drought compared to wild type plants. This suggests that sucrose also plays an important role in drought acclimation in addition to seasonal wood formation [60].

In addition to its influence on wood anatomy, drought has also major impact on the genes putatively involved cell wall biosynthesis. For instance, in *P. trichocarpa*, 119 genes related to cell wall organization, carbohydrate metabolism, and lipid metabolism were significantly differentially expressed. Among these genes four master switch regulators (*PtrWND1*, *PtrWND1B*, *PtrWND2A*, *PtrWND2B*) which are orthologs of the Arabidopsis ANAC043 (Table 1), as well as the secondary level regulators *PtrMYB2*, *PtrMYB21* and Potri.009G061500 (an ortholog of Arabidopsis *AtMYB83*, Table 1) were downregulated [61], indicating major changes in the transcription of cell wall biosynthesis genes. Additionally, 29 laccases-encoding genes were also found to be differentially regulated [61]. Except for one laccase all other laccases were downregulated under drought [61]. These findings corroborate major changes in cell wall composition, though biochemical analyses were not conducted.

It is clear that the supply with energy and precursors of cell wall components, especially carbohydrates, has to be sustained under drought since earlier studies showed strong correlations between growth and cambial carbohydrate concentrations [62]. However, transcriptome analyses also identified many genes involved in lipid metabolism [10,61]. There is now accumulating evidence for the role lipids in SCW formation: metabolomic and transcriptomic analyses revealed links between decreasing concentrations of sterols and fatty acids with increasing transcript levels of genes involved in β-oxidation and the glyoxylate cycle required for the production of carbohydrate precursors for SCW [63,64]. Transcriptome analysis of Eucalyptus xylem revealed a putative UDP-glucose:sterol glucosyltransferase, which produces sitosterol-cellodextrin. This molecule serves as a primer the elongation of β-1,4-glucan strands by the cellulose synthases [63,65]. These lipoglucan primers are discussed as being responsible for elongation by a special CesA isoform [66], forming a possible indirect link to channel carbon from lipids towards cellulose.

Other players, whose role in regulating drought acclimation has not yet been understood, are micro RNAs (miRNAs), which are widely known to function as regulatory molecules in gene expression under stress conditions [67,68]. A recent study identified five up- and seven down-regulated miRNAs in *P. trichocarpa* under drought stress, which in total regulate 72 targets according to degradome analysis. Among these were four UDP-glucosyl transferases, which transfer sugars to target molecules and have been shown to build up β-1,4-glucan from UDP-activated glucose [69,70].

It is important to mention that trees can recover from xylem embolism to a certain extent. To re-establish the water flow in the xylem it is necessary to refill the vessels [45]. Although this process is not yet fully understood, there is substantial evidence that paratracheal parenchyma cells also designated as vessel-associated cells (VACs) play an important role in this process [71]. As stated earlier, VACs are living cells surrounding the xylem vessels, which makes them ideal candidates to facilitate the refilling process [15]. To refill the vessels with water, a change in the osmotic gradient between VACs and vessel cells is indispensable. The most important osmotic compound in this regard is most likely sucrose [72]. Experimental evidence has been provided by mimicking xylem embolism in *P. trichocarpa* [72], which resulted in enhanced expression of α-amylases (*PtrAMY1*, *PtrAMY2*, *PtrAMY3*) leading to a reduction in starch content in the VACs [72]. Simultaneously, the expression of sucrose transporters (*PtrSUT2a* and *PtrSUT2b*) as well as aquaporins (*PtrPIP1.2*, *PtrPIP1.4*, *PtrPIP1.5*, *PtrPIP2.2*) was upregulated. These findings support that upon xylem embolism, starch is degraded to sucrose, which is transported from the VACs to the vessel cells leading to enhanced water uptake of the vessel cells via aquaporins. Upregulation of aquaporins of the PIP1 and PIP2 family in VACs has been observed in several tree species (*P. trichocarpa* [72–74], *Populus* x '*Okanese*' [75], *Juglas regia* [76]), substantiating the hypothesis of VAC-mediated refilling of vessels upon embolism. It is still under discussion as to whether the VACs store sufficient carbohydrates for this process or if they are replenished by carbohydrate export from the phloem during embolism recovery [77].

3.3. Phytohormones Mediate Xylem Changes under Drought

Although transcriptomic studies have strongly enhanced our knowledge on plant drought responses and growth regulation, they cannot elucidate the underlying signals. In this regard phytohormones such as cytokinins, auxin, abscisic acid (ABA) and many others play crucial roles. The major plant hormone mediating drought stress is ABA [78,79]. Although drought has drastic effects on xylem formation, it is still unclear if ABA plays a direct role in regulating wood formation process. Circumstantial evidence suggests a role of ABA in fall for late wood formation because the concentration of ABA in the cambium increases in fall when cells with small lumina and thick cell walls are produced [80]. Reduction of stem radial growth under drought stress is accompanied by a strong increase in cambial ABA concentrations [81]. Therefore, ABA might play a negative role in secondary xylem development similar to that found for its function regulating bud dormancy [82], but no direct evidence on cellular or molecular level has been reported so far. An alternative suggestion was that enhanced ABA concentrations in fall might influence stomatal aperture, thereby leading to a decline in carbon fixation and consequently in a reduction in stem growth [83]. It is also possible that ABA interacts with the activity of auxin inhibited xylem initiation [84]. To understand the effect of ABA on wood formation in different environments it is necessary to get more insight on the molecular basis of wood formation and to identify the genes in ABA signaling which may affect the regulation of wood cell development. The first evidence for this idea was found in Arabidopsis, where ABA-dependent upregulation of the vessel specific master switch genes VND2, VND4 and VND6 was observed resulting in enhanced xylem differentiation [85].

Another important class of phytohormones involved in drought acclimation are cytokinins, trans-zeatin, kinetin and 6-benzylaminopurine in particular. Cytokinins are generally referred to as growth promoting phytohormones by initiating cell division (reviewed by [86]). Together with ethylene and auxin, cytokinins regulate cambial activity. In poplar, overexpression of the Arabidopsis *CYTOKININ OXIDASE 2*, an enzyme involved in cytokinin catabolism, resulted in non-detectable levels of the cytokinin *trans*-zeatin and its storage form zeatin-*O*-glucoside and lead to a reduced number of cambial cells, resulting in impaired wood formation [87]. Moreover, cytokinin activity was high in primary meristems supporting its function in early wood formation processes [88] and reduced in the cambial zone of drought stressed poplar, indicating a regulatory function of cytokinins under drought stress [89]. There is now increasing evidence that cytokinins interact with jasmonic acid. A reverse genetic approach showed that application of 10 μM methyl-jasmonic acid promotes xylem development in Arabidopsis roots [90]. In addition, exogenous application of jasmonic acid reduces the cytokinin responses in regard to xylem formation. A follow-up study provided evidence that drought stress also induces xylem development by increasing jasmonic acid responses and diminishing cytokinin responses [91]. These results imply that cytokinin activity is a negative regulator of xylem development under drought stress while jasmonic acid promotes xylogenesis. The molecular mechanism how cytokinins mediate xylem development is not yet fully understood, but it was shown that external application of 50 ng/mL kinetin downregulates the expression of the vessel specific master switch genes *VND6 and VND7*, which are important for proto- and meta-xylem differentiation [28]. Cytokinins are known to induce cell wall loosening expansion genes [92,93]. Application of 6-benzylaminopurine has a negative effect on cell wall thickness [94]. These findings corroborate the function of cytokinins as negative regulators of secondary xylem, yet there is still plenty of research necessary to identify the role of specific cytokinins in these processes.

Cytokinin levels in plants are affected by nitrogen supply and are implicated in stress responses mboxciteB95-forests-465031,B96-forests-465031. It is possible that they also mediate nitrogen effects in wood. Under nitrogen starvation, wood properties resemble those of wood produced under drought with smaller vessel lumina and thicker fiber cell walls [97]. Under these conditions, secondary cell wall related transcription factors, like members of *MYB*, *bHLH*, *WRKY* and *WD40* gene families, are up-regulated, leading to increased cell wall formation [97–99]. In contrast, under high nitrogen availability the secondary cell walls of fibers and vessels are thinner and the vessel lumina are expanded [97,99,100].

Notably, changes in nitrogen supply affected ABA and JA [101], which are known to mediate stress responses. Synergistic effects of drought stress and nitrogen availability, which both influence wood formation [101–105] have been noted [106–108]. For example, when water and nitrogen are sufficiently available, water supply increases nitrogen uptake, leading to larger vessel lumina and in turn to higher susceptibility to drought stress [109–111]. SCW-related genes are not suppressed by drought under high nitrogen supply, only under nitrogen limitation [112]. Consequently, water and nitrogen deprivation has a negative synergistic effect on secondary growth compared to either limitation alone [112,113]. These examples show that drought stress impacts on xylem development at different levels, clearly as an osmotic stress as demonstrated by PEG studies but also via its influence on tree nutrition. To generate or select more drought-resistant trees, it will be necessary to disentangle the complex hormonal network with emphasis on the mechanism of phytohormone accumulation, the transition of active to inactive forms and the turnover to identify the key mechanisms how water limitation affects xylem anatomy and cell wall composition.

3.4. Salt Severely Affects Wood Formation

Survival and productivity of trees are affected by salt through different processes, mainly related to osmotic stress, nutritional imbalances or ion toxicity [114,115]. High salt concentration within the plant itself can be toxic, resulting in the inhibition of many physiological and biochemical processes. Saline environments restrict the ability of plant to take up water and, therefore, reduce plant growth [116]. Uptake of salt facilitates water uptake and has been identified as a tolerance mechanism in poplars, *P. euphratica* in particular, which are acclimated to salt deserts [117]. Hence, the xylem originating from salinity affected meristem also exhibits modifications in its anatomical and chemical aspects [118]. It has been shown that salt stressed poplars have a reduced number of cell layers in cambial zone compared to control plants, indicating growth reduction [119]. Similar to drought-stressed poplars, vessel lumina of salt-exposed trees are reduced [119,120] and the vessel frequency is increased [121]. The sensitivity of these anatomical alterations to salt exposure depends on the salt-sensitivity of the species. For example, the salt-sensitive poplar species *P.* x *canescens* reacts strongly to moderate NaCl levels, while the salt-tolerant *P. euphratica* shows diminished cambial activity and decreased vessel lumina only after long-term exposure to higher salinity levels, i.e., 150 mM NaCl [120,121]. Among other reasons, diminished nutrient supply, in particular lower calcium and potassium supply caused decreased xylem radial growth under salt stress [119].

Salinity impairs the tree water status and may increase tension in the water-conducting system, promoting cavitation and subsequently, embolisms [118]. However, trees can reinforce the wall strength of conducting cells so as to prevent against vessel collapse under osmotic stress [122]. This phenomenon is also found in poplar vessel cell walls, which show a significant increase in strength when exposed to salinity [120]. Furthermore, increased vessel frequencies can at least partly compensate negative effects on hydraulic efficiency [121]. Cell wall reinforcement under salt stress goes along with drastic alterations in major wood compounds such as cellulose, hemicelluloses, and lignin [121]. The transcriptional changes found under these conditions were opposite to those found during tension wood formation in poplar [121]. Therefore, the type of wood developed under salinity was termed "pressure wood" [121].

Since the anatomical changes in secondary xylem under salt and drought stress might be of similar nature, the question arises if there are similar hormones involved in these processes. The activity of the vascular cambium largely determines the rate of wood formation and is strictly controlled by the interplay of several phytohormones [123]. Early studies demonstrate that a concentration gradient of auxin across the cambial zone regulates cell division and expansion of the xylem elements [124–126]. Osmotic stress in plants is known to influence auxin transport [127]. Using an auxin responsive reporter gene construct (*GH3:GUS*), an impact of salt on auxin activity was shown in poplar wood [128]. Associated with wood alteration, free-auxin levels decrease in developing xylem under salt stress. In contrast to salt susceptible grey poplar (*P.* x *canescens*), the salt tolerant *P. euphratica* shows only little reduction in auxin and moreover, an increase in indole-3-acetic acid-amido conjugates, which can be used as a source of auxin [120]. Overexpression of auxin amidohydrolase (ILL3) from poplar in Arabidopsis rendered the plants more salt tolerant [120], indicating an important function of auxin in salt acclimation. The signaling pathways show how auxin and other plant hormones affect gene regulation remains elusive.

Recent transcriptome comparisons of salt treated and non-stressed eucalypt trees uncovered responses of a large number of transcription factors (TFs) including members of the MYB TF family [129] to salt. Upregulation genes encoding transcription factors (*EgrMYB20*, *EgrMYB47* and *EgrMYB36*) involved as master switches in secondary xylem development were observed in several eucalyptus genotypes tested, in particular under the higher (125 mM) salt treatment [130]. *BplMYB46*, a MYB gene from *Betula platyphylla* (birch), has also been reported to be involved in both abiotic stress tolerance and secondary wall biosynthesis [131]. *BplMYB46* is predominantly expressed in stems and its expression is induced by NaCl [131]. In *BplMYB46* overexpressing lines a positive effect on lignin and cellulose content but a negative effect on hemicellulose content was found. Correspondingly, the expression of genes related to lignin and cellulose biosynthesis (phenylalanine ammonia lyase (PAL), caffeoyl-CoA O-methyltransferase (CCoAOMT), 4-coumarate-coa ligase (4CL), POD, Laccase (LAC), cinnamoyl-CoA reductase (CCR), cellulose synthase (CesA) was significantly upregulated, and that of genes related to hemicellulose biosynthesis (e.g., fragile fiber (FRA) and irregular xylem (IRX)) was significantly down-regulated in *BplMYB46* overexpressing lines. Hence, *BplMYB46* functions in linking stress responses and cell wall properties [131]. Further functional analyses by forward and reverse genetic approaches are required to understand the acclimatory responses of trees to salt stress in more detail. While a number of candidate genes has already been studied for their involvement in enhancing salt tolerance in leaves or roots [47], systematic genetic studies targeting at enhanced xylem salt resistance are lacking.

4. Conclusions

Trees respond to environmental stress factors like drought and salt with the adjustment of xylem formation primarily to maintain the hydraulic function of the wood. These responses are strongly affected by nutrients, especially by major cations (K^+, Ca^{2+}) and by nitrogen availability. Interestingly, the anatomical changes in response to different environmental cues are quite similar and characterized by a reduction in xylem vessel lumina and fortification of secondary cell walls in vessels and fibers. However, the molecular regulations that act between the initial stimulus and the output, i.e., modified wood, are just poorly understood. It has become apparent that phytohormones play a crucial role in mediating those stress stimuli by modulating xylogenesis related genes like the NAC and MYB master switch regulators, which lead to stress specific gene transcription. Yet, this review shows that our picture of the processes leading to wood acclimation is far from being complete. In order to put the puzzle together, it will be important to not only continue investigating specific stimulus responses, but also to integrate the information of all processes to get a complete picture of wood stress acclimation. Transcriptome analyses helped us a lot in getting a better picture of the processes during stress acclimation, yet they mostly focus on gene clusters build by GO term or KEGG analyses. It will be crucial to exploit this powerful resource more in-depth in the future to get to the key genes involved in those regulatory mechanisms. Identification and functional characterization of candidate genes improving stress tolerance and/or growth under nutrient-limiting conditions will promote the generation of new genotypes either by improving trees by genetic engineering or by smart breeding. Significant advances have been made over recent years to build up an effective toolbox for the characterization of candidate genes, especially in poplar. While overexpression of candidate genes has been well established for years, the generation of loss-of-function mutants has always been difficult. However, recent advances in the CRISPR/Cas9 technology have provided researchers with a powerful tool to investigate gene function and to generate genetically modified trees.

Author Contributions: C.E., S.S., L.K., D.Y., G.J.S., A.K. and A.P. wrote the draft manuscript, C.E. and A.P. reviewed, edited and finalized the manuscript.

Funding: We are grateful for financial support in the frame of WATBIO (Development of improved perennial biomass crops for water stressed environments), which is a collaborative research project funded from the European Union's Seventh Programme for research, technological development and demonstration under grant agreement No. 311929. S.S. acknowledges financial support by the DAAD, D.Y. by the C.S.C., and A.K. and G.J.S. by the graduate programme MaFO-Holz funded by the Lichtenberg programme of the country Lower Saxony.

Acknowledgments: We thank Merle Fastenrath for technical support with the anatomical studies used to prepare the figures in this review.

Conflicts of Interest: The authors declare no conflict of interest.

References

1.   Boer, A.H.D.; Volkov, V. Logistics of water and salt transport through the plant: Structure and functioning of the xylem. *Plant Cell Environ.* **2003**, *26*, 87–101. [CrossRef]
2.   Myburg, A.A.; Lev-Yadun, S.; Sederoff, R.R. Xylem Structure and Function. In *Encycolpedia of Life Science*; American Cancer Society: London, UK, 2013; ISBN 978-0-470-01590-2.
3.   Tyree, M.T.; Sperry, J.S. Vulnerability of Xylem to Cavitation and Embolism. *Annu. Rev. Plant. Physiol. Plant. Mol. Biol.* **1989**, *40*, 19–36. [CrossRef]
4.   Sperry, J.S.; Hacke, U.G.; Pittermann, J. Size and function in conifer tracheids and angiosperm vessels. *Am. J. Bot.* **2006**, *93*, 1490–1500. [CrossRef] [PubMed]
5.   Alder, N.N.; Sperry, J.S.; Pockman, W.T. Root and stem xylem embolism, stomatal conductance, and leaf turgor in *Acer grandidentatum* populations along a soil moisture gradient. *Oecologia* **1996**, *105*, 293–301. [CrossRef] [PubMed]
6.   Choat, B.; Cobb, A.R.; Jansen, S. Structure and function of bordered pits: New discoveries and impacts on whole-plant hydraulic function. *New Phytol.* **2008**, *177*, 608–626. [CrossRef]
7.   Zwieniecki, M.A.; Holbrook, N.M. Confronting Maxwell's demon: Biophysics of xylem embolism repair. *Trends Plant Sci.* **2009**, *14*, 530–534. [CrossRef] [PubMed]
8.   Awad, H.; Barigah, T.; Badel, E.; Cochard, H.; Herbette, S. Poplar vulnerability to xylem cavitation acclimates to drier soil conditions. *Physiologia Plantarum* **2010**, *139*, 208–288. [CrossRef]
9.   Rita, A.; Cherubini, P.; Leonardi, S.; Todaro, L.; Borghetti, M. Functional adjustments of xylem anatomy to climatic variability: Insights from long-term *Ilex aquifolium* tree-ring series. *Tree Physiol.* **2015**, *35*, 817–828. [CrossRef]
10. Wildhagen, H.; Paul, S.; Allwright, M.; Smith, H.K.; Malinowska, M.; Schnabel, S.K.; Paulo, M.J.; Cattonaro, F.; Vendramin, V.; Scalabrin, S.; et al. Genes and gene clusters related to genotype and drought-induced variation in saccharification potential, lignin content and wood anatomical traits in *Populus nigra. Tree Physiol* **2018**, *38*, 320–339. [CrossRef] [PubMed]
11. Bogeat-Triboulot, M.-B.; Brosché, M.; Renaut, J.; Jouve, L.; Thiec, D.L.; Fayyaz, P.; Vinocur, B.; Witters, E.; Laukens, K.; Teichmann, T.; et al. Gradual soil water depletion results in reversible changes of gene expression, protein profiles, ecophysiology, and growth performance in *Populus euphratica*, a poplar growing in arid regions. *Plant Physiol.* **2007**, *143*, 876–892. [CrossRef] [PubMed]
12. Aichinger, E.; Kornet, N.; Friedrich, T.; Laux, T. Plant Stem Cell Niches. *Annu. Rev. Plant Biol.* **2012**, *63*, 615–636. [CrossRef] [PubMed]
13. Plomion, C.; Leprovost, G.; Stokes, A. Wood Formation in Trees. *Plant Physiol.* **2001**, *127*, 1513–1523. [CrossRef] [PubMed]
14. Furuta, K.M.; Hellmann, E.; Helariutta, Y. Molecular control of cell specification and cell differentiation during procambial development. *Annu. Rev. Plant Biol.* **2014**, *65*, 607–638. [CrossRef] [PubMed]
15. Słupianek, A.; Kasprowicz-Maluśki, A.; Myśkow, E.; Turzańska, M.; Sokołowska, K. Endocytosis acts as transport pathway in wood. *New Phytol.* **2018**. [CrossRef]
16. Růžička, K.; Ursache, R.; Hejátko, J.; Helariutta, Y. Xylem development – from the cradle to the grave. *New Phytol.* **2015**, *207*, 519–535. [CrossRef]
17. Déjardin, A.; Laurans, F.; Arnaud, D.; Breton, C.; Pilate, G.; Leplé, J.-C. Wood formation in Angiosperms. *Comptes Rendus Biologies* **2010**, *333*, 325–334. [CrossRef] [PubMed]
18. McFarlane, H.E.; Döring, A.; Persson, S. The cell biology of cellulose synthesis. *Annu. Rev. Plant Biol.* **2014**, *65*, 69–94. [CrossRef] [PubMed]
19. Pauly, M.; Gille, S.; Liu, L.; Mansoori, N.; de Souza, A.; Schultink, A.; Xiong, G. Hemicellulose biosynthesis. *Planta* **2013**, *238*, 627–642. [CrossRef]
20. Scheller, H.V.; Ulvskov, P. Hemicelluloses. *Annu. Rev. Plant Biol.* **2010**, *61*, 263–289. [CrossRef]
21. Voxeur, A.; Wang, Y.; Sibout, R. Lignification: Different mechanisms for a versatile polymer. *Curr. Opin. Plant Biol.* **2015**, *23*, 83–90. [CrossRef]
22. Liu, Q.; Luo, L.; Zheng, L. Lignins: Biosynthesis and Biological Functions in Plants. *Int. J. Mol. Sci.* **2018**, *19*, 335. [CrossRef] [PubMed]
23. Yamamoto, E.; Bokelman, G.H.; Lewis, N.G. Phenylpropanoid Metabolism in Cell Walls. In *Plant Cell Wall Polymers*; ACS Symposium Series; American Chemical Society: Washington, DC, USA, 1989; Volume 399, pp. 68–88. ISBN 978-0-8412-1658-7.

24. Zhao, Q. Lignification: Flexibility, biosynthesis and regulation. *Trends Plant Sci.* **2016**, *21*, 713–721. [CrossRef]
25. Barros, J.; Serk, H.; Granlund, I.; Pesquet, E. The cell biology of lignification in higher plants. *Ann. Bot.* **2015**, *115*, 1053–1074. [CrossRef] [PubMed]
26. Schuetz, M.; Smith, R.; Ellis, B. Xylem tissue specification, patterning, and differentiation mechanisms. *J. Exp. Bot.* **2013**, *64*, 11–31. [CrossRef]
27. Ye, Z.-H.; Zhong, R. Molecular control of wood formation in trees. *J. Exp. Bot.* **2015**, *66*, 4119–4131. [CrossRef]
28. Kubo, M.; Udagawa, M.; Nishikubo, N.; Horiguchi, G.; Yamaguchi, M.; Ito, J.; Mimura, T.; Fukuda, H.; Demura, T. Transcription switches for protoxylem and metaxylem vessel formation. *Genes Dev.* **2005**, *19*, 1855–1860. [CrossRef]
29. Yamaguchi, M.; Mitsuda, N.; Ohtani, M.; Ohme-Takagi, M.; Kato, K.; Demura, T. Vascular-related NAC-domain 7 directly regulates a broad range of genes for xylem vessel differentiation. *BMC Proc.* **2011**, *5*, O37. [CrossRef]
30. Zhong, R.; Lee, C.; Ye, Z.-H. Functional characterization of poplar wood-associated NAC domain transcription factors. *Plant Physiol.* **2010**, *152*, 1044–1055. [CrossRef] [PubMed]
31. Mariconti, L.; Pellegrini, B.; Cantoni, R.; Stevens, R.; Bergounioux, C.; Cella, R.; Albani, D. The E2F family of transcription factors from *Arabidopsis thaliana*. *J. Biol. Chem.* **2002**, *277*, 9911–9919. [CrossRef]
32. Du, J.; Miura, E.; Robischon, M.; Martinez, C.; Groover, A. The *Populus* class III HD ZIP transcription factor *POPCORONA* affects cell differentiation during secondary growth of woody stems. *PLoS ONE* **2011**, *6*, e17458. [CrossRef]
33. McCarthy, R.L.; Zhong, R.; Fowler, S.; Lyskowski, D.; Piyasena, H.; Carleton, K.; Spicer, C.; Ye, Z.-H. The poplar MYB transcription factors, PtrMYB3 and PtrMYB20, are involved in the regulation of secondary wall biosynthesis. *Plant Cell Physiol.* **2010**, *51*, 1084–1090. [CrossRef] [PubMed]
34. Zhong, R.; McCarthy, R.L.; Haghighat, M.; Ye, Z.-H. The poplar MYB master switches bind to the SMRE site and activate the secondary wall biosynthetic program during wood formation. *PLoS ONE* **2013**, *8*, e69219. [CrossRef] [PubMed]
35. Pear, J.R.; Kawagoe, Y.; Schreckengost, W.E.; Delmer, D.P.; Stalker, D.M. Higher plants contain homologs of the bacterial *celA* genes encoding the catalytic subunit of cellulose synthase. *Proc. Natl. Acad. Sci. USA* **1996**, *93*, 12637–12642. [CrossRef] [PubMed]
36. Djerbi, S.; Lindskog, M.; Arvestad, L.; Sterky, F.; Teeri, T.T. The genome sequence of black cottonwood (*Populus trichocarpa*) reveals 18 conserved cellulose synthase (*CesA*) genes. *Planta* **2005**, *221*, 739–746. [CrossRef] [PubMed]
37. Suzuki, S.; Li, L.; Sun, Y.-H.; Chiang, V.L. The Cellulose Synthase Gene Superfamily and Biochemical Functions of Xylem-Specific Cellulose Synthase-Like Genes in *Populus trichocarpa*. *Plant Physiol.* **2006**, *142*, 1233–1245. [CrossRef] [PubMed]
38. Watanabe, Y.; Schneider, R.; Barkwill, S.; Gonzales-Vigil, E.; Hill, J.L., Jr.; Samuels, A.L.; Persson, S.; Mansfield, S.D. Cellulose synthase complexes display distinct dynamic behaviors during xylem transdifferentiation. *Proc. Natl. Acad. Sci. USA* **2018**, *115*, E6366–E6374. [CrossRef] [PubMed]
39. Polle, A.; Janz, D.; Teichmann, T.; Lipka, V. Poplar genetic engineering: Promoting desirable wood characteristics and pest resistance. *Appl. Microbiol. Biotechnol.* **2013**, *97*, 5669–5679. [CrossRef]
40. Liu, C.-J.; Miao, Y.-C.; Zhang, K.-W. Sequestration and transport of lignin monomeric precursors. *Molecules* **2011**, *16*, 710–727. [CrossRef]
41. Alejandro, S.; Lee, Y.; Tohge, T.; Sudre, D.; Osorio, S.; Park, J.; Bovet, L.; Lee, Y.; Geldner, N.; Fernie, A.R.; Martinoia, E. AtABCG29 Is a Monolignol Transporter Involved in Lignin Biosynthesis. *Curr. Biol.* **2012**, *22*, 1207–1212. [CrossRef]
42. Miao, Y.-C.; Liu, C.-J. ATP-binding cassette-like transporters are involved in the transport of lignin precursors across plasma and vacuolar membranes. *Proc. Natl. Acad. Sci. USA* **2010**, *107*, 22728–22733. [CrossRef]
43. Sandhu, A.P.S.; Randhawa, G.S.; Dhugga, K.S. Plant cell wall matrix polysaccharide biosynthesis. *Mol. Plant* **2009**, *2*, 840–850. [CrossRef]
44. Savi, T.; Bertuzzi, S.; Branca, S.; Tretiach, M.; Nardini, A. Drought-induced xylem cavitation and hydraulic deterioration: Risk factors for urban trees under climate change? *New Phytol.* **2015**, *205*, 1106–1116. [CrossRef]
45. Nardini, A.; Savi, T.; Trifilò, P.; Gullo, M.A.L. Drought stress and the recovery from xylem embolism in woody plants. In *Progress in Botany*; Progress in Botany; Springer: Cham, Switzerland, 2017; Volume 79, pp. 197–231. ISBN 978-3-319-71412-7.
46. Fischer, U.; Polle, A. *Populus* responses to abiotic stress. In *Genetics and Genomics of Populus*; Plant Genetics and Genomics: Crops and Models; Springer: New York, NY, USA, 2010; pp. 225–246. ISBN 978-1-4419-1540-5.

47. Polle, A.; Chen, S.L.; Eckert, C.; Harfouche, A. Engineering drought resistance in forest trees. *Front. Plant Sci.* **2019**, *9*. [CrossRef]

48. Arend, M.; Fromm, J. Seasonal change in the drought response of wood cell development in poplar. *Tree Physiol.* **2007**, *27*, 985. [CrossRef] [PubMed]

49. Beniwal, R.S.; Langenfeld-Heyser, R.; Polle, A. Ectomycorrhiza and hydrogel protect hybrid poplar from water deficit and unravel plastic responses of xylem anatomy. *Environ. Exp. Bot.* **2010**, *69*, 189–197. [CrossRef]

50. Schreiber, S.G.; Hacke, U.G.; Chamberland, S.; Lowe, C.W.; Kamelchuk, D.; Bräutigam, K.; Campbell, M.M.; Thomas, B.R. Leaf size serves as a proxy for xylem vulnerability to cavitation in plantation trees. *Plant Cell Environ.* **2016**, *39*, 272–281. [CrossRef]

51. Fonti, P.; Heller, O.; Cherubini, P.; Rigling, A.; Arend, M. Wood anatomical responses of oak saplings exposed to air warming and soil drought. *Plant Biol.* **2013**, *15*, 210–219. [CrossRef]

52. Colangelo, M.; Camarero, J.J.; Borghetti, M.; Gazol, A.; Gentilesca, T.; Ripullone, F. Size matters a lot: Drought-affected italian oaks are smaller and show lower growth prior to tree death. *Front. Plant Sci.* **2017**, *8*. [CrossRef]

53. Fichot, R.; Barigah, T.S.; Chamaillard, S.; Le Thiec, D.; Laurans, F.; Cochard, H.; Brignolas, F. Common trade-offs between xylem resistance to cavitation and other physiological traits do not hold among unrelated *Populus deltoides* x *Populus nigra* hybrids: Xylem resistance to cavitation and water relations in poplar. *Plant Cell Environ.* **2010**, *33*, 1553–1568. [CrossRef] [PubMed]

54. Rathgeber, C.B.K.; Cuny, H.E.; Fonti, P. Biological Basis of Tree-Ring Formation: A Crash Course. *Front. Plant Sci.* **2016**, *7*. [CrossRef] [PubMed]

55. Langer, K.; Ache, P.; Geiger, D.; Stinzing, A.; Arend, M.; Wind, C.; Regan, S.; Fromm, J.; Hedrich, R. Poplar potassium transporters capable of controlling K^+ homeostasis and K^+-dependent xylogenesis. *Plant J.* **2002**, *32*, 997–1009. [CrossRef]

56. Wind, C.; Arend, M.; Fromm, J. Potassium-Dependent Cambial Growth in Poplar. *Plant Biol.* **2004**, *7*, 30–37. [CrossRef]

57. Larisch, C.; Dittrich, M.; Wildhagen, H.; Lautner, S.; Fromm, J.; Polle, A.; Hedrich, R.; Rennenberg, H.; Müller, T.; Ache, P. Poplar wood rays are involved in seasonal remodeling of tree physiology. *Plant Physiol.* **2012**, *160*, 1515–1529. [CrossRef] [PubMed]

58. Danielsen, L.; Polle, A. Poplar nutrition under drought as affected by ectomycorrhizal colonization. *Environ. Exp. Bot.* **2014**, *108*, 89–98. [CrossRef]

59. Giovannelli, A.; Emiliani, G.; Traversi, M.L.; Deslauriers, A.; Rossi, S. Sampling cambial region and mature xylem for non structural carbohydrates and starch analyses. *Dendrochronologia* **2011**, *29*, 177–182. [CrossRef]

60. Xue, L.-J.; Frost, C.J.; Tsai, C.-J.; Harding, S.A. Drought response transcriptomes are altered in poplar with reduced tonoplast sucrose transporter expression. *Sci. Rep.* **2016**, *6*. [CrossRef] [PubMed]

61. Tang, S.; Dong, Y.; Liang, D.; Zhang, Z.; Ye, C.-Y.; Shuai, P.; Han, X.; Zhao, Y.; Yin, W.; Xia, X. Analysis of the drought stress-responsive transcriptome of black cottonwood (*Populus trichocarpa*) using deep RNA sequencing. *Plant Mol. Biol. Rep.* **2015**, *33*, 424–438. [CrossRef]

62. Deslauriers, A.; Giovannelli, A.; Rossi, S.; Castro, G.; Fragnelli, G.; Traversi, L. Intra-annual cambial activity and carbon availability in stem of poplar. *Tree Physiol.* **2009**, *29*, 1223–1235. [CrossRef]

63. Paux, E.; Tamasloukht, M.; Ladouce, N.; Sivadon, P.; Grima-Pettenati, J. Identification of genes preferentially expressed during wood formation in *Eucalyptus*. *Plant Mol. Biol.* **2004**, *55*, 263–280. [CrossRef]

64. Druart, N.; Johansson, A.; Baba, K.; Schrader, J.; Sjödin, A.; Bhalerao, R.R.; Resman, L.; Trygg, J.; Moritz, T.; Bhalerao, R.P. Environmental and hormonal regulation of the activity-dormancy cycle in the cambial meristem involves stage-specific modulation of transcriptional and metabolic networks. *Plant J.* **2007**, *50*, 557–573. [CrossRef]

65. Doblin, M.S.; Kurek, I.; Jacob-Wilk, D.; Delmer, D.P. Cellulose biosynthesis in plants: From genes to rosettes. *Plant Cell Physiol.* **2002**, *43*, 1407–1420. [CrossRef]

66. Peng, L.; Kawagoe, Y.; Hogan, P.; Delmer, D. Sitosterol-β-glucoside as primer for cellulose synthesis in plants. *Science* **2002**, *295*, 147–150. [CrossRef]

67. Leung, A.K.L.; Sharp, P.A. MicroRNA functions in stress responses. *Mol. Cell* **2010**, *40*, 205–215. [CrossRef]

68. Sunkar, R.; Li, Y.-F.; Jagadeeswaran, G. Functions of microRNAs in plant stress responses. *Trends Plant Sci.* **2012**, *17*, 196–203. [CrossRef] [PubMed]

69. Shuai, P.; Liang, D.; Zhang, Z.; Yin, W.; Xia, X. Identification of drought-responsive and novel *Populus trichocarpa* microRNAs by high-throughput sequencing and their targets using degradome analysis. *BMC Genomics* **2013**, *14*, 233. [CrossRef] [PubMed]

70. Brown, C.; Leijon, F.; Bulone, V. Radiometric and spectrophotometric in vitro assays of glycosyltransferases involved in plant cell wall carbohydrate biosynthesis. *Nat. Protoc.* **2012**, *7*, 1634–1650. [CrossRef]

71. Secchi, F.; Pagliarani, C.; Zwieniecki, M.A. The functional role of xylem parenchyma cells and aquaporins during recovery from severe water stress. *Plant Cell Environ.* **2017**, *40*, 858–871. [CrossRef] [PubMed]

72. Secchi, F.; Zwieniecki, M.A. Sensing embolism in xylem vessels: The role of sucrose as a trigger for refilling. *Plant Cell Environ.* **2011**, *34*, 514–524. [CrossRef] [PubMed]

73. Secchi, F.; Maciver, B.; Zeidel, M.L.; Zwieniecki, M.A. Functional analysis of putative genes encoding the PIP2 water channel subfamily in *Populus trichocarpa*. *Tree Physiol.* **2009**, *29*, 1467–1477. [CrossRef]

74. Secchi, F.; Zwieniecki, M.A. Patterns of PIP gene expression in *Populus trichocarpa* during recovery from xylem embolism suggest a major role for the PIP1 aquaporin subfamily as moderators of refilling process. *Plant Cell Environ.* **2010**, *33*, 1285–1297. [CrossRef] [PubMed]

75. Almeida-Rodriguez, A.M.; Hacke, U.G. Cellular localization of aquaporin mRNA in hybrid poplar stems. *Am. J. Bot.* **2012**, *99*, 1249–1254. [CrossRef] [PubMed]

76. Sakr, S.; Alves, G.; Morillon, R.; Maurel, K.; Decourteix, M.; Guilliot, A.; Fleurat-Lessard, P.; Julien, J.-L.; Chrispeels, M.J. Plasma membrane aquaporins are involved in winter embolism recovery in walnut tree. *Plant Physiol.* **2003**, *133*, 630–641. [CrossRef] [PubMed]

77. Nardini, A.; Lo Gullo, M.A.; Salleo, S. Refilling embolized xylem conduits: Is it a matter of phloem unloading? *Plant Sci.* **2011**, *180*, 604–611. [CrossRef] [PubMed]

78. Fujita, Y.; Fujita, M.; Shinozaki, K.; Yamaguchi-Shinozaki, K. ABA-mediated transcriptional regulation in response to osmotic stress in plants. *J. Plant Res.* **2011**, *124*, 509–525. [CrossRef] [PubMed]

79. Yoshida, T.; Fernie, A.R. Remote Control of Transpiration via ABA. *Trends Plant Sci.* **2018**, *23*, 755–758. [CrossRef] [PubMed]

80. Lachaud, S. Participation of auxin and abscisic acid in the regulation of seasonal variations in cambial activity and xylogenesis. *Trees-Struct. Funct.* **1989**, *3*, 125–137. [CrossRef]

81. Luisi, A.; Giovannelli, A.; Traversi, M.L.; Anichini, M.; Sorce, C. Hormonal responses to water deficit in cambial tissues of *Populus alba* L. *J. Plant Grow. Regul.* **2014**, *33*, 489–498. [CrossRef]

82. Tylewicz, S.; Petterle, A.; Marttila, S.; Miskolczi, P.; Azeez, A.; Singh, R.K.; Immanen, J.; Mähler, N.; Hvidsten, T.R.; Eklund, D.M.; et al. Photoperiodic control of seasonal growth is mediated by ABA acting on cell-cell communication. *Science* **2018**, *360*, 212–215. [CrossRef]

83. Sorce, C.; Giovannelli, A.; Sebastiani, L.; Anfodillo, T. Hormonal signals involved in the regulation of cambial activity, xylogenesis and vessel patterning in trees. *Plant Cell Rep.* **2013**, *32*, 885–898. [CrossRef]

84. Popko, J.; Haensch, R.; Mendel, R.R.; Polle, A.; Teichmann, T. The role of abscisic acid and auxin in the response of poplar to abiotic stress. *Plant Biol.* **2010**, *12*, 242–258. [CrossRef]

85. Jensen, M.K.; Kjaersgaard, T.; Nielsen, M.M.; Galberg, P.; Petersen, K.; O'Shea, C.; Skriver, K. The *Arabidopsis thaliana* NAC transcription factor family: Structure–function relationships and determinants of ANAC019 stress signalling. *Biochem. J.* **2010**, *426*, 183–196. [CrossRef]

86. Keshishian, E.A.; Rashotte, A.M. Plant cytokinin signalling. *Essays Biochem.* **2015**, *58*, 13–27. [CrossRef]

87. Nieminen, K.; Immanen, J.; Laxell, M.; Kauppinen, L.; Tarkowski, P.; Dolezal, K.; Tähtiharju, S.; Elo, A.; Decourteix, M.; Ljung, K.; et al. Cytokinin signaling regulates cambial development in poplar. *PNAS* **2008**, *105*, 20032–20037. [CrossRef]

88. Paul, S.; Wildhagen, H.; Janz, D.; Teichmann, T.; Hänsch, R.; Polle, A. Tissue- and cell-specific cytokinin activity in *Populus x canescens* monitored by *ARR5::GUS* reporter lines in summer and winter. *Front. Plant Sci.* **2016**, *7*. [CrossRef]

89. Paul, S.; Wildhagen, H.; Janz, D.; Polle, A. Drought effects on the tissue- and cell-specific cytokinin activity in poplar. *AoB Plants* **2018**, *10*. [CrossRef]

90. Jang, G.; Chang, S.H.; Um, T.Y.; Lee, S.; Kim, J.-K.; Choi, Y.D. Antagonistic interaction between jasmonic acid and cytokinin in xylem development. *Sci. Rep.* **2017**, *7*. [CrossRef]

91. Jang, G.; Choi, Y.D. Drought stress promotes xylem differentiation by modulating the interaction between cytokinin and jasmonic acid. *Plant Signal. Behav.* **2018**, *13*. [CrossRef]

92. Brenner, W.G.; Ramireddy, E.; Heyl, A.; Schmülling, T. Gene regulation by cytokinin in Arabidopsis. *Front Plant Sci.* **2012**, *3*. [CrossRef]

93. Nafisi, M.; Fimognari, L.; Sakuragi, Y. Interplays between the cell wall and phytohormones in interaction between plants and necrotrophic pathogens. *Phytochemistry* **2015**, *112*, 63–71. [CrossRef]

94. Jung, K.W.; Oh, S.-I.; Kim, Y.Y.; Yoo, K.S.; Cui, M.H.; Shin, J.S. *Arabidopsis* histidine-containing phosphotransfer factor 4 (AHP4) negatively regulates secondary wall thickening of the anther endothecium during flowering. *Mol. Cells* **2008**, *25*, 294–300.

95. Wilkinson, S.; Kudoyarova, G.R.; Veselov, D.S.; Arkhipova, T.N.; Davies, W.J. Plant hormone interactions: Innovative targets for crop breeding and management. *J. Exp. Bot.* **2012**, *63*, 3499–3509. [CrossRef] [PubMed]

96. Reguera, M.; Peleg, Z.; Abdel-Tawab, Y.M.; Tumimbang, E.B.; Delatorre, C.A.; Blumwald, E. Stress-induced cytokinin synthesis increases drought tolerance through the coordinated regulation of carbon and nitrogen assimilation in rice. *Plant Physiol.* **2013**, *163*, 1609–1622. [CrossRef] [PubMed]

97. Euring, D.; Bai, H.; Janz, D.; Polle, A. Nitrogen-driven stem elongation in poplar is linked with wood modification and gene clusters for stress, photosynthesis and cell wall formation. *BMC Plant Biol.* **2014**, *14*, 391. [CrossRef] [PubMed]

98. Van den Broeck, H.C.; Maliepaard, C.; Ebskamp, M.J.M.; Toonen, M.A.J.; Koops, A.J. Differential expression of genes involved in C1 metabolism and lignin biosynthesis in wooden core and bast tissues of fibre hemp (*Cannabis sativa* L.). *Plant Sci.* **2008**, *174*, 205–220. [CrossRef]

99. Camargo, E.L.O.; Nascimento, L.C.; Soler, M.; Salazar, M.M.; Lepikson Neto, J.; Marques, W.L.; Alves, A.; Teixeira, P.J.P.L.; Mieczkowski, P.; Carazzolle, M.F.; et al. Contrasting nitrogen fertilization treatments impact xylem gene expression and secondary cell wall lignification in Eucalyptus. *BMC Plant Biol.* **2014**, *14*. [CrossRef] [PubMed]

100. Plavcová, L.; Hacke, U.G.; Almeida-Rodriguez, A.M.; Li, E.; Douglas, C.J. Gene expression patterns underlying changes in xylem structure and function in response to increased nitrogen availability in hybrid poplar. *Plant Cell Environ.* **2013**, *36*, 186–199. [CrossRef] [PubMed]

101. Luo, J.; Zhou, J.; Li, H.; Shi, W.; Polle, A.; Lu, M.; Sun, X.; Luo, Z.-B. Global poplar root and leaf transcriptomes reveal links between growth and stress responses under nitrogen starvation and excess. *Tree Physiol.* **2015**, *35*, 1283–1302. [CrossRef] [PubMed]

102. Benson, M.L.; Myers, B.J.; Raison, R.J. Dynamics of stem growth of *Pinus radiata* as affected by water and nitrogen supply. *For. Ecol. Manag.* **1992**, *52*, 117–137. [CrossRef]

103. Dobbertin, M. Tree growth as indicator of tree vitality and of tree reaction to environmental stress: A review. *Eur. J. For. Res.* **2006**, *125*, 89. [CrossRef]

104. Novaes, E.; Osorio, L.; Drost, D.R.; Miles, B.L.; Boaventura-Novaes, C.R.D.; Benedict, C.; Dervinis, C.; Yu, Q.; Sykes, R.; Davis, M.; et al. Quantitative genetic analysis of biomass and wood chemistry of *Populus* under different nitrogen levels. *New Phytol.* **2009**, *182*, 878–890. [CrossRef] [PubMed]

105. Galle, A.; Esper, J.; Feller, U.; Ribas-Carbo, M.; Fonti, P. Responses of wood anatomy and carbon isotope composition of *Quercus pubescens* saplings subjected to two consecutive years of summer drought. *Ann. For. Sci.* **2010**, *67*, 809. [CrossRef]

106. Liu, Z.; Dickmann, D.I. Abscisic acid accumulation in leaves of two contrasting hybrid poplar clones affected by nitrogen fertilization plus cyclic flooding and soil drying. *Tree Physiol.* **1992**, *11*, 109–122. [CrossRef]

107. Liu, Z.; Dickmann, D.I. Responses of two hybrid *Populus* clones to flooding, drought, and nitrogen availability. I. Morphology and growth. *Can. J. Bot.* **1992**, *70*, 2265–2270. [CrossRef]

108. Ibrahim, L.; Proe, M.F.; Cameron, A.D. Interactive effects of nitrogen and water availabilities on gas exchange and whole-plant carbon allocation in poplar. *Tree Physiol.* **1998**, *18*, 481–487. [CrossRef] [PubMed]

109. Mazzoleni, S.; Dickmann, D.I. Differential physiological and morphological responses of two hybrid *Populus* clones to water stress. *Tree Physiol.* **1988**, *4*, 61–70. [CrossRef]

110. Harvey, H.P.; Van Den Driessche, R. Nutrition, xylem cavitation and drought resistance in hybrid poplar. *Tree Physiol.* **1997**, *17*, 647–654. [CrossRef]

111. Zimmermann, M.H. Hydraulic architecture of some diffuse-porous trees. *Can. J. Bot.* **2011**, *56*, 2286–2295. [CrossRef]

112. Euring, D. Nitrogen responsive wood formation in poplar (*Populus* sp.). In *Chapter 3: N-Responsive Network*; Georg-August-Universität Göttingen: Göttingen, Germany, 2014.

113. Tarighaleslami, M.; Zarghami, R.; Mashhadi, M.; Boojar, A.; Oveysi, M. Effects of drought stress and different nitrogen levels on morphological traits of proline in leaf and protein of corn seed (*Zea mays* L.). *Am.-Euras. J. Agric. Environ. Sci.* **2012**, *12*, 49–56.

114. Chen, S.; Polle, A. Salinity tolerance of *Populus*. *Plant Biol.* **2010**, *12*, 317–333. [CrossRef] [PubMed]

115. Polle, A.; Chen, S. On the salty side of life: Molecular, physiological and anatomical adaptation and acclimation of trees to extreme habitats. *Plant Cell Environ.* **2015**, *38*, 1794–1816. [CrossRef] [PubMed]

116. Munns, R.; Tester, M. Mechanisms of Salinity Tolerance. *Annu. Rev. Plant Biol.* **2008**, *59*, 651–681. [CrossRef] [PubMed]

117. Ottow, E.A.; Brinker, M.; Teichmann, T.; Fritz, E.; Kaiser, W.; Brosché, M.; Kangasjärvi, J.; Jiang, X.; Polle, A. *Populus euphratica* displays apoplastic sodium accumulation, osmotic adjustment by decreases in calcium and soluble carbohydrates, and develops leaf succulence under salt stress. *Plant Physiol.* **2005**, *139*, 1762–1772. [CrossRef]

118. Lautner, S. Wood formation under drought stress and salinity. In *Cellular Aspects of Wood Formation*; Fromm, J., Ed.; Springer: Berlin, Germany, 2013; Volume 20, pp. 187–202. ISBN 978-3-642-36490-7.

119. Escalante-Pérez, M.; Lautner, S.; Nehls, U.; Selle, A.; Teuber, M.; Schnitzler, J.-P.; Teichmann, T.; Fayyaz, P.; Hartung, W.; Polle, A.; et al. Salt stress affects xylem differentiation of grey poplar (*Populus* x *canescens*). *Planta* **2009**, *229*, 299–309. [CrossRef]

120. Junghans, U.; Polle, A.; Düchting, P.; Weiler, E.; Kuhlman, B.; Gruber, F.; Teichmann, T. Adaptation to high salinity in poplar involves changes in xylem anatomy and auxin physiology. *Plant Cell Environ.* **2006**, *29*, 1519–1531. [CrossRef] [PubMed]

121. Janz, D.; Lautner, S.; Wildhagen, H.; Behnke, K.; Schnitzler, J.-P.; Rennenberg, H.; Fromm, J.; Polle, A. Salt stress induces the formation of a novel type of 'pressure wood' in two *Populus* species. *New Phytol.* **2012**, *194*, 129–141. [CrossRef]

122. Hacke, U.G.; Sperry, J.S. Functional and ecological xylem anatomy. *Perspect. Plant Ecol. Evol. Syst.* **2001**, *4*, 97–115. [CrossRef]

123. Bhalerao, R.P.; Fischer, U. Environmental and hormonal control of cambial stem cell dynamics. *J. Exp. Bot.* **2017**, *68*, 79–87. [CrossRef]

124. Uggla, C.; Moritz, T.; Sandberg, G.; Sundberg, B. Auxin as a positional signal in pattern formation in plants. *PNAS* **1996**, *93*, 9282–9286. [CrossRef]

125. Uggla, C.; Mellerowicz, E.J.; Sundberg, B. Indole-3-Acetic Acid Controls Cambial Growth in Scots Pine by Positional Signaling. *Plant Physiol.* **1998**, *117*, 113–121. [CrossRef]

126. Nilsson, J.; Karlberg, A.; Antti, H.; Lopez-Vernaza, M.; Mellerowicz, E.; Perrot-Rechenmann, C.; Sandberg, G.; Bhalerao, R.P. Dissecting the Molecular Basis of the Regulation of Wood Formation by Auxin in Hybrid Aspen. *Plant Cell* **2008**, *20*, 843–855. [CrossRef]

127. Sheldrake, A.R. Effects of Osmotic Stress on Polar Auxin Transport in Avena Mesocotyl Sections. *Planta* **1979**, *145*, 113–117. [CrossRef] [PubMed]

128. Teichmann, T.; Bolu-Arianto, W.H.; Olbrich, A.; Langenfeld-Heyser, R.; Gobel, C.; Grzeganek, P.; Feussner, I.; Hansch, R.; Polle, A. GH3::GUS reflects cell-specific developmental patterns and stress-induced changes in wood anatomy in the poplar stem. *Tree Physiol.* **2008**, *28*, 1305–1315. [CrossRef]

129. Oh, J.E.; Kwon, Y.; Kim, J.H.; Noh, H.; Hong, S.-W.; Lee, H. A dual role for MYB60 in stomatal regulation and root growth of *Arabidopsis thaliana* under drought stress. *Plant Mol. Biol.* **2011**, *77*, 91–103. [CrossRef]

130. Sixto, H.; González-González, B.D.; Molina-Rueda, J.J.; Garrido-Aranda, A.; Sanchez, M.M.; López, G.; Gallardo, F.; Cañellas, I.; Mounet, F.; Grima-Pettenati, J.; et al. *Eucalyptus* spp. and *Populus* spp. coping with salinity stress: An approach on growth, physiological and molecular features in the context of short rotation coppice (SRC). *Trees* **2016**, *30*, 1873–1891. [CrossRef]

131. Guo, H.; Wang, Y.; Wang, L.; Hu, P.; Wang, Y.; Jia, Y.; Zhang, C.; Zhang, Y.; Zhang, Y.; Wang, C.; Yang, C. Expression of the MYB transcription factor gene *BplMYB46* affects abiotic stress tolerance and secondary cell wall deposition in *Betula platyphylla*. *Plant Biotechnol. J.* **2017**, *15*, 107–121. [CrossRef] [PubMed]

Growth and Needle Properties of Young *Pinus koraiensis* Sieb. et Zucc. Trees across an Elevational Gradient

Ying Fan [1], W. Keith Moser [2] and Yanxia Cheng [3,*]

1 College of Forest Science, Beijing Forestry University, No.35 Qinghua Donglu, Haidian District, Beijing 100083, China; yingjanefan@gmail.com
2 Rocky Mountain Research Station, Forest Service, United States Department of Agriculture, 2500 S. Pine Knoll Drive, Flagstaff, AZ 86001-6381, USA; wkmoser@fs.fed.us
3 College of Science, Beijing Forestry University, No.35 Qinghua Donglu, Haidian District, Beijing 100083, China
* Correspondence: chyx@bjfu.edu.cn; Tel.: +86-10-62336189

Received: 5 November 2018; Accepted: 7 January 2019; Published: 11 January 2019

Abstract: A better understanding of the response of plant growth to elevational gradients may shed light on how plants respond to environmental variation and on the physiological mechanisms underlying these responses. This study analyzed whole plant growth and physiological and morphological properties of needles in young *Pinus koraiensis* Sieb. et Zucc. trees at thirteen points along an elevational gradient ranging from 750 to 1350 m above sea level (a.s.l.) at the end of a growing season on Changbai Mountain in northeastern China. Sampling and analyses indicated the following; (1) many needle properties of *P. koraiensis* varied with forest type along the elevational gradient though some needle properties (e.g., intrinsic water use efficiency, concentration of chlorophyll, and leaf mass per area) did not change with elevation and forest types; (2) growth was significantly influenced by both forest type and elevation and growth of saplings in *P. koraiensis* and mixed broadleaved forests was greater than that in evergreen forests and increased with elevation in both forest types; (3) in *P. koraiensis* and mixed broadleaved forests, there were significant correlations between growth properties and light saturation point, leaf water potential, mean within-crown humidity, annual precipitation, cumulative temperature ($\geq$5 °C), within-crown air temperature, and atmospheric pressure; while in evergreen forests, the leaf C, leaf P content, net rate of light saturation in photosynthesis, water content of soil, within-crown humidity, annual precipitation, cumulative temperature ($\geq$5 °C), within-crown air temperature, and total soil P content displayed a significant relationship with plant growth. These results may help illuminate how *P. koraiensis* responds to environmental variation and evaluate the adaptive potential of *Pinus koraiensis* to climate change. Data presented here could also contribute to the more accurate estimation of carbon stocks in this area and to refinement of a plant trait database.

Keywords: elevation gradient; forest type; growth; leaf properties; *Pinus koraiensis* Sieb. et Zucc.

---

1. Introduction

Variation in biomass accumulation, biomass allocation, and leaf physiological and morphological responses to elevational gradients may reflect adaptations of plants to environmental variation and indicate potential responses to future climatic variability. Ecologists and biogeographers have long been aware of the value of elevational gradients for understanding how plants respond to changes in macroclimate [1]. Elevation affects environmental factors, such as temperature [2], precipitation [3], partial carbon dioxide (CO_2) pressure [4], and ultraviolet (UV) irradiance [5] over long periods of time. Hence, elevational gradients can serve as powerful case studies for understanding longer-term, larger-scale plant responses to environmental changes as a complement to controlled experiments [6,7].

Changes in temperature, precipitation, partial CO_2 pressure and UV irradiance affect biomass accumulation [8], biomass allocation [9], photosynthetic leaf characteristics [10,11], plant water status [12], chlorophyll concentrations in leaves [13], leaf mass per area [14], and leaf nutrients [15] along elevational gradients. Annual biomass accumulation reflects the growth of plants directly. Biomass allocation tends to optimize allocation to components where resources are the most limited. For example, plants growing in nutrient-poor soils may preferentially allocate biomass to roots: where light is a limiting factor, biomass preferentially accrues to leaves [16]. Photosynthesis is one of the most important physiological parameters for all aspects of plant growth [17,18]. Many indices quantify processes related to photosynthesis. The apparent quantum yield (AQY, $mol \cdot mol^{-1}$) may indicate weak light utilization efficiency [19]. The light saturation point (LSP, $\mu mol \cdot m^{-2} \cdot s^{-1}$) measures the ability of a plant to adapt to strong light [20,21]. The net rate of light saturation in photosynthesis (Asat, $\mu mol\ CO_2 \cdot m^{-2} \cdot s^{-1}$) indicates maximum photosynthetic potential [22]. Stomatal conductance (Cond, $mol\ H_2O \cdot m^{-2} \cdot s^{-1}$) is related to both carbon fixation and leaf transpiration [23]. Intrinsic water use efficiency (iWUE, $\mu mol\ CO_2 \cdot mmol^{-1}\ H_2O$), calculated as Pn/Cond, is reported to stimulate tree growth in some high elevational areas [24]. Many studies have shown that the spectral water band index (WI) can be used to predict plant water status: leaf water potential [25]. Composition and concentration of chlorophyll influence the photosynthetic capacity of leaves [26]. Leaf mass per area (LMA, $g \cdot m^{-2}$) is a measure of the light harvesting surface for a given amount of dry matter investment, also closely related to plant growth [27]. On the other hand, leaf nutrients are important to plant ecophysiological processes; for example, relative and absolute leaf nitrogen content are highly correlated with plant photosynthetic capacity [28]. Therefore, responses of photosynthetic characteristics, water relations, chlorophyll concentrations, leaf mass per area, and leaf nutrient levels to elevational gradients may reflect responses and adaptations of plants to environmental variation.

The primeval forest on Changbai Mountain is a well preserved temperate mountain forest ecosystem in northeast China [29]. This region is at the ecological limit of many species which are thus very sensitive to climatic changes but also has experienced climate warming during the past 50 years [30]. Hence, Changbai Mountain has become an important area for study and conservation in China. The Changbai Mountain ecosystem has characteristics typical of montane forests at similar ranges of elevation. At elevations from 750 to 1100 m, *Pinus koraiensis* and mixed broadleaved forests (PBMF) are dominated by *Pinus koraiensis* Sieb. et Zucc., *Tilia amurensis* Rupr., *Populus davidiana* Dode, and *Betula platyphylla* Suk. From 1100 to 1700 m, evergreen coniferous forests (ECF) are dominated by *Larix olgensis* Henry, *Picea jezoensis* Carr., and *Abies nephrolepis* (Trautv.) Maxim., but *Pinus koraiensis* is also present. Between 1700 and 1950 m, *Betula ermanii* is the dominant species. Beyond this elevation and up to 2691 m, the landscape consists of tundra [31,32]. The incline in the topography has made this region an ideal place for studying the ecophysiology of alpine trees along an elevational gradient.

Pinus koraiensis is one of the three major five-needle pines in the northern hemisphere, and is the most abundant coniferous species in the Changbai Mountain landscape [33]. It plays an important role in maintaining biodiversity and providing ecosystem services in East Asia. This native pine is also an economically valuable species because of its edible pine nuts and high-quality wood products [34]. The essential oil of *Pinus koraiensis* needles contains properties valuable in treating high cholesterol and colorectal cancer [35,36]. In recent decades, extreme freezing, forest fires, pests, and diseases caused by extreme weather events and overutilization have degraded the natural *Pinus koraiensis* forest, so considerable efforts have been made to preserve *Pinus koraiensis* [37,38]. Studies of *Pinus koraiensis* saplings report that natural regeneration of *Pinus koraiensis*, considered to be shade tolerant, takes place even under dense canopies, where young plants may survive under such conditions for 10 to 20 years [39]. Long-term survival and establishment of forests are often associated with the young trees' leaf properties in the understory [40]. To date, no research has dealt with growth and needle property responses to an elevational gradient in wild *Pinus koraiensis* saplings.

To explore these relationships, *Pinus koraiensis* saplings of the same age and under the same light condition (represented by canopy openness) were sampled along an elevational gradient from 750 to 1350 m a.s.l. by 50 m increments on Changbai Mountain (above 1350 m a.s.l., few *Pinus koraiensis* saplings were found). We focused on three subjects: (1) needle properties of young *Pinus koraiensis* trees across an elevational gradient in two forest types; (2) effects of elevation on growth of *Pinus koraiensis* in two forest types; and (3) needle characteristics, soil properties, and atmospheric indices that may influence the growth of *Pinus koraiensis* in two forest types. Results of this study may help in illuminating how *Pinus koraiensis* saplings respond to environmental variation and which physiological and morphological mechanisms are at work.

2. Materials and Methods

2.1. Study Area

The sampling area of this study was on the north slope of the Changbai Mountain Natural Reserve (42°25′–42°09′, 128°04′–128°55′) in Jilin Province, northeastern China. The study transected the *Pinus koraiensis* distribution (750–1350 m a.s.l.) and crossed two vegetation types (PBMF and ECF) free from anthropogenic disturbance. The climate is a continental monsoon type with long, cold winters; hot, rainy summers; mean annual precipitation (RIA) of 644.25–820.89 mm; and cumulative temperature (≥ 5 °C, ACT5) of 1682.26 to 2388.28 °C [41]. The mean within-crown temperatures (WCT) decreased by 0.68 °C and the mean within-crown levels of humidity (WCH) increased by 0.93% with each 100 m increase in elevation during the growing season [32] (Table 1). Soil properties are summarized in Table 2.

Table 1. General description of plots at various elevations on Changbai Mountain.

Elevation (m)	Type of Forests	Slope Aspect	Slope (°)	WCT # (°C)	ACT5 ## (°C)	RIA ## (mm)	WCH # (%)	Pair (kPa)
750	PBMF	N	<10	15.55	2388.28	644.25	77.17	91.80 ± 0.07
800	PBMF	N	<10	15.21	2329.44	658.97	77.63	90.60 ± 0.02
850	PBMF	N	<10	14.87	2270.61	673.69	78.10	90.75 ± 0.11
900	PBMF	N	<10	14.53	2211.77	688.41	78.56	90.88 ± 0.03
950	PBMF	N	<10	14.19	2152.94	703.13	79.03	90.78 ± 0.02
1000	PBMF	N	<10	13.85	2094.10	717.85	79.49	90.03 ± 0.06
1050	PBMF	N	<10	13.51	2035.27	732.57	79.96	89.05 ± 0.02
1100	ECF	N	<10	13.17	1976.43	747.29	80.42	88.49 ± 0.01
1150	ECF	N	<10	12.83	1917.60	762.01	80.89	87.85 ± 0.04
1200	ECF	N	<10	12.49	1858.76	776.73	81.35	88.02 ± 0.04
1250	ECF	N	<10	12.15	1799.93	791.45	81.82	86.99 ± 0.09
1300	ECF	N	<10	11.81	1741.09	806.17	82.28	86.94 ± 0.04
1350	ECF	N	<10	11.47	1682.26	820.89	82.75	85.81 ± 0.01

Note: PBMF = *Pinus koraiensis* and mixed broadleaved forests, ECF = evergreen coniferous forests, WCT = within-crown air temperature, WCH = within-crown relative humidity, RIA = annual precipitation, ACT5 = cumulative temperature (≥ 5 °C), Pair = atmospheric pressure, # data from [28], ## data from [41].

Table 2. General description of soil properties at various elevations on Changbai Mountain.

Elevation (m)	N Soil (%)	C Soil (%)	P Soil (%)	pH	WC Soil (%)
750	0.2148 ± 0.0389 de	5.8475 ± 0.4568 ab	0.3492 ± 0.0258 a	5.4492 ± 0.1319 c	27.7338 ± 1.6653 a
800	0.0674 ± 0.0017 a	4.8923 ± 0.4933 a	0.3008 ± 0.0114 a	5.2173 ± 0.0715 bc	43.3591 ± 1.3379 c
850	0.1046 ± 0.0098 ab	6.0790 ± 1.0523 ab	0.3005 ± 0.0072 a	5.5433 ± 0.2471 c	47.1886 ± 3.1609 b
900	0.1624 ± 0.0203 bcd	4.1960 ± 0.9026 a	0.3228 ± 0.0241 a	5.1678 ± 0.0570 abc	50.7534 ± 0.6221 a
950	0.2551 ± 0.0271 e	8.1750 ± 0.5909 b	0.3264 ± 0.0181 a	5.0633 ± 0.0840 ab	47.9172 ± 2.2047 b
1000	0.1691 ± 0.0115 cd	4.7742 ± 0.2660 a	0.2957 ± 0.0109 a	5.2323 ± 0.0293 bc	50.7662 ± 0.3353 c
1050	0.1154 ± 0.0047 abc	3.6204 ± 1.2684 a	0.3159 ± 0.0104 a	4.8427 ± 0.0378 a	50.5732 ± 0.4940 b
1100	0.1343 ± 0.0063 CD	2.4571 ± 0.1983 AB	0.2519 ± 0.0103 BC	5.1591 ± 0.0367 D	32.8224 ± 4.0644 D
1150	0.0919 ± 0.0027 A	2.2594 ± 0.6608 AB	0.2327 ± 0.0041 AB	4.9967 ± 0.0194 C	33.0481 ± 2.5072 AB
1200	0.1107 ± 0.0038 B	3.9636 ± 0.1945 CD	0.2419 ± 0.0044 BC	4.8632 ± 0.0620 B	31.4864 ± 1.7371 BC
1250	0.1199 ± 0.0024 BC	4.8995 ± 0.1732 D	0.1951 ± 0.0261 A	4.5588 ± 0.0385 A	40.8705 ± 4.3609 C
1300	0.0868 ± 0.0067 A	1.5724 ± 0.2830 A	0.2733 ± 0.0130 BC	4.9167 ± 0.0254 BC	51.0298 ± 0.5400 C
1350	0.1500 ± 0.0113 D	3.3030 ± 0.5744 BC	0.2794 ± 0.0118 C	4.9364 ± 0.0037 BC	44.5513 ± 1.0680 A

Note: For each elevation, the sample size was 5. Significant differences among elevations are denoted by lowercase letters in PBMF (*Pinus koraiensis* Sieb. et Zucc. and mixed broadleaved forests) and uppercase letters in ECF (evergreen coniferous forests) (both at the $p < 0.05$ level).

2.2. Plant Materials

Five healthy *Pinus koraiensis* saplings ($n = 5$; $N = 65$; >300 m apart from each other at each elevation) up to 97 ± 30 cm in height, without any evident damage or deformed shape, at seemingly the same age (35 years $\pm$ 5) were selected randomly as sample trees in the understory of this natural forest at each elevation from 750 to 1350 m at thirteen 50 m intervals of the transect. All saplings were located under a closed canopy layer with approximately the same light conditions. Tree height, basal stem diameter, crown length (S–N), crown width (E–W), and canopy openness were recorded and are summarized in Table 3.

Table 3. Information at each elevation of plants sampled.

Elevation (m)	Height (cm)	BSD (cm)	Height/BSD	Crown Length (cm)	Crown Width (cm)	Canopy Openness (%)
750	99.21 $\pm$ 4.79	1.79 $\pm$ 0.12	56.03 $\pm$ 2.90 ns	95.00 $\pm$ 8.34	86.40 $\pm$ 10.33	77.93 $\pm$ 1.05 ns
800	97.82 $\pm$ 4.69	1.84 $\pm$ 0.10	53.28 $\pm$ 1.74 ns	95.60 $\pm$ 9.57	85.20 $\pm$ 8.06	76.93 $\pm$ 1.09 ns
850	97.72 $\pm$ 5.13	1.85 $\pm$ 0.11	52.82 $\pm$ 0.84 ns	95.00 $\pm$ 9.99	84.60 $\pm$ 10.08	78.77 $\pm$ 0.57 ns
900	95.73 $\pm$ 3.20	1.83 $\pm$ 0.14	53.08 $\pm$ 2.51 ns	93.60 $\pm$ 9.24	83.60 $\pm$ 7.54	76.82 $\pm$ 0.68 ns
950	99.36 $\pm$ 4.35	1.80 $\pm$ 0.11	55.72 $\pm$ 2.43 ns	97.00 $\pm$ 9.32	86.60 $\pm$ 9.63	78.83 $\pm$ 0.80 ns
1000	99.89 $\pm$ 3.43	1.85 $\pm$ 0.13	54.82 $\pm$ 3.03 ns	98.00 $\pm$ 10.26	85.80 $\pm$ 9.65	77.86 $\pm$ 0.51 ns
1050	98.46 $\pm$ 4.85	1.79 $\pm$ 0.15	55.98 $\pm$ 3.30 ns	108.20 $\pm$ 9.48	95.20 $\pm$ 10.28	78.28 $\pm$ 1.08 NS
1100	97.96 $\pm$ 4.48	1.83 $\pm$ 0.13	54.12 $\pm$ 2.21 ns	84.40 $\pm$ 4.55	73.80 $\pm$ 4.85	79.36 $\pm$ 0.72 NS
1150	98.36 $\pm$ 3.62	1.82 $\pm$ 0.14	54.82 $\pm$ 2.79 ns	94.20 $\pm$ 9.93	86.40 $\pm$ 9.56	78.28 $\pm$ 0.87 NS
1200	99.50 $\pm$ 4.55	1.82 $\pm$ 0.13	55.13 $\pm$ 2.16 ns	97.40 $\pm$ 9.35	85.20 $\pm$ 10.51	77.56 $\pm$ 0.57 NS
1250	97.30 $\pm$ 5.17	1.82 $\pm$ 0.14	54.04 $\pm$ 2.02 ns	97.40 $\pm$ 9.52	84.40 $\pm$ 10.37	78.49 $\pm$ 1.08 NS
1300	94.04 $\pm$ 4.10	1.90 $\pm$ 0.13	50.11 $\pm$ 2.36 ns	94.80 $\pm$ 9.56	85.60 $\pm$ 9.14	78.71 $\pm$ 0.98 NS
1350	97.85 $\pm$ 5.37	1.79 $\pm$ 0.12	54.77 $\pm$ 1.22 ns	96.00 $\pm$ 9.36	83.40 $\pm$ 8.13	77.56 $\pm$ 1.18 ns

Note: BSD = basal stem diameter. Lowercase letters ns mean not statistically significant in PBMF (*Pinus koraiensis* and mixed broadleaved forests) and uppercase letters NS mean not statistically significant in ECF (evergreen coniferous forests) (both at the $p < 0.05$ level). Sample size at each elevation was 5.

2.3. Canopy Openness

Understory light conditions were measured by hemispheric photography. Canopy openness was defined as the fraction of open sky in a hemisphere, visible from a point beneath the canopy [42]. A camera (Nikon Coolpix 950 with FC-E8 fisheye; Nikon Corp., Tokyo, Japan) was leveled and oriented above each sapling and then pictures were taken with its fisheye lens. The photographs were analyzed using the Adobe Photoshop® (San Jose, CA, USA) software to calculate the ratio of the canopy area to the entire sky area as a measure of canopy openness [43]. The pictures were taken during the morning on the same day as the photosynthetic measurements.

2.4. Soil Properties

Soils near each sapling were collected with soil corers (6 soil cores for each sapling). Organic and mineral layers of each core were separated and then the organic layer was screened. After the stones and plant residues were set aside, the organic layer of each sapling was thoroughly mixed and spread on shallow aluminum trays, and then air-dried. Next, the air-dried samples were finely ground and homogenized for the nutrient determination. Soil pH (pH soil) was determined from a deionized water-based saturated paste extract. The total soil carbon content (C soil) was determined by a simplified colorimetric method [44]. The soil total nitrogen content (N soil) was determined by the semimicro Kjeldahl method [45]. Total soil phosphorus content (P soil) was determined by a colorimetric method [46]. Soil water content (WC soil) is defined as the dry weight divided by the wet weight of the soil.

2.5. Needle Gas Exchange

Because we used saplings, that photosynthesis could be measured without cutting any branches from trees. We also decided to take needle samples at the end of the growing season in 2014, as suggested by Shi et al. [47]. Photosynthetic light response curves for each young tree were obtained from fully expanded, mature, and visibly healthy one-year-old needles. Five young trees were measured on clear and sunny days at each elevation. Determinations were taken in situ with a portable photosynthetic system (LI 6400, Li-Cor, Lincoln, NE, USA) with a built-in red-blue light source (LI-6400-02B). Measurements were taken from low to high elevations of CO_2 levels maintained at 380 $m \cdot mol \cdot mol^{-1}$, temperature at 25 ± 1 °C, water vapor within leaf chamber at 21 ± 1 $m \cdot mol \cdot mol^{-1}$, and a constant flow rate at 500 $mL \cdot min^{-1}$.

A pre-experiment was conducted to determine the saturation light intensity. Before measurements for the light response curves, needles were given a dose of photosynthetically active radiation (PAR) of 1000 $\mu mol \cdot m^{-2} \cdot s^{-1}$ (approximately equal to the saturation light intensity) for at least 10 min in the chamber until CO_2 uptake was steady-state, where steady-state in practice means that CO_2 uptake shows no systematic increase or decrease ($\pm 2\%$) over a 5 min period. This is important to ensure a steady-state activation of Rubisco. Measurements were taken at 14 different light levels (2100, 1800, 1500, 1200, 900, 600, 400, 200, 150, 100, 50, 20, 10, and 0 $\mu mol \cdot m^{-2} \cdot s^{-1}$) and recorded automatically. The minimum waiting time was 90 s and the maximum 120 s at each light level. After the measurements, projected areas of measured needles were determined with a scanner using leaf area analysis software. Since the needles are three-sided, the red-blue light illuminated the largest needle side. All rates of gas exchange were based on the projected needle area. Predicted nonlinear light response curves were constructed and fit the observations well ($R^2 > 0.99$). Last, we calculated the photosynthetic parameters (AQY, LSP, Asat, WUE, and Cond) from the light response curve data. According to the relationship between PPFD and Pn, light-response curves of *Pinus koraiensis* at different elevations were fitted by a nonlinear model [48]:

$$P(I) = \alpha \frac{1 - 2\beta I - \beta \gamma I^2 + (\gamma + \beta) Ic}{(1 + \gamma I)^2}$$

2.6. Needle Chlorophyll Concentration

The five groups of one-year-old needles mentioned earlier were used to determine the difference in the amount of needle chlorophyll. The chlorophyll from these fresh needles was extracted with an 80% (v/v) acetone solution by grinding and centrifuging methods and their concentrations quantified, using a 722 spectrometer (the 3rd Analytical Instrument Company of Shanghai, Shanghai, China). The amounts of chlorophyll a (Chl a, $mg \cdot g^{-1}$), chlorophyll b (Chl b, $mg \cdot g^{-1}$), and total chlorophyll (Chl, $mg \cdot g^{-1}$) were expressed as the amount of chlorophyll per fresh needle mass. In our calculations, we used the equation by Arnon [49]:

$$\text{Chl a} = 12.7 \times OD663 - 2.69 \times OD645; \text{Chl b} = 22.9 \times OD645 - 4.86 \times OD663; \text{Chl} = \text{Chl a} + \text{Chl b}$$

where OD645 and OD663 are the optical densities at wavelengths of 645 nm and 662 nm, respectively.

2.7. Needle Morphological Trait

Projected areas of the five groups of one-year-old needles from five saplings at each elevation were determined. Needles were laid out and scanned separately (CanoScan LiDE 120, Tokyo, Japan), with a reference object for scale. The total projected needle area for each sapling was calculated with image processing software (Image J; National Institute of Mental Health, Bethesda, MD, USA) [50]. Then, these needles were oven-dried (75 °C, 72 h) to a constant weight to obtain their dry weight and the LMA ($g \cdot m^{-2}$) of each sapling was calculated.

2.8. Needle Water Band Index

Thirty one-year-old needles from each of the five similar saplings at each elevation were selected. They were sampled evenly in five directions, south, east, north, west, and central. Spectral reflectance, at wavelengths from 310 to 1130 nm, was measured using a UniSpec Spectral Analysis System (PP Systems, Haverhill, MA, USA) with a 0.5 mm diameter optical fiber and an internal 5 V halogen lamp. Last, spectral water band index (WI = R900/R970) was calculated to characterize the plant water potential [51], where R900 and R970 are spectral reflectance at a wavelength of 900 nm and 970 nm, respectively.

2.9. Growth Properties

We harvested the 65 saplings and divided each into current-year needles, current-year branches, one-year-old needles, one-year-old branches, stems, and roots after measuring the gas exchange, chlorophyll concentration, spectral reflectance, and LMA of needles. Biomass samples were oven-dried (75 °C, 72 h) to a constant moisture level to obtain the dry weight of their current-year needle biomass (CN, g), their current-year branch biomass (CB, g), their one-year-old needle biomass (ON, g), and their one-year-old branch biomass (OB, g). The current-year needle biomass/total biomass ratio (CNR) and one-year-old needle biomass/total biomass ratio (ONR) of each sapling were then calculated.

2.10. Needle Nutrients

We finely ground and homogenized the dried one-year-old needles for the nutrient determination. Leaf C, N, and P content per unit leaf mass were determined by the same methods as the soil.

2.11. Statistical Analyses

Before our analyses, we tested for homogeneity of variances and normality of the data. All data were found to meet the requirements of the one-way analysis of variance (ANOVA) assumptions. Differences among means were determined by a Duncan's test at a significance level of <0.05. ANOVAs were used to test the effects of elevation on the various variables. Relationships between photosynthetic parameters, growth parameters, and elevational gradient were determined using a general linear regression model for each forest type. Pearson's correlation coefficients were used to detect the relationships between growth and other variables. Multivariate analysis of variance was used to help determine whether changes in elevation and forest type had significant effects on growth parameters. Redundancy analysis was used to identify the relationship between the leaf characteristic parameters, atmospheric index, soil factors, and growth data using the R vegan software package.

3. Results

3.1. Needle Properties of Young Pinus koraiensis Trees along an Elevational Gradient in Two Different Forest Types

The AQY increased with elevation in PBMF, i.e., when elevation was below 1100 m ($R^2 = 0.90$, $p < 0.01$) and also followed a monotonic trend with respect to elevation in ECF, ($R^2 = 0.33$, $p < 0.05$) (Figure 1). The LSP increased with elevation in both PBMF ($R^2 = 0.90$, $p < 0.01$) and ECF ($R^2 = 0.84$, $p < 0.01$) (Figure 1). Area-based Asat was also observed to increase with elevation in PBMF ($R^2 = 0.78$, $p < 0.01$) and ECF ($R^2 = 0.79$, $p < 0.01$) (Figure 1). The results showed no significant differences ($p > 0.05$) between iWUE at various elevations in either PBMF or ECF (Figure 1). In addition, Cond increased with elevation in PBMF ($R^2 = 0.74$, $p < 0.01$) (Figure 1).

There were no significant differences ($p > 0.05$) between the average needle Chl a, Chl b and total needle chlorophyll from various elevations. Taking all elevations into account, Chl a in needles ranged between 1.02 and 1.08 mg·g^{-1}, Chl b ranged between 0.41 and 0.45 mg·g^{-1} and total needle chlorophyll ranged between 1.42 and 1.53 mg·g^{-1}. Chl a, Chl b, and total needle chlorophyll were observed to be not significantly ($p > 0.05$) affected by forest type (Figure 2).

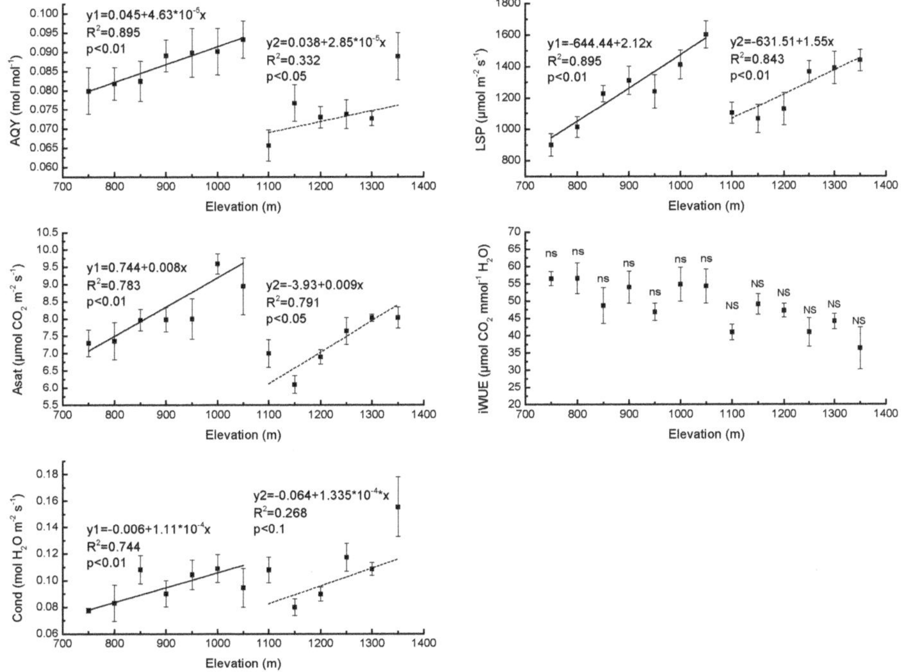

Figure 1. Regression equations of the mean values of apparent quantum yield (AQY), light saturation point (LSP), light saturation net photosynthetic rate (Asat), intrinsic water use efficiency (iWUE) and stomatal conductance (Cond) of needles in *Pinus koraiensis* Sieb. et Zucc. as a function of elevation on Changbai Mountain. The sample size was 7 for *Pinus koraiensis* and mixed broadleaved forests (solid lines) and 6 for evergreen coniferous forests (dotted lines). For each elevation, the sample size was 5. The abbreviations "ns" and "NS" mean not statistically significant in PBMF (*Pinus koraiensis* and mixed broadleaved forests) and ECF (evergreen coniferous forests), respectively (both at the $p < 0.05$ level).

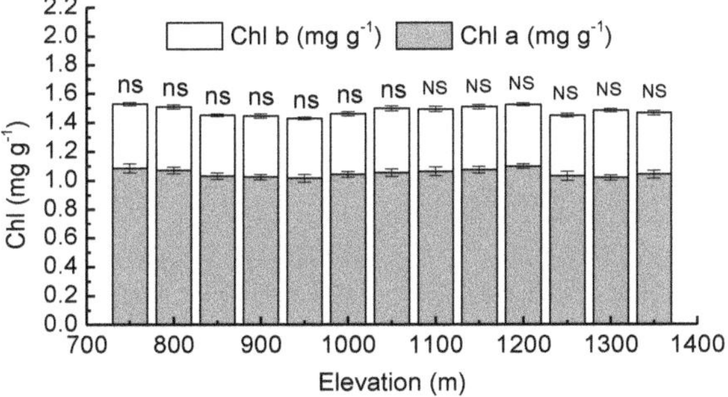

Figure 2. Mean values of chlorophyll a concentration (Chl a) and chlorophyll b concentration (Chl b) in needles of *Pinus koraiensis* at different elevations, Changbai Mountain. The abbreviations "ns" and "NS" mean not statistically significant in PBMF (*Pinus koraiensis* and mixed broadleaved forests) and ECF (evergreen coniferous forests), respectively (both at the $p < 0.05$ level). For each elevation, the sample size was 5.

Water band index was greatest at 1100 m (WI = 0.9838 ± 0.0007), and smallest at 1350 m (WI = 0.9509 ± 0.0378). With increasing elevation, the WI changed frequently (Figure 3).

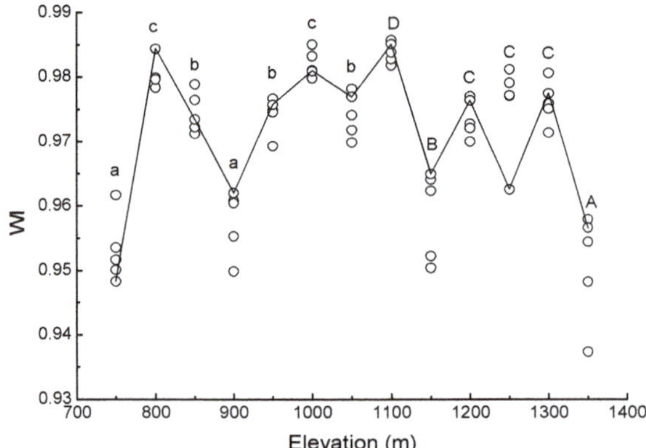

Figure 3. Scatter plot of water band index (WI) of needles in *Pinus koraiensis* at each elevation on Changbai Mountain. For each elevation, the sample size was 5. Lowercase letters denote significant differences among elevations in PBMF (*Pinus koraiensis* and mixed broadleaved forests). Uppercase letters denote significant differences in ECF (evergreen coniferous forests) (both at the $p < 0.05$ level).

Differences in LMA among the populations from different elevations were not significant ($p > 0.05$), either in PBMF or in ECF. LMA was also observed to be not significantly ($p > 0.05$) affected by forest type; mean values were 58.8 and 59.5 g·m^{-2} in PBMF and ECF, respectively (Figure 4).

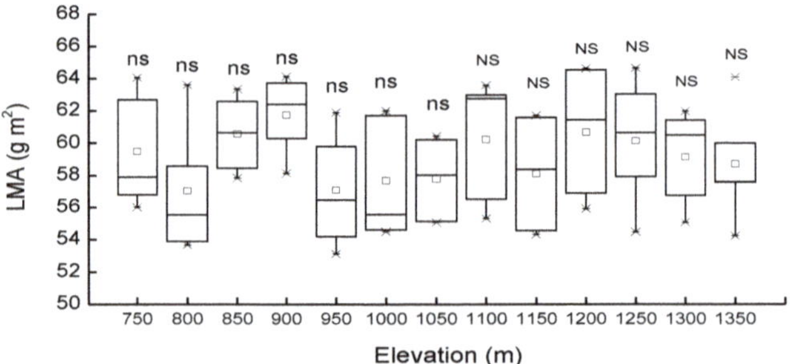

Figure 4. Mean values of leaf mass per area (LMA) in *Pinus koraiensis* at different elevations, Changbai Mountain. The abbreviations "ns" and "NS" means not statistically significant in PBMF (*Pinus koraiensis* and mixed broadleaved forests) and ECF (evergreen coniferous forests), respectively (at the $p < 0.05$ level). For each elevation, the sample size was 5.

There was a significant difference in Nleaf at different elevations in each forest type. Values ranged from 1.09% (± 0.04) to 1.31% (± 0.01) in PBMF and from 1.02% (± 0.01) to 1.24% (± 0.006) in ECF (Figure 5). In both PBMF and ECF, Pleaf increased with elevation ($R^2 = 0.94$, $p < 0.05$ for PBMF, $R^2 = 0.44$, $p < 0.05$ for ECF) (Figure 5). The lowest Cleaf (27.73% $\pm$ 1.67) occurred at an elevation of 750 m. At an elevation of 1050 m, Cleaf was 45% higher than at 750 m. In ECF, Cleaf increased by 38% from the lowest to the highest elevation (Figure 5).

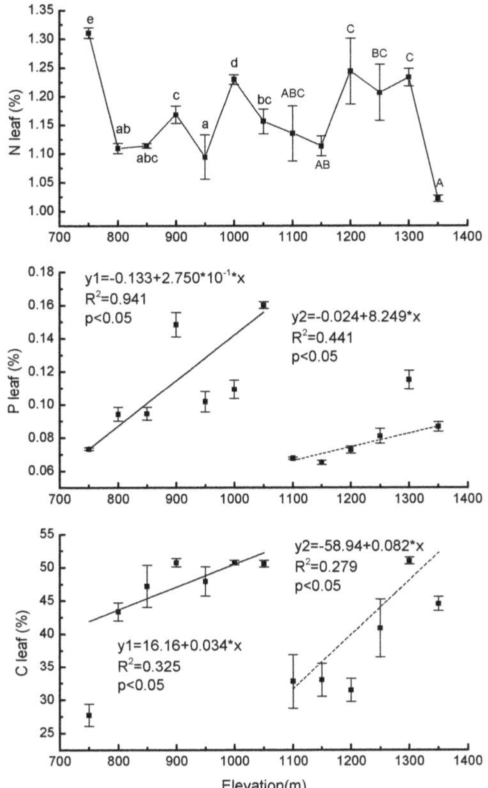

Figure 5. Regression equations of the mean values of total leaf N, total leaf P, and total leaf C content in *Pinus koraiensis* as a function of elevation on Changbai Mountain. The sample size was 7 for *Pinus koraiensis* and mixed broadleaved forests (solid lines) and 6 for evergreen coniferous forests (dotted lines). For each elevation, the sample size was 5. Lowercase letters denote significant differences among elevations in PBMF (*Pinus koraiensis* and mixed broadleaved forests). Uppercase letters denote significant differences in ECF (evergreen coniferous forests) (both at the $p < 0.05$ level).

3.2. Growth Properties of Young Pinus koraiensis Trees along an Elevational Gradient in Two Different Forest Types

Forest type and elevation had significant effects on the growth properties CN, CB, ON, OB, CNR, and ONR (Table 4).

Table 4. Results of multivariate ANOVA showing the effects of forest type and elevation on growth properties.

		Total	CN	CB	ON	OB	CNR	ONR
Forest Type	Pr(>F)	<2.2×10^{-16} ***	3.2×10^{-15} ***	7.533×10^{-13} ***	1.598×10^{-13} ***	2.516×10^{-14} ***	<2.2×10^{-16} ***	6.209×10^{-14} ***
	F	39.208	107.150	80.318	87.467	96.466	140.900	92.001
Elevation	Pr(>F)	1.805×10^{-9} ***	4.699×10^{-15} ***	1.895×10^{-14} ***	3.697×10^{-16} ***	<2.2×10^{-16} ***	<2.2×10^{-16} ***	5.336×10^{-16} ***
	F	2.691	19.262	17.976	21.788	120.520	22.425	21.408

Note: CN = biomass of current-year needles, CB = biomass of current-year branches, ON = biomass of one-year-old needles, OB = biomass of one-year-old branches, CNR = current-year needle biomass/total biomass ratio, ONR = one-year-old needle biomass/total biomass ratio. *** Correlation is significant at the <0.01 level (two-tailed). The sample size was 65.

Our results showed that CN, CB, ON, OB, CNR, and ONR increased significantly with increasing elevation (R^2 = 0.84, $p < 0.01$; R^2 = 0.55, $p < 0.05$; R^2 = 0.66, $p < 0.05$; R^2 = 0.49, $p < 0.05$; R^2 = 0.91, $p < 0.01$ and R^2 = 0.73, $p < 0.01$, respectively) in PBMF (Figure 6). In ECF, OB, CNR, and ONR also increased significantly with increasing elevation (R^2 = 0.65, $p < 0.05$) (Figure 6). There were also significantly higher ($p < 0.05$) CN, CB, ON, OB, CNR, and ONR values of *Pinus koraiensis* in PBMF when compared to the saplings in ECF (Table 5).

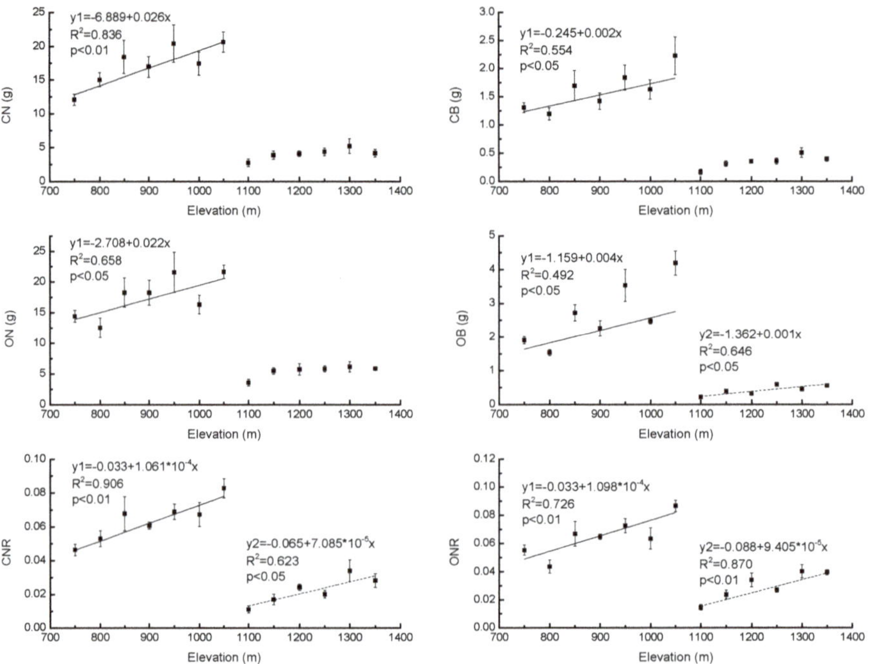

Figure 6. Regression equations of biomass of current-year needles (CN), biomass of current-year branches (CB), biomass of one-year-old needles (ON), biomass of one-year-old branches (OB), current-year needle biomass/total biomass ratio (CNR), and one-year-old needle biomass/total biomass ratio (ONR) of *Pinus koraiensis* at different elevations, Changbai Mountain. The sample size was 7 for *Pinus koraiensis* and mixed broadleaved forests (solid lines) and 6 for evergreen coniferous forests (dotted lines). For each elevation, the sample size was 5.

Results of redundancy analysis showed that the needle properties could explain 42.26% of the differences in growth in PBMF. The first two RDA axes explained 93.05% of the total variance of the relationship between the growth indices and the needle properties. The first RDA axis primarily reflected the changing trend of Nleaf, LMA and WI values. The correlation coefficients between these three factors and the first RDA axis were −0.41, −0.52, and 0.66, respectively. The second RDA axis primarily reflected the changing trend of Chl a, TC, and LSP, and correlation coefficients with the sorting axis were −0.39, −0.45, and 0.48. Growth index CN was positively correlated with WI, Chlb, and TC, and growth index OB was positively correlated with LSP (Figure 7). Correlation analysis indicated that the growth index CN was positively correlated with Pleaf, Cleaf, LSP, and WI, and the growth index OB was positively correlated with Pleaf, Cleaf, and LSP (Table 5). These showed that WI and LSP were factors that significantly affect growth of *Pinus koraiensis* in PBMF. RDA and correlation analysis showed that growth of *Pinus koraiensis* was significantly correlated with Pleaf, Cleaf, and Asat in ECF (Figure 7, Table 5).

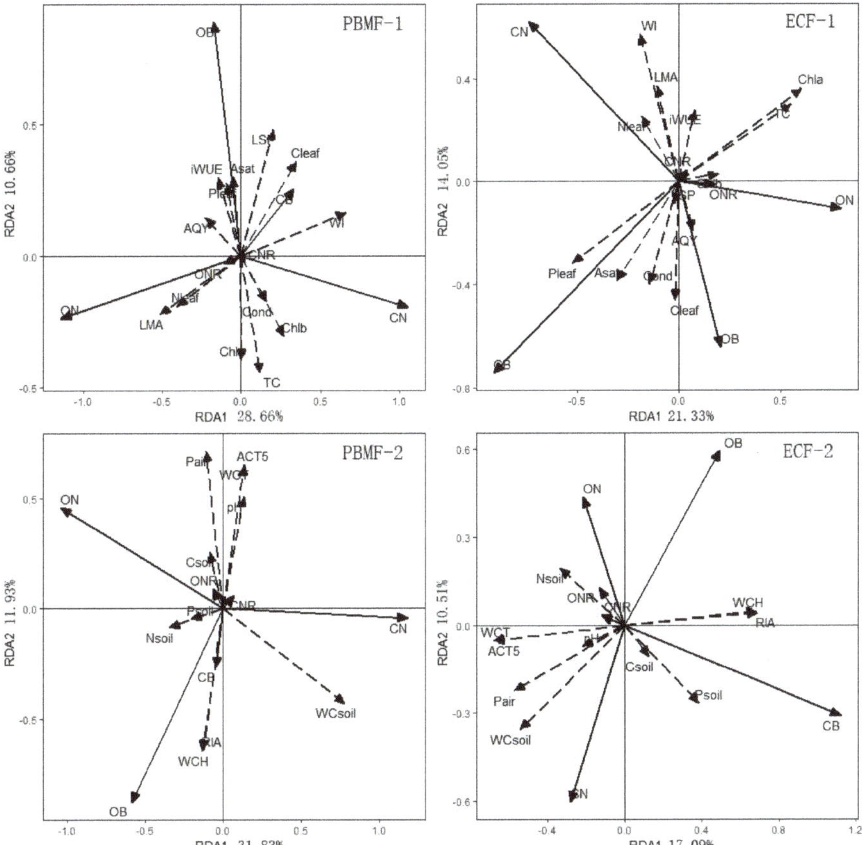

Figure 7. PBMF-1: RDA analysis between growth and needle properties in *Pinus koraiensis* and mixed broadleaved forests; PBMF-2: RDA analysis between growth properties and environmental factors in *Pinus koraiensis* and mixed broadleaved forests. ECF-1: RDA analysis between growth and needle properties in evergreen coniferous forests; ECF-2: RDA analysis between growth properties and environmental factors in evergreen coniferous forests.

The redundancy analysis of growth properties and environmental factors revealed that environmental factors could explain 45.11% of growth changes in PBMF, and 97.03% of the total variance of the relationship between growth and environmental factors was explained. The first sorting axis primarily reflected the changing trend of WC soil, N soil, and P soil. The correlation coefficients of these three factors and the first sorting axis were 0.78, −0.35, and −0.21, respectively. The second RDA axis primarily reflected the changing trend of Pair, RIA, ACT5, WCH, and WCT, and their correlation coefficients with the RDA axis were 0.72, −0.64, 0.65, −0.65, and 0.65. As can be seen from the RDA diagram, growth indicator OB was positively correlated with WCH and RIA, and negatively correlated with ACT5, WCT, and pH soil in PBMF (Figure 7). Correlation analysis also showed that growth index OB was positively correlated with WCH and RIA, and negatively correlated with WCT, ACT5, and Pair (Table 6). These results indicate that growth indicators were significantly influenced by WCH, RIA, WCT, ACT5, and Pair in PBMF; WCT, WCH, ACT5, RIA, WC soil, and P soil were the most relevant environmental factors limiting the growth of *Pinus koraiensis* in ECF (Figure 7, Table 6).

Table 5. Correlation coefficients of growth properties with leaf indicators at each forest type.

	Chl a	Chl b	TC	N Leaf	P Leaf	C Leaf	LMA	AQY	LSP	Asat	Cond	iWUE	WI
						PBMF							
CN	−0.031	0.187	0.056	−0.232	0.482 **	0.396 *	−0.234	0.228	0.522 **	0.175	0.310	−0.285	0.337 *
CB	−0.111	0.249	0.019	−0.153	0.440 **	0.325	−0.258	0.171	0.513 **	0.121	0.153	−0.134	0.214
ON	−0.020	0.115	0.034	−0.114	0.530 **	0.305	−0.090	0.296	0.491 **	0.203	0.259	−0.250	0.127
OB	−0.141	0.038	−0.098	−0.212	0.594 **	0.416 *	−0.201	0.291	0.607 **	0.266	0.106	−0.081	0.230
CNR	−0.019	0.173	0.060	−0.252	0.452 **	0.384 *	−0.202	0.191	0.518 **	0.206	0.356 *	−0.312	0.337 *
ONR	0.004	0.067	0.033	−0.061	0.488 **	0.224	0.012	0.274	0.434 **	0.231	0.251	−0.228	0.022
						ECF							
CN	−0.130	0.244	0.004	0.155	0.646 **	0.377 *	−0.005	0.268	0.512 **	0.300	0.174	−0.052	−0.112
CB	−0.322	0.169	−0.179	0.131	0.713 **	0.450 *	−0.100	0.326	0.458 *	0.384 *	0.343	−0.197	−0.227
ON	0.001	0.289	0.127	0.132	0.611 **	0.377 *	−0.027	0.324	0.538 **	0.288	0.176	−0.006	−0.230
OB	−0.112	0.182	−0.008	−0.108	0.593 **	0.535 **	0.007	0.492 **	0.441 *	0.441 *	0.357	−0.133	−0.422
CNR	−0.119	0.242	0.012	0.147	0.644 **	0.388 *	0.024	0.262	0.521 **	0.306	0.179	−0.054	−0.102
ONR	0.042	0.303	0.166	0.106	0.602 **	0.412 *	−0.023	0.327	0.565 **	0.295	0.197	−0.015	−0.260

Note: * Correlation is significant at the 0.05 level (two-tailed). ** Correlation is significant at the 0.01 level (two-tailed). The sample size was 35 for PBMF, *Pinus koraiensis*, and mixed broadleaved forests and 30 for ECF, evergreen coniferous forests.

Table 6. Growth properties of each forest type and their correlation coefficients with environment factors.

	Mean ± S.E.	WCT	WCH	ACT5	RIA	Pair	N Soil	C Soil	P Soil	pH Soil	WC Soil
						PBMF					
CN	18.15 ± 6.64 a	−0.647 **	0.648 **	−0.647 **	0.647 **	−0.611 **	−0.040	−0.185	−0.198	−0.185	0.226
CB	1.71 ± 0.77 a	−0.629 **	0.630 **	−0.629 **	0.629 **	−0.631 **	−0.032	−0.287	−0.126	−0.292	0.087
ON	18.33 ± 6.58 a	−0.665 **	0.666 **	−0.665 **	0.665 **	−0.555 **	0.039	−0.125	−0.133	−0.233	−0.058
OB	2.79 ± 1.28 a	−0.795 **	0.796 **	−0.795 **	0.795 **	−0.747 **	0.045	−0.175	−0.060	−0.450 **	0.035
CNR	0.06 ± 0.02 a	−0.616 **	0.616 **	−0.616 **	0.616 **	−0.563 **	−0.033	−0.139	−0.209	−0.132	0.238
ONR	0.06 ± 0.02 a	−0.599 **	0.599 **	−0.599 **	0.599 **	−0.446 **	0.074	−0.059	−0.114	−0.182	−0.170
						ECF					
CN	4.62 ± 2.89 b	−0.605 **	0.604 **	−0.605 **	0.605 **	−0.497 **	−0.116	−0.022	0.215	−0.214	−0.257
CB	0.40 ± 0.24 b	−0.734 **	0.734 **	−0.734 **	0.734 **	−0.594 **	−0.100	0.047	0.423 *	−0.217	−0.397 *
ON	6.02 ± 3.16 b	−0.683 **	0.682 **	−0.683 **	0.683 **	−0.569 **	−0.048	0.029	0.285	−0.231	−0.401 *
OB	0.46 ± 0.25 b	−0.874 **	0.874 **	−0.874 **	0.874 **	−0.845 **	0.112	0.072	0.348	−0.304	−0.555 **
CNR	0.02 ± 0.01 b	−0.617 **	0.617 **	−0.617 **	0.617 **	−0.509 **	−0.105	0.025	0.201	−0.244	−0.254
ONR	0.03 ± 0.01 b	−0.724 **	0.723 **	−0.724 **	0.724 **	−0.609 **	−0.031	0.083	0.281	−0.274	−0.439 *

Note: * Correlation is significant at the 0.05 level (two-tailed). ** Correlation is significant at the 0.01 level (two-tailed). The sample size was 35 for PBMF, *Pinus koraiensis*, and mixed broadleaved forests and 30 for ECF, evergreen coniferous forests. Lowercase letters denote significant differences between PBMF and ECF.

4. Discussion

4.1. Response of Needle Properties to Elevational Gradient in Two Different Forest Types

Differences in photosynthetic characteristics of *Pinus koraiensis* between high and low elevation plants have been documented in our study. We showed that AQY, Asat, and LSP increased significantly with the increase in elevation for both PBMF and ECF. These results indicate that the ability to utilize weak light, the ability to adapt to strong illumination and maximum photosynthetic potential increase with elevation in both PBMF and ECF. Further, the AQY and Asat of PBMF are greater than those of ECF. There was no significant elevational difference in iWUE in either PBMF or ECF, suggesting a simultaneous decrease in photosynthesis with water deficit-induced lower stomatal conductance along the elevational gradient in each forest type. The maximum photosynthetic potential in plants at a given elevation has been variously found to be equal, lower, or higher at higher elevations compared to lower elevations [52]. A previous study showed that conifer populations from high elevations have evolved to exhibit higher maximum rates of CO_2 assimilation than trees from low elevations [53]. There is also some previous evidence that photosynthetic capacity decreases with increasing elevation [54,55], which may be caused by lower activity of Rubisco [52]. These inconsistent results may be explained by differences in plant material, equipment, elevational range, and conditions [55]. Photosynthesis is dependent on stomata for its supply of CO_2 [56]. It is generally accepted that photosynthesis increases when the stomata open and decreases when the stomata close [57,58]. However, a study by Farquhar and Sharkey (1982) [59] provided no evidence of a positive correlation between photosynthesis and Cond. Our study shows that Asat is affected primarily by stomatal factors in each forest type. Kumar et al. [60] reported that the Cond displayed some degree of plasticity in response to elevation and suggested that this could be one of the adaptive features allowing a wider elevational distribution of plants.

The chlorophyll concentration may be an important nonstomatal factor that affects the photosynthetic characteristics of plants. Studies have shown steady declines in total chlorophyll concentrations over an elevational range from 650 m a.s.l. to 1950 m a.s.l. [61]. Chlorophyll degradation is generally related to stress [62]. With increasing elevation, plants are often exposed to low temperatures, large diurnal temperature fluctuations, high UV-B radiation, and low partial CO_2 pressures: these conditions are harmful to chlorophyll formation and conducive to chlorophyll degradation, thus serving as negative factors for plant growth [63,64]. It has also been reported that high elevation populations tend to have higher leaf chlorophyll concentrations than those from low elevations [65]. Higher chlorophyll concentrations could increase leaf photosynthetic efficiency as an important protective strategy for a harsher alpine environment [66]. In our study, the chlorophyll concentration in needles did not change significantly with increasing elevation in PBMF or ECF, nor was it correlated with AQY, LSP, and Asat. The elevational disparity may not be large enough to affect the needle chlorophyll concentration of *Pinus koraiensis* saplings.

Stomatal closure occurs when water availability is reduced [67]. Stomatal conductance may be a good indicator of plant water status, but it indicates only a short-term response [68]. The water band index (WI) can track variation in stomatal aperture [69], so it has utility for predicting components of plant water status including leaf water potential [25], relative water content [70] and water content as a percentage of dry mass [71]. The present study shows that WI varied significantly with elevation, and this result based on the soil moisture content for each forest type at the time of study.

LMA is a widely used index in functional ecology because it is thought to reflect relative growth and important physiological traits, such as photosynthetic rate. An increase in LMA with increasing elevation has been generally reported in natural populations of herbs and shrubs [72], conifers [73], and broadleaved tree species [74,75] and across a wide spectrum of tree taxa reaching the tree line [76]. LMA was also found to remain constant along narrow elevational gradients [77–79] or even to decrease [80]. Additionally, earlier studies have suggested that greater photosynthetic capacity is often related to higher levels of LMA, which is enhanced by ambient irradiance leading to facilitated growth [81–83]. Differences in the accumulation of starch or the number of palisade cell layers [84] are likely to result in inconsistent findings. Plants growing under low irradiance have thinner leaves, consequently having lower LMA [85]. Here, we found no elevational differences in LMA, possibly because the plants we sampled spend long periods of time in the understory. This would be an advantage for the survival of *Pinus koraiensis* seedlings and saplings [83,86], although further investigation is needed.

The C leaf and P leaf ranges of needles in *Pinus koraiensis* were consistent with the results of previous studies [87]. The range of total nitrogen content in leaves of *Pinus koraiensis* was also consistent with that of previous measurements [88]. Significant differences in N leaf were detected in both forest types, but these responses to elevational gradients varied nonlinearly with increasing elevation, which was similar to previous studies on spruce [89]. The content of total leaf nitrogen in samples at the lowest elevation was the highest and the content in the highest elevation was the lowest, because N is relatively less limiting due to high rates of plant turnover and decomposition as well as abundant N fixers at low elevations [90]. The increase in P leaf in sampled saplings with increasing elevation in both forest types may result from leaching of highly weathered soils, as well as chemical processes that immobilize P within the soils [91]. Our results suggested increased N limitation and decreased P limitation with increasing elevation, results consistent with studies by Fisher et al. [15]. Additionally, our elevational variation of leaf photosynthetic characteristics explained the elevational variation of leaf carbon content well although further research is needed in this area due to our small sample size.

There are striking differences in many leaf traits between sun-exposed and shade leaves, such as leaf photosynthesis and chlorophyll concentration [92], LMA [93], leaf water potential [94], and leaf nutrients [95]. Elevational variation observed in this study could be related to changes in sun-exposed leaves, in shade leaves or in both. As leaves in this study were not differentiated between sun and shade, we cannot distinguish any effect of type of leaves upon these traits. Future investigations should account for these potential differences.

4.2. Response of Growth Properties to Elevational Gradient in Two Different Forest Types

Elevational gradient and forest type affect the growth of plants. Studies have shown that growth does not follow a simple linear trend as a function of elevation. Based on data from a tropical forest transect in the Peruvian Andes, Girardin et al. [96] demonstrated that net primary productivity does not vary linearly with elevation. Analyzing data from 2400 trees across a 1650 m elevational gradient in Kosnipata Valley, Peru, Rapp et al. [97] also found that growth does not show a consistent trend with elevation within species, although higher-elevation species had lower growth rates than lower-elevation species. In our study, the absolute annual biomass accumulation (CN, CB, ON, and OB) of *Pinus koraiensis* increased with increasing elevation in both PBMF and ECF. However, an inflection point was observed at the junction of these two forest types; furthermore, the average CN, CB, ON, and OB in PBMF were significantly higher than those in ECF.

Growth in PBMF increased with increasing elevation within a certain temperature range, a pattern that may be largely determined by precipitation. Based on data from ring-width chronologies, Yu et al. [98] found that precipitation in the previous September and the current June is the primary limiting factor for growth of *Pinus koraiensis* at low-elevation sites. Higher precipitation in September results in a greater amount of moisture available in the soil, a condition advantageous to the growth of *Pinus koraiensis* in the next year [30]. Precipitation (including that in the previous September and the June) increases with an increase in elevation [41], which would be the main reason why the absolute biomass and relative growth rate increase with increasing elevation of PBMF. Wang et al. (2013) [30] built regression equations to predict the future growth of *Pinus koraiensis* under future climate change scenarios. They found that the radial growth of *Pinus koraiensis* will increase at higher elevations relative to lower elevations, which is consistent with our findings. Results of redundancy and correlation analysis indicated that the growth of *Pinus koraiensis* was significantly affected by LSP as well as WI, WCH, and RIA in PBMF. This may be because a large part of light used for photosynthesis of understory plants comes from the sunflecks. Therefore, plants should have higher light saturation points to adapt to strong sunflecks that may appear at any time.

When temperatures drop below a threshold value (the threshold value in this study is ACT5 = 2000 °C), growth abruptly slows, even with an increase in precipitation. This is probably the result of a sudden decline in the rate of metabolism when the temperature is below a certain range [99], coupled with the decreasing number of growing degree-days [100]. As is generally known, frost damage to growth increases with increasing elevation [101]. Other studies have shown that the current July temperature (decreasing with rising elevation) is the main limiting factor for the growth of *Pinus koraiensis*, because a decrease in temperature may lead to a delay in the onset of the growth period and the termination of growth before the end of the normal growing season [30]. In addition, decreases in temperature along an elevational gradient result in changes of forest type, i.e., from PBMF to ECF, leading to a variety of stand compositions and developmental stages; soil types [102] and rates of mineralization of dead organic matter and nutrient cycling [103] may also differ, all of which affect plant growth. As the result of the interaction of many factors, high elevation saplings (ECF) showed much more stress in growth when compared with saplings growing at a low elevation (PBMF); hence the greater growth in PBMF than in ECF.

Within a certain range of low temperatures in ECF, growth increases again with increasing elevation. The increase in Asat caused by increasing precipitation is a likely explanation for this pattern and represents an important strategy for plant survival under adverse environmental conditions. Asat is an important photosynthetic parameter representing maximum photon utilization capacity in plants; hence, Asat reflects the net assimilation rate [104] and a relatively high photosynthetic rate would result in a higher growth rate. Our results also showed that there was a significant positive correlation between plant growth and P soil and P leaf in ECF. According to some reports, P may be more limiting than N [105]. The amount of phosphorus in the soil is related to the high degree of weathering of apatite by physical and chemical functions and also related to the chemical process of fixing phosphorus within the soil. Due to the high rate of regeneration and decomposition by plants and animals, and the abundant nitrogen-fixing microorganisms present, N is much less a limiting factor than P in soil. The increase of leaf phosphorus with increasing elevation is another important reason for the increasing growth in ECF.

Low temperatures may explain why few *Pinus koraiensis* saplings were found above their current upper elevational limits (1350 m). At very low temperatures, plants cannot perform normal physiological activities and cells cannot differentiate properly. Low temperatures there also limit both the survival and activity of squirrels (*Sciurus* spp.), which are the primary seed dispersers of *Pinus koraiensis* [33] and facilitators of germination of *Pinus koraiensis* seeds [106].

5. Conclusions

The responses of needle and growth properties to elevational gradients (low elevation to upper elevational limits) and the possible underlying morphological and physiological mechanisms in natural young *Pinus koraiensis* trees were studied for the first time on Changbai Mountain. These responses and adaptations to elevation are related to forest type. The growth of young *Pinus koraiensis* trees increases with rising elevations in each vegetation belt. Redundancy and correlation analysis has shown that the growth of plants is closely related to the light saturation point, leaf water potential, mean within-crown humidity, annual precipitation, cumulative temperature ($\geq$5 °C), within-crown air temperature, and atmospheric pressure in *Pinus koraiensis* and mixed broadleaved forests; in evergreen forests, the leaf C, leaf P content, net rate of light saturation in photosynthesis, water content of soil, within-crown humidity, annual precipitation, cumulative temperature ($\geq$5 °C), within-crown air temperature, and total soil P content displayed a significant relationship with growth of *Pinus koraiensis*. This study will help us to understand better the ecophysiological processes that may enable young *Pinus koraiensis* trees to adapt to varying environments and to evaluate the species' adaptive potential in scenarios of climate change, although further studies and demonstrations are needed. These data could also provide baseline data for future work on exploring the contribution of young trees to estimated forest carbon stocks and refining a leaf trait database. Further investigation of these issues is required before these results can be used in a broader range of species and conditions.

Author Contributions: Conceptualization, Y.F., W.K.M., and Y.C.; Data Curation, Y.F.; Formal Analysis, Y.F.; Funding Acquisition, Y.C.; Investigation, Y.F. and Y.C.; Methodology, Y.F., W.K.M., and Y.C.; Project administration, Y.C.; Writing—Original Draft, Y.F.; Writing—Review and Editing, Y.F., W.K.M., and Y.C.

Funding: This research was funded by (the Key Project of National Key Research and Development Plan) grant number (2017YFC050400101) and (the Program of National Natural Science Foundation of China) grant number (31670643).

Acknowledgments: We thank Zhen Huang, Junwei Wang, Shuai Liu, Shasha Song, Xian Wu, Yizhi Zhou, and Lei Yi, for field work assistance.

Conflicts of Interest: The authors declare no conflict of interest.

References

1. Sundqvist, M.K.; Sanders, N.J.; Wardle, D.A. Community and ecosystem responses to elevational gradients: Processes, mechanisms, and insights for global change. *Ann. Rev. Ecol. Evol. Syst.* **2013**, *44*, 261–280. [CrossRef]
2. Panek, J.A.; Waring, R.H. Stable carbon isotopes as indicators of limitations to forest growth imposed by climate stress. *Ecol. Appl.* **1997**, *7*, 854–863. [CrossRef]
3. Miller, J.M.; Farquhar, G.D. Carbon isotope discrimination by a sequence of eucalyptus species along a subcontinental rainfall gradient in Australia. *Funct. Ecol.* **2001**, *15*, 222–232. [CrossRef]
4. Kouwenberg, L.L.R.; Kürschner, W.M.; Mcelwain, J.C. Stomatal frequency change over altitudinal gradients: Prospects for paleoaltimetry. *Rev. Mineral. Geochem.* **2007**, *66*, 215–241. [CrossRef]
5. Korner, C. *Alpine Plant Life: Functional Plant Ecology of High Mountain Ecosystems*; Springer: Berlin/Hamburg, Germany, 1999; p. 1501.
6. Fukami, T.; Wardle, D.A. Long-term ecological dynamics: Reciprocal insights from natural and anthropogenic gradients. *Proc. R. Soc. Lond. B Biol. Sci.* **2005**, *272*, 2105–2115. [CrossRef] [PubMed]
7. Walker, L.R.; Wardle, D.A.; Bardgett, R.D.; Clarkson, B.D. The use of chronosequences in studies of ecological succession and soil development. *J. Ecol.* **2010**, *98*, 725–736. [CrossRef]
8. Malhi, Y.; Silman, M.; Salinas, N.; Bush, M.; Meir, P.; Saatchi, S. Introduction: Elevation gradients in the tropics: Laboratories for ecosystem ecology and global change research. *Glob. Chang. Biol.* **2010**, *16*, 3171–3175. [CrossRef]
9. Malhi, Y.; Girardin, C.A.; Goldsmith, G.R.; Doughty, C.E.; Salinas, N.; Metcalfe, D.B.; Huaraca, H.W.; Silvaespejo, J.E.; Del, A.J.; Farfán, A.F. The variation of productivity and its allocation along a tropical elevation gradient: A whole carbon budget perspective. *New Phytol.* **2017**, *214*, 1019–1032. [CrossRef]
10. Terashima, I.; Masuzawa, T.; Ohba, H.; Yokoi, Y. Is photosynthesis suppressed at higher elevations due to low CO_2 pressure? *Ecology* **1995**, *76*, 2663–2668. [CrossRef]
11. Wang, Q.; Iio, A.; Tenhunen, J.; Kakubari, Y. Annual and seasonal variations in photosynthetic capacity of fagus crenata along an elevation gradient in the naeba mountains, Japan. *Tree Physiol.* **2008**, *28*, 277–285. [CrossRef]
12. Körner, C.; Cochrane, P.M. Stomatal responses and water relations of eucalyptus pauciflora in summer along an elevational gradient. *Oecologia* **1985**, *66*, 443–455. [CrossRef] [PubMed]
13. Chen, T.; Zhao, Z.; Zhang, Y.; Qiang, W.; Feng, H.; An, L.; Li, Z. Physiological variations in chloroplasts of rhodiola coccinea along an altitudinal gradient in tianshan mountain. *Acta Physiol. Plant.* **2012**, *34*, 1007–1015. [CrossRef]
14. Reinhardt, K.; Castanha, C.; Germino, M.J.; Kueppers, L.M. Ecophysiological variation in two provenances of pinus flexilis seedlings across an elevation gradient from forest to alpine. *Tree Physiol.* **2011**, *31*, 615–625. [CrossRef] [PubMed]
15. Fisher, J.B.; Malhi, Y.; Cuba Torres, I.; Metcalfe, D.B.; van de Weg, M.J.; Meir, P.; Silva-Espejo, J.E.; Huaraca Huasco, W. Nutrient limitation in rainforests and cloud forests along a 3,000-m elevation gradient in the peruvian andes. *Oecologia* **2013**, *172*, 889–902. [CrossRef] [PubMed]
16. Chapin, F.S.I.; Matson, P.A.I.; Mooney, H.A. Principles of terrestrial ecosystem eology. In *Terrestrial Decomposition*; Springer: Berlin/Hamburg, Germany, 2002.

17. Kramer, P.J.; Kozlowski, T.T. 17–environmental and cultural factors affecting growth. In *Physiology of Woody Plants*; Academic Press: Cambridge, MA, USA, 1979; pp. 628–702.

18. Jensen, A.M.; Gardiner, E.S.; Vaughn, K.C. High-light acclimation in quercus robur l. Seedlings upon over-topping a shaded environment. *Environ. Exp. Bot.* **2012**, *78*, 25–32. [CrossRef]

19. Wang, Q.; Zhang, Q.; Fan, D.; Lu, C. Photosynthetic light and CO_2 utilization and c4 traits of two novel super-rice hybrids. *Plan. Physiol.* **2006**, *163*, 529–537. [CrossRef] [PubMed]

20. Kneeshaw, D.D.; Kobe, R.K.; Coates, K.D.; Messier, C. Sapling size influences shade tolerance ranking among southern boreal tree species. *J. Ecol.* **2010**, *94*, 471–480. [CrossRef]

21. Pastur, G.M.; Lencinas, M.V.; Peri, P.L.; Arena, M. Photosynthetic plasticity of nothofagus pumilio seedlings to light intensity and soil moisture. *For. Ecol. Manag.* **2007**, *243*, 274–282. [CrossRef]

22. Roberntz, P.; Stockfors, J. Effects of elevated cO_2 concentration and nutrition on net photosynthesis, stomatal conductance and needle respiration of field-grown norway spruce trees. *Tree Physiol.* **1998**, *18*, 233–241. [CrossRef] [PubMed]

23. Ziska, L.H.; Sullivan, J.H. Physiological sensitivity of plants along an elevational gradient to uv-b radiation. *Am. J. Bot.* **1992**, *79*, 863–871. [CrossRef]

24. Huang, R.; Zhu, H.; Liu, X.; Liang, E.; Grießinger, J.; Wu, G.; Li, X.; Bräuning, A. Does increasing intrinsic water use efficiency (iwue) stimulate tree growth at natural alpine timberline on the southeastern tibetan plateau? *Glob. Planet. Chang.* **2017**, *148*, 217–226. [CrossRef]

25. Peñuelas, J.; Filella, I.; Biel, C.; Serrano, L.; Savé, R. The reflectance at the 950–970 nm region as an indicator of plant water status. *Int. J. Remote Sens.* **1993**, *14*, 1887–1905. [CrossRef]

26. Li, Y.; Yang, D.; Xiang, S.; Li, G. Different responses in leaf pigments and leaf mass per area to altitude between evergreen and deciduous woody species. *Aust. J. Bot.* **2013**, *61*, 424–435. [CrossRef]

27. Wright, I.J.; Reich, P.B.; Westoby, M.; Ackerly, D.D.; Baruch, Z.; Bongers, F.; Cavenderbares, J.; Chapin, T.; Cornelissen, J.H.; Diemer, M. The worldwide leaf economics spectrum. *Nature* **2004**, *428*, 821–827. [CrossRef]

28. Reich, P.B.; Walters, M.B.; Kloeppel, B.D.; Ellsworth, D.S. Different photosynthesis-nitrogen relations in deciduous hardwood and evergreen coniferous tree species. *Oecologia* **1995**, *104*, 24–30. [CrossRef]

29. Fang, O.; Wang, Y.; Shao, X. The effect of climate on the net primary productivity (npp) of Pinus koraiensis in the changbai mountains over the past 50 years. *Trees* **2016**, *30*, 281–294. [CrossRef]

30. Wang, H.; Shao, X.; Jiang, Y.; Fang, X.; Wu, S. The impacts of climate change on the radial growth of Pinus koraiensis along elevations of changbai mountain in northeastern china. *For. Ecol. Manag.* **2013**, *289*, 333–340. [CrossRef]

31. Zhang, Y.; Drobyshev, I.; Gao, L.; Zhao, X.; Bergeron, Y. Disturbance and regeneration dynamics of a mixed korean pine dominated forest on changbai mountain, north-eastern China. *Dendrochronologia* **2014**, *32*, 21–31. [CrossRef]

32. Yu, D.; Wang, Q.; Liu, J.; Zhou, W.; Qi, L.; Wang, X.; Zhou, L.; Dai, L. Formation mechanisms of the alpine erman's birch (*Betula ermanii*) treeline on changbai mountain in northeast china. *Trees* **2014**, *28*, 935–947. [CrossRef]

33. Hutchins, H.E.; Hutchins, S.A.; Liu, B.W. The role of birds and mammals in korean pine (*Pinus koraiensis*) regeneration dynamics. *Oecologia* **1996**, *107*, 120–130. [CrossRef]

34. Kang, K.S.; Choi, W.Y.; Han, S.U.; Kim, C.S. Effective number and seed production in a clonal seed orchard of pznus korazenszs'. *For. Genet.* **2004**, *11*, 277–280.

35. Kim, J.H.; Lee, H.J.; Jeong, S.J.; Lee, M.H.; Kim, S.H. Essential oil of Pinus koraiensis leaves exerts antihyperlipidemic effects via up-regulation of low-density lipoprotein receptor and inhibition of acyl-coenzyme a: Cholesterol acyltransferase. *Phytother. Res.* **2012**, *26*, 1314–1319. [CrossRef] [PubMed]

36. Cho, S.M.; Lee, E.O.; Kim, S.H.; Lee, H.J. Essential oil of Pinus koraiensis inhibits cell proliferation and migration via inhibition of p21-activated kinase 1 pathway in hct116 colorectal cancer cells. *BMC Complement. Altern. Med.* **2014**, *14*, 275. [CrossRef] [PubMed]

37. Li, W.H. Degradation and restoration of forest ecosystems in china. *For. Ecol. Manag.* **2004**, *201*, 33–41.

38. Zhu, J.; Mao, Z.; Hu, L.; Zhang, J. Plant diversity of secondary forests in response to anthropogenic disturbance levels in montane regions of northeastern china. *J. For. Res.* **2007**, *12*, 403–416. [CrossRef]

39. Sun, Y.; Zhu, J.; Sun, O.J.; Yan, Q. Photosynthetic and growth responses of Pinus koraiensis seedlings to canopy openness: Implications for the restoration of mixed-broadleaved korean pine forests. *Environ. Exp. Bot.* **2016**, *129*, 118–126. [CrossRef]

40. Royo, A.A.; Carson, W.P. On the formation of dense understory layers in forests worldwide: Consequences and implications for forest dynamics, biodiversity, and succession. *Can. J. For. Res.* **2006**, *36*, 1345–1362. [CrossRef]

41. Fan, B.; Sang, W.; Axmacher, J.C. Forest vegetation responses to climate and environmental change: A case study from changbai mountain, ne china. *For. Ecol. Manag.* **2012**, *262*, 2052–2060.

42. Jennings, S.B.; Brown, N.D.; Sheil, D. Assessing forest canopies and understorey illumination: Canopy closure, canopy cover and other measures. *Forestry* **1999**, *72*, 59–74. [CrossRef]

43. Frazer, G.W.; Fournier, R.A.; Trofymow, J.; Hall, R.J. A comparison of digital and film fisheye photography for analysis of forest canopy structure and gap light transmission. *Agric. For. Meteorol.* **2001**, *109*, 249–263. [CrossRef]

44. Schollenberger, C.J. A rapid approximate method for determining soil organic matter. *Soil Sci.* **1927**, *24*, 65–68. [CrossRef]

45. Mitchell, A.K. Acclimation of pacific yew (*Taxus brevifolia*) foliage to sun and shade. *Tree Physiol.* **1998**, *18*, 749–757. [CrossRef] [PubMed]

46. Isaac Berenblum, E.C. An improved method for the colorimetric determination of phosphate. *Biochem. J.* **1938**, *32*, 295–298. [CrossRef]

47. Shi, P.; Körner, C.; Hoch, G. End of season carbon supply status of woody species near the treeline in western china. *Basic Appl. Ecol.* **2005**, *7*, 370–377. [CrossRef]

48. Ye, Z.-P. A new model for relationship between irradiance and the rate of photosynthesis in oryza sativa. *Photosynthetica* **2007**, *45*, 637–640. [CrossRef]

49. Arnon, D.I. Copper enzymes in isolated chloroplasts. Polyphenoloxidase in beta vulgaris. *Plant Physiol.* **1949**, *24*, 1. [CrossRef] [PubMed]

50. Blackman, C.J.; Brodribb, T.J.; Jordan, G.J. Leaf hydraulics and drought stress: Response, recovery and survivorship in four woody temperate plant species. *Plant Cell Environ.* **2009**, *32*, 1584–1595. [CrossRef] [PubMed]

51. Sims, D.A.; Gamon, J.A. Estimation of vegetation water content and photosynthetic tissue area from spectral reflectance: A comparison of indices based on liquid water and chlorophyll absorption features. *Remote Sens. Environ.* **2003**, *84*, 526–537. [CrossRef]

52. Korner, C.; Diemer, M. In situ photosynthetic responses to light, temperature and carbon dioxide in herbaceous plants from low and high altitude. *Funct. Ecol.* **1987**, *1*, 179–194. [CrossRef]

53. Saxe, H.; Cannell, M.G.R.; Johnsen, B.; Ryan, M.G.; Vourlitis, G. Tree and forest functioning in response to global warming. *New Phytol.* **2001**, *149*, 369–399. [CrossRef]

54. Cabrera, H.; Rada, F.; Cavieres, L. Effects of temperature on photosynthesis of two morphologically contrasting plant species along an altitudinal gradient in the tropical high andes. *Oecologia* **1998**, *114*, 145–152. [CrossRef] [PubMed]

55. Rada, F.; Azocar, A.; Gonzalez, J.; Briceño, B. Leaf gas exchange in espeletia schultzii wedd, a giant caulescent rosette species, along an altitudinal gradient in the venezuelan andes. *Acta Oecol.* **1998**, *19*, 73–79. [CrossRef]

56. Morison, J.I.L. Stomatal response to increased CO_2 concentration. *J. Exp. Bot.* **1998**, *49*, 443–452. [CrossRef]

57. Heber, U.; Neimanis, S.; Lange, O.L. Stomatal aperture, photosythesis and water fluxes in mesophyll cells as affected by the abscission of leaves. Simultaneous measurements of gas exchange, light scattering and chlorphyll fluorescence. *Planta* **1986**, *167*, 554–562. [CrossRef] [PubMed]

58. Bunce, J.A. Effects of boundary layer conductance on substomatal pressures of carbon dioxide. *Plant Cell Environ.* **1988**, *11*, 205–208. [CrossRef]

59. And, G.D.F.; Sharkey, T.D. Stomatal conductance and photosynthesis. *Ann. Rev. Plant Physiol.* **1982**, *33*, 317–345.

60. Kumar, N.; Kumar, S.; Ahuja, P.S. Photosynthetic characteristics of hordeum, triticum, rumex, and trifolium species at contrasting altitudes. *Photosynthetica* **2005**, *43*, 195–201. [CrossRef]

61. Liu, W.; Fan, X.; Wang, J.; Zhang, C.; Lu, W.; Gadow, K.V. Spectral reflectance response of fraxinus mandshurica leaves to above-and belowground competition. *Int. J. Remote Sens.* **2012**, *33*, 5072–5086. [CrossRef]

62. Curran, P.J. Exploring the relationship between reflectance red edge and chlorophyll content in slash pine. *Tree Physiol.* **1990**, *7*, 33–48. [CrossRef]

63. Haldimann, P. Effects of changes in growth temperature on photosynthesis and carotenoid composition in zea mays leaves. *Physiol. Plant.* **1996**, *97*, 554–562. [CrossRef]

64. Lefsrud, M.G.; Kopsell, D.A. Biomass production and pigment accumulation in kale grown under different radiation cycles in a controlled environment. *HortScience* **2006**, *41*, 1412–1415.

65. Oleksyn, J.; Modrzýnski, J.; Tjoelker, M.; Reich, P.; Karolewski, P. Growth and physiology of picea abies populations from elevational transects: Common garden evidence for altitudinal ecotypes and cold adaptation. *Funct. Ecol.* **1998**, *12*, 573–590. [CrossRef]

66. Ran, F.; Zhang, X.; Zhang, Y.; Korpelainen, H.; Li, C. Altitudinal variation in growth, photosynthetic capacity and water use efficiency of abies faxoniana rehd. Et wils. Seedlings as revealed by reciprocal transplantations. *Trees* **2013**, *27*, 1405–1416. [CrossRef]

67. Lovisolo, C.; Perrone, I.; Carra, A.; Ferrandino, A.; Flexas, J.; Medrano, H.; Schubert, A. Drought-induced changes in development and function of grapevine (*Vitis* spp.) organs and in their hydraulic and non-hydraulic interactions at the whole-plant level: A physiological and molecular update. *Funct. Plant Biol.* **2010**, *37*, 98–116. [CrossRef]

68. Chaves, M.M.; Zarrouk, O.; Francisco, R.; Costa, J.M.; Santos, T.; Regalado, A.P.; Rodrigues, M.L.; Lopes, C.M. Grapevine under deficit irrigation: Hints from physiological and molecular data. *Ann. Bot.* **2010**, *105*, 661–676. [CrossRef] [PubMed]

69. Dzikiti, S.; Verreynne, J.S.; Stuckens, J.; Strever, A.; Verstraeten, W.W.; Swennen, R.; Coppin, P. Determining the water status of satsuma mandarin trees [citrus unshiu marcovitch] using spectral indices and by combining hyperspectral and physiological data. *Agric. For. Meteorol.* **2010**, *150*, 369–379. [CrossRef]

70. Cibula, W.G.; Zetka, E.F.; Rickman, D.L. Response of thematic mapper bands to plant water stress. *Int. J. Remote Sens.* **1992**, *13*, 1869–1880. [CrossRef]

71. Ustin, S.L.; Roberts, D.A.; Pinzón, J.; Jacquemoud, S.; Gardner, M.; Scheer, G.; Castañeda, C.M.; Palacios-Orueta, A. Estimating canopy water content of chaparral shrubs using optical methods. *Remote Sens. Environ.* **1998**, *65*, 280–291. [CrossRef]

72. Korner, C.; Allison, A.; Hilscher, H. Altitudinal variation of leaf diffusive conductance and leaf anatomy in heliophytes of montane new guinea and their interrelation with microclimate. *Flora* **1983**, *174*, 91–135. [CrossRef]

73. Hultine, K.; Marshall, J. Altitude trends in conifer leaf morphology and stable carbon isotope composition. *Oecologia* **2000**, *123*, 32–40. [CrossRef]

74. Piper, F.I. Intraspecific trait variation and covariation in a widespread tree species (*Nothofagus pumilio*) in southern chile. *New Phytol.* **2011**, *189*, 259–271.

75. Bresson, C.C.; Vitasse, Y.; Kremer, A.; Delzon, S. To what extent is altitudinal variation of functional traits driven by genetic adaptation in european oak and beech? *Tree Physiol.* **2013**, *31*, 1164–1174. [CrossRef]

76. Smith, M. Alpine treelines: Functional ecology of the global high elevation tree limits. *Mt. Res. Dev.* **2013**, *33*, 357. [CrossRef]

77. Sveinbjornsson, B.; Nordell, O.; Kauhanen, H. Nutrient relations of mountain birch growth at and below the elevational tree-line in swedish lapland. *Funct. Ecol.* **1992**, *6*, 213–220. [CrossRef]

78. Kudo, G. Altitudinal effects on leaf traits and shoot growth of betulaplatyphyl. *Can. J. For. Res.* **2011**, *25*, 1881–1885. [CrossRef]

79. Birmann, K.; Körner, C. Nitrogen status of conifer needles at the alpine treeline. *Plant Ecol. Divers.* **2009**, *2*, 233–241. [CrossRef]

80. Schoettle, A.W.; Rochelle, S.G. Morphological variation of pinus flexilis (pinaceae), a bird-dispersed pine, across a range of elevations. *Am. J. Bot.* **2000**, *87*, 1797–1806. [CrossRef]

81. Ellsworth, D.S.; Reich, P.B. Leaf mass per area, nitrogen content and photosynthetic carbon gain in acer saccharum seedlings in contrasting forest light environments. *Funct. Ecol.* **1992**, *6*, 423–435. [CrossRef]

82. Le, R.X.; Walcroft, A.S.; Sinoquet, H.; Chaves, M.M.; Rodrigues, A.; Osorio, L. Photosynthetic light acclimation in peach leaves: Importance of changes in mass: Area ratio, nitrogen concentration, and leaf nitrogen partitioning. *Tree Physiol.* **2001**, *21*, 377–386.

83. Pollastrini, M.; Stefano, V.D.; Ferretti, M.; Agati, G.; Grifoni, D.; Zipoli, G.; Orlandini, S.; Bussotti, F. Influence of different light intensity regimes on leaf features of vitis vinifera l. In ultraviolet radiation filtered condition. *Environ. Exp. Bot.* **2011**, *73*, 108–115. [CrossRef]

84. Luomala, E.M.; Laitinen, K.; Sutinen, S.; Kellomäki, S.; Vapaavuori, E. Stomatal density, anatomy and nutrient concentrations of scots pine needles are affected by elevated co 2 and temperature. *Plant Cell Environ.* **2005**, *28*, 733–749. [CrossRef]

85. Björkman, O. *Responses to Different Quantum Flux Densities*; Springer: Berlin/Heidelberg, Germany, 1981; pp. 57–107.

86. Reich, P.B.; Ellsworth, D.S.; Walters, M.B. Leaf structure (specific leaf area) modulates photosynthesis-nitrogen relations: Evidence from within and across species and functional groups. *Funct. Ecol.* **1998**, *12*, 948–958. [CrossRef]

87. Park, B.-B.; Byun, J.-K.; Park, P.-S.; Lee, S.-W.; Kim, W.-S. Growth and tissue nutrient responses of fraxinus rhynchophylla, fraxinus mandshurica, Pinus koraiensis, and abies holophylla seedlings fertilized with nitrogen, phosphorus, and potassium. *J. Korean Soc. For. Sci.* **2010**, *99*, 186–196.

88. Cai-Feng, Y.; Shi-Jie, H.; Yu-Mei, Z.; Cun-Guo, W.; Guan-Hua, D.; Wen-Fa, X.; Mai-He, L. Needle-age related variability in nitrogen, mobile carbohydrates, and δ13c within Pinus koraiensis tree crowns. *PLoS ONE* **2012**, *7*, e35076.

89. Luo, J.; Zang, R.; Li, C. Physiological and morphological variations of picea asperata populations originating from different altitudes in the mountains of southwestern china. *For. Ecol. Manag.* **2006**, *221*, 285–290. [CrossRef]

90. Hedin, L.O.; Brookshire, E.J.; Menge, D.N.; Barron, A.R. The nitrogen paradox in tropical forest ecosystems. *Ann. Rev. Ecol. Evol. Syst.* **2009**, *40*, 613–635. [CrossRef]

91. Vitousek, P.M.; Porder, S.; Houlton, B.Z.; Chadwick, O.A. Terrestrial phosphorus limitation: Mechanisms, implications, and nitrogen-phosphorus interactions. *Ecol. Appl.* **2010**, *20*, 5–15. [CrossRef]

92. Lichtenthaler, H.K.; Ač, A.; Marek, M.V.; Kalina, J.; Urban, O. Differences in pigment composition, photosynthetic rates and chlorophyll fluorescence images of sun and shade leaves of four tree species. *Plant Physiol. Biochem.* **2007**, *45*, 577–588. [CrossRef]

93. Hendrik, P.; Ulo, N.; Lourens, P.; Wright, I.J.; Rafael, V. Causes and consequences of variation in leaf mass per area (lma): A meta-analysis. *New Phytol.* **2010**, *182*, 565–588.

94. Guiying Ben, C.B.O.; Sharkey, T.D. Comparisons of photosynthetic responses of xanthium strumarium and helianthus annuus to chronic and acute water stress in sun and shade. *Plant Physiol.* **1987**, *84*, 476–482.

95. Rosati, A.; Esparza, G.; Dejong, T.M.; Pearcy, R.W. Influence of canopy light environment and nitrogen availability on leaf photosynthetic characteristics and photosynthetic nitrogen-use efficiency of field-grown nectarine trees. *Tree Physiol.* **1999**, *19*, 173–180. [CrossRef]

96. Girardin, C.A.J.; Malhi, Y.; Aragão, L.E.O.C.; Mamani, M.; Huasco, W.H.; Durand, L.; Feeley, K.J.; Rapp, J.; Silva-Espejo, J.E.; Silman, M. Net primary productivity allocation and cycling of carbon along a tropical forest elevational transect in the peruvian andes. *Glob. Chang. Biol.* **2010**, *16*, 3176–3192. [CrossRef]

97. Rapp, J.M.; Silman, M.R.; Clark, J.S.; Girardin, C.A.; Galiano, D.; Tito, R. Intra- and interspecific tree growth across a long altitudinal gradient in the peruvian andes. *Ecology* **2012**, *93*, 2061–2072. [CrossRef]

98. Yu, D.; Wang, Q.; Wang, Y.; Zhou, W.; Ding, H.; Fang, X.; Jiang, S.; Dai, L. Climatic effects on radial growth of major tree species on changbai mountain. *Ann. For. Sci.* **2011**, *68*, 921–933. [CrossRef]

99. Gillooly, J.F.; Charnov, E.L. Effects of size and temperature on metabolic rate. *Science* **2001**, *293*, 2248–2251. [CrossRef]

100. Reiners, W.A.; Hollinger, D.Y.; Lang, G.E. Temperature and evapotranspiration gradients of the white mountains, new hampshire, USA. *Arct. Alp. Res.* **1984**, *16*, 31–36. [CrossRef]

101. Dittmar, C.; Fricke, W.; Elling, W. Impact of late frost events on radial growth of common beech (*Fagus sylvatica* L.) in southern germany. *Eur. J. For. Res.* **2006**, *125*, 249–259. [CrossRef]

102. Yan, C.; Han, S.; Zhou, Y.; Zheng, X.; Yu, D.; Zheng, J.; Dai, G.; Li, M.-H. Needle δ13c and mobile carbohydrates in Pinus koraiensis in relation to decreased temperature and increased moisture along an elevational gradient in ne china. *Trees* **2012**, *27*, 389–399. [CrossRef]

103. Salinas, N.; Malhi, Y.; Meir, P.; Silman, M.; Cuesta, R.R.; Huaman, J.; Salinas, D.; Huaman, V.; Gibaja, A.; Mamani, M.; et al. The sensitivity of tropical leaf litter decomposition to temperature: Results from a large-scale leaf translocation experiment along an elevation gradient in peruvian forests. *New Phytol.* **2011**, *189*, 967–977. [CrossRef]

104. Surabhi, G.K.; Reddy, K.R.; Singh, S.K. Photosynthesis, fluorescence, shoot biomass and seed weight responses of three cowpea (*Vigna unguiculata* (L.) Walp.) cultivars with contrasting sensitivity to uv-b radiation. *Environ. Exp. Bot.* **2009**, *66*, 160–171. [CrossRef]

105. Vitousek, P.M.; Sanford, R.L. Nutrient cycling in moist tropical forest. *Ann. Rev. Ecol. Syst.* **1986**, *17*, 137–167. [CrossRef]
106. Asakawa, S. Further investigation on hastening the germination of Pinus koraiensis seeds. *J. Jpn. For. Soc.* **1956**, *38*, 1–4.

 *forests*

Article

Environmental Controls on the Seasonal Variation in Gas Exchange and Water Balance in a Near-Coastal Mediterranean *Pinus halepensis* Forest

Mariangela N. Fotelli [1,*], Evangelia Korakaki [2], Spyridon A. Paparrizos [3], Kalliopi Radoglou [4], Tala Awada [5] and Andreas Matzarakis [6,7]

[1] Forest Research Institute, Hellenic Agricultural Organization Demeter, 57006 Vassilika, Thessaloniki, Greece
[2] Institute of Mediterranean and Forest Ecosystems, Hellenic Agricultural Organization Demeter, P. O. Box 14180, Terma Alkmanos, 11528 Ilisia, Athens, Greece; e.korakaki@fria.gr
[3] LSCE/IPSL, CEA-CNRS-UVSQ, Université Paris-Saclay, Orme des Merisiers, 91191 Gif-sur-Yvette, France; spyridon.paparrizos@lsce.ipsl.fr
[4] Department of Forestry & Management of the Environment & Natural Resources, Democritus University of Thrace, Pantazidou 193, 68200 Nea Orestiada, Greece; kradoglo@fmenr.duth.gr
[5] School of Natural Resources, University of Nebraska, 807 Hardin Hall, 3310 Holdrege Street, Lincoln, NE 68583-0968, USA; tawada2@unl.edu
[6] Chair of Environmental Meteorology, Faculty of Environment and Natural Resources, University of Freiburg, Werthmannstr. 10, D-79085 Freiburg, Germany; andreas.matzarakis@meteo.uni-freiburg.de
[7] Research Center Human Biometeorology, German Meteorological Service, D-79104 Freiburg, Germany
* Correspondence: fotelli@fri.gr; Tel.: +30-2310-461172 (ext. 238)

Received: 22 February 2019; Accepted: 3 April 2019; Published: 5 April 2019

Abstract: Aleppo pine (*Pinus halepensis* Mill.) is widespread in most countries of the Mediterranean area. In Greece, Aleppo pine forms natural stands of high economic and ecological importance. Understanding the species' ecophysiological traits is important in our efforts to predict its responses to ongoing climate variability and change. Therefore, the aim of this study was to assess the seasonal dynamic in Aleppo pine gas exchange and water balance on the leaf and canopy levels in response to the intra-annual variability in the abiotic environment. Specifically, we assessed needle gas exchange, water potential and $\delta^{13}C$ ratio, as well as tree sap flow and canopy conductance in adult trees of a mature near-coastal semi-arid Aleppo pine ecosystem, over two consecutive years differing in climatic conditions, the latter being less xerothermic. Maximum photosynthesis (A_{max}), stomatal conductance (g_s), sap flow per unit leaf area (Q_l), and canopy conductance (G_s) peaked in early spring, before the start of the summer season. During summer drought, the investigated parameters were negatively affected by the increasing potential evapotranspiration (PET) rate and vapor pressure deficit (VPD). Aleppo pine displayed a water-saving, drought avoidance (isohydric) strategy via stomatal control in response to drought. The species benefited from periods of high available soil water, during the autumn and winter months, when other environmental factors were not limiting. Then, on the leaf level, air temperature had a significant effect on A_{max}, while on the canopy level, VPD and net radiation affected Q_l. Our study demonstrates the plasticity of adult Aleppo pine in this forest ecosystem in response to the concurrent environmental conditions. These findings are important in our efforts to predict and forecast responses of the species to projected climate variability and change in the region.

Keywords: Aleppo pine; Greece; photosynthesis; water potential; $\delta^{13}C$; sap flow; canopy conductance; climate

1. Introduction

Aleppo pine (*Pinus halepensis* Mill.) is widespread in most countries of the Mediterranean area [1,2]. In Greece, Aleppo pine forms forests of economic importance (e.g., wood, resin, medicinal, and honey products) [1], comprising 26% of the coniferous forests in the country.

Aleppo pine can reach heights of 10–20 m, depending on the precipitation regime in the area, and has relatively shallow roots, usually not exceeding depths of 5 m [3]. The species is adapted to the xerothermic conditions (high temperatures and droughts) of the Mediterranean, due to its drought avoidance strategy of reducing stomatal conductance under water shortage [4]. This isohydric response allows Aleppo pine to limit the reduction of needle water potential and xylem cavitation, to which it is quite vulnerable [5,6]. Moreover, Aleppo pine has adapted its physiological activity to the seasonally changing climatic regime in the region. It actively grows during two periods of the growing season (spring and autumn) when temperatures are favorable and water is available. This behavior is more pronounced in the coastal regions than in continental forest sites [7]. On the other hand, extreme winter and summer temperatures and intensive summer droughts may cause growth activity to cease [8] and lead to extensive dieback and growth declines in Aleppo pine forests [9].

Improving our understanding of the driving factors that control Aleppo pine responses to climatic conditions is important for managing the species and forecasting its responses to climate variability, extremes, and change. Relationships between various physiological traits and the abiotic environment have been reported in the literature for Aleppo pine [10–12]. Studies have demonstrated a strong stomatal regulation in the species and coordination between foliage water potential and stomatal conductance to balance water loss [13]. However, needle and canopy stomatal responses to changes in evaporative demands, especially in combination with high temperatures remain unclear. Aleppo pine populations vary greatly in their response to extreme weather events across their distribution range [9,14]. Populations growing under the driest environments seem to be most impacted by extreme droughts and are prone to growth decline, but recover quickly. It is, however, unknown how the species will respond to the drier and hotter conditions forecasted for the Mediterranean basin under climate change [15], particularly in the eastern part [16]. Studying the seasonal dynamics of physiological traits in Aleppo pine in response to the concurrent climatic conditions will advance our understanding of the drivers that control growth and performance in the species and resilience of these forests. Few studies on the ecophysiological responses of Aleppo pine to drought regimes included Greek provenances and have focused on ecotypic variability assessed in plantations [17–20], not on adult trees of natural Aleppo pine forests. Thus, information on the ecophysiological responses of natural Aleppo pine forests in Greece is scarce.

In the present study, we assessed the seasonal dynamics of physiological traits of a mature near-coastal Aleppo pine ecosystem in Sani, Chalkidiki, northern Greece, over two consecutive years. We measured foliage gas exchange, water potential and stable carbon isotopic ratio, and tree sap flow rate and canopy conductance to characterize water balance dynamics of the species in response to climatic variability. Our specific aims were to (a) describe the seasonal variation in the physiological traits of Aleppo pine trees and (b) determine the climatic factors that control the observed seasonal trends. The combination of selected complementary techniques provides vital information from the needle to the stand level for assessing the performance of this dominant Mediterranean forest species under the prevailing climate change.

2. Materials and Methods

2.1. Site Description

The study was conducted at the peninsula of Kassandra, Chalkidiki, Greece. The experimental site is located at the Stavronikita forest (latitude: 40°06′22″ N, longitude: 23°18′80″ E, altitude 15 m.a.s.l., slope 1%, c. 300 m distance from the coast). The site is in a natural Aleppo pine (*Pinus halepensis*) stand with a mean tree height of 16 m, a mean diameter at breast height of 45 cm, a mean tree basal area of

0.19 m^2, and a stand basal area of 23.68 m^{2}ha^{-1}. The understorey consists of a maquis shrub vegetation, dominated with *Pistacia lentiscus* L., *Phyllirea media* L., and *Quercus coccifera* L. The soil has a high pH (7.5–8.2) and, according to European soil classification, it lies at the boundary between Calcari-chromic Vertisols and Chromic Luvisols [21,22].

2.2. Environmental Conditions

The climate on site is Mediterranean (Csa), according to Köppen-Geiger's classification, and is characterized by rainy winters and semi-arid growing seasons [23]. Micrometeorological data are available for the period 1978–1997 and from 2007 to present, from a fully automated weather station operating at a c. 50 m distance from the forest stand. Air temperature and air relative humidity (RHT2nl, Delta-T Devices Ltd., Cambridge, UK), photosynthetically active radiation (SKP215; Skye Instruments Ltd., Llandrindod Wells, UK), solar radiation (SKS1110, Skye Instruments, UK), wind speed (model 4.3515.30.000, THIES CLIMA, Göttingen, Germany), wind direction (WD4, Delta-T Devices Ltd., UK), precipitation (AR100 and RGB1, EM UK), and soil temperature at a depth of 15 cm (ST1, Delta-T Devices Ltd., UK) were continuously recorded. All parameters were data-logged on a 1-h basis (DL2e Delta-T Logger, Delta-T Devices Ltd., Cambridge, UK). Missing data due to a short-term malfunction of the meteorological station were completed after extrapolation from the respective data from the closest meteorological station of Loutra Thermis (latitude 40°30′ N, longitude 23°04′ E, 30 m.a.s.l.). The filling of the missing data gaps was performed by using the double-mass curve technique [24] followed by a t-test [25]. Moreover, vapor pressure deficit (VPD) was estimated using the RayMan model [26,27], while potential stand evapotranspiration (PET) and available soil water capacity (aSWC) of the study site were calculated with the water balance model WBS3. WBS3 is a forest–hydrological model that requires daily mean values of air temperature and daily total precipitation as meteorological inputs [28] and takes into account several forest stand parameters as input, as described in detail in a previous study [29].

An aridity index (AI) [30] was selected to estimate aridity conditions prevailing at the study area. A number of aridity indices have been proposed; these indicators serve to identify, locate, or delimit regions that suffer from a deficit of water availability [31]. The aridity index is estimated as follows (Equation (1)):

$$AI = P/PET, \tag{1}$$

where P is precipitation (mm), which in our study is equal to rainfall, and PET is potential evapotranspiration (mm). The boundaries that define the various degrees of aridity are shown in Table S1 [32].

2.3. Measurement Campaigns

Four dominant, non-neighboring Aleppo pine trees were selected for measurements and needle collection. Attention was paid to choosing healthy individuals, since infestation by the insect *Marhalina hellenica* (Genn.) is spread in *Pinus halepensis* forests of Chalkidiki. Three sun-exposed branches of the lower canopy (approximately three meters above ground) were marked and were thereafter used for measurements of gas exchange and midday water potential. After completion of each set of gas exchange measurements, the needles were sampled for carbon isotopic ratio analysis, as described below. Neighboring needles of the same branches were used for water potential measurements.

Measurements were conducted over two consecutive years on a monthly basis; gas exchange was measured from December 2007 to November 2009, while needle midday water potential was measured from January 2008 to October 2009. Needle δ^{13}C was determined from January 2008 to May 2009 due to technical limitations.

2.4. Gas Exchange and Needle Water Potential

For gas exchange measurement, we used the Li-6400 open path infra-red gas analyzer with a Li 6400-40 fluorescence chamber (Li-Cor, Lincoln, NE, USA). Maximum photosynthesis (A_{max}) and stomatal conductance (g_s) measurements were conducted on current year, fully expanded, and sun exposed needles between 10:00 and 13:00. The needles were carefully arranged in the 2-cm^2 cuvette in a way to exclude overlapping and to fully cover the area of the cuvette and they were acclimated for c. 10 min in the chamber at a CO_2 concentration of 400 ppm, under a photosynthetically active radiation (PAR) level of 1000 $\mu molm^{-2}s^{-1}$ from November to March and 1500 $\mu molm^{-2}s^{-1}$ from April to October. CO_2 flow rate was set to 300 $\mu mols^{-1}$ and temperature inside the chamber was controlled within the range of 17–28 °C, depending on the seasonal fluctuation of ambient air temperature.

Midday water potential (midday Ψ) measurements were conducted between 13:00 to 14:00 using a portable pressure chamber (model PMS 1003, PMS Instruments, Corvallis, OR, USA). The needles' water potential was measured after gas exchange measurements.

2.5. Needle $\delta^{13}C$ Signature

At the study site, new needles were fully expanded until the end of May each year. New needles (<1 year old), fully developed from the preceding May, were collected on a monthly basis from the same dominant Aleppo pine trees for the determination of the needle carbon isotopic ratio ($\delta^{13}C$). Samples were oven-dried at 65 °C until at a constant weight and then sent to the University of Nebraska Water Sciences Laboratory for analysis (https://watersciences.unl.edu/). Samples were finely ground and $\delta^{13}C$ was determined using mass spectrometry. The carbon isotope ratio ($\delta^{13}C$) of each sample was then determined as $\delta^{13}C$ (‰) = [(Rsample/Rstandard) − 1] × 1000, where Rsample is the $^{13}C/^{12}C$ of the sample and Rstandard is the $^{13}C/^{12}C$ ratio of the Vienna Pee Dee Belemnite (VPDB) standard.

2.6. Tree Sap Flow and Canopy Stomatal Conductance

Xylem sap flux was monitored using the thermal dissipation method [33,34]. In July 2008, 2-cm long Granier-type sensors and measurements were taken until early November 2009. Probe pairs were inserted radially into the stem of five dominant Aleppo pine trees averaging 43.5 cm diameter at breast height (DBH) with a vertical separation between the probes of approximately 12.0 cm. Probes were installed in the outer sapwood of the north-facing side of the stem and both probes and stems were insulated to minimize natural temperature gradients.

The temperature difference between the Granier-type probes was recorded at 10 second intervals and stored as 15 min averages on a data logger (CR10X Campbell Scientific, Logan, UT, USA) and used to obtain sap flux density by means of the equation derived empirically by Granier [33]. The daily maximum temperature difference was used as an estimate of the temperature difference under zero flow conditions. This variable was approximately constant over the study period (average coefficient of variation ± SE = 1.24% ± 0.04%).

Natural temperature gradients in the stem can interfere with sap flow measurements. These were measured over 60 days, but as values were consistently <5.2% of the sap flow signal, no corrections were applied [35].

The thickness of active sapwood was estimated using an allometric relationship obtained from a close by Aleppo pine site (in Peukochori, Chalkidiki; Radoglou K, unpublished data). For this purpose, 20 wood slices were used to estimate sapwood and heartwood areas. The equation best fitted to our data ($r^2 = 0.999$) was $A_s = 0.077 \times (DBH^{1.9905})$, where A_s stands for sapwood area (in m^2) and DBH for diameter at breast height (in m). Allometric relationships were also applied to estimate total tree leaf area [36] and used to calculate sap flow per unit leaf area (Q_l; $kgm^{-2}day^{-1}$).

Canopy stomatal conductance (G_s; mms^{-1}) was derived from sap flow measurements as described by [37].

Mean daily values of Q_l and G_s corresponding to days with mean daily values of VPD <0.1 kPa were excluded [38].

2.7. Statistical Analysis

Statistical analysis was performed with SPSS 23.0 (IBM Corp., SPSS for Windows, NY, USA) and OriginPro 8.0 (OriginLab Corp., Northampton, MA, USA). Relationships between physiological traits, as well as between physiological and single or combined environmental parameters, were examined using linear and non-linear regression analyses and coefficients of determination (adjusted R^2). The physiological traits tested were A_{max}, g_s, Ψ_{mid}, needle $\delta^{13}C$, Q_l, and G_s, while the respective environmental parameters were rainfall, air relative humidity, VPD, mean, maximum and minimum air temperature (T_{mean}, T_{max} and T_{min}, respectively), PET, net radiation, daytime net radiation, and aSWC of the actual day the physiological parameters were measured, or averaged over (a) the respective month, (b) the preceding month, (c) one week prior to measurements, and (d) two weeks prior to measurements. For the regression model between Q_l, VPD, and net radiation, mean hourly values for the wetter period (November to March of each study year, when data were available) were considered after excluding the ones corresponding to VPD <0.1 KPa. For all analyses, the environmental parameter(s) having an insignificant effect on each regression model ($p > 0.05$) were excluded from the model. When the combined effect of more than one environmental parameter on physiological traits was tested, only the independent environmental parameters were entered into the regression model. All tested significant regression models are presented in Table S2. The regression models with the highest adjusted R^2 and the highest significance level are presented in figures. The level of significance of each relationship ($p < 0.05$, $p < 0.01$, $p < 0.001$) is given in the respective plot.

3. Results

3.1. Climatic Conditions

The seasonal fluctuation of T_{mean} cumulative precipitation, mean aSWC, and mean VPD during the two-year study is presented in Figure 1. On average, 2009 was characterized by a combination of both higher average air temperatures and rainfall relative to 2008, resulting in lower VPD, higher aSWC, and higher aridity index (less xerothermic conditions) in 2009 compared to 2008 and to averages from the previous decade (Table 1).

Table 1. Annual cumulative rainfall and mean annual aridity index, aSWC and VPD during the study years and the previous decade.

Period	Rainfall (mm)	Aridity Index	aSWC (%)	VPD (KPa)
2008	542.1	0.77	36.9	0.55
2009	680.2	1.16	46.6	0.46
2008–2017	544.9	0.84	37.5	0.48

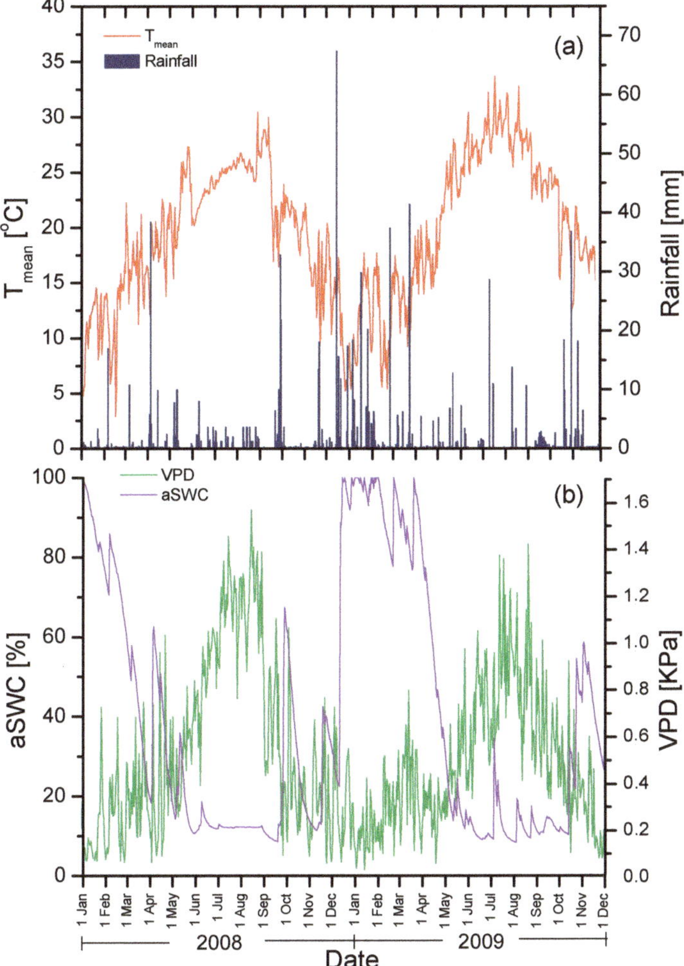

Figure 1. Daily values of: (**a**) Mean air temperature (T_{mean}) and rainfall and (**b**) available soil water capacity (aSWC) and vapor pressure deficit (VPD) during the study period.

3.2. Seasonal Patterns of Gas Exchange, Needle Water Potential, and $\delta^{13}C$ Composition

Gas exchange rates were high at the beginning of April in both years of measurements (Figure 2a,b). In 2008, a second pick in gas exchange was recorded in early July, before the summer drought was intensified, whereas in 2009 a second pick was evident in early October, after the completion of the drought season.

The combined effects of daytime net radiation and T_{mean} on sampling dates largely explains the variation in A_{max} between October and March ($R_{adj}^2 = 0.62$, $p < 0.05$; Figure 3). The particularly low A_{max} values observed in October and November 2008, compared to the same period in 2009, could be due to the substantially lower air temperatures of the former period compared to the latter. T_{mean} ranged from 8.5 to 15.5 °C in October–November measuring days of 2008 vs. 19.6 to 21.7 °C in 2009, while the T_{min} of the preceding nights was, similarly, lower in these measuring days of 2008 (5.4–11.2 °C) vs. 2009 (15.7–18.4 °C).On the other hand, the higher T_{mean}, increased aSWC, and decreased VPD during October–December of 2009, compared to the same period in 2008, resulted in a substantial increase in gas exchange, which reached values comparable to those observed during the spring in Aleppo pine. The strong and significant relationship between A_{max} and g_s ($R_{adj}^2 = 0.59$, $p < 0.01$; Figure 4a) indicated a close stomatal regulation of photosynthesis during the study period.

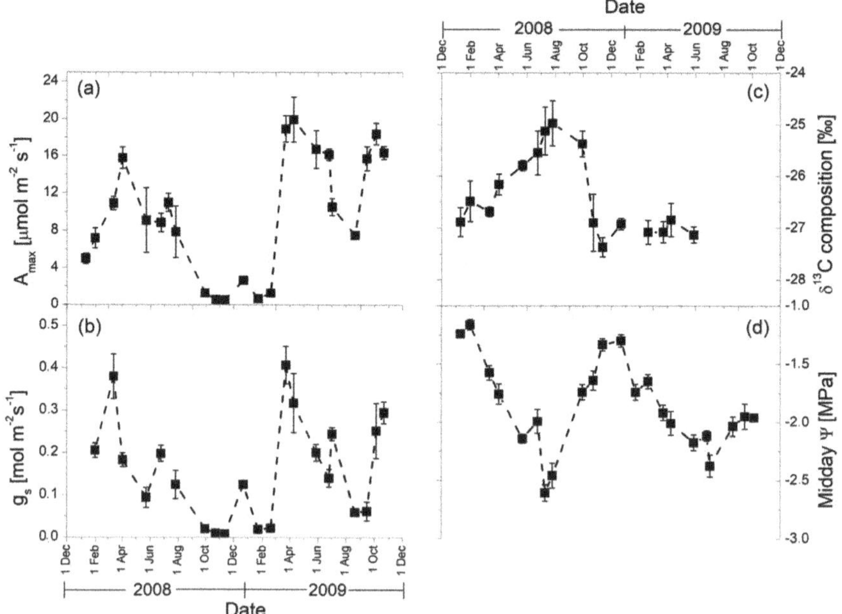

Figure 2. Seasonal pattern of monthly (**a**) maximum photosynthesis (A_{max}), (**b**) stomatal conductance (g_s), (**c**) $\delta^{13}C$ ratio, and (**d**) midday water potential (Ψ) during two consecutive years (2008 and 2009). $n = 4$ trees $\pm$ SE.

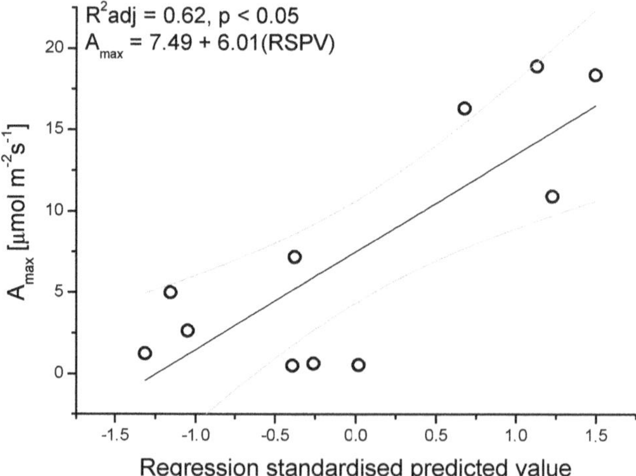

Figure 3. Regression model describing the combined effect of net radiation [kWm^{-2}] and mean air temperature [$°C$] on needle A_{max} during the period October–March. Mean daily values of net radiation and air temperature corresponding to the days of A_{max} measurements were used. For net radiation, only daytime values were used. RSPV stands for regression standardized predicted value. The confidence levels of the models are depicted by grey upper and lower bands.

The δ^{13}C ratio and midday Ψ of the current year needles displayed seasonal variability (Figure 2c,d) that was more pronounced in 2008 than 2009, consistent with the observed lower VPD and higher aSWC in 2009 vs. 2008 (Figure 1; Table 1). A significant negative linear relationship was recorded between δ^{13}C and Ψ ($R_{adj}^2 = 0.38$, $p < 0.05$; Figure 4b), with the highest δ^{13}C and the lowest midday Ψ values reported in August. Among all tested environmental parameters, the average VPD over two weeks period prior to sampling had the strongest effect on the δ^{13}C ratio of the needles ($R_{adj}^2 = 0.64$, $p < 0.001$; Figure 4a). The PET of the day of measurements was found to be the strongest predictor of midday Ψ ($R_{adj}^2 = 0.72$, $p < 0.001$; Figure 5b).

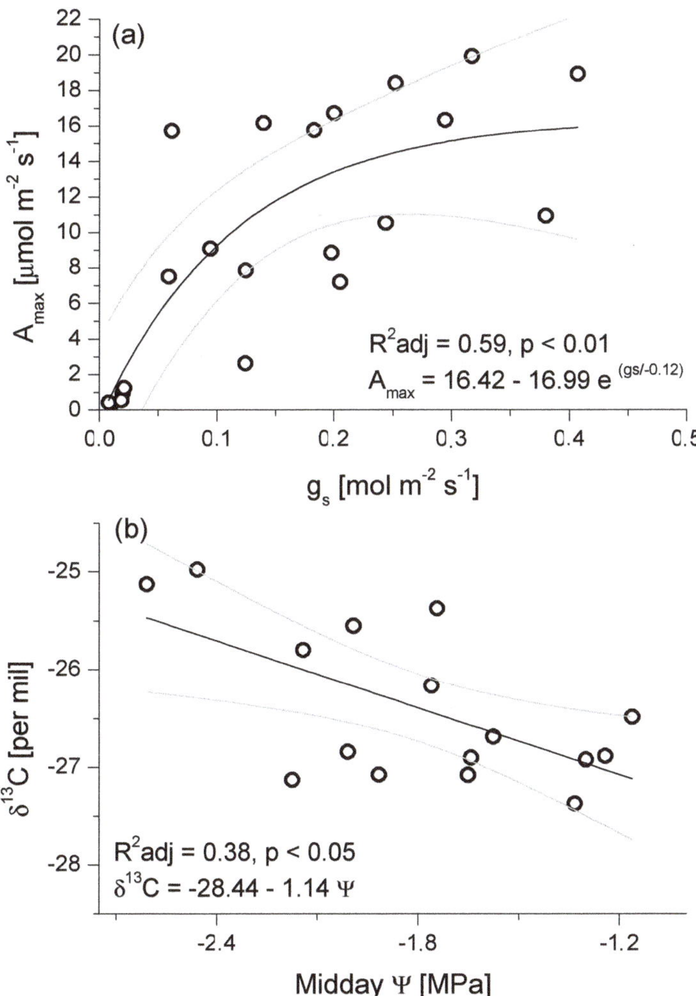

Figure 4. Regression models describing the relationship between needle (**a**) A_{max} and g_s and (**b**) δ^{13}C and midday Ψ. The confidence levels of the models are depicted by grey upper and lower bands.

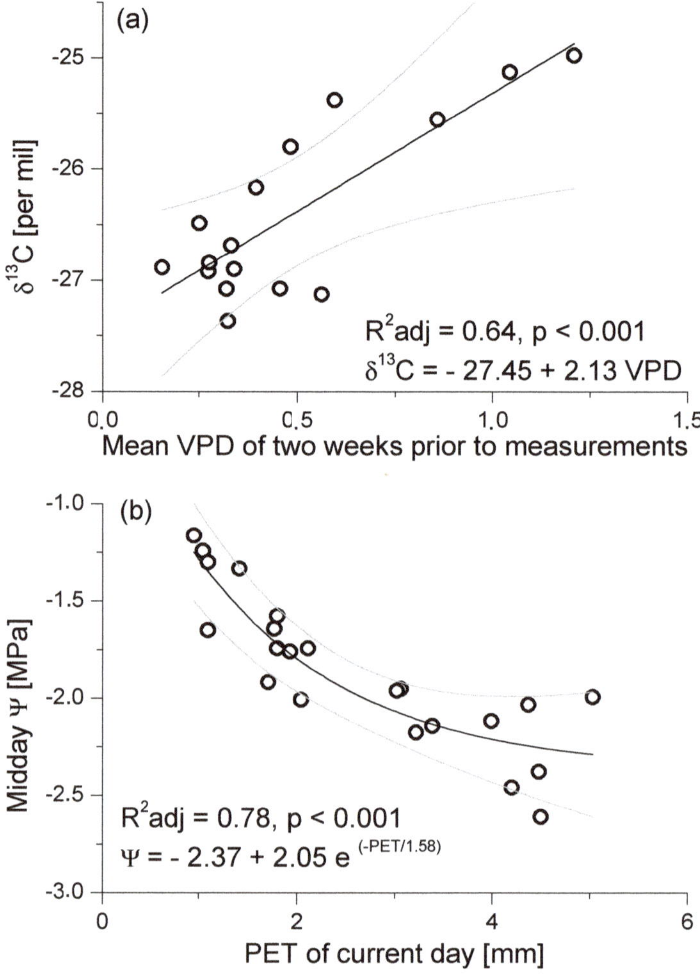

Figure 5. Regression models describing the relationship between (**a**) needle δ^{13}C and mean vapor pressure deficit (VPD) averaged over the preceding two weeks prior to measurements and (**b**) needle midday Ψ and potential stand evapotranspiration (PET) of the current day of measurements. The confidence levels of the models are depicted by grey upper and lower bands.

3.3. Seasonal Patterns of Sap Flow and Canopy Conductance

In 2009, maximum Q_l values were reached early in the growing season, when mean daily aSWC was still quite high (Figure 6a). Comparison with the same period in 2008 was not possible, since sap flow measurements were initiated in July 2008. In addition, datalogger malfunctioning resulted in missing values in April 2009, thus, not allowing comparison with the gas exchange maximum values in April. During the dry months, from July to September of both study years, similar declining trends were apparent in Q_l, with higher overall rates in less xerothermic 2009 compared to 2008 (Figure 6a). Q_l increased in October 2008, with the increase in aSWC, from 13% (in the second half of September) to 49% (first half of October), while mean daily values of VPD and net radiation were still not limiting. Q_l declined until December 2008, before increasing again to reach maximum values in March 2009 (Figure 6a), when VPD started to increase and aSWC was still very high (>85%). The combined effect of net radiation and VPD during the wet season (November–March) of the study period largely explained the variation in Q_l (R_{adj}^2 = 0.64, p < 0.001; Figure 7). Thus, when water availability was not a limiting factor, Q_l was mainly controlled by VPD and net radiation.

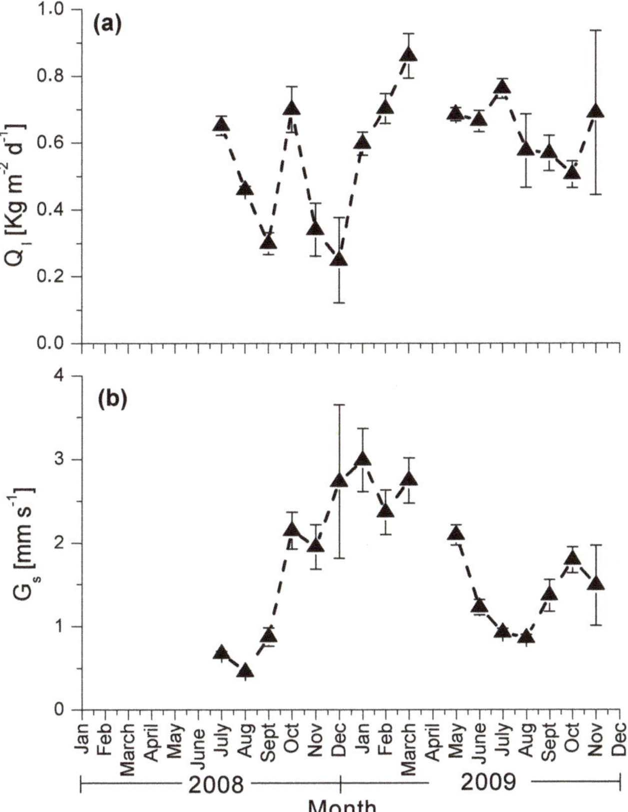

Figure 6. Seasonal patterns of mean monthly values of: (**a**) Sap flow per unit leaf area Q_l and (**b**) canopy stomatal conductance (G_s) measured over two consecutive years (2008 and 2009). n = 5 trees ± SE.

To investigate the causes of the drought-induced decrease in Q_l, canopy stomatal conductance (G_s) was derived from Q_l. The seasonal pattern of G_s was strongly controlled by PET (R_{adj}^2 = 0.69, p < 0.001; Figure 8a) and VPD (R^2 = 0.62, p < 0.001; Figure 8c) of the respective period, indicating increased stomatal control as drought progressed. G_s was also strongly related to aSWC (R_{adj}^2 = 0.72, p < 0.001; Figure 8b) and tracked its seasonal changes, which explains the high values in October and December 2008, as well as in January 2009 (Figure 6b).

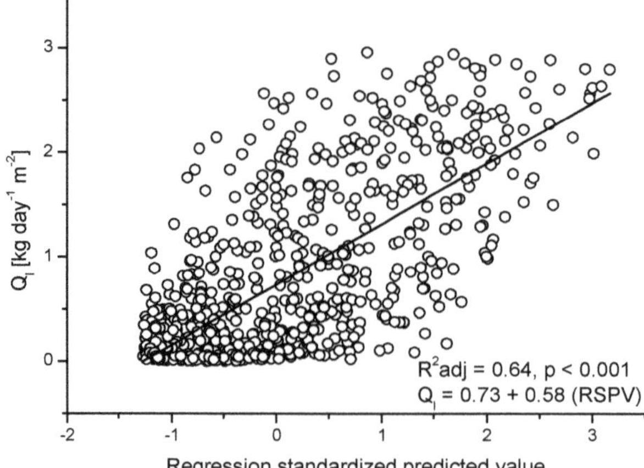

Figure 7. Regression model describing the combined effect of vapor pressure deficit (VPD) [kPa] and net radiation [kWm^{-2}] on Q$_l$. Mean hourly values of VPD and net radiation corresponding to the days' sap flow was monitored during the wet period (November to March) were used. RSPV stands for regression standardized predicted value.

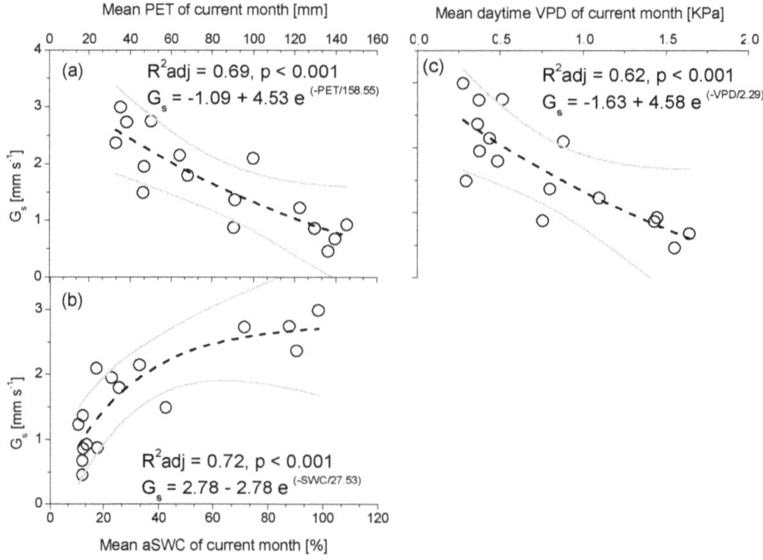

Figure 8. Regression models describing the relationship between mean monthly canopy stomatal conductance (G$_s$) and (**a**) potential stand evapotranspiration (PET), (**b**) available soil water capacity (aSWC), and (**c**) vapor pressure deficit (VPD). The confidence levels of the models are depicted by grey upper and lower bands.

4. Discussion

The seasonal variation in key physiological traits as impacted by the environmental conditions were investigated in a mature near-coastal Aleppo pine forest in Sani, Chalkidiki, northern Greece. The study site is characterized by semi-arid growing seasons (mean aridity index of April to October during the last decade was 0.38) and compared to other Mediterranean Aleppo pine forests, it falls within the average rainfall range but has relatively high mean air temperatures (Table S3). Assessments were conducted during two consecutive years (2008 and 2009) differing in climatic conditions. The latter was characterized by considerably higher water availability, while the former was comparable or even drier than the last decade's average (Figure 1; Table 1). This enabled monitoring of the ecophysiological responses of Aleppo pine under a wider range of climatic conditions. Within this frame, the effects of key environmental parameters on the gas exchange and water balance of *Pinus halepensis* were tested.

Aleppo pine exhibited a bimodal pattern of A_{max} which peaked twice in each year, when conditions were favorable (Figure 2a), corresponding with the species growth activity in the spring and autumn [6]. Stomatal regulation over A_{max} was observed throughout the year (Figure 4a), which appeared to respond to the limiting environmental factors. During the xerothermic summer conditions, PET exceeded that of aSWC and caused a decline in midday Ψ to values comparable to those reported in adult Aleppo pine trees (c. −2.6 MPa) [6]. As a result, g_s also declined, thus limiting A_{max} (Figure 2a,b), in line with the isohydric water-saving strategy of Aleppo pine [4,8,39]. This reduction of g_s is probably a response to minimize conductivity loss. Similar midday Ψ levels caused a c. 30% loss of conductivity in Aleppo pine seedlings subjected to drought [4], indicating its relatively high vulnerability to xylem embolism [6].

Air temperature and net radiation appear to be key controllers of A_{max} during the period of October to March (Figure 3). High A_{max} rates were reached during the warm days of October–November 2009, but not during the substantially colder measuring days between October 2008 and March 2009 (Figure 2a). It has similarly been reported [40] that the decrease of night temperatures below 10 °C, accompanied by a photoperiod below 12 h, results in low photosynthetic rates in seedlings of *Pinus strobus* during autumn. The A_{max} of Mediterranean pines was also found to be controlled by the preceding night temperatures and internal factors during autumn and winter months [41], as well as by extreme preceding summer droughts [10]. Any photoinhibition effects on A_{max} during the colder months of 2008 can be excluded since Fv/Fm values remained high (above 0.84; data not shown) and the air temperature was not low enough to account for such a response [42].

The quite xerothermic summer of 2008 resulted in a low midday Ψ (Figure 2d) that was reflected in the $\delta^{13}C$ isotopic ratio (Figure 2c), which increased linearly with decreasing midday Ψ (Figure 4b). This has been observed in other forest species during water deficits [43], owing to decreased Rubisco discrimination against ^{13}C under stomatal closure due to abiotic stresses [44]. Midday Ψ was found to immediately respond to short-term changes in air evaporative demand, as similarly shown by [11], since current day evapotranspiration explains 72% of its variation ($p < 0.001$; Figure 5b). On the contrary, the air vapor pressure deficit over the last two weeks strongly affected the needle $\delta^{13}C$ (R^2 adj = 0.64, $p < 0.001$, Figure 5a), while this effect was less pronounced when shorter or longer time intervals were examined (Table S2). Foliar $\delta^{13}C$ being affected by recent environmental conditions has been previously reported [29,45] and may apply for Aleppo pine as well, as the isotopic signature of recently produced assimilates could be detected quicker in a conifer that maintains an active C metabolism through most of the year.

Sap flow per unit leaf area (Q_l) showed a seasonal water saving pattern, with maximum values in early spring and a gradual decline as summer drought progressed (Figure 6a) and stomatal control increased (Figure 2b) to prevent water loss. A second peak in Q_l occurred in autumn 2008 (Figure 6a) associated with aSWC increase (from 13% in the last half of September to 49% in the first half of October). It is also evident that, when water availability was not a limiting factor, Q_l was mainly controlled by VPD and net radiation (Figure 7). Our results indicate that, in accordance to the findings from other Mediterranean type ecosystems [37,46], interannual variability in sap flow of Aleppo pines can be substantial, to avoid periods of prolonged drought and high transpirational demands.

Consistent with the patterns of leaf-level responses and sap flow, canopy conductance (G_s) decreased during the summer drought months (Figure 6b) being strongly affected by the gradual increases in PET, VPD (Figure 8a,c) and air temperature (Table S2), as similarly reported in numerous other studies [37,47]. As PET increased, midday Ψ decreased (Figure 5b) and g_s followed the same pattern to prevent conductivity losses [48]. On the other hand, g_s responded positively when water availability increased (Figure 8b), possibly due a gradual refill of previously cavitated tracheids [49].

The studied Aleppo pine stand exhibited plasticity to environmental conditions and showed the ability to recover from the effects induced by drought when climatic factors were improved. A similar response is reported for Aleppo pine growth when precipitation is increased [9]. However, during the period of increased water availability (October to March), varying patterns may be observed on the leaf and on the canopy level, responding to different parameters that seem to have a limiting effect. Thus, photosynthesis at the level of the lower canopy was greatly affected by air temperature and net radiation and increased when temperature was optimal (in October–November 2009), while sap flow and conductance at the canopy level responded positively to the favorable VPD, net radiation, and evapotranspiration during the same period of both years.

The results of the present study demonstrate the potential of this dominant Mediterranean forest tree species, to overcome the adverse conditions during summer droughts and to take advantage of more favorable water regimes occurring in early spring and occasionally also during autumn and winter, if other environmental parameters are not limiting, in a semi-arid ecosystem in Chalkidiki, Greece. Furthermore, some light is shed on the environmental controls over key physiological traits at the leaf and the canopy level.

5. Conclusions

By assessing the ecophysiological responses of the studied Aleppo pine forest in Northern Greece within a range of varying climatic conditions during a two year period, different but complementary patterns were revealed. During xerothermic periods, a typical isohydric behavior was exhibited by Aleppo pine; A_{max}, Q_l and g_s declined through stomatal control to limit Ψ reduction and loss of conductivity. However, in periods when water availability was not a limiting factor, the species was able to maximize its carbon gain if other controlling parameters, such as air temperature and net radiation, simultaneously ensured a favorable environmental regime. In conclusion, a high plasticity of the *Pinus halepensis* at the studied forest site to concurrent environmental conditions is indicated. Continuing studies are needed across Aleppo pine locations in the Mediterranean, particularly at its eastern part which is expected to be more prone to climate change, to improve our understanding of the species responses to ongoing climate variability.

Supplementary Materials: The following are available online at http://www.mdpi.com/1999-4907/10/4/313/s1, Table S1: Classification of aridity index (AI) categories; Table S2: Significant regression models describing the relationship between physiological and single or combined environmental parameters over different time intervals. The adjusted R^2 values and the levels of significance are presented. The models in bold are the most significant ones with the highest adjusted R^2, explaining the variation in the respective physiological parameter, which are presented in figures; Table S3: Characterization of the study site, in comparison to other Aleppo pine sites of Mediterranean countries.

Author Contributions: Conceptualization, M.N.F. and E.K.; Formal analysis, M.N.F., E.K. and S.A.P.; Investigation, M.N.F. and E.K.; Methodology, M.N.F. and E.K.; Project administration, K.R. and T.A.; Resources, K.R. and T.A.; Validation, M.N.F., E.K. and S.A.P.; Writing—original draft, M.N.F. and E.K.; Writing—review & editing, M.N.F., E.K., S.A.P., K.R., T.A. and A.M.

Funding: This research was funded by the General Secretariat for Research and Technology, Greece, grant number 05NON-EU-230 and the FoResMit LIFE14 CCM/IT/000905 project.

Acknowledgments: We wish to thank Gavriil Spyroglou and Nikolaos Fyllas for the biometrical characterization of the site, as well as George Halyvopoulos and Grigoris Morakis for their contribution in field measurements.

Conflicts of Interest: The authors declare no conflict of interest. The funders had no role in the design of the study; in the collection, analyses, or interpretation of data; in the writing of the manuscript, or in the decision to publish the results.

References

1. Chambel, M.R.; Climent, J.; Pichot, C.; Ducci, F. Mediterranean Pines (*Pinus halepensis* Mill. and *brutia* Ten.). In *Forest Tree Breeding in Europe: Current State-of-the-Art and Perspectives*; Pâques, L.E., Ed.; Springer: Dordrecht, The Netherlands, 2013; Volume 25, pp. 229–265, ISBN 978-2-11128054-0.
2. Euro + Med PlantBase—The Information Resource for Euro-Mediterranean Plant Diversity. Available online: http://ww2.bgbm.org/EuroPlusMed/ (accessed on 12 February 2019).
3. Dafis, S.; Moulopoulos, C. Einwirkung der Bodenfeuchtigkeit auf das Wachstum und Wurzelbildung von Samlingen der Aleppo und Hartkiefer (*Pinus halepensis* Mill und *P. brutia* Ten). *Beih. Z. Schweiz. Forstver.* **1969**, *46*, 225–260.
4. Klein, T.; Cohen, S.; Yakir, D. Hydraulic adjustments underlying drought resistance of *Pinus halepensis*. *Tree Physiol.* **2011**, *31*, 637–648. [CrossRef] [PubMed]
5. Klein, T.; Cohen, S.; Paudel, I.; Preisler, Y.; Rotenberg, E.; Yakir, D. Diurnal dynamics of water transport, storage and hydraulic conductivity in pine trees under seasonal drought. *iForest* **2016**, *9*, e1–e10. [CrossRef]
6. Oliveras, I.; Martinez Vilalta, J.; Jimenez-Ortiz, T.; Lledo, J.; Escarre, A.; Piñol, J. Hydraulic properties of *Pinus halepensis, Pinus pinea* and *Tetraclinis articulata* in a dune ecosystem of Eastern Spain. *Plant Ecol.* **2003**, *169*, 131–141. [CrossRef]
7. Pacheco, A.; Camarero, J.J.; Ribas, M.; Gazol, A.; Gutierrez, E.; Carrer, M. Disentangling the climate-driven bimodal growth pattern in coastal and continental Mediterranean pine stands. *Sci. Total Environ.* **2017**, *615*, 1518–1526. [CrossRef]
8. Prislan, P.; Gričar, J.; de Luis, M.; Novak, K.; del Castillo, M.E.; Schmitt, U.; Koch, G.; Štrus, J.; Mrak, P.; Žnidarič, M.T.; et al. Annual cambial rhythm in *Pinus halepensis* and *Pinus sylvestris* as indicator for climate adaptation. *Front. Plant Sci.* **2016**, *7*, 1923. [CrossRef]
9. Gazol, A.; Ribas, M.; Gutiérrez, E.; Camarero, J.J. Aleppo pine forests from across Spain show drought-induced growth decline and partial recovery. *Agric. For. Meteorol.* **2017**, *232*, 186–194. [CrossRef]
10. Schiller, G.; Cohen, Y. Water regime of a pine forest under a Mediterranean climate. *Agric. For. Meteorol.* **1995**, *74*, 181–193. [CrossRef]
11. Klein, T.; Shpringer, I.; Fikler, B.; Elbaz, G.; Cohen, S.; Yakir, D. Relationships between stomatal regulation, water-use, and water-use efficiency of two coexisting key Mediterranean tree species. *For. Ecol. Manag.* **2013**, *302*, 34–42. [CrossRef]
12. Birami, B.; Gattmann, M.; Heyer, A.G.; Grote, R.; Arneth, A.; Ruehr, N.K. Heat Waves Alter Carbon Allocation and Increase Mortality of Aleppo Pine Under Dry Conditions. *Front. For. Glob. Chang.* **2018**, *1*, 8. [CrossRef]
13. Anderegg, W.R.L.; Wolf, A.; Arango-Velez, A.; Choat, B.; Chmura, D.J.; Jansen, S.; Kolb, T.; Li, S.; Meinzer, F.C.; Pita, P.; et al. Woody plants optimise stomatal behaviour relative to hydraulic risk. *Ecol. Lett.* **2018**, *21*, 968–977. [CrossRef] [PubMed]
14. Del Castillio, J.; Voltas, J.; Ferrio, J.P. Carbon isotope discrimination, radial growth, and NDVI share spatiotemporal responses to precipitation in Aleppo pine. *Trees Struct. Funct.* **2015**, *29*, 223–233. [CrossRef]
15. García-Ruiz, J.M.; López-Moreno, J.I.; Vicente-Serrano, S.M.; Lasanta-Martínez, T.; Beguería, S. Mediterranean water resources in a global change scenario. *Earth-Sci. Rev.* **2011**, *105*, 121–139. [CrossRef]
16. Sarris, D.; Christodoulakis, D.; Körner, C. Recent decline in precipitation and tree growth in the eastern Mediterranean. *Glob. Chang. Biol.* **2007**, *13*, 1187–1200. [CrossRef]
17. Grünwald, C.; Schiller, G. Needle xylem water potential and water saturation deficit in provenances of *Pinus halepensis* Mill. and *P. brutia* Ten. *Forêt Méditerranéenne* **1988**, *10*, 407–414.
18. Michelozzi, M.; Loreto, F.; Colom, R.; Rossi, F.; Calamassi, R. Drought responses in Aleppo pine seedlings from two wild provenances with different climatic features. *Photosynthetica* **2011**, *49*, 564–572. [CrossRef]
19. Klein, T.; Di Matteo, G.; Rotenberg, E.; Cohen, S.; Yakir, D. Differential ecophysiological response of a major Mediterranean pine species across a climatic gradient. *Tree Physiol.* **2012**, *33*, 26–36. [CrossRef]
20. David-Schwartz, R.; Paudel, I.; Mizrachi, M.; Delzon, S.; Cochard, H.; Lukyanov, V.; Badel, E.; Capdeville, G.; Shklar, G.; Cohen, S. Indirect Evidence for Genetic Differentiation in Vulnerability to Embolism in *Pinus halepensis*. *Front. Plant Sci.* **2016**, *7*, 768. [CrossRef]
21. European Soil Bureau Network—European Commission. *Soil Atlas of Europe*, 11th ed.; Office for Official Publications of the European Communities: Luxembourg, 2005; p. 128, ISBN 92-894-8120-X.
22. Orfanoudakis, M.; Democritus University of Thrace, Nea Orestiada, Greece. Personal communication, 2019.

23. Peel, M.C.; Finlayson, B.L.; McMahon, T.A. Updated world map of the Köppen-Geiger climate classification. *Hydrol. Earth Syst. Sci.* **2007**, *11*, 1633–1644. [CrossRef]
24. Dingman, S.L. *Physical Hydrology*, 2nd ed.; Prentice Hall: Upper Saddler River, NJ, USA, 2002.
25. Paparrizos, S. The Effect of Climate on the Hydrological Regime of Selected Greek Areas with Different Climate Conditions. Ph.D. Thesis, Faculty of Environment and Natural Resources, Albert-Ludwigs-University of Freiburg, Freiburg, Germany, 2016.
26. Matzarakis, A.; Rutz, F.; Mayer, H. Modelling radiation fluxes in simple and complex environments—Application of the RayMan model. *Int. J. Biometeorol.* **2007**, *51*, 323–334. [CrossRef]
27. Matzarakis, A.; Rutz, F.; Mayer, H. Modelling radiation fluxes in simple and complex environments: Basics of the RayMan model. *Int. J. Biometeorol.* **2010**, *54*, 131–139. [CrossRef] [PubMed]
28. Matzarakis, A.; Mayer, H.; Schindler, D.; Fritsch, J. Simulation des Wasserhaushaltes eines Buchenwaldes mit dem forstlichen Wasserhaushaltsmodell WBS3. *Bericht Des Meteorologischen Instituts der Universität Freiburg* **2000**, *5*, 137–146.
29. Fotelli, M.N.; Nahm, M.; Radoglou, K.; Rennenberg, H.; Matzarakis, A. Seasonal and interannual ecophysiological responses of beech (*Fagus sylvatica*) at its south-eastern distribution limit in Europe. *For. Ecol. Manag.* **2009**, *257*, 1157–1164. [CrossRef]
30. United Nations Environment Programme (UNEP). *World Atlas of Desertification*; Edward Arnold: London, UK, 1992; p. 69, ISBN 0 340 55512 2.
31. Paparrizos, S.; Maris, F.; Matzarakis, A. Integrated analysis and mapping of aridity over Greek areas with different climate conditions. *Glob. NEST J.* **2016**, *18*, 131–145.
32. Food and Agriculture Organization (FAO). *Forest Resources Assessment 1990: Tropical Countries*; FAO-UN: Rome, Italy, 1993; ISBN 92-5-103390-0.
33. Granier, A. A new method of sap flow measurement in tree stems. *Ann. Sci. For.* **1985**, *42*, 193–200. [CrossRef]
34. Granier, A. Evaluation of transpiration in a Douglas-fir stand by means of sap flow measurements. *Tree Physiol.* **1987**, *3*, 309–319. [CrossRef] [PubMed]
35. Do, F.; Rocheteau, A. Influence of natural temperature gradients on measurements of xylem sap flow with thermal dissipation probes. 1. Field observations and possible remedies. *Tree Physiol.* **2002**, *22*, 641–648. [CrossRef]
36. Mitsopoulos, I.D.; Dimitrakopoulos, A.P. Allometric equations for crown fuel biomass of Aleppo pine (*Pinus halepensis* Mill.) in Greece. *Int. J. Wildland Fire* **2007**, *16*, 642–647. [CrossRef]
37. Martínez-Vilalta, J.; Mangirón, M.; Ogaya, R.; Sauret, M.; Serrano, L.; Peñuelas, J.; Piñol, J. Sap flow of three co-occurring Mediterranean woody species under varying atmospheric and soil water conditions. *Tree Physiol.* **2003**, *23*, 747–758. [CrossRef]
38. Phillips, N.; Oren, R. A comparison of daily representations of canopy conductance based on two conditional time-averaging methods and dependence of daily conductance on environmental factors. *Ann. For. Sci.* **1998**, *55*, 217–235. [CrossRef]
39. Salazar-Tortosa, D.; Castro, J.; Rubio de Casas, R.; Viñegla, B.; Sánchez-Cañete, E.P.; Villar-Salvador, P. Gas exchange at whole plant level shows that a less conservative water use is linked to a higher performance in three ecologically distinct pine species. *Environ. Res. Lett.* **2018**, *13*, 045004. [CrossRef]
40. Chang, C.Y.; Unda, F.; Zubilewich, A.; Mansfield, S.D.; Ensminger, I. Sensitivity of cold acclimation to elevated autumn temperature in field-grown *Pinus strobus* seedlings. *Front. Plant Sci.* **2015**, *6*, 165. [CrossRef]
41. Awada, T.; Radoglou, K.; Fotelli, M.N.; Constantinidou, H.-I.A. Ecophysiology of three Mediterranean pine species under contrasting light regimes. *Tree Physiol.* **2002**, *23*, 33–41. [CrossRef]
42. Pflug, E.; Brüggemann, W. Frost-acclimation of photosynthesis in overwintering Mediterranean holm oak, grown in Central Europe. *Int. J. Plant Biol.* **2012**, *3*, e1. [CrossRef]
43. Fotelli, M.N.; Geßler, A.; Peuke, A.D.; Rennenberg, H. Drought affects the competition between *Fagus sylvatica* L. seedlings and an early successional species (*Rubus fruticosus*): Growth, water status and δ13C composition. *New Phytol.* **2001**, *151*, 427–435. [CrossRef]
44. Farquhar, G.D.; Ehleringer, J.R.; Hubick, K.T. Carbon isotope discrimination and photosynthesis. *Annu. Rev. Plant Physiol. Plant Mol. Biol.* **1989**, *40*, 503–537. [CrossRef]
45. Keitel, C.; Matzarakis, A.; Rennenberg, H.; Geßler, A. Carbon isotopic composition and oxygen isotopic enrichment in phloem and total leaf organic matter of European beech (*Fagus sylvatica* L.) along a climate gradient. *Plant Cell Environ.* **2006**, *29*, 1492–1507. [CrossRef] [PubMed]

46. Sánchez-Costa, E.; Poyatos, R.; Sabaté, S. Contrasting growth and water use strategies in four co-occurring Mediterranean tree species revealed by concurrent measurements of sap flow and stem diameter variations. *Agric. For. Meteorol.* **2015**, *207*, 24–37. [CrossRef]

47. Oren, R.; Phillips, N.; Ewers, B.E.; Pataki, D.E.; Megonigal, J.P. Sap- flux scaled transpiration response to light, vapor pressure deficit, and leaf area reduction in a flooded *Taxodium distichum* forest. *Tree Physiol.* **1999**, *19*, 337–347. [CrossRef]

48. Sperry, J.S. Hydraulic constraints on plant gas exchange. *Agric. For. Meteorol.* **2000**, *104*, 13–23. [CrossRef]

49. Tognetti, R.A.; Michelozzi, M.; Giovanelli, A. Geographical variation in water relations, hydraulic architecture and terpene composition of Aleppo pine seedlings from Italian provenances. *Tree Physiol.* **1997**, *17*, 241–250. [CrossRef]

Article

Pretreatment with High-Dose Gamma Irradiation on Seeds Enhances the Tolerance of Sweet Osmanthus Seedlings to Salinity Stress

Xingmin Geng *, Yuemiao Zhang, Lianggui Wang and Xiulian Yang

College of Landscape Architecture, Nanjing Forestry University, Nanjing 210037, China;
lgwang2017@163.com (Y.Z.); wlg@njfu.com.com (L.W.); yangxl339@sina.com (X.Y.)
* Correspondence: xmgeng@njfu.edu.cn; Tel.: +86-15951902586

Received: 16 April 2019; Accepted: 7 May 2019; Published: 10 May 2019

Abstract: The landscape application of sweet osmanthus (*Osmanthus fragrans* (Thunb.) Lour.) with flower fragrance and high ornamental value is severely limited by salinity stress. Gamma irradiation applied to seeds enhanced their tolerance to salinity stress as reported in other plants. In this study, *O. fragrans* 'Huangchuang Jingui' seeds were pretreated with different doses of gamma irradiation, and tolerance of the seedlings germinated from the irradiated seeds to salinity stress and the changes of reactive oxygen species (ROS) production and ROS scavenging systems induced by gamma irradiation were observed. The results showed that seed pretreatment with different doses of gamma irradiation enhanced the tolerance of sweet osmanthus seedlings to salinity stress, and the positive effect induced by gamma irradiation was more remarkable with the increase of radiation dose (50–150 Gy). The pretreatment with high-dose irradiation decreased O_2^- production under salinity stress and mitigated the oxidative damage marked by a lower malondialdehyde (MDA) level, which could be related to the significant increase of superoxide dismutase (SOD), peroxidase (POD) and catalase (CAT) activities in the seedlings germinated from the irradiated seeds compared to the corresponding control seedlings. In addition, the accumulation of proline in the irradiated seedlings may contribute to enhancing their tolerance to salt stress by the osmotic adjustment. The study demonstrated the importance of regulating plant ROS balance under salt stress and provided a potential approach to improve the tolerance of sweet osmanthus to salt stress.

Keywords: plant tolerance; reactive oxygen species; antioxidant activity; proline

1. Introduction

Salinity has been threatening more and more land in the world [1,2]. Soil salinity greatly affects plant growth, development and productivity, thus posing a serious threat to agricultural and landscape plants in many regions of the world. Salt stress increases the concentration of toxic ions in plant cells, causes ion homeostasis disruption and then results in oxidative damage with excess generation of reactive oxygen species (ROS) [3].

ROS, including free radicals like O_2^- and OH·, and non-radicals like H_2O_2 and $^1O_2$, can be generated in the process of aerobic metabolism. Under favorable conditions, ROS production is controlled at basal levels and is beneficial to plants by supporting cellular proliferation, physiological function and viability [4]. However, the accumulation of ROS was accelerated by various environmental stresses such as salinity, drought, heat and high light [5–8], which caused damage to protein, DNA and lipid and thereby affecting normal cellular function [9,10]. Plants possess specific mechanisms to detoxify ROS which include enzymatic antioxidants such as multiple superoxide dismutase (SOD), catalase (CAT), peroxidase (POD), ascorbate peroxidase (APX), glutathione reductase (GR), dehydroascorbate reductase (DHAR) and non-enzymatic antioxidants such as ascorbic acid (AA) and glutathione (GSH).

SOD enzymatically disproportionate O_2^- (the primary product of oxygen reduction) into H_2O_2 and O_2 [11], and the SOD-catalyzed reaction provides the initial defense against ROS in plant cells. POD and CAT catalyze the conversion of H_2O_2 to water and O_2. APX, GR, DHAR, AA and GSH regulate mainly ROS by the ascorbate-glutathione (ASC-GSH) cycle. Under stresses, keeping higher activities of enzymatic antioxidants or the level of non-enzymatic antioxidants contributes to plant tolerance to environmental stresses [4,6].

Gamma rays belong to ionizing radiation and interact to atoms or molecules to produce free radicals in cells, which affects plant cellular structure and metabolism, e.g., the dilation of thylakoid membranes, alteration in photosynthesis, modulation of the antioxidative system, and accumulation of phenolic compounds [12–14]. It is possible that ^{60}Co-γ radiation could keep more complete cellular structures of the irradiation-induced mutants under chilling treatment than wild-type plants [14]. It has been reported that the activities of scavenging enzymes, such as SOD, POD and CAT, are generally increased in various plant species by the treatment of ionizing radiation [15–17]. ROS scavenging systems are important inducible factors for improving stress resistance through gamma-irradiation [18–20]. For example, gamma irradiation could help plants acclimate to lethal salinity by increasing antioxidant enzymes in irradiation-pretreated plants and alleviating oxidative damage [19,20].

Sweet osmanthus is a representative species of the genus *Osmanthus*, evergreen ornamental plants with excellent fragrance, color and shape. Sweet osmanthus has been cultivated at least for 2500 years in China. It was introduced into Japan, then to England in 18th century, and is now cultivated in Southeast Asia and many European countries [21]. However, soil salinity adversely affects the landscape application and geographic distribution of sweet osmanthus. The objective of this study was to explore a new approach to improve the salt tolerance of sweet osmanthus and determine how seed pretreatment with gamma irradiation improves the tolerance of sweet osmanthus through influencing antioxidant activity and ROS balance.

2. Materials and Methods

2.1. Seed Collection, Gamma Irradiation and Storage

Seeds of *O. fragrans* 'Huangchuan Jingui' (a cultivated variety) were collected on the campus of Nanjing Forestry University (Nanjing, Jiangsu Province) in May 2013. The seeds were randomly divided into four groups: Group 1 as non-irradiated controls and Group 2, 3 and 4 exposed to gamma irradiation with three irradiation doses. Gamma irradiation was conducted using a ^{60}Co-γ gamma source at a dose rate of 1.3 Gy/min, and the doses of exposure in this study were 50, 100 and 150 Gy. After irradiation, the non-irradiated and irradiated seeds were pretreated with 0.1% gibberellin (GA_3) for 24 h in order to break seed dormancy quickly and stored in wet sand at 4 °C before germination. The germinated seeds were sown in soil (1:1, v/v, mixture of garden soil and peat moss) in October 2013 and grown in greenhouse. Two years later, these plants were used in the following experiments of salt stress and tolerance assays.

2.2. Salt Stress and Tolerance Assays

Salinity stress was carried out using the method of water culture. Ten seedlings from each treatment (0, 50, 100, 150 Gy irradiation) with uniform size were chosen and grown in glass bottles (10 cm diameter and 8 cm high, one seedling in one bottle) filled with Hoagland solution with sodium chloride at different concentration, 20, 40, 60, 80, 100 and 120 mmol/L, sequentially. All seedlings were exposed to each concentration for three days before being moved to the next level.

Morphological changes of the seedlings due to salt stress were observed in the process of salt stress treatment, and the injury index was recorded at each concentration of salt stress. The extent of salinity injury was divided into five grades as shown in Table 1. The salinity injury index was calculated according to the following formula:

Table 1. Degree of leaf damage under salinity stress.

Grade	Salinity Damage Level	Description
0	No damage	No morphological damage symptoms of the whole plant
1	Mild	Less than 20% of the leaves have scorched margin and dehydration symptoms
2	Moderate	Nearly 50% of the leaves scorch, yellow with rust or wither
3	Severe	More than 50% of the leaves scorch, yellow with rust or wither
4	Very severe	More than 90% of the leaves scorch and wither or even the whole plant dies

Injury index = (Σ injury grades $\times$ corresponding number of seedlings)/(the highest grade $\times$ total number of seedlings); Injury rate = (Number of seedlings with morphological symptoms of salinity injury/total number of seedlings) $\times$ 100%. The plants were continuously monitored and observed for 1 week after salt stress. This same experiment was conducted twice again to confirm the results.

2.3. Sampling for Physiological Index

As described above, ten non-irradiated and ten irradiated seedlings were sequentially exposed to salt stress with different concentrations of NaCl solution for three days. Leaves from the ten individual non-irradiated and ten irradiated seedlings were collected after every concentration of salinity treatment and were stored in a −80 °C freezer for the measurement of O_2^- content, antioxidant enzyme activity, proline, malondialdehyde (MDA) and soluble proteins. This experiment was conducted three times to confirm the results.

2.3.1. Measurement of O_2^- Content and MDA Level

The content of O_2^- in leaves was determined by the hydroxylamine hydrochloride method according to Ke et al. [22] with minor modifications. Frozen leaves (0.3 g of fresh weight) were ground in 6 mL of potassium phosphate buffer (65 mM, pH 7.8), and centrifuged for 15 min at 10,000 rpm. The supernatant (1 mL) was mixed with 0.2 mL of 10 mmol/L hydroxylamine hydrochloride, then incubated for 20 min at 25 °C. After 1 mL of 17 mmol/L sulfanilic amide and 1 mL of 7 mmol/L α-naphthylamine were added, the mixture was further incubated for 20 min at 25 °C. The reaction mixtures were extracted with the same volume of chloroform (3.2 mL), and the absorbance was determined at 530 nm. Sodium nitrite was used to make a standard solution for calculating the content of O_2^-. MDA level was determined by the thiobarbituric acid (TBA) method as previously described [23].

2.3.2. Determination of Antioxidant Enzyme Activity and Proline Level

Leaves (0.3 g of fresh weight) were homogenized in a mortar and pestle with 2 mL of ice-cold phosphate buffer (50 mM, pH 7.8). The homogenate was centrifuged at 9000 rpm and 4 °C for 20 min. The supernatant was used for measuring the activities of SOD, POD and CAT. The procedure was conducted at 4 °C.

SOD activity was assayed by monitoring its ability to inhibit the photochemical reduction of nitro blue tetrazolium (NBT) using the method of Dhindsa et al. [24]. The reaction mixture (3 mL) contained 1.5 mL of phosphate buffer (50 mM, pH 7.8), 0.5 mL of 0.1 mM EDTA, 0.5 mL of 130 mM methionine, 0.5 mL of 0.5 mM NBT, 0.5 mL of 0.02 mM riboflavin, and 0.05 mL of enzyme extract. Riboflavin was added last, and the reaction mixtures were illuminated under 4000 Lm/m^2 for 30 min. The reaction was stopped by switching off the light and the tubes were covered with a black cloth. Non-illuminated and illuminated reactions without the supernatant served as calibration standards. One unit of SOD activity was defined as the amount of enzyme required to cause 50% inhibition of the reduction of NBT as monitored at 560 nm.

POD activity was measured following the method of Zhang and Kirkham [25] with some modifications. The enzyme extract (0.02 mL) was added to the reaction mixture containing 0.3% guaiacol solution and 3% hydrogen peroxide solution. The reaction was started by adding the enzyme extract and the absorbance increase at 470 nm in 5 min was recorded.

CAT activity was assayed in a reaction mixture containing phosphate buffer (pH 7.8), 0.1 mol/L H_2O_2 and enzyme extract. The reaction was initiated by adding the enzyme extract. The decrease of H_2O_2 was monitored at 240 nm in at least 4 min.

Proline was determined using colorimetric methods [26] as previously described [27].

2.4. Data Analyses

Statistical analyses were carried out using SPSS17.0 software (SPSS Company, Chicago, IL, USA) to calculate the mean and standard deviation. The Duncan's multiple range test (DMRT) was applied to test the significance in differences among treatments ($P < 0.05$).

3. Results

3.1. Tolerance in Irradiated Seedlings to Salinity Stress

To determine the level of enhanced tolerance to salt stress induced by gamma irradiation, we investigated the morphological changes of two-year-old sweet osmanthus seedlings developed from the irradiated seeds. The injury degree induced by salinity stress was represented by the injury index and injury rate. As shown in Table 2, the injury degree of sweet osmanthus seedlings was gradually aggravated when exposed to successively increasing NaCl concentration of 20, 40, 60, 80, 100 and 120 mmol/L, and the injury degree in non-irradiated seedlings was significantly different from the irradiated ones. The non-irradiated and irradiated seedlings at lower dose appeared to show slight injury with yellowing of basic leaves under 20 mmol/L salt stress. The similar injury symptom in 150-Gy gamma irradiated seedlings was not observed until exposed to 40 mmol/L NaCl. The non-irradiated seedlings suffered from moderate injury when NaCl concentration was increased to 40 mmol/L: nearly 50% of the leaves scorched, yellowed with rust or withered, and the salt injury rate reached 100%. Similar injury symptom in 50-Gy and 100-Gy gamma irradiated seedlings was observed when exposed to 60 mmol/L NaCl. However, 150-Gy gamma irradiated seedlings suffered from moderate injury under 80 mmol/L NaCl. When salt concentration was increased to 100 mmol/L, all seedlings appeared to show very severe injury symptom: 100% of the leaves scorched, withered, or the whole plant died.

Table 2. Effects of different dose of gamma irradiation on the salt injury index and injury rate of *O. fragrans* 'Huangchuan Jingui' under salt stress.

NaCl Concentration (mmol/L)	Salt Injury Index				Salt Injury Rate (%)			
	0 Gy	50 Gy	100 Gy	150 Gy	0 Gy	50 Gy	100 Gy	150 Gy
0	0.00 a	0.00 a	0.00 a	0.00 a	0.00 a	0.00 a	0.00 a	0.00 a
20	0.08 ± 0.01 a	0.07 ± 0.01 a	0.07 ± 0.01 a	0.00 ± 0.00 b	63.33 a	30.00 b	26.67 c	0.00 d
40	0.55 ± 0.05 a	0.37 ± 0.03 b	0.36 ± 0.01 b	0.28 ± 0.04 c	100.0 a	90.00 b	70.00 c	33.33 d
60	0.70 ± 0.05 a	0.56 ± 0.06 b	0.52 ± 0.06 b	0.40 ± 0.04 c	100.0 a	100.0 a	96.67 b	80.00 c
80	0.90 ± 0.02 a	0.74 ± 0.05 b	0.67 ± 0.05 c	0.54 ± 0.04 c	100.0 a	100.0 a	100.0 a	100.0 a
100	0.90 ± 0.01 a	0.89 ± 0.01 a	0.78 ± 0.06 c	0.85 ± 0.03 b	100.0 a	100.0 a	100.0 a	100.0 a
120	0.98 ± 0.01 a	0.98 ± 0.01 a	0.95 ± 0.01 b	0.97 ± 0.01 a	100.0 a	100.0 a	100.0 a	100.0 a

Note: Different normal letters within the same row indicate significant difference among different doses of gamma irradiation at 0.05 level ($P < 0.05$).

3.2. Effects of High-Doses Gamma Irradiation on ROS Level and Lipid Peroxidation in Response to Salt Stress

ROS production can be accelerated by various environmental stresses. As shown in Figure 1A, O_2^- production was accelerated as the increase of salt concentrations, and the O_2^- level in gamma-irradiated seedlings was lower than non-irradiated seedlings, except in response to 20 mmol/L salt stress (Figure 1A).

As shown in Figure 1B, MDA level in the gamma-irradiated seedlings was higher than the level observed in the non-irradiated seedlings under control condition, which indicated that gamma-ray irradiation applied on the seeds resulted in oxidative damage of cell membrane in sweet osmanthus seedlings to some extent. The MDA level in the gamma-irradiated seedlings under 20~60 mmlo/L and 120 mmol/L salt stress was lower than in the non-irradiated seedlings, and no significant difference in MDA level between the gamma-irradiated seedlings and the control seedlings was observed under 80~100 mmlo/L salt stress.

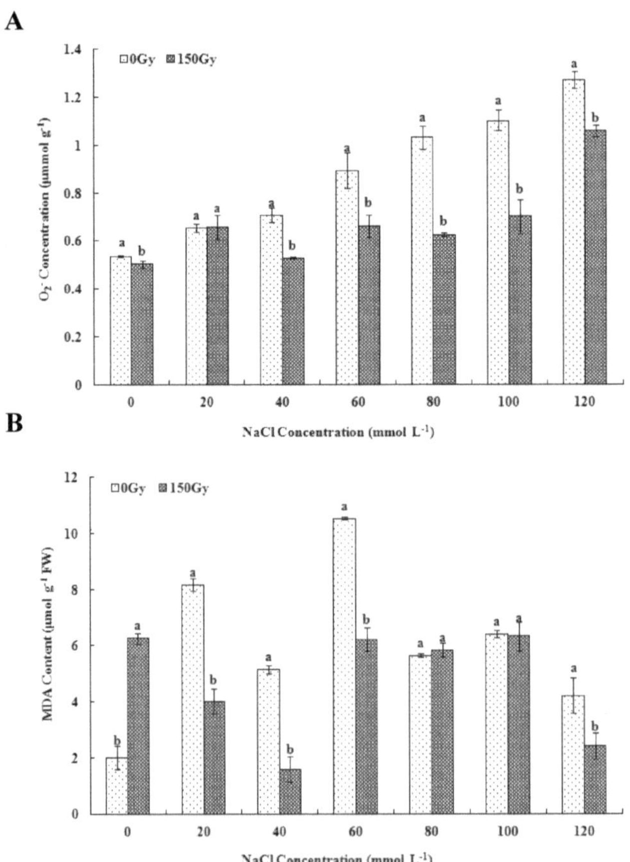

Figure 1. Effects of high-dose gamma irradiation on O_2^- production (**A**) and MDA level (**B**) in *Osmanthus fragrans* (Thunb.) Lour. seedlings in response to salt stress. Note: 0 Gy was used as the control; 150 Gy was the seedlings germinated from 150-Gy gamma irradiated seeds. Means with different letters above bars were significantly different at $P < 0.05$ between the control and the irradiation treatment.

3.3. Effects of High-Doses Gamma Irradiation on ROS Scavenging Systems in Response to Salt Stress

SOD activity in the control and the irradiated seedlings under salt stress was gradually enhanced along with the increasing of salt concentrations and reached the peak in the control seedlings under 60 mmol/L NaCl or in the irradiated seedlings under 80 mmol/L NaCl (Figure 2A). The pretreatment with 150-Gy gamma irradiation increased the SOD activities in seedlings under salt stress compared with the levels observed in the corresponding control seedlings, and the increase was statistically significant ($P < 0.05$) under stress with 40 or 80~120 mmlo/L NaCl.

POD activity increased initially and thereafter declined along with the increasing of salt concentrations. Maximum value in the control seedlings was observed under 40 mmol/L NaCl, but the peak in the irradiated seedlings appeared under 60 mmol/L NaCl. POD activity in the irradiated seedlings was significantly enhanced under higher concentration of NaCl ($\geq$60 mmol/L) (Figure 2B). Exposed to successively increasing salt stress, the activity of CAT in the irradiated seedlings was increased compared with the non-irradiated seedlings, except in response to 80 mmlo/L NaCl (Figure 2C).

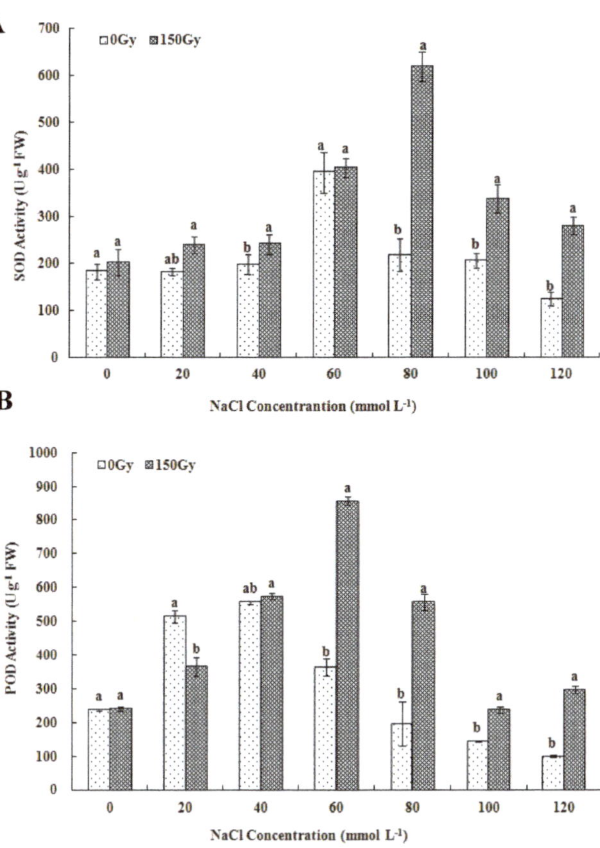

Figure 2. *Cont.*

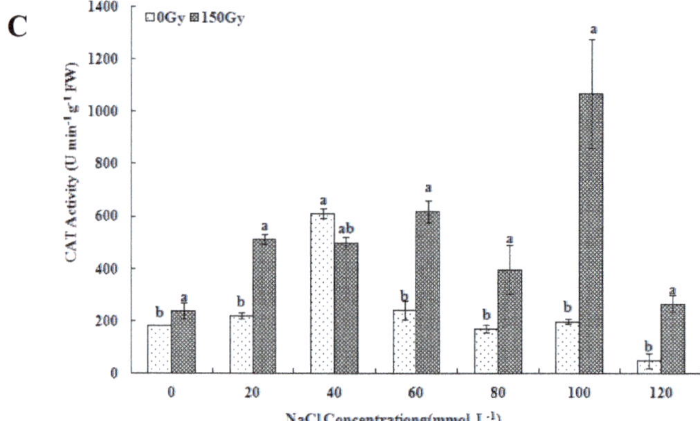

Figure 2. Effects of high-dose gamma irradiation on the activity of SOD (**A**), POD (**B**) and CAT (**C**) in *Osmanthus fragrans* seedlings in response to salt stress. Note: 0 Gy was used as the control; 150 Gy was the seedlings germinated from 150-Gy gamma irradiated seeds. Means with different letters above bars were significantly different at $P < 0.05$ between the control and the irradiation treatment.

3.4. Effects of High-Doses Gamma Irradiation on Proline Level in Response to Salt Stress

The proline level in the irradiated seedlings was lower than that in the non-irradiated seedlings in control and 20 mmol/L NaCl condition (Figure 3). More accumulation of proline in the irradiated seedlings was observed when exposed to higher concentration of salt stress. And the proline content in irradiated seedlings increased with the increase of salt stress.

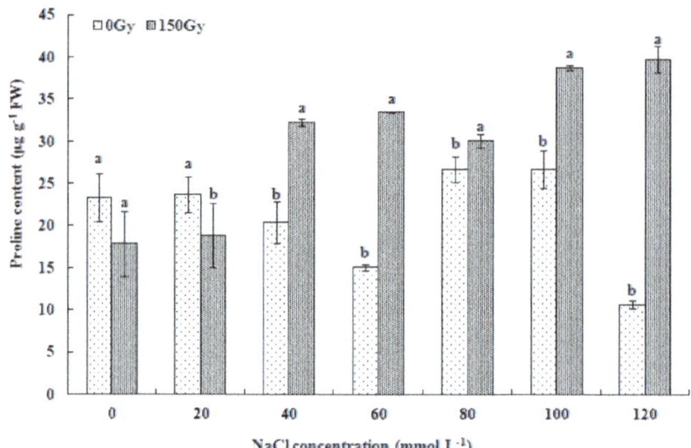

Figure 3. Effects of high-dose gamma irradiation on proline level in *Osmanthus fragrans* seedlings in response to salt stress. Note: 0 Gy was used as the control; 150 Gy was the seedlings germinated from 150-Gy gamma irradiated seeds. Means with different letters above bars were significantly different at $P < 0.05$ between the control and the irradiation treatment.

4. Discussion and Conclusions

The effects of gamma irradiation on the morphological changes and biological responses of plants are dependent on radiation doses [13]. Low-dose gamma irradiation improves seed germination and seedling growth, but adverse effects are induced by high-dose [28]. We previously reported that 50-Gy gamma irradiation displayed the maximal positive effects on seed germination and seedling growth of sweet osmanthus [29], and a high dose resulted in the decrease of seed germination and the poor growth of seedlings, even leading seedlings to death. In the present study, the seed pretreatment with different doses of gamma irradiation enhanced the tolerance of sweet osmanthus seedlings to salinity stress, and the survival of seedlings developed from the seeds irradiated with higher doses of gamma-ray showed stronger tolerance to salinity stress (Table 2). The result is different from the previous reports in rice [30], sweet potato [31] and *arabidopsis* seedlings [20], in these studies gamma irradiation at relative low dose improved the seedling growth under salinity stress. The differences of the response to irradiation dose were hypothesized to be related to the sensitivity to radiation. It has been well documented that there are great differences of sensitivity to irradiation between the taxa from the level of varieties of the same species to main plant divisions [32]. The resistance to irradiation of sweet osmanthus may be stronger than other plants mentioned in front. In addition, in the present study salt tolerance experiments were carried out after two years of seed radiation and normal cultivation. As reported by Zaka et al. [33], the effects induced by irradiation were often not reproducible or only transitory. The higher radiation dose may maintain the radiation effect for a longer time. According to the current research results, it is uncertain whether higher doses of radiation cause genotype variation and alter plant resistance genetically and stabilize radiation effects, which requires further research in the later stage.

Salinity stress resulted in the accumulation of excess O_2^- (Figure 1). If excessive O_2^- cannot be removed in time, it will produce $\cdot OH$ and $^1O_2$ with strong activity and toxicity, and result in membrane lipid peroxidation and oxidative damage to the cell membrane [34]. The pretreatment with high-dose gamma irradiation suppressed ROS production in salinity-stressed conditions and mitigated lipid peroxidation and the damage of cell membrane as manifested by the lower O_2^- and MDA level in the irradiated seedlings compared with the controls (Figure 1). The enhanced defense against ROS damage under high-dose gamma irradiation was probably due to the increased activities of antioxidant enzymes. POD, SOD and CAT activities were all significantly increased to different extents in response to salinity stress (Figure 2). The responses of antioxidant enzymes to salinity stress in the irradiated seedlings are consistent with other reported studies [20,35]. We also found another *O. fragrans* "Zi Yin gui", whose tolerance to salinity stress was enhanced similarly to "Huangchuan Jingui" and the defense response against ROS damage was induced by pretreatment with different doses of gamma irradiation; although the up-regulated extent of every single antioxidant enzyme was not exactly the same (data not shown). As reported by Ashraf [6], the extent to which the activities of enzymatic antioxidants and the level of non-enzymatic antioxidants were up-regulated under salinity stress is highly variable among plant species and even between two cultivars of the same species. The response and activation of antioxidant systems induced by irradiation exposure are dependent upon the γ-ray dose rate, the doses of gamma rays and the developmental stages [36,37]. The ASC-GSH cycle system were the main ROS-scavenging systems in the chloroplasts [11], the activities of APX and GR were found to be higher in the salt-tolerant cultivars of potato than those in the salt-sensitive cultivars [38]. ASC-GSH cycle also plays an important role in plant salt tolerance, hence further research on this system is necessary in the future in order to fully understand the effect of gamma irradiation on salinity tolerance of sweet osmanthus.

The proline level was significantly increased in the irradiated sweet osmanthus seedlings when exposed to higher concentration of salt stress (Figure 3). Proline is one of the major osmolytes regulating osmotic adjustment in plants exposed to osmotic stresses, and active accumulation of proline is associated with salinity tolerance in various plant species [5]. Proline contributes to membrane stability and mitigates the effect of NaCl on cell membrane disruption [39]. The proline content was also increased by irradiation in sweet potato [31], *arabidopsis* [20] and sugarcane [40]. Proline, other than being an osmoprotectant, can act as a singlet oxygen quencher and scavenger of hydroxyl radicals [41], and may be important in preventing oxidative damage caused by ROS when it accumulates during the exposure of the plants to adverse environmental conditions. Hossain and Fujita [42] provide evidence for the role of proline, which can protect against salt-induced oxidative damage by reducing H_2O_2 and lipid peroxidation levels, and by enhancing antioxidant defense and methylglyoxal detoxification systems.

In summary, our study provides a new approach to improve the tolerance of sweet osmanthus to salt stress by the application of irradiation, and proves that enhanced tolerance to salinity stress is closely related to the ROS-scavenging system.

Author Contributions: X.G. directed the project, drafted the first version of the manuscript, finalized figures and manuscript revision; Y.Z. conducted salinity stress experiments, collected morphological and physiological data on salinity tolerance, prepared draft figures; L.W. initiated the project, contributed to manuscript writing and finalizing; X.Y. supervised the experiments on physiological analyses and contributed to the manuscript writing.

Funding: This work was funded by the Agriculture-Forestry Plant Germplasm Resources Exploration, Innovation and Utilization Project of the National Twelve-Five Science and Technology Support Program (2013BA001B).

Acknowledgments: The authors thank the NJFU Greenhouse staff for supporting the cultivation of sweet osmanthus seedlings and graduate students Na Li for assistance in radiation treatment methods and concentrations.

Conflicts of Interest: The authors declare no conflict of interest.

References

1. Murtaza, G.; Ghafoor, A.; Owens, G.; Qadir, M.; Kahlon, U.Z. Environmental and economic benefits of saline-sodic soil reclamation using low-quality water and soil amendments in conjunction with a rice-wheat cropping system. *J. Agron. Crop Sci.* **2009**, *195*, 124–136. [CrossRef]
2. Rozema, J.; Flowers, T. Ecology crops for a salinized world. *Science* **2008**, *322*, 1478–1480. [CrossRef] [PubMed]
3. Zhu, J.K. Regulation of ion homeostasis under salt stress. *Curr. Opin. Plant Biol.* **2003**, *6*, 441–445. [CrossRef]
4. Mittler, R. ROS are good! *Trends Plant Sci.* **2017**, *22*, 11–19. [CrossRef] [PubMed]
5. Ashraf, M.P.J.C.; Harris, P.J.C. Potential biochemical indicators of salinity tolerance in plants. *Plant Sci.* **2004**, *166*, 3–16. [CrossRef]
6. Ashraf, M. Biotechnological approach of improving plant salt tolerance using antioxidants as markers. *Biotechnol. Adv.* **2009**, *27*, 84–93. [CrossRef]
7. Carvalho, M.D. Drought stress and reactive oxygen species, production, scavenging and signaling. *Plant Signal Behav.* **2008**, *3*, 156–165. [CrossRef]
8. Suzuki, N.; Koussevitzky, S.; Mittler, R.; Miller, G. ROS and redox signalling in the response of plants to abiotic stress. *Plant Cell Environ.* **2012**, *35*, 259–270. [CrossRef] [PubMed]
9. Apel, K.; Hirt, H. Reactive oxygen species, metabolism, oxidative stress and signal transduction. *Annu. Rev. Plant Biol.* **2004**, *55*, 373–399. [CrossRef]
10. Kai, H.; Hirashima, K.; Matsuda, O.; Ikegami, H.; Winkelmann, T.; Nakahara, T.; Iba, K. Thermotolerant cyclamen with reduced acrolein and methyl vinyl ketone. *J. Exp. Bot.* **2012**, *63*, 4143–4150. [CrossRef]
11. Das, K.; Roychoudhury, A. Reactive oxygen species (ROS) and response of antioxidants as ROS-scavengers during environmental stress in plants. *Front. Environ. Sci.* **2014**, *2*, 1–13. [CrossRef]
12. Wi, S.G.; Chung, B.Y.; Kim, J.S.; Kim, J.H.; Baek, M.H.; Lee, J.W.; Kim, Y.S. Ultrastructural changes of cell organelles in *Arabidopsis* stem after gamma irradiation. *J. Plant Biol.* **2005**, *48*, 195–200. [CrossRef]
13. Wi, S.G.; Chung, B.Y.; Kim, J.S.; Kim, J.H.; Baek, M.H.; Lee, J.W.; Kim, Y.S. Effects of gamma irradiation on morphological changes and biological responses in plants. *Micron* **2007**, *38*, 553–564. [CrossRef] [PubMed]
14. Wang, S.; Yang, R.; Shu, C.; Zhang, X.C. Screening for cold-resistant tomato under radiation mutagenesis and observation of the submicroscopic structure. *Acta Physiol. Plant* **2016**, *38*, 1–12. [CrossRef]
15. Cho, H.S.; Lee, H.S.; Pai, H.S. Expression patterns of diverse genes in response to gamma irradiation in Nicotiana tabacum. *J. Plant Biol.* **2000**, *43*, 82–87. [CrossRef]
16. Moussa, H.R. Gamma irradiation effects on antioxidant enzymes and G6PDH activities in *Vicia Faba* plants. *J. New Seeds* **2008**, *9*, 89–99. [CrossRef]
17. Zaka, R.; Vandecasteele, C.M.; Misset, M.T. Effects of low chronic doses of ionizing radiation on antioxidant enzymes and G6PDH activities in *Stipa capillata* (Poaceae). *J. Exp. Bot.* **2002**, *53*, 1979–1987. [CrossRef] [PubMed]

18. Kim, J.H.; Chung, B.Y.; Kim, J.S.; Wi, S.G. Effects of in planta gamma-irradiation on growth, photosynthesis, and antioxidative capacity of red pepper (*Capsicum annuum* L.). *J. Plant Biol.* **2005**, *48*, 47–56. [CrossRef]

19. Chen, S.; Chai, M.L.; Jia, Y.F.; Gao, Z.S.; Zhang, L.; Gu, M.X. In Vitro Selection of salt tolerant variants following ^{60}Co gamma irradiation of long-term callus cultures of *zoysia matrella* [L.] merr. *Plant Cell Tiss. Organ Cult.* **2011**, *107*, 493–500. [CrossRef]

20. Qi, W.C.; Zhang, L.; Xu, H.B.; Wang, L.; Jiao, Z. Physiological and molecular characterization of the enhanced salt tolerance induced by low-dose gamma irradiation in *Arabidopsis* seedlings. *Biochem. Biophys. Res. Commun.* **2014**, *450*, 1010–1015. [CrossRef]

21. Xiang, Q.B.; Liu, Y.L. *An Illustrated Monograph of the Sweet Osmanthus Cultivars in China*; Zhejiang Science and Technology Press: Hangzhou, China, 2008.

22. Ke, D.S.; Wang, A.G.; Sun, G.C.; Dong, L.F. The effect of active oxygen on the activity of ACC synthase induced by exogenous IAA. *Acta Bot. Sin.* **2002**, *44*, 551–556.

23. Geng, X.M.; Liu, J.; Lu, J.G.; Hu, F.R.; Okubo, H. Effects of cold storage and different pulsing treatments on postharvest quality of cut OT Lily 'Mantissa' flowers. *J. Facul. Agric. Kyushu Univ.* **2009**, *54*, 41–45.

24. Dhindsa, R.A.; Plumb, D.P.; Thorpe, T.A. Leaf senescence, correlated with increased permeability and lipid peroxidation, and decreased levels of superoxide dismutase and catalase. *Exp. Bot.* **1981**, *126*, 93–101. [CrossRef]

25. Zhang, J.; Kirkham, M.B. Drought-stress-induced changes in activities of superoxide dismutase, catalase, and peroxidase in wheat species. *Plant Cell Physiol.* **1994**, *35*, 785–791. [CrossRef]

26. Bates, L.S.; Waldren, R.P.; Teare, I.D. Rapid determination offree proline for water-stress studies. *Plant Soil* **1973**, *39*, 205–207. [CrossRef]

27. Geng, X.M.; Liu, X.; Ji, M.; Hoffmann, W.A.; Grunden, A.; Xiang, Q.Y.J. Enhancing heat tolerance of the little dogwood *cornus canadensis* L. f. with introduction of a superoxide reductase gene from the hyperthermophilic archaeon *pyrococcus furiosus*. *Front. Plant Sci.* **2016**, *7*, 1–7. [CrossRef]

28. Marcu, D.; Cristea, V.; Daraban, L. Dose-dependent effects of gamma radiation on lettuce (*Lactuca sativa* var. *capitata*) seedlings. *Int. J. Radiat. Biol.* **2013**, *89*, 219–223. [CrossRef]

29. Geng, X.M.; Wang, L.G.; Li, N.; Yang, X.L. Study on the seed germination and seedling growth of *Osmanthus fragrans* under $^{6\circ}$Co-γ irradiation. *J. Nuclear Agric. Sci.* **2016**, *30*, 0216–0223, (In Chinese with English abstract).

30. Shereen, A.; Ansari, R.; Mumtaz, S.; Bughio, H.R.; Mujtaba, S.M.; Shirazi, M.U.; Khan, M.A. Impact of gamma irradiation induced changes on growth and physiological responses of rice under saline conditions. *Pakist. J. Bot.* **2009**, *41*, 2487–2495.

31. He, S.Z.; Han, Y.F.; Wang, Y.P.; Zhai, H.; Liu, Q.C. In vitro selection and identification of sweet potato (*Ipomoea batatas* (L.) Lam.) plants tolerant to NaCl. *Plant Cell Tiss. Organ Cult.* **2009**, *96*, 69–74. [CrossRef]

32. Haruhiko, W.; Koshiba, T.; Matsui, T.; Satô, M. Involvement of peroxidase in differential sensitivity to γ-radiation in seedlings of two *Nicotiana* species. *Plant Sci.* **1998**, *132*, 109–119.

33. Zaka, R.; Chenal, C.; Misset, M.T. Effects of low doses of short-term gamma irradiation on growth and development through two generations of *Pisum sativum*. *Sci. Total Environ.* **2004**, *320*, 121–129. [CrossRef]

34. Halliwell, B. Reactive species and antioxidants. Redox biology is a fundamental theme of aerobic life. *Plant Physiol.* **2006**, *141*, 312–322. [CrossRef]

35. Helaly, M.N.M.; El-Hosieny, A.M.R. Effectiveness of gamma irradiated protoplasts on improving salt tolerance of lemon (*Citrus limon* L. Burm. f.). *Am. J. Plant Physiol.* **2011**, *6*, 190–208. [CrossRef]

36. Qi, W.C.; Zhang, L.; Feng, W.S.; Xu, H.B.; Wang, L.; Jiao, Z. ROS and ABA Signaling Are Involved in the Growth Stimulation Induced by Low-Dose Gamma Irradiation in *Arabidopsis* Seedling. *Appl. Biochem. Biotechnol.* **2015**, *175*, 1490–1506. [CrossRef]

37. Goh, E.J.; Kim, J.B.; Kim, W.J.; Ha, B.K.; Kim, S.H.; Kang, S.Y.; Seo, Y.W.; Kim, D.S. Physiological changes and anti-oxidative responses of *Arabidopsis* plants after acute and chronic γ-irradiation. *Radiat. Environ. Biophys.* **2014**, *53*, 677–693. [CrossRef]

38. Aghaei, K.; Ehsanpour, A.A.; Komatsu, S. Potato responds to salt stress by increased activity of antioxidant enzymes. *J. Integr. Plant Biol.* **2009**, *51*, 1095–1103. [CrossRef]

39. Nikam, A.A.; Devarumath, R.M.; Ahuja, A.; Babu, H.; Shitole, M.G.; Suprasanna, P. Radiation-induced in vitro mutagenesis system for salt tolerance and other agronomic characters in sugarcane (*Saccharum officinarum* L.). *Crop J.* **2015**, *3*, 46–56. [CrossRef]

40. Mansour, M.M.F. Protection of plasma membrane of onion epidermal cells by glycinebetaine and proline against NaCl stress. *Plant Physiol. Biochem.* **1998**, *36*, 767–772. [CrossRef]
41. Rejeb, K.B.; Abdelly, C.; Savouré, A. How reactive oxygen species and proline face stress together. *Plant Physiol. Biochem.* **2014**, *80*, 278–284. [CrossRef]
42. Hossain, M.A.; Fujita, M. Evidence for a role of exogenous glycinebetaine and proline inantioxidant defense and methylglyoxal detoxification systems in mung bean seedlings under salt stress. *Physiol. Mol. Biol. Plants* **2010**, *16*, 19–29. [CrossRef] [PubMed]

 *forests*

Article

Spatial and Temporal Calcium Signaling and Its Physiological Effects in Moso Bamboo under Drought Stress

Xiong Jing, Chunju Cai *, Shaohui Fan, Lujun Wang and Xianli Zeng

International Center for Bamboo and Rattan, State Forestry and Grassland Administration Key Laboratory of Bamboo and Rattan, Beijing 100102, China; jxicbr@163.com (X.J.); fansh@icbr.ac.cn (S.F.); wuatang@icbr.ac.cn (L.W.); zengxl723@163.com (X.Z.)
* Correspondence: caicj@icbr.ac.cn; Tel.: +86-010-8478-9806

Received: 25 January 2019; Accepted: 22 February 2019; Published: 2 March 2019

Abstract: Elevations in cytosolic free calcium concentration constitute a fundamental signal transduction mechanism in plants; however, the particular characteristics of calcium ion (Ca^{2+}) signal occurrence in plants is still under debate. Little is known about how stimulus-specific Ca^{2+} signal fluctuations are generated. Therefore, we investigated the identity of the Ca^{2+} signal generation pathways, influencing factors, and the effects of the signaling network under drought stress on *Phyllostachys edulis* (Carrière) J. Houz. Non-invasive micro testing and laser confocal microscopy technology were used as platforms to detect and record Ca^{2+} signaling in live root tip and leaf cells of *P. edulis* under drought stress. We found that Ca^{2+} signal intensity (absorption capacity) positively correlated with degree of drought stress in the *P. edulis* shoots, and that Ca^{2+} signals in different parts of the root tip of *P. edulis* were different when emitted in response to drought stress. This difference was reflected in the Ca^{2+} flux and in regional distribution of Ca^{2+}. Extracellular Ca^{2+} transport requires the involvement of the plasma membrane Ca^{2+} channels, while abscisic acid (ABA) can activate the plasma membrane Ca^{2+} channels. Additionally, Ca^{2+} acted as the upstream signal of H_2O_2 in the signaling network of *P. edulis* under drought stress. Ca^{2+} was also involved in the signal transduction process of ABA, and ABA can promote the production of Ca^{2+} signals in *P. edulis* leaves. Our findings revealed the physiological role of Ca^{2+} in drought resistance of *P. edulis*. This study establishes a theoretical foundation for research on the response to Ca^{2+} signaling in *P. edulis*.

Keywords: Ca^{2+} signal; drought stress; living cell; Moso Bamboo (*Phyllostachys edulis*); plasma membrane Ca^{2+} channels; signal network

---

1. Introduction

Calcium ions (Ca^{2+}) are a primary signaling element for diverse cell processes in response to environmental cues. Ca^{2+} is a vital regulatory molecule for response to stress in plant growth and development [1,2]. When plants are affected by various physical stimuli, such as temperature, drought, salt, light, gravity, or chemical substances, such as plant hormones and pathogenic inducers, extracellular and intracellular sources of calcium release Ca^{2+} into the cytoplasm through Ca^{2+} channels. The spatio-temporal activity of membrane-localized Ca^{2+} channels or transporters causes an increase in cytosolic free calcium ion concentration (Ca^{2+})cyt, resulting in specific signals [3,4]. Almost all of the extracellular stimuli can lead to changes in intracellular (Ca^{2+})cyt. However, different stress stimuli can lead to variations in the pattern of Ca^{2+} spatial-temporal changes. There are significant differences in time, frequency, amplitude, and regional distribution. Studies have illustrated that plants may rely on different forms of Ca^{2+} to reflect the specificity of different stimulation signals to achieve signal transduction [5].

Although Ca^{2+} signaling has been extensively studied in other gramineous plants, such as rice, little is known about the Ca^{2+} signal identities and functions of the clonal plant *Phyllostachys edulis* (Carrière) J. Houz. Therefore, a better understanding of the calcium signal characteristics of bamboo under drought stress is an important prerequisite for the study of clonal habits of bamboo using signal transduction methods.

At present, research has shown that root tip cells of *P. edulis* seedlings transport Ca^{2+} from the extracellular region, cell walls, nucleus, and other calcium stores under drought stress. With increased duration of drought stress, the distribution of and changes in Ca^{2+} will produce regular fluctuations [6]. The more pronounced the drought stress, the greater the Ca^{2+} distribution in the root tip. Exogenous application of calcium fertilizer can relieve the physiological effects of drought stress and improve drought resistance in *P. edulis* [7]. Although roots constitute the most direct organ of water absorption, both Ca^{2+} uptake velocity and density of the different organ changed under drought stress [8]. Moreover, stomatal opening and closing behaviors were also regulated by Ca^{2+} signals at the cellular scale [9].

When there are no environmental pressures present, the majority of the Ca^{2+} in plant cells is distributed in the extracellular compartments, cell wall, vacuoles, and endoplasmic reticulum, with less concentrated distribution in the cytoplasm to prevent the precipitation of calcium and phosphoric acid [10]. To regulate this lower Ca^{2+} concentration in the cytoplasm, plant cells will actively export Ca^{2+}. When a stimulus signal reaches a cell, plasma Ca^{2+} channels transiently increase Ca^{2+} permeability. When the cytoplasmic Ca^{2+} concentration increases to a certain threshold, it binds to calmodulin (CaM) to form Ca-CaM compounds, and thus activate CaM. Activated CaM further activates various key enzymes in the plant, which further phosphorylates and dephosphorylates phospholipase, nicotinamide adenine dinucleotide (NAD) kinase, and Ca^{2+}-ATPase. In addition, it amplifies the initial stimulation signal and subsequently causes the cells to produce a physiological response corresponding to the signal, such as cell division, material synthesis, etc. [2,11].

Under drought stress, a complex signaling network is formed by a communication mechanism between regulatory signals. Ca^{2+} signaling can be combined with calcium receptors, such as CaM, to amplify the signal and transmit the oscillation to initiate stomatal closure and production of reactive oxygen species [12,13]. Previously, reactive oxygen species were considered toxic byproducts of plant metabolism. However, recent studies have shown that reactive oxygen species also have an important part to play in cellular signal transduction and regulation networks [14–16]. Abscisic acid (ABA) is a root chemical signal that plays an important role in regulating stomatal movement of plants under drought stress. H_2O_2 can be used as a downstream signal of ABA to activate Ca^{2+} channels in the plasma membrane. Involvement of ABA can induce stomatal closure by increasing Ca^{2+} concentration in guard cells [17]. Sha et al. used 25% PEG-6000 to simulate water stress in maize plants, inducing CaM gene expression in the leaves [18]. The study found that exogenous ABA treatment can also induce significant CaM gene expression, and that H_2O_2 is involved in ABA-induced CaM gene expression in the late regulation period.

The emergence of new technologies has made it possible to study changes in Ca^{2+} signaling in response to environmental stress in plants. Non-invasive microelectrode technology (NMT) and laser confocal microscopy are effective techniques for detecting Ca^{2+} signals. Antoine et al. demonstrated the influx of Ca^{2+} over the course of fertilization in maize with non-invasive microelectrode technology [19]. In addition, a live-cell Ca^{2+} imaging platform has been used to detect Ca^{2+} signals in the cytoplasm and nucleus of *Arabidopsis thaliana* (L.) Heynh. This technique was also used to observe the spatial-temporal distribution of Ca^{2+} in living cells of *A. thaliana* under stimulated adverse environmental conditions [20]. These studies use a method to generate Ca^{2+} signaling pathways by treating experimental materials with Ca^{2+} inhibitors. Ethylene glycol-bis (2-aminoethylether)-N,N,N′,N′-tetraacetic acid (EGTA), lanthanum chloride ($LaCl_3$), and chlorpromazine (CPZ) are the most commonly used reagents. At present, laser confocal scanning microscopy has been used to study Ca^{2+} fluorescence localization in *P. edulis* cells under drought stress; however, most of these studies have been limited to the root tip [7] and few studies have been conducted on Ca^{2+} localization in cells located in the leaves of *P. edulis* using this technique.

To bridge this scientific gap, in this study we investigated the spatial-temporal location and flux velocity of Ca^{2+} ions by inducing drought stress in *P. edulis* with 20% PEG-6000. Using non-invasive micro-test technology (NMT) and laser confocal microscopy, we demonstrated the regularity of cellular Ca^{2+} dynamics in response to drought stress and provided measures of Ca^{2+} signaling in *P. edulis* leaf cells. We further studied the Ca^{2+} signaling pathway and analyzed the communication of the signal network pathway between Ca^{2+}, H_2O_2, and ABA in leaf cells using Ca^{2+} inhibitors (Ca^{2+} channel blockers) and ABA. The aim of this study was to reveal the physiological role of Ca^{2+} in drought resistance and establish a theoretical foundation for the cellular response to Ca^{2+} signaling in *P. edulis*.

2. Materials and Methods

2.1. Plant Materials and Treatment

The sprouting seed materials used in this study were taken from the parent *P. edulis* from Guilin, Guangxi in September 2017. The thousand seed weight of these seed materials is equal to 22.75 ± 0.35 g. The *P. edulis* seeds were treated in November 2017. The seeds were soaked in warm water for 24 h at 50 °C, and then removed and disinfected with 5‰ potassium permanganate solution for 5 min. The sterilized seeds were then repeatedly washed with distilled water and germinated in an incubator in a dark environment set to a constant temperature of 25 °C. After the seeds germinated, the seedlings were selected for regular, even growth and placed in a Petri dish with pad disinfectant (lower layer) and filter paper (upper layer). The dishes were cultured at a constant temperature of 28 °C in an illumination incubator (PRX-1000B, Safe, Ningbo, China). The Petri dishes were set to point the roots vertically downwards. Proper humidity was maintained in the Petri dishes and any seeds found to be growing mold were removed. When the vertical root length reached approximately 3 cm, the seedlings were transferred to a Seed Germination Pouch (Phytotc CYG-98LB, size: 30 cm × 25 cm, Beijing Bioconsumable Tech., Ltd., Beijing, China) and cultured with a 12-h light period with a light intensity of 120–150 mmol m^{-2} s^{-1} and a temperature of 22 ± 2 °C. Figure 1 gives more details about the performance of the experimental materials.

Figure 1. Cultivation of *P. edulis sprouting* seedlings.

For each experimental treatment, ten strains of *P. edulis* seedlings were selected in duplicate. To substitute drought stress, 20% polyethylene glycol-6000 (PEG, Coolaber, Beijing, China) was used for 5 min, 10 min, 15 min, and 30 min. Distilled water was used for the control (CK). In addition, Ca^{2+} inhibitors and ABA (Sigma-Aldrich, Shanghai, China) were used to treat *P. edulis* seedlings. The Ca^{2+} production pathway and its distribution were studied, and the effects of Ca^{2+} signaling on the H_2O_2/ABA signaling network were analyzed under drought stress. The Ca^{2+} inhibitors used in the experiment were extracellular Ca^{2+} chelating agent EGTA, Ca^{2+} channel blocker $LaCl_3$, and CaM antagonist CPZ (Sigma-Aldrich, Shanghai, China). The dosage of additives was added according to the method of Lu [21].

2.2. Laser Confocal Microscopy Luminescence Imaging

Here, we present detailed instructions for laser confocal microscopy luminescence imaging of cytosolic Ca^{2+} and H_2O_2 concentration and distribution in root tip and leaf cells of *P. edulis* seedlings.

2.2.1. Esterified Fluorescent Probe Stock Solution Configuration

The calcium ion fluorescent probe Fluo-8 (AAT Bioquest, Sunnyvale, CA, USA) was fully dissolved in dimethyl sulfoxide (DMSO, Sigma-Aldrich, Shanghai, China) to a concentration of 1 mmol/L, $-20\,^{\circ}C$ dark storage reserve.

$2',7'$-dichloro fluorescin diacetate (H2DCFDA, Sigma-Aldrich, Shanghai, China) was made up of 50 mmol/L DMSO mother liquor, which was stored in a separate container and frozen.

2.2.2. Fluorescent Labeling

As to the root tip cells and leaf cells Ca^{2+} fluorescent labeling, we put the lower epidermis of the leaves of *P. edulis* seedlings flat on transparent tape and used a surgical blade to gently scrape the upper epidermis off the leaves. Then we placed the samples of lower epidermis in Hank's balanced salt solution (without calcium ions) containing 20 µmol/L Fluo-8. We incubated them for 40 min in the dark at room temperature and rinsed the lower epidermis of *P. edulis* leaves with a buffer several times. We then incubated them again at room temperature for 20 min, ensuring that the esterification probe was fully dissolved. Finally we placed them on a glass slide and added 0.5 mL of Hank's buffer to complete slice preparation.

As to the H_2O_2 fluorescent labeling of *P. edulis* leaf cells, the lower epidermis of *P. edulis* seedling leaves was first laid flat on transparent tape, and then the upper epidermis of the leaves was removed using a surgical blade and placed in MES, Free acid, monohydrate (Coolaber, Beijing, China) buffer to a final concentration of 50 µmol/L and incubated at room temperature in the dark for 15 min. We rinsed the lower epidermis samples several times with a buffer and placed them on glass slides. We then added 0.5 mL of buffer to complete slice preparation.

2.2.3. Laser Confocal Microscope Observation

Localization of Ca^{2+} fluorescence in *P. edulis* root tips and leaves: The prepared in vivo test slices were placed on a laser confocal microscope (LSM510, LeicaDM4, Berlin, Germany) for observation and scanning. Fluorescence intensity was controlled and all parameters were kept constant during the test (parameter settings: excitation wavelength 488 nm, BP 505-530, Pinhole 280, DG 581, AO 0.1, AG 1.34). At least three fields from different repeats were selected for each test. After the results were stable, one field of view was used for the analysis.

Localization of H_2O_2 fluorescence in root tips and leaves of *P. edulis*: The prepared in vivo test slices were placed on a laser confocal microscope (LSM510) for observation and scanning (parameter settings: excitation wavelength 488 nm, BP 505-530, Pinhole 386, DG 768, AO 0, AG 1). At least three fields from different repeats were selected for each test. After the results were stable, one field of view was used for analysis.

2.3. Measurement of Ca^{2+} Flux

Net Ca^{2+} flux was measured using non-invasive micro-test technology (Physiolyzer, Younger USA LLC, Amherst, MA 01002, USA; Xuyue (Beijing) Sci. & Tech. Co., Ltd., Beijing, China). NMT non-invasively measures Ca^{2+} fluxes with a high temporal and spatial resolution. It measures the concentration gradient of Ca^{2+} by means of selective microsensor oscillation between two points in the root tip of *P. edulis* seedlings (Figure 2).

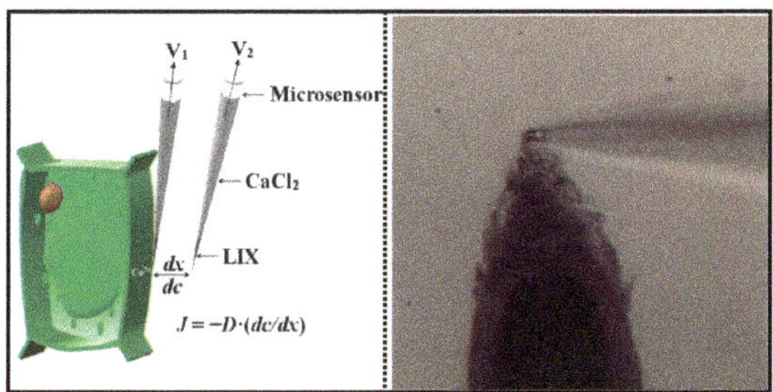

Figure 2. Schematic diagram and measured photo of Ca^{2+} flux analysis by non-invasive micro-tests. The diagram is to demonstrate the principles behind NMT testing and is not to scale. The tip diameter of the microsensor is about 5 μm.

After different test treatments, the roots were rinsed with redistilled water and immediately incubated in measuring solution to equilibrate for 10 min. Then, the roots were transferred to a measuring chamber containing 10−15 mL of a fresh measuring solution. Ions were monitored in the following solutions: 0.1 mM KCl, 0.1 mM $CaCl_2$, 0.1 mM $MgCl_2$, 0.5 mM NaCl, 0.3 mM MES, and 0.2 mM Na_2SO_4, following adjustment of the pH to 6.0. The measuring chamber was mounted on the micromanipulator, and the flux microsensor was positioned close to the root tip at four points: at the pileorhiza, meristematic zone, elongation zone, and mature zone (0 μm, 200 μm, 500 μm, and 800 μm from the root apex) of *P. edulis* seedlings.

The system setup parameters in the experiment are as follows. The Ca^{2+} flux microsensor ($\Phi 4.5 \pm 0.5$ μm, XY-CGQ-01, Xuyue (Beijing) Sci. &Tech. Co., Ltd., Beijing, China) was filled with a backfilling solution (100 mM $CaCl_2$) to a length of approximately 1.0 cm from the tip. The micropipettes were front filled with 40–50 μm columns of selective liquid ion-exchange cocktails (Ca^{2+} LIX, XY-SJ-Ca, YoungerUSA LLC, Amherst, MA, USA). An Ag/AgCl wire microsensor holder YG003-Y11 (Younger USA) was inserted in the back of the microsensor to make electrical contact with the electrolyte solution. YG003-Y11 (Younger USA) was used as the reference microsensor. Prior to the flux measurement, the flux microsensor was calibrated with a measuring solution having different concentrations of Ca^{2+} 0.1 mM and 0.01 mM. The electrodes with a Nernstian slope > 22 mV per decade were used in this study. Only the same flux microsensor was calibrated again according to the same procedure and standards after each test. Data was discarded if the post-test calibrations failed. The following figure shows the schematic and actual measurement of Ca^{2+} flux.

The data for Ca^{2+} fluxes were calculated by Fick's law of diffusion as follows:

$$J = -D \cdot (dc/dx) \tag{1}$$

where dx (30 μm) is the distance the flux microsensor moved repeatedly from one point to another perpendicular to the surfaces of the samples at a frequency of ca. 0.3 Hz.

2.4. Statistical Analysis

Data were analyzed by single factorial analysis of variance and statistical correlation analysis. The significance of differences among means was evaluated using the least significant difference test, with a family wise error rate of 0.05, using the Statistical Package for Social Sciences, v18.0 (SPSS Inc., Chicago, IL, USA). Significant differences are marked with alphabet.

3. Results and Discussion

3.1. Variation in the Flux and Distribution of Ca^{2+} among Different Parts of the Root Tips of P. edulis Seedlings under Drought Stress

The transient net Ca^{2+} flux was measured from different regions along the root axis using NMT in four different areas, the root apex and root hair zone, including pileorhiza, the meristematic zone, elongation zone, and mature zone (Figure 3a). Responses in the root apex Ca^{2+} flux are shown in Figure 3b. The results showed that there was a significant difference in the ability of *P. edulis* seedlings to absorb Ca^{2+} from different parts of the root tip after drought stress. The Figure 3b showed the most uptake of Ca^{2+} in pileorhiza. The pileorhiza was the area with the strongest Ca^{2+} signal response ability, and the Ca^{2+} concentration in the root pileorhiza was the highest. The elongation zone was the region with the strongest Ca^{2+} efflux. The intensity of Ca^{2+} uptake capacity at different parts of the root tip from strong to weak was: pileorhiza, mature area, meristematic zone, and elongation zone. The Figure 3c shows the fluorescence localization of *P. edulis* root tip cells by laser confocal microscopy. To better observe the distribution of Ca^{2+} in the pileorhiza, the corresponding heat map is shown in Figure 3d. Combined with Figure 3b, it can be concluded that there were differences in the responsiveness of different parts of the root tip of *P. edulis* to drought stress, which is reflected in both the Ca^{2+} flux and regional distribution.

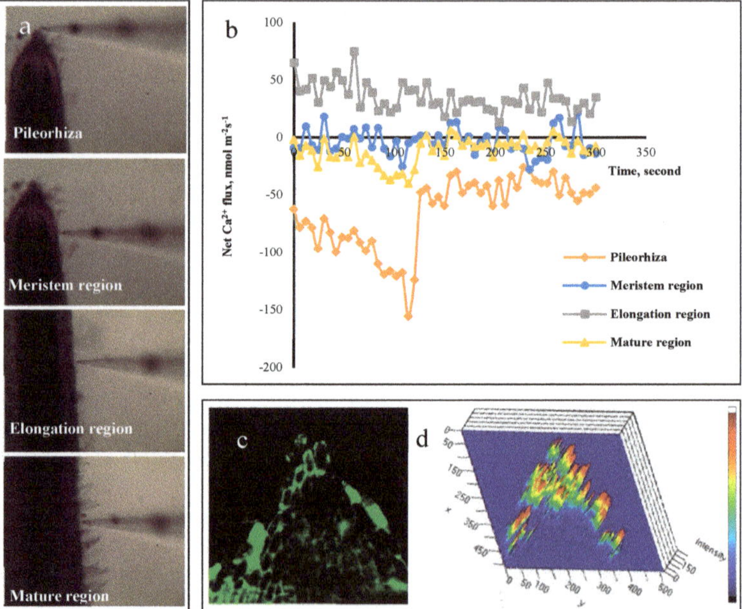

Figure 3. The flux and distribution of Ca^{2+} in different parts of *Phyllostachys edulis* root tips under drought stress. (**a**,**b**) Net Ca^{2+} fluxes in different parts of root tip of *P. edulis* seedlings under drought stress induced by 20% PEG for 10 min. At each position, an average Ca^{2+} flux was measured for 5 min before the electrode was repositioned (Data Repetition: 6 replicates). (**c**) Green fluorescence intensity is positively correlated with the Ca^{2+} concentration. The Ca^{2+} fluorescence localization at the root tip of the *P. edulis* seedlings treated with PEG-simulated drought stress for 10 min. (**d**) A heat map of a further analysis for (**c**), which clearly presented that the concentrated distribution of Ca^{2+} was in the pileorhiza of the *P. edulis* root tip cells under drought stress.

We confirmed this difference in the results of Ying [7], which used laser confocal microscopy on fluorescence localization in the root tip. This study showed that Ca^{2+} absorption intensity in the elongation zone was weakest under drought stress, while Ying's study showed that under drought stress, the Ca^{2+} in the root tip of *P. edulis* was mainly distributed in the pileorhiza and elongation zone; the distribution in the meristem area was relatively lower. This was due to the presence of a large number of small vacuoles in the cells of the elongation zone, which were not present in meristem zone. The vacuoles, as a calcium bank in plant cells, may provide a large amount of Ca^{2+} for *P. edulis* cytoplasm under drought stress. This leads to a greater concentration of Ca^{2+} in the elongate zone.

3.2. Effects of Drought Stress Duration on Ca^{2+} Absorption Regularity in the Pileorhiza of P. edulis

Under different durations of PEG-induced drought stress, we observed different conductivity, which represents leaf cell permeability (Figure 4a). An increase in drought stress duration caused electrolyte leakage in the cells, which in turn led to an increase in leaf conductivity. The cell membrane permeability of *P. edulis* seedlings increased, and there was a positive correlation between time and membrane permeability. The difference in conductivity increase between the 10- and 15-min treatments was higher than that in the 5–10 min period (difference in conductivity = 15.2%). The relative electrical conductivity of *P. edulis* seedlings under drought stress was 40.4% higher in the 30-min treatment than in the 15 min treatment. We observed significant correlation between Ca^{2+} flux and drought stress time ($p < 0.01$, Pearson correlation coefficient (r) = 0.967) (Figure 4b). All the flux data in Figure 4b represents stable and optimal real-time flux that can respond to Ca^{2+} concentration in response to drought stress. Relative to the control group, Ca^{2+} flux in pileorhiza of *P. edulis* seedlings gradually shifted from efflux to influx. Ten minutes of PEG treatment was found to be a "threshold", after which efflux became influx. There was also a positive correlation between leachate conductivity and time as Ca^{2+} absorption intensity in the pileorhiza ($p < 0.01$, $r = 0.976$), and leaf membrane permeability increased.

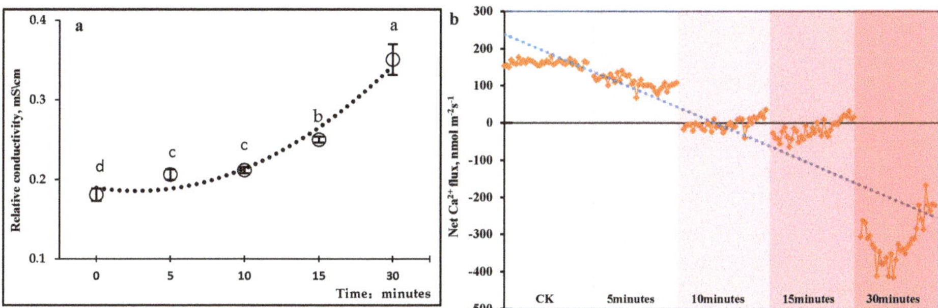

Figure 4. Ca^{2+} flux of the pileorhiza cells of *P. edulis* seedlings under different durations of PEG-treatment. (**a**) The relative conductivity of leaves from *P. edulis* seedlings treated with PEG-simulated drought stress for 5 min, 10 min, 15 min, and 30 min (Data Repetition: 6 replicates). The control (CK) is represented by 0, which was a control group incubated with distilled water. The dashed curve ($y = 0.0132x^2 - 0.0406x + 0.2171$, $R^2 = 0.9615$) represents the linear regression model for relative conductivity and PEG-simulated drought stress. (**b**) The depth of the red color represents the length of treatment time with 20% PEG simulating drought stress. From left to right in Figure 4b: the CK group, PEG-simulated drought stress treatment for 5 min, 10 min, 15 min, and 30 min, respectively. The flux data for each treatment in the figures reflected the steady-state real-time flux of maximum response to Ca^{2+} under drought stress (Data Repetition: 3 replicates); each segment of real-time data comprises flux data for three minutes.

Ca^{2+} flux in the pileorhiza of *P. edulis* seedlings in the control group (CK) maintained an out-of-range value of approximately 152 nmol m^{-2} s^{-1}. Figure 5 showed that Ca^{2+} flux in the pileorhiza was only slightly changed in the experimental group after 5 min of PEG treatment, indicating that the slight degree of drought stress did not significantly induce the stress response leading to Ca^{2+} signaling in such a short time. When treated with PEG for 30 min, the Figure 5 shows a "V-shaped" pattern of ion uptake, with net flux first increasing and then decreasing. As drought stress time increased, Ca^{2+} flux gradually returned to control group standard of efflux. It is evident that there is a positive correlation between Ca^{2+} signal and drought stress time (Figure 5). Under drought stress, there is a "stress threshold" for the Ca^{2+} signal response. When the stress level reached this threshold, the Ca^{2+} flux at the pileorhiza underwent significant changes, which may be related to the plant's ability to withstand drought stress.

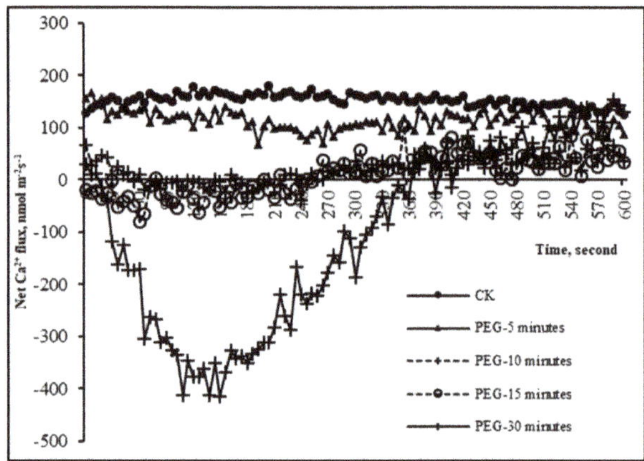

Figure 5. Changes in net Ca^{2+} flux in the pileorhiza of *P. edulis* seedlings after PEG-induced drought stress. Real-time Ca^{2+} flux by the pileorhiza of by PEG-induced drought stress for 5 min, 10 min, 15 min, 30 min, and CK, respectively (Data Repetition: 3 replicates).

Thirty minutes of PEG simulated drought stress showed obvious Ca^{2+} absorption in the pileorhiza, and Ca^{2+} flux oscillation increased. With the increase in the duration of drought stress, Ca^{2+} absorption increased. *P. edulis* seedlings were able to absorb Ca^{2+} from the extracellular environment under drought stress conditions, suggesting that while the intracellular calcium stores provide Ca^{2+} to the cytoplasm, Ca^{2+} absorbed from extracellular environment was also an important way to accumulate calcium signaling ions in the pileorhiza. It is suggested that Ca^{2+}, as an important signaling molecule in response to drought stress, participated in the transmission of drought signaling to the roots of *P. edulis* seedlings.

3.3. Analysis of Ca^{2+} Signal Transport Patterns in the Pileorhiza of P. edulis under Drought Stress

The study of the Ca^{2+} signaling system in plant cells was mainly carried out by pretreatment of experimental materials with Ca^{2+} signal inhibitors [21]. Regulation of Ca^{2+} channels is vital. Plasma membrane Ca^{2+}-permeable channels interact with Ca^{2+} activated nicotinamide adenine dinucleotide phosphate (NADPH) to form a self-amplifying system—a ROS-Ca^{2+} hub [22]. This system could provide the transduction and amplification of the initial Ca^{2+} or reactive oxygen species (ROS) stimuli into a more sustainable response, with implications for cell growth, hormonal signaling, and stress response [23,24]. To obtain further evidence of the role of plasma membrane Ca^{2+} channels in *P. edulis* root tips, different Ca^{2+} inhibitors and ABA were used to further determine the Ca^{2+} signaling pathways. Ca^{2+} signal intensity varied with time in the pileorhiza of *P. edulis* seedlings treated with $LaCl_3$ (Ca^{2+} channel blocker), EGTA (extracellular Ca^{2+} chelating agent), and exogenous ABA.

To determine the factors affecting Ca^{2+} transport, Ca^{2+} flux in *P. edulis* root tip treated with a calcium antagonist and ABA was measured using NMT. As shown in Figure 6a, $LaCl_3$ was applied to the *P. edulis* seedlings treated with PEG-induced drought stress. The $LaCl_3$ treatment significantly impeded extracellular Ca^{2+} influx at the pileorhiza compared to the experimental group without $LaCl_3$. Extracellular Ca^{2+} channels were involved in Ca^{2+} fluid transport. The extracellular Ca^{2+} in seedlings could enter cells through Ca^{2+} channels under drought stress, which was one of the reasons for the increase of cytoplasmic Ca^{2+} concentration.

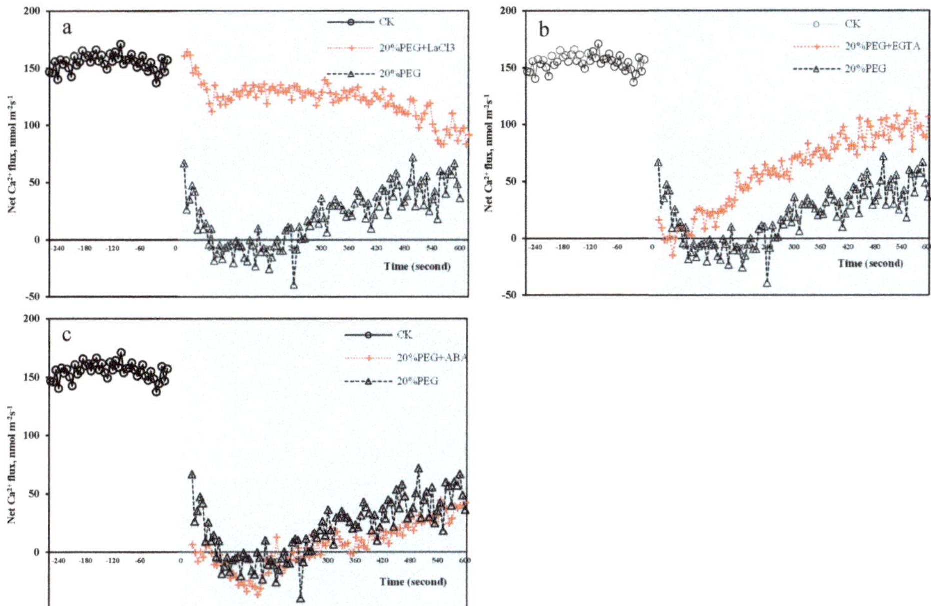

Figure 6. Factor analysis of Ca^{2+} flux of *P. edulis* seedlings under drought stress (Data Repetition: 3 replicates). The white area represents Ca^{2+} oscillation in the pileorhiza after treatment of the *P. edulis* seedlings with distilled water (CK), and the black line graph in the blue area indicates the Ca^{2+} oscillation in the pileorhiza treated with 20% PEG-induced drought stress for 10 min. The red line in the blue region indicates Ca^{2+} oscillation under 0.5 mmol/L $LaCl_3$ (**a**), 3 mmol/L EGTA (**b**), and 30 μmol/L ABA (**c**) in the 20% PEG treatment for 10 min.

In addition, EGTA was used to treat the roots of *P. edulis* seedlings under PEG-simulated drought stress. Ca^{2+} flux at the pileorhiza was almost always in a state of efflux (Figure 6b), while net Ca^{2+} flux of without EGTA treatment saw a pattern of "efflux-absorption-efflux". The EGTA-chelating extracellular Ca^{2+} increased, resulting in decreases of extracellular Ca^{2+} concentration and a decrease in the ability of pileorhiza to absorb Ca^{2+}; this caused an increase in Ca^{2+} excretion. Ca^{2+} uptake in the pileorhiza with the addition of EGTA was significantly lower than in the CK. It is worth noting that treating the pileorhiza with EGTA and $LaCl_3$ while under drought stress had the opposite effects on Ca^{2+} uptake curves. Ca^{2+} efflux rate from the pileorhiza of $LaCl_3$ decreased as EGTA treatment increased. $LaCl_3$ treatment hindered the Ca^{2+} channel in the pileorhiza, which led to a reduction of Ca^{2+} excretion from the pileorhiza.

Figure 6c shows Ca^{2+} absorption in the pileorhiza when treated with 30 μmol/L ABA for 10 min as PEG-induced drought stress treatment. Compared with the control, the ABA treatment showed a smaller oscillation in Ca^{2+} flux in the pileorhiza, and Ca^{2+} net flux reflected its strong absorption capacity. Therefore, ABA can promote the increase in Ca^{2+} absorption intensity in the pileorhiza of *P. edulis* seedlings. We conclude that ABA may activate the extracellular Ca^{2+} channel in the pileorhiza and promote Ca^{2+} uptake (Figure 6a,b). The efflux of ABA-induced Ca^{2+} spikes remained high among all treatments. These results further confirm that plasma membrane Ca^{2+} channels activity is involved in Ca^{2+} signaling in PEG-induced drought stress by controlling (Ca^{2+})cyt through Ca^{2+} influxes. Under drought stress, adding the Ca^{2+} channel blocker $LaCl_3$ and extracellular Ca^{2+} chelating agent EGTA could significantly inhibit extracellular Ca^{2+} influx. Exogenous application of ABA could increase the ability of the pileorhiza to absorb Ca^{2+}.

3.4. Analysis of Ca^{2+} Signaling Pathway in Leaves under Drought Stress

In addition to chemical signal root transduction, drought stress triggered Ca^{2+} movement in leaf cells. Ca^{2+} may regulate leaf stomatal movement to control transpiration and respiration, as a self-protection and adaption measure in response to environmental stress. To simultaneously monitor cell-specific Ca^{2+} *P. edulis* seedling leaves, we used a confocal microscope to record Ca^{2+} fluorescence localization (Figure 7).

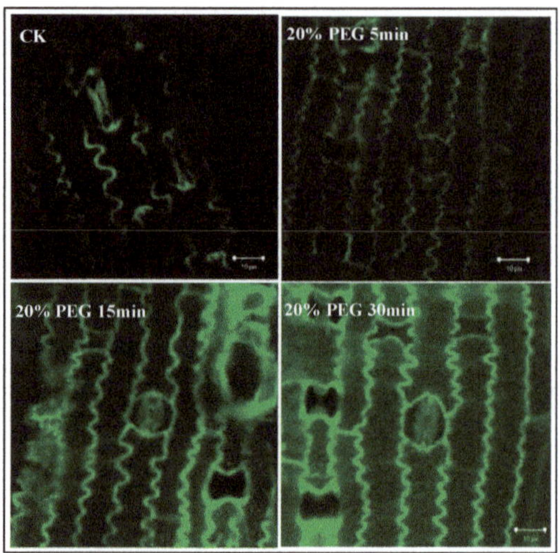

Figure 7. Localization of cellular Ca^{2+} dynamics in leaves of *P. edulis* seedlings. Calcium-fluorescence localization of the leaves of the *P. edulis* seedlings under drought stress for 5 min, 15 min, and 30 min induced by 20% PEG, respectively. The green fluorescence intensity represents the Ca^{2+} concentration and is positively correlated with the Ca^{2+} concentration.

CK is the Ca^{2+} fluorescence map of the lower epidermal cells treated with distilled water. Under normal water conditions, Ca^{2+} was mainly localized within the cell walls of guard cells, accessory cells, and long cells in *P. edulis* seedling leaves. Ca^{2+} was less commonly found to be distributed in the cytoplasm. As drought stress duration increased, Ca^{2+} concentration in the cytoplasm increased, except for in tethered cells. Among them, Ca^{2+} in long cells and guard cells increased significantly. The fluorescence intensity in the cytoplasm was also higher than that of the CK. Fluorescence intensity in the cytoplasm of accessory cells did not significantly increase. There were particularly pronounced differences in Ca^{2+} distribution in the 30 min PEG treatment. In addition to in tethered cells, the cytoplasmic fluorescence in all living cells, including guard cells, increased. Our results suggest that Ca^{2+} signals in response to PEG stress reflect distinct cellular Ca^{2+} dynamics.

With increased duration of drought stress, the fluorescence intensity in the cytoplasm of long cells increased. Only when a certain degree of drought stress was reached would Ca^{2+} in the cytoplasm of leaf guard cells escape from the cell wall and diffuse into the cytoplasm, resulting in a significant increase in the fluorescence intensity of the cytoplasm. In comparison, the Ca^{2+} in the cytoplasm of accessory cells was not obviously enhanced. To further verify that Ca^{2+} in the leaf cells of *P. edulis* was also transmitted through the cytoplasmic Ca^{2+} channels, three Ca^{2+} inhibitors were used to treat the seedlings under drought conditions. Ca^{2+} fluorescence localization under these conditions is shown in Figure 8.

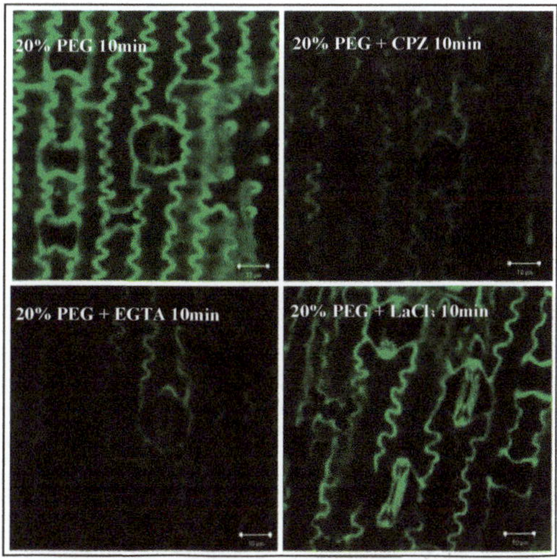

Figure 8. Localization of cellular Ca^{2+} dynamics in leaves of *P. edulis* seedlings treated with different Ca^{2+} inhibitors and drought stress. Ca^{2+} fluorescence localization in leaves of plants treated with 20% PEG-induced drought stress and 0.1 mmol/L CPZ, 3 mmol/L EGTA, or 0.5 mmol/L $LaCl_3$ for 10 min.

To confirm the effect of calcium channels on Ca^{2+}, we analyzed the Ca^{2+} fluorescence under three Ca^{2+} inhibitor treatments. The addition of CPZ inhibited the binding of Ca^{2+} to CAM in the seedlings; thus, the Ca^{2+} signal was not transmitted further. The results showed that after treatment with 0.1 mmol/L CPZ in the seedlings under PEG-simulated drought stress for 10 min, the brightness of the fluorescence in the cell wall and cytoplasm of the leaf cells was less than under the treatment without CPZ. Under drought stress, the Ca^{2+} in leaf cells of the *P. edulis* seedlings treated with 0.1 mmol/L CPZ were mainly distributed in the cell wall. The fluorescence brightness in the cell wall decreased more obviously than in other parts of the leaf.

We analyzed the distribution of Ca^{2+} in leaves treated with 3 mmol/L EGTA. Fluorescence intensity in the cell wall and cytoplasm reduced, indicating that Ca^{2+} concentration in the cell wall and cytoplasm decreased compared with the control group. This indicates that *P. edulis* seedling leaves can still take up Ca^{2+} from extracellular pathways under drought stress to produce calcium signals. Extracellular Ca^{2+} was chelated in the leaves under EGTA treatment, resulting in a decrease in extracellular Ca^{2+} uptake.

Treatment with $LaCl_3$ prevents Ca^{2+} from extracellular entry, and Ca^{2+} in the cytoplasm of leaf cells significantly reduced. Ca^{2+} was mainly concentrated in the cell walls in this treatment (Figure 8). The results of Ca^{2+} fluorescence localization in leaf cells after treatment with 0.5 mmol/L $LaCl_3$ for 10 min in drought-stressed *P. edulis* seedlings showed a decrease in Ca^{2+} fluorescence intensity in the cytoplasm of the cells. This was caused by the $LaCl_3$ blocking the cytoplasmic Ca^{2+} channel, blocking extracellular Ca^{2+} entry through the cell wall. This shows that Ca^{2+} in the leaves of *P. edulis* can be transported through Ca^{2+} channels of the plasma membrane, absorbing Ca^{2+} from outside the cell, and using the potential difference to generate calcium signals.

3.5. Analysis of the Effects of Ca^{2+} Signals on H_2O_2 and ABA Signaling Pathways under Drought Stress

The H_2O_2 fluorescence localization map of *P. edulis* leaves is shown in Figure 9. H_2O_2 concentration in the leaf cytoplasm increased with drought stress time (Figure 9a). *P. edulis* seedling leaves treated with different Ca^{2+} inhibitors under PEG-simulated drought stress were subjected to laser confocal technology for H_2O_2 fluorescence localization to study the relationship between Ca^{2+} signaling and H_2O_2 in the stress signaling pathway (Figure 9b). Compared with CK, plants treated with 0.1 mmol/L CPZ showed lower leaf fluorescence, indicating that CPZ could also prevent H_2O_2 signal transduction. The calcium signals were unable to be transmitted normally, resulting in the decrease of H_2O_2 in the leaf cells. This indicates that the regulation of H_2O_2 activity requires the participation of Ca^{2+}. Both 0.5 mmol/L $LaCl_3$ and 3 mmol/L EGTA inhibited the production of H_2O_2 in guard cells, accessory cells, and long cells in leaves of *P. edulis* seedlings under drought stress. It is inferred that the Ca^{2+} signal is generated upstream of the active oxygen signal in the drought stress signaling network of *P. edulis*, and that H_2O_2 activity in leaf cells requires Ca^{2+} participation.

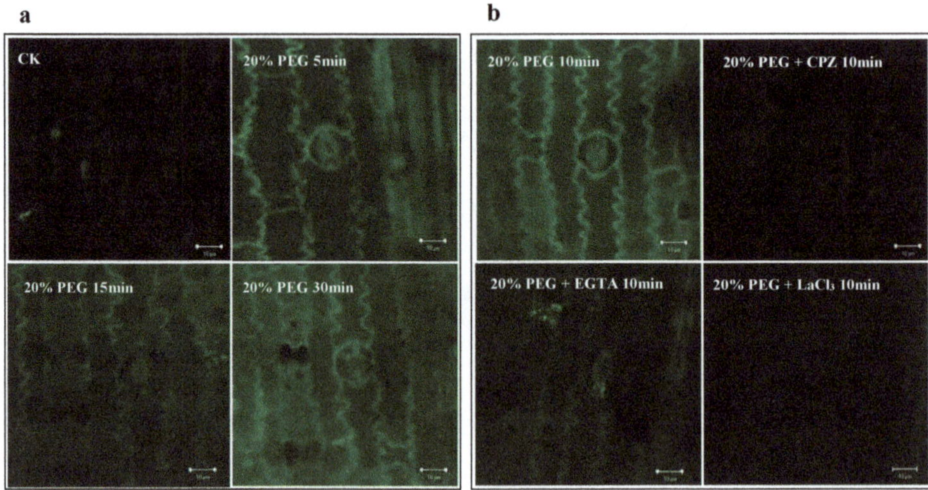

Figure 9. H_2O_2 fluorescence in *P. edulis* seedling leaves under drought stress. The green fluorescence intensity in the figure represents concentration of hydrogen peroxide, with more fluorescence indicating a higher concentration of H_2O_2. (**a**) H_2O_2 fluorescence localization at different treatment times (5 min, 15 min, and 30 min) using 20% PEG-induced drought stress. (**b**) The samples were treated with different Ca^{2+} inhibitors (CPZ, EGTA, and $LaCl_3$) for 10 min under 20% PEG-induced drought stress.

Laser confocal microscopy was used to detect the Ca^{2+} and H_2O_2 fluorescence signal in living leaf cells. The results indicated that the exogenous application of ABA under drought stress led to an increase in the concentration of Ca^{2+} and H_2O_2 in the mesophyll cells of *P. edulis* seedlings. Interestingly, there was a positive correlation between Ca^{2+} and H_2O_2 dynamics. ABA might activate Ca^{2+} channels of the plasma membrane and promote the production of Ca^{2+} signals in the pileorhiza of *P. edulis* seedlings.

H_2O_2 fluorescence intensity in guard cells was significantly higher in the ABA-treated group than in the group not treated with ABA (Figure 10). In plants, ABA is involved in many stress responses [25–27]. Regulatory systems of reactive oxygen species (ROS) are known to be integrated with other pathways involving Ca^{2+} signaling, protein kinases, and hormones pathways to regulate the defense mechanism in plants [28]. The existence of ABA-induced genes, which are expressed in stomatal guard cells, allows us to closely examine the role of Ca^{2+} [29]. In rice, the Ca^{2+}/CaM-dependent protein kinase OsDMI3 has been shown to be required for ABA-induced antioxidant defense [30]. These results demonstrated that ABA also plays a key role in the signal transduction of Ca^{2+} in the signaling network of the leaves of *P. edulis* treated with drought stress. Ca^{2+} not only acted as the upstream signal of H_2O_2, but was also involved in the signal transduction process of ABA. ABA could promote the production of Ca^{2+} signaling in leaves and stimulate the burst of H_2O_2, a reactive oxygen species, in the guard cells. *P. edulis* may enhance drought tolerance via ABA-induced stomatal closure by ROS production.

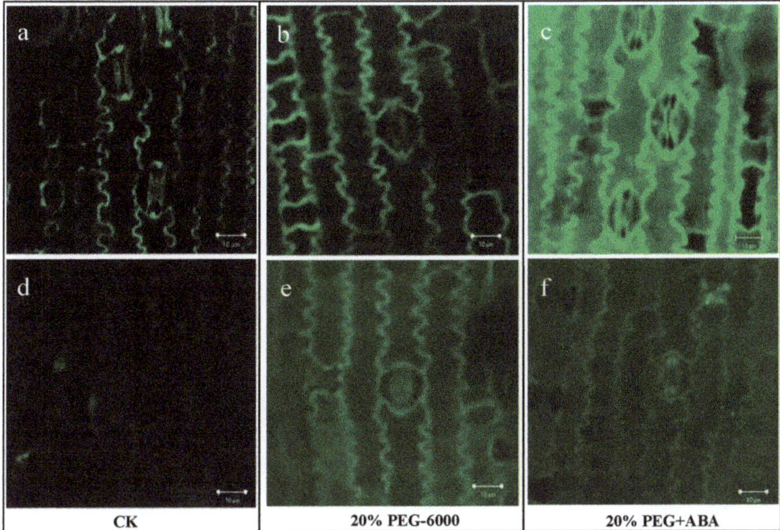

Figure 10. Localization of fluorescence of Ca^{2+} and H_2O_2 in mesophyll cells after application of exogenous ABA under drought stress. (**a**) Calcium-fluorescence localization of leaves of *P. edulis* seedlings with distilled water. (**b**) Calcium-fluorescence localization of leaves of *P. edulis* seedlings under 20% PEG 6000-induced stress for 10 min. (**c**) Calcium-fluorescence localization of leaves of *P. edulis* seedlings with the application of 30μmol/L ABA to 20% PEG-induced stress for 10 min. (**d–f**) H_2O_2 fluorescence localization under the same treatment as above.

4. Conclusions

This study indicated that the conductivity of *P. edulis* leaves increased with the degree of drought stress induced. Ca^{2+} is an important signaling molecule in response to drought stress in the root tip of *P. edulis*, especially in the transmission of stress signals and resistance to drought stress. Under drought stress, root tip cells of *P. edulis* can be used to transport Ca^{2+} from the extracellular environment to the cytoplasm, Ca^{2+} channels participate in extracellular Ca^{2+} transportation, and ABA may activate Ca^{2+} channels in the plasma membrane and promote Ca^{2+} signal production in the pileorhiza of *P. edulis*. In *P. edulis* leaves, Ca^{2+} can also be transported through the Ca^{2+} channels of the plasma membrane under drought stress, absorbing Ca^{2+} from outside the cell and generating Ca^{2+} signals through potential difference. The responsiveness of Ca^{2+} signals to drought stress in leaves of *P. edulis* from strong to weak was shown as follows: (1) long cells; (2) guard cells; (3) accessory cells; and (4) plug cells. Ca^{2+} acts as the upstream signal of H_2O_2 in the signal network of the mesophyll cells of the *P. edulis* under drought stress. It is also involved in ABA signal transduction process. ABA could promote Ca^{2+} signal production and stimulate H_2O_2 bursts in *P. edulis* leaves.

This study provides a method for the spatial and temporal localization of Ca^{2+} signaling and flux in *P. edulis*. Further research on Ca^{2+} signaling is essential, as it may help shed light on the physiology of *P. edulis* under stress. There are also areas for improvement in this study. The first is that processing the mesophyll samples may damage the cells or put them under stress conditions, thus affecting the Ca^{2+} fluorescence of the leaves under the laser confocal microscope. Autofluorescence of lignin can also have an effect on Ca^{2+} fluorescence. It is particularly important to improve this methodology in the future.

Author Contributions: X.J., C.C., and S.F. conceived and designed the experiments; X.J. performed the experiments; X.J. analyzed the data; L.W. and X.Z. contributed reagents/materials/analysis tools; X.J. wrote the paper, and C.C. revised it.

Funding: This research was funded by the International Centre for Bamboo and Rattan Center. Effects of Drought Stress on Water Physiology and Productivity of *Phyllostachys edulis* (grant number 1632018008).

Acknowledgments: The authors acknowledge the financial supports of the Foundation of International Centre for Bamboo and Rattan (No. 1632018008).

Conflicts of Interest: The authors declare no conflict of interest.

References

1. Hepler, P.K. Calcium: A central regulator of plant growth and development. *Plant. Cell* **2005**, *17*, 2142–2155. [CrossRef] [PubMed]

2. Reddy, A.S.N.; Ali, G.S.; Celesnik, H.; Day, I.S. Coping with stresses: Roles of calcium-and calcium/calmodulin-regulated gene expression. *Plant. Cell* **2011**, *23*, 2010–2032. [CrossRef] [PubMed]

3. Reddy, A.S.N. Calcium: Silver bullet in signaling. *Plant Sci.* **2001**, *160*, 381–404. [CrossRef]

4. Michard, E. Glutamate receptor-like genes form Ca^{2+} channels in pollen tubes and are regulated by pistil D-serine. *Science* **2011**, *332*, 434–437. [CrossRef] [PubMed]

5. Miedema, H.; Bothwell, J.H.F.; Brownlee, C.; Davies, J.M. Calcium uptake by plant cells-channels and pumps acting in concert. *Trends Plant. Sci.* **2001**, *6*, 514–519. [CrossRef]

6. Jiang, Q. Effect of calcium ions in the process of responses of *Phyllostachys edulis*. Master's Thesis, Zhejiang A. & F. University, Hangzhou, China, 2012.

7. Ying, Y.Q.; Du, X.H.; Jiang, Q.; Xu, C.M.; Wu, J.S. Distribution of Ca^{2+} at the Tip of *Phyllostachys edulis* Root under Drought Stress and Physiological Functions of Exogenous Ca^{2+}. *Scientia Silvae Sinicae* **2013**, *49*, 141–146.

8. Wilkins, K.A.; Elsa, M.; Swarbreck, S.M.; Davies, J.D. Calcium-Mediated Abiotic Stress Signaling in Roots. *Front. Plant Sci.* **2016**, *7*, 1269. [CrossRef] [PubMed]

9. Li, F.; Zhang, H.; Liu, L.; Guo, L.L.; Wang, L.S.; Zheng, Y.P. Effects of Exogenous Ca^{2+} on Stomatal Traits and Gas Exchange Parameters of Anthurium scherzerianum. *North. Hortic.* **2017**, *8*, 80–85.

10. Edel, K.H.; Marchadier, E.; Brownlee, C.; Kudla, J.; Hetherington, A.M. The Evolution of Calcium-Based Signalling in Plants. *Curr. Biology* **2017**, *27*, 667–679. [CrossRef] [PubMed]

11. Yuan, P.; Jauregui, E.; Du, L.; Tanaka, K. Calcium signatures and signaling events orchestrate plant-microbe interactions. *Curr. Opin. Plant. Biology* **2017**, *38*, 173–183. [CrossRef] [PubMed]

12. Knight, H.; Trewavas, A.J.; Knight, M.R. Calcium signalling in Arabidopsis thaliana responding to drought and salinity. *Plant. J.* **1997**, *12*, 1067. [CrossRef] [PubMed]

13. Shinozaki, K.; Shinozaki, Y. Gene Expression and Signal Transduction in Water-Stress Response. *Plant. Physiol.* **1997**, *115*, 327–334. [CrossRef] [PubMed]

14. Tamás, L.; Mistrík, I.; Huttová, J.; Haluskova, L.; Valentovicova, K.; Zelinova, V. Role of reactive oxygen species-generating enzymes and hydrogen peroxide during cadmium, mercury and osmotic stresses in barley root tip. *Planta* **2010**, *231*, 221–231. [CrossRef] [PubMed]

15. Oracz, K.; Elmaaroufbouteau, H.; Kranner, I.; Bogatek, R.; Corbineau, F.; Bailly, C. The mechanisms involved in seed dormancy alleviation by hydrogen cyanide unravel the role of reactive oxygen species as key factors of cellular signaling during germination. *Plant. Physiol.* **2009**, *150*, 494–505. [CrossRef] [PubMed]

16. Yoshioka, H.; Asai, S.; Yoshioka, M.; Kobayashi, M. Molecular mechanisms of generation for nitric oxide and reactive oxygen species, and role of the radical burst in plant immunity. *Mol. Cell* **2009**, *28*, 321–329. [CrossRef] [PubMed]

17. Kwak, J.M.; Mori, I.C.; Pei, Z.M.; Leonhardt, N.; Torres, M.A.; Dangl, J.L.; Bloom, R.E.; Bodde, S.; Jones, J.D.G.; Schroeder, J.I. NADPH oxidase AtrbohD and AtrbohF genes function in ROS-dependent ABA signaling in Arabidopsis. *EMBO J.* **2003**, *22*, 2623–2633. [CrossRef] [PubMed]

18. Sha, Q.; Jiang, M.Y.; Lin, F.; Wang, J.X. The expression of calmodulin genes induced by water stress is associated with ABA and H_2O_2. *J. Nanjing Agric. Univ.* **2009**, *3*, 52–57.

19. Antoine, A.F.; Faure, J.E.; Dumas, C.; Feijo, J.A. Differential contribution of cytoplasmic Ca^{2+} and Ca^{2+} influx to gamete fusion and egg activation in maize. *Nat. Cell Biology* **2001**, *3*, 1120–1123. [CrossRef] [PubMed]

20. Zhu, X.; Taylor, A.; Zhang, S.; Zhang, D.; Feng, Y.; Liang, G. Measuring spatial and temporal Ca^{2+} signals in Arabidopsis plants. *J. Vis. Exp.* **2015**, *91*, 1–13.
21. Lu, Y. The Analysis of Calcium Signal Characteristics and Its Effect on *Phyllostachys edulis* Seeding under Drought Stress. Master's Thesis, Zhejiang A. & F. University, Hangzhou, China, 2014.
22. Gelli, A.; Higgins, V.J.; Blumwald, E. Activation of Plant Plasma Membrane Ca^{2+}-Permeable Channels by Race-Specific Fungal Elicitors. *Plant. Physiol.* **1997**, *113*, 269–279. [CrossRef] [PubMed]
23. Demidchik, V.; Shabala, S.; Isayenkov, S.; Cuin, T.A.; Pottosin, I. Calcium transport across plant membranes: Mechanisms and functions. *New Phytol.* **2018**, *220*, 49–69. [CrossRef] [PubMed]
24. Zhu, J.K. Salt and drought stress signal transduction in plants. *Annu. Rev. Plant. Biol.* **2002**, *53*, 247–273. [CrossRef] [PubMed]
25. Tuteja, N.; Sopory, S.K. Chemical signaling under abiotic stress environment in plants. *Plant. Signal. Behav.* **2008**, *3*, 525–536. [CrossRef] [PubMed]
26. Alqurashi, M.; Thomas, L.; Gehring, C.; Marondedze, C. A Microsomal Proteomics View of H$_2$O$_2$- and ABA-Dependent Responses. *Proteome* **2017**, *5*, 22. [CrossRef] [PubMed]
27. Hetherington, A.M. Guard cell signaling. *Cell* **2001**, *107*, 711–714. [CrossRef]
28. Suzuki, N.; Katano, K. Coordination Between ROS Regulatory Systems and Other Pathways Under Heat Stress and Pathogen Attack. *Front. Plant. Sci.* **2018**, *9*, 490. [CrossRef] [PubMed]
29. Taylor, J.E.; Renwick, K.F.; Webb, A.A.R.; McAinsh, M.R.; Furini, A. ABA-regulated promoter activity in stomatal guard cells. *Plant. J.* **1995**, *7*, 129–134. [CrossRef] [PubMed]
30. Huang, L.; Zhang, M.; Jia, J.; Zhao, X.; Huang, X.; Ji, E.; Ni, L.; Jiang, M. An Atypical Late Embryogenesis Abundant Protein OsLEA5 Plays A Positive Role in ABA-Induced Antioxidant Defense in *Oryza sativa* L. *Plant. Cell Physiol.* **2018**, *59*, 916–929. [CrossRef] [PubMed]

 *forests*

Article

Natural and Synthetic Hydrophilic Polymers Enhance Salt and Drought Tolerance of *Metasequoia glyptostroboides* Hu and W.C.Cheng Seedlings

Jing Li [1,2], Xujun Ma [1], Gang Sa [1], Dazhai Zhou [3], Xiaojiang Zheng [3], Xiaoyang Zhou [1], Cunfu Lu [1], Shanzhi Lin [1], Rui Zhao [1] and Shaoliang Chen [1,*]

[1]   Beijing Advanced Innovation Center for Tree Breeding by Molecular Design,
      College of Biological Sciences and Technology, Beijing Forestry University, Qinghua East Road 35,
      Beijing 100083, China; kaka19832008@163.com (J.L.); maxujun@lzb.ac.cn (X.M.); sg_1214@126.com (G.S.);
      zhouxiaoyang@bjfu.edu.cn (X.Z.); lucunfu@bjfu.edu.cn (C.L.); szlin@bjfu.edu.cn (S.L.);
      ruizhao926@126.com (R.Z.)
[2]   School of Architectural and Artistic Design, Henan Polytechnic University, Jiaozuo 454000, China
[3]   Key Laboratory of Biological Resources Protection and Utilization in Hubei Province,
      Hubei University for Nationalities, Enshi 445000, China; sws0048@163.com (D.Z.); hbzxj123@126.com (X.Z.)
*   Correspondence: Lschen@bjfu.edu.cn; Tel.: +86-(0)10-6233-8129

Received: 6 September 2018; Accepted: 11 October 2018; Published: 15 October 2018

Abstract: We compared the effects of hydrophilic polymer amendments on drought and salt tolerance of *Metasequoia glyptostroboides* Hu and W.C.Cheng seedlings using commercially available Stockosorb and Luquasorb synthetic hydrogels and a biopolymer, Konjac glucomannan (KGM). Drought, salinity, or the combined stress of both drought and salinity caused growth retardation and leaf injury in *M. glyptostroboides*. Under a range of simulated stress conditions, biopolymers and synthetic hydrogels alleviated growth inhibition and leaf injury, improved photosynthesis, and enhanced whole-plant and unit transpiration. For plants subjected to drought conditions, Stockosorb hydrogel amendment specifically caused a remarkable increase in water supply to roots due to the water retention capacity of the granular polymer. Under saline stress, hydrophilic polymers restricted Na^+ and Cl^- concentrations in roots and leaves. Moreover, root K^+ uptake resulted from K^+ enrichment in Stockosorb and Luquasorb granules. Synthetic polymers and biopolymers increased the ability of *M. glyptostroboides* to tolerate combined impacts of drought and salt stress due to their water- and salt-bearing capacities. Similar to the synthetic polymers, the biopolymer also enhanced *M. glyptostroboides* drought and salt stress tolerance.

Keywords: hydrophilic polymers; Stockosorb; Luquasorb; Konjac glucomannan; photosynthesis; ion relation

---

1. Introduction

Soil salinity and drought pose major problems in agriculture and forestry [1–4]. Soil salinization often accompanies drought due to evaporative salt accumulation in upper soil layers. Together, these cause soil degradation and erosion [3]. Molecular physiology indicates that multiple stress signaling networks are involved in the plant response to dehydration and saline conditions. These networks specifically include the abscisic acid-activated signaling pathway, mitogen-activated protein kinase (MAPK) cascades, extracellular adenosine triphosphate (ATP) signaling, and hydrogen peroxide catabolic process [3,5–8]. Gene transformation, mycorrhization, and polymer amendments to soil can enhance drought and salt tolerance at the tissue and cellular level [1–3,9]. These interventions can increase osmotic adjustment, antioxidative defense, water use efficiency, and ionic homeostasis in herbaceous and woody plants [2,3,6,9,10].

Hydrophilic polymers are commonly used as soil conditioners. They aid plant growth and development in drying soils [11,12] by increasing the plant's ability to absorb and retain large volumes of water and maintain osmotic balance. Amendments of hydrophilic polymers have been shown to increase available moisture levels around the root zone [13–16], thereby improving plant survival under drought stress [14,17–19]. Additional evidence shows that the presence of polymers increased the water holding capacity [14,20,21], decreased water percolation rates [22], and reduced the need for frequent irrigation [23].

In addition to enhancing plant survival in arid soils, hydrophilic polymers also aid plant growth in saline soils [24–29]. The water- and salt-retentive capacities of Stockosorb and Luquasorb polymers were shown to limit the accumulation of toxic ions in the plants [26–28]. Stockosorb hydrogel amendments to saline soil (potassium mine refuse) improved Ca^{2+} uptake in salt-resistant *Populus euphratica* Oliv. due to the polymer's cation exchange character [26]. We have previously shown that the exchangeable K^+ contained in Stockosorb and Luquasorb enables *Populus simonii* × (*P. pyramidalis* + *Salix matsudana*) (*Populus popularis* cv. 35-44) (a salt-sensitive poplar species) to maintain K^+/Na^+ homeostasis under saline conditions [27,28]. Hydrophilic polymers may also improve soil pore water quality available to plant roots, which enabled poplars under investigation to tolerate a combined stress of drought and salinity [28].

Most synthetic organic polymers (also known as plastic) may degrade under natural conditions, but typically persist in the environment for extended periods and are thus the subject of general environmental concern (Environmental issues and concern of synthetic polymers; https://prezi. com/5hthiclu4uwh/environmental-issues-and-concern-of-synthetic-polymers/). To consider the effectiveness of plastic alternatives, this study also investigated a biopolymer's effects on plant growth under adverse conditions. Konjac glucomannan (KGM) is a natural macromolecule made of β-1,4 linked D-mannose and D-glucose residues derived from *Amorphophallus konjac* K. Koch ex N.E. Br. The strong hydrogen bonds formed by hydroxyl groups in solution give this polysaccharide a high water absorbency [30,31]. We investigated KGM along with the synthetic Stockosorb 500 XL and Luquasorb hydrogels as a soil conditioner in soils subjected to drought and saline conditions. Amended soils were planted with seedlings of *Metasequoia glyptostroboides*, a valuable tree species widely used for coastal shelter, farmland protection, city greening, and as an ornamental. We compared the biopolymer and synthetic hydrogels in terms of their effects on plant growth, photosynthesis, water status, and ion relations under drought or saline conditions, or both drought and saline conditions.

2. Materials and Methods

2.1. Hydrophilic Polymers

This study used hydrophilic polymers Stockosorb 500 XL (granular type, cross-linked poly potassium-co-(acrylic resin polymer)-co-polyacrylamide hydrogel, Stockhausen GmbH, Krefeld, Germany), Luquasorb® product (powder type, potassium polyacrylate, BASF Corporation, Ludwigshafen, Germany), and Konjac flour-derived glucomannan (KGM purity >70%, Key Laboratory of Biological Resources Protection and Utilization in Hubei Province, Enshi, China).

2.2. Plant Material and Treatments

One-year-old seedlings of *M. glyptostroboides* (Hu and W.C.Cheng) were obtained from Xingdoushan Nature Reserve, Hubei province, China. In March, the seedlings were planted within individual 2.5 L pots containing sandy soil (sand:soil = 1:1, *v/v*). The potted plants were kept well-watered and received 500 mL of Hoagland's nutrient solution every two weeks. All plants were placed in a greenhouse at Beijing Forestry University prior to the initiation of salt and drought treatments. In early June, plants that exhibited healthy, uniform appearance were transferred to 5 L pots filled with either control soil (no salt) or saline soil (soil was pretreated with 1 L 50 mM NaCl).

These were left as controls or amended with one of the three polymer types (0.5% by dry weight). Plants were then divided into four groups, with each subjected to the following treatments:

(1) Control (Non-polymer), Control + Stockosorb, Control + Luquasorb, Control + KGM;
(2) NaCl (Non-polymer), NaCl + Stockosorb, NaCl + Luquasorb, NaCl + KGM;
(3) Drought (Non-polymer), Drought + Stockosorb, Drought + Luquasorb, Drought + KGM;
(4) Drought + NaCl (Non-polymer), Drought + NaCl + Stockosorb, Drought + NaCl + Luquasorb, Drought + NaCl + KGM.

Drought treatment consisted of withholding water after the initiation of water stress. Control plants were kept well-watered during the period of the experiment.

2.3. Shoot Height Measurement

The shoot height of three to five plant replicants (per treatment) was measured every seven days for a total 46-day exposure to saline and drought treatments (*M. glyptostroboides* leaves started to wilt after 46 days). Shoot height was measured from the growing tip to the base of the stem.

2.4. Whole-Plant Water Consumption

To simulate drought conditions (water stress), each pot containing one plant was covered with a plastic bag secured around the stem base. The whole-plant water consumption was measured under natural sunlight by the daily weight loss of the pot together with the plant over a 12 h period (07:00–19:00) [28,32]. At each sampling interval (day 7, 14, 21, 28, 35, and 42), three to five individual pots were examined for each treatment.

2.5. Leaf Gas-Exchange

Leaf transpiration rates (TRN), stomatal conductance (Gs), and net photosynthetic rates (Pn) of upper mature leaves were measured each week with a CIRAS-2 Portable Photosynthesis System (PP systems, Amesbury, MA, USA). TRN, Gs, and Pn were always measured between 8:30–11:00 a.m. under natural conditions, where photosynthetically active radiation (PAR) was ca. 1000 μmol m^{-2} s^{-1}. During the measurements, leaf temperature (T_{leaf}) ranged from 25 °C to 33 °C. Three to five individual seedlings per-treatment were measured at each interval.

2.6. Chlorophyll a Fluorescence

Chlorophyll *a* fluorescence was measured on a weekly basis using a PAM-2100 Fluorometer (Heinz Walz GmbH, Effeltrich, Germany) and methods described in Wang et al. (2007) [33]. The maximal efficiency of PSII photochemistry (Fv/Fm) was calculated based on measured fluorescence parameters [33].

2.7. Leaf Membrane Permeability

Leaf membrane permeability was examined after 20 days of saline and drought treatment. For each plant, 30 fresh leaf cubes (0.2 × 0.2 cm) were immersed in 10 mL of distilled water and then vacuumed for 30 min. Electrical conductivity (E_1) was then measured with a DDS-307 conductivity meter (INESA Scientific Instrument Co., Ltd., Shanghai, China) at room temperature. Following this, leaf samples were incubated in boiling water (95 °C–100 °C) for 30 min and subjected to subsequent electrical conductivity (E_2) measurements at room temperature. Leaf membrane permeability was calculated as $E_1/E_2 \times 100\%$.

2.8. Plant Harvest

Destructive harvests were conducted after 46 days of exposure to saline and drought treatments. Three to five replicated seedlings were harvested for each treatment. The roots were thoroughly rinsed free of soil with deionized water. All sampled materials (root, leaf, and stem) were then oven-dried at 60 °C for five to eight days to determine the dry weight. Dried samples were ground into powder and stored for compositional analysis.

2.9. Ion Analysis of Leaves and Roots

Dried samples (0.5 g) of leaves and roots were extracted with 1 M HNO_3 as described by Storey (1995) [34]. Concentrations of Na^+ and K^+ were measured by an atomic absorption spectrometer (PerkinElmer 2280, Perkin-Elmer Corporation, Norwalk, CT, USA). Cl^- concentrations were determined by a modified silver titration method [35].

2.10. Ion Analysis of Soil

Soil pore water composition was analyzed immediately following plant harvest. Concentrations of Na^+, Cl^-, and K^+ were measured from aqueous soil extracts (dried soil rinsed in deionized water = 1:5, w/v). Na^+ and K^+ concentrations were measured by atomic absorption spectrophotometry (PerkinElmer 2280) at 589.0 and 766.5 nm, respectively, while Cl^- concentrations were measured by silver titration [35].

2.11. Data Analysis

The software program SPSS (SPSS Statistics 17.0, 2008, SPSS Inc., Chicago, IL, USA) was used to calculate basic statistical parameters for measured data. Unless otherwise stated, differences are interpreted as statistically significant for the $p < 0.05$ level.

3. Results

3.1. Occurrence of Leaf Injury under Water and Salt Stress

Observation during the experiment revealed that drought and/or salt caused leaf injury in plants without a polymer amendment after 16–26 day of the stress treatment (Table 1; Figure 1). Leaves exhibited chlorosis or necrosis prior to abscission (Figure 1). Hydrogels appear to have delayed stress-induced leaf injury until 30–46 day (Table 1; Figure 1). Plants amended with the biopolymer exhibited a pronounced delay of leaf injury relative to that of plants treated with synthetic polymer (Table 1).

Table 1. Effect of drought and saline stress on the timing (in days) of leaf injuries in *Metasequoia glyptostroboides* Hu and W.C.Cheng amended with synthetic (Stockosorb, Luquasorb) or natural (Konjac glucomannan, KGM) hydrophilic polymers and those not amended with polymers (Non-Polymer).

Treatment	Non-Polymer	Stockosorb	Luquasorb	KGM
Control	NI	NI	NI	NI
NaCl	26 ± 2	NI	36 ± 2	NI
Drought	17 ± 1	34 ± 1	30 ± 2	46 ± 1
Drought + NaCl	16 ± 1	34 ± 2	30 ± 2	46 ± 1

Legend: NI, non-injury. Each value (±SE) is the mean of three to five individual plants.

3.2. Leaf Membrane Permeability

Membrane permeability (MP) was examined when the stress symptoms were visible in plants subjected to drought and saline conditions. Relative to control plants, leaf MP significantly increased in non-polymer-treated plants after 20 day of drought or combined drought and saline stress (Figure 2). However, the stress-induced increase in MP for these plants was markedly reduced by all polymers (Stockosorb, Luquasorb, or KGM) (Figure 2). Saline treatment did not cause a significant increase in MP, irrespective of polymer treatments at the observation time (Figure 2).

Figure 1. Images of *Metasequoia glyptostroboides* Hu and W.C.Cheng test plants subjected to drought and saline stress after 20 days. Columns indicate soils amended without (Non-Polymer) or with synthetic (Stockosorb, Luquasorb) and natural (Konjac glucomannan, KGM) hydrophilic polymers. Rows indicate *M. glyptostroboides* seedlings subjected to control, salt exposure, drought, and combined salt and drought exposure (see Materials and Methods section).

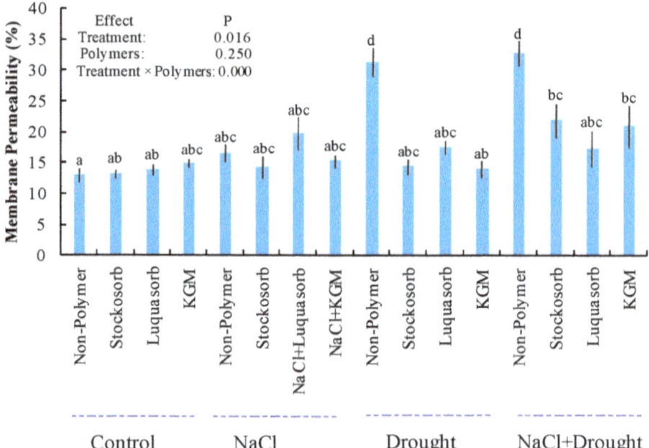

Figure 2. Effect of drought and salt stress on leaf membrane permeability in *Metasequoia glyptostroboides* seedlings supplemented with synthetic (Stockosorb, Luquasorb) and natural (Konjac glucomannan, KGM) hydrophilic polymers along with plants not amended with polymers (Non-Polymer). Each column represents the mean of three to five individual plants and bars represent the standard error of the mean. Columns labelled with different letters (a, b, c, and d) are significantly different at the $p < 0.05$ level.

3.3. Shoot Height Increment

Amendment of 0.5% (by weight) Stockosorb, Luquasorb, or KGM caused no significant effect on height increment in control *M. glyptostroboides* plants (Figure 3A). Drought, salt, and combined drought and salt stress reduced shoot elongation in the absence of polymers (Figure 3A). The presence of hydrophilic polymers alleviated growth inhibition under stress conditions (Figure 3A). Plants subjected to salt stress and amended with Stockosorb exhibited a more pronounced effect in height enhancement than those amended with Luquasorb and KGM (Figure 3A). Plants subjected to combined drought and saline stress and amended with Luquasorb retained shoot growth better than those amended with Stockosorb and KGM (Figure 3A).

3.4. Plant Dry Weight

For plants not amended with polymers, dry weights of roots, stem, and leaves decreased significantly after 46 days of drought, salt, or combined drought and salt stress. These caused respective declines of 44%, 62%, and 65% in whole-plant biomass (Figure 3B). *M. glyptostroboides* roots and leaves were more sensitive to these stresses relative to stems (Figure 3B). Stockosorb, Luquasorb, and KGM alleviated whole-plant biomass inhibition by salt and drought stress, although specific tissue types did not show pronounced effects (Figure 3B). KGM also improved plant biomass for plants subjected to drought or salt stress (Figure 3B). For samples subjected to combined drought and salt stress, these effects were not as pronounced as those exhibited by plants treated with the two synthetic polymers (Figure 3B).

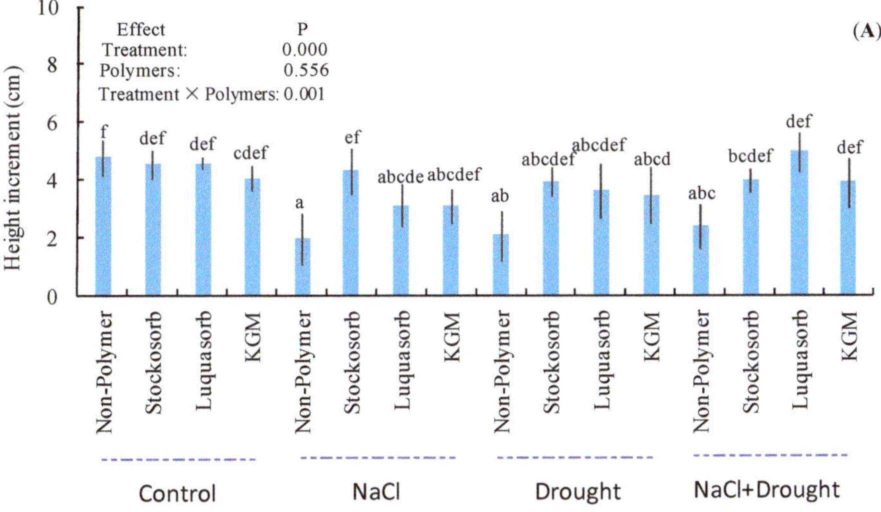

Figure 3. *Cont.*

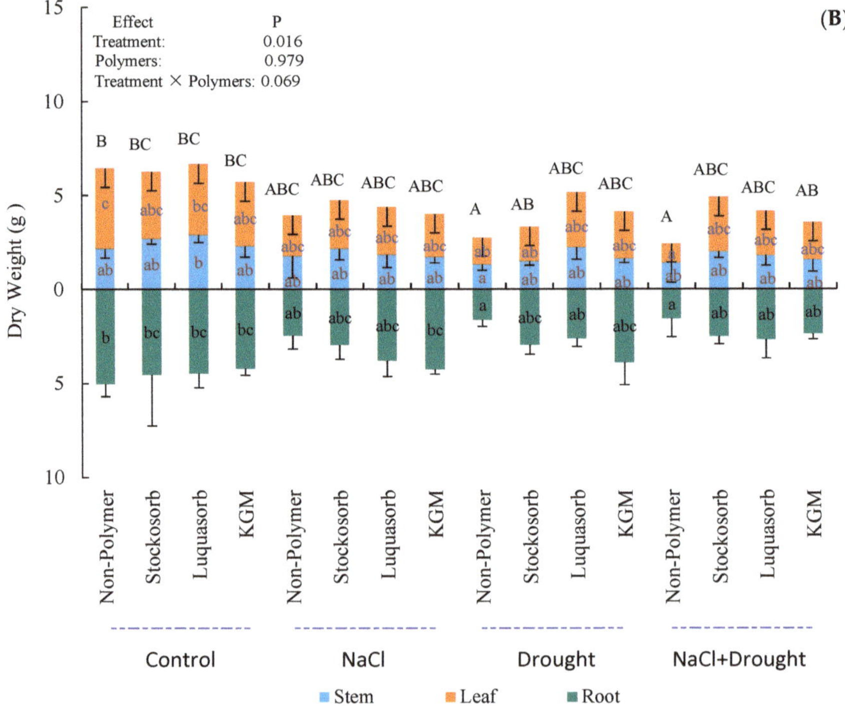

Figure 3. Effect of drought and salt stress on height increment and biomass in *Metasequoia glyptostroboides* seedlings. (**A**): Shoot height increment. (**B**): Plant dry weight. Columns show samples amended with synthetic (Stockosorb, Luquasorb) and natural (Konjac glucomannan, KGM) hydrophilic polymers and those not amended (Non-Polymer). Each column represents mean analysis of three to five individual plants and bars represent the standard error of the mean. Columns labelled with different letters (a–f) are significantly different at the $p < 0.05$ level. (Note: Columns labelled with A, B, and C, indicate significant difference in whole-plant dry weight between treatments).

3.5. Whole-Plant Water-Consumption

Drought and/or salt stress decreased the daily water-consumption of plants not treated with polymers (Figure 4). Synthetic and natural polymer amendments increased the daily water-loss for plants subjected to stress (Figure 4). For plants subjected to drought and salt stress, water-consumption of plants treated with Stockosorb exceeded that of those treated with Luquasorb or KGM (Figure 4). For plants subjected to drought or combined drought and salt stress, plants treated with KGM exhibited less water loss than those treated with Stockosorb or Luquasorb (Figure 4).

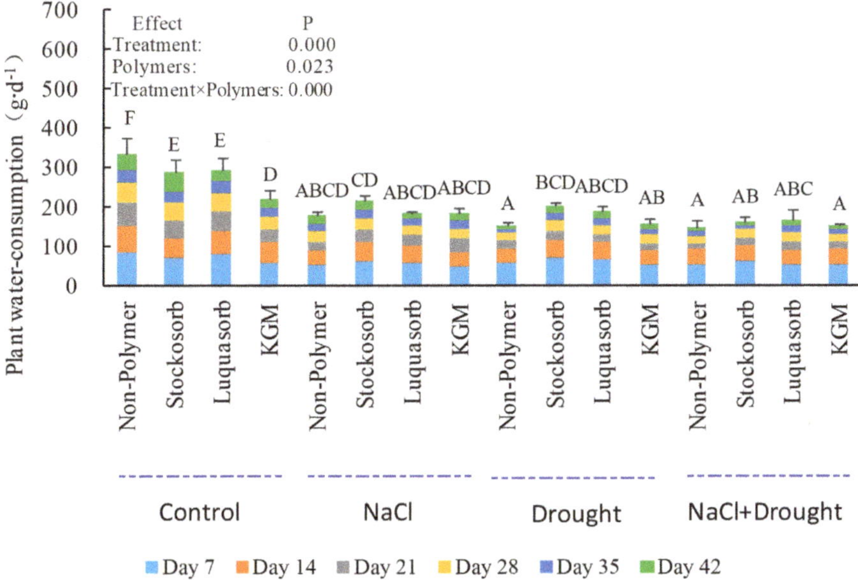

Figure 4. Effect of drought and salt stress on whole-plant water consumption in *Metasequoia glyptostroboides* seedlings. Columns show samples amended with synthetic (Stockosorb, Luquasorb) and natural (Konjac glucomannan, KGM) hydrophilic polymers and those not amended (Non-Polymer). Whole-plant water loss was measured after 7, 14, 21, 28, 35, and 42 days of salt and drought treatment. Each column represents the mean of three to five individual plants and bars represent the standard error of the mean. Columns labelled with different letters (A–F) are significantly different at the $p < 0.05$ level.

3.6. Leaf Gas-Exchange

In the absence of hydrogels, drought and salt stress caused declines in net photosynthetic rates (Pn), transpiration rates (TRN), and stomatal conductance (Gs) (Figure 5). Hydrogel amendments increased Gs, Pn, and TRN under these stress conditions (relative to rates for plants not treated with polymers) (Figure 5). Relative to plants treated with Luquasorb or KGM, plants treated with Stockosorb showed the highest rates of gas exchange under drought or salt stress (Figure 5). The Pn of KGM-treated plants was 56%–60% higher than that of plants treated with synthetic polymers under combined drought and salt stress (Figure 5A). However, these combined-stressed plants treated with each polymer showed similar TRN and Gs values (Figure 5B,C).

3.7. Chlorophyll a Fluorescence

Seedlings subjected to drought and/or salt stress but not treated with polymers showed lower maximum values for PSII photochemistry (Fv/Fm) efficiency (Figure 6). This finding supports the photosynthetic response in water- and salt-stressed plants (Figure 5). Polymer amendments alleviated the drought and salt effects irrespective of powder and granular type of polymer (Figure 6).

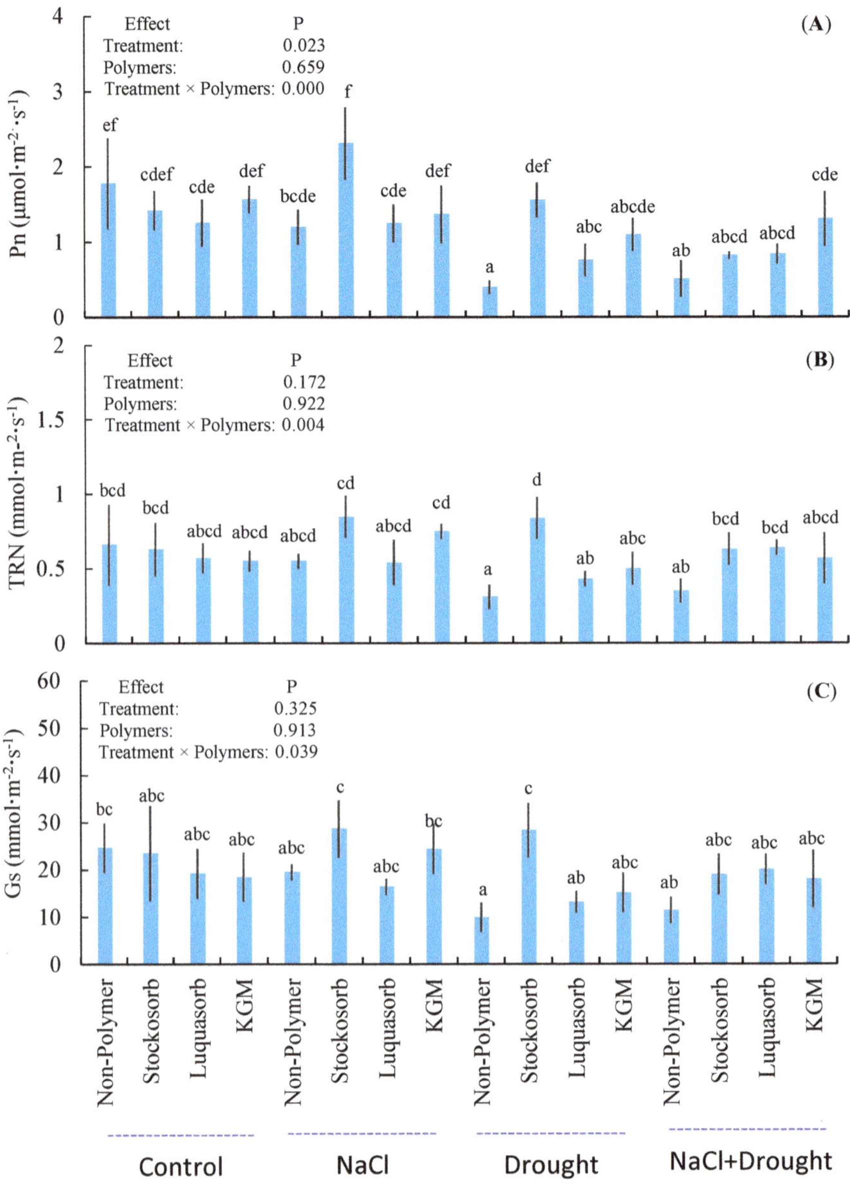

Figure 5. Effect of drought and salt stress on leaf net photosynthetic rates (Pn), transpiration rates (TRN), and stomatal conductance (Gs) in *Metasequoia glyptostroboides* seedlings amended with synthetic (Stockosorb, Luquasorb) and natural (Konjac glucomannan, KGM) hydrophilic polymers along with those not amended with polymers (Non-Polymer). (**A**): Net photosynthetic rate. (**B**): Transpiration rate. (**C**): Stomatal conductance. Mean gas exchange values were measured on a weekly basis. Each column represents mean values for three to five individual plants and bars represent the standard error of the mean. Columns labelled with different letters (a–f) are significantly different at the $p < 0.05$ level.

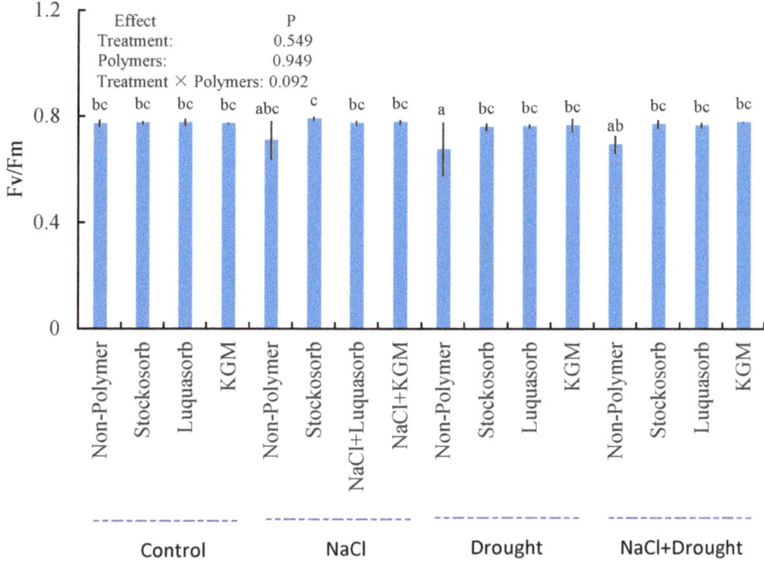

Figure 6. Effect of drought and salt stress on the maximum photochemical efficiency of photosystem II in *Metasequoia glyptostroboides* seedlings amended with synthetic (Stockosorb, Luquasorb) and natural (Konjac glucomannan, KGM) hydrophilic polymers along with those not amended (Non-Polymer). Chlorophyll *a* fluorescence was measured on a weekly basis. Each column represents the mean value for three to five individual plants and bars represent the standard error of the mean. Columns labelled with different letters (a, b, and c) are significantly different at the $p < 0.05$ level.

3.8. Ion Concentrations in Roots and Leaves

Salt (NaCl) treatment significantly increased Na^+ and Cl^- levels in roots and leaves for plants not treated with polymers. However, leaves retained markedly higher salt concentrations than roots (Figure 7A,B). Samples treated with polymers and subjected to saline stress showed lower salt concentrations in roots and leaves (Figure 7A,B). Plants subjected to combined drought and salt stress showed more pronounced salt accumulation in leaves and roots relative to that measured from plants only subjected to salt stress (Figure 7A,B). For plants subjected to combined stresses, KGM was more effective than synthetic polymers at restricting Na^+ and Cl^- buildup in roots and leaves (Figure 7A,B).

Plants subjected to salt stress but not treated with polymers exhibited lower K^+ levels in roots, but not in leaves, regardless of drought stress (Figure 7C). In plants subjected to salt stress and amended with polymers, this decline in root K^+ was reduced, and the effect was most pronounced in plants amended with Luquasorb (Figure 7C). In plants subjected to combined drought and salt stress, Stockosorb, and Luquasorb exhibited an increase in root K^+ concentrations (relative to plants not amended with polymers), but KGM had no such effect (Figure 7C).

3.9. Ion Concentrations in Soils

Relative to control soils, soils subjected to saline and drought stress exhibited higher Na^+ and Cl^- concentrations (Figure 8A,B). Hydrogel amendments lowered Na^+ and Cl^- concentrations in salt-treated soils, regardless of drought stress (Figure 8A,B). Soils subjected to combined drought and salt stress and treated with Luquasorb showed the most significant decrease in Na^+ and Cl^- levels relative to soils treated with Stockosorb and KGM (Figure 8A,B). Soils amended with synthetic polymers, in particular Luquasorb, exhibited increased K^+ concentrations relative to soils treated with KGM, irrespective of control or stress treatment (Figure 8C).

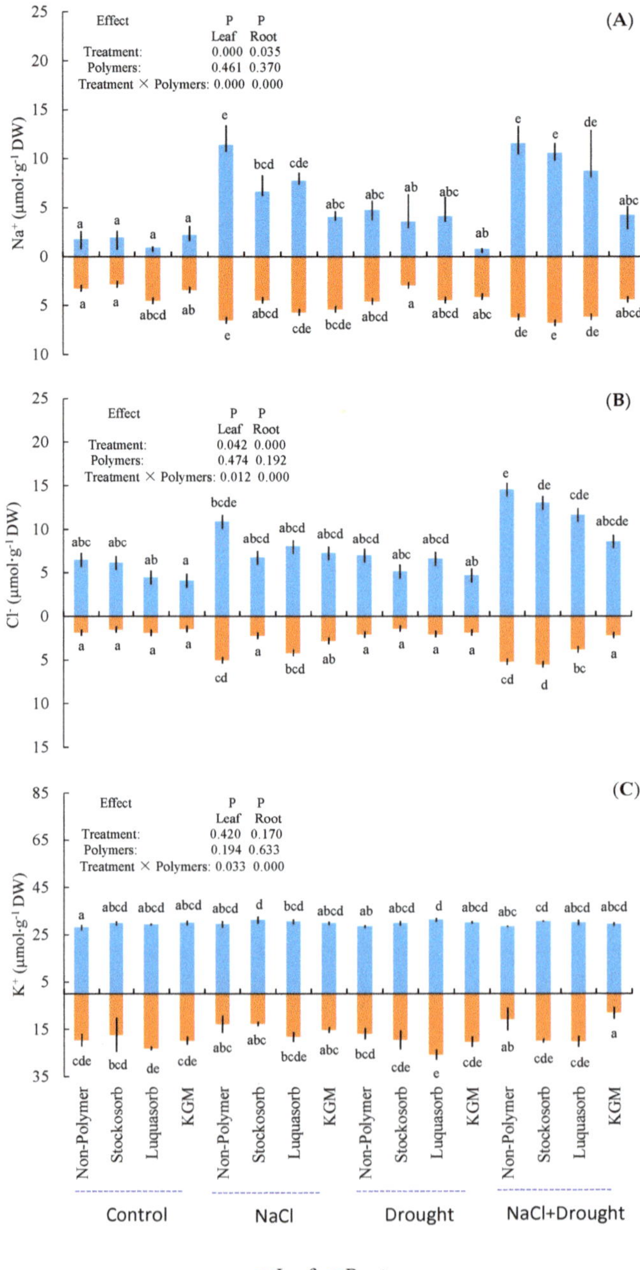

Figure 7. Effects of drought and salt stress on Na$^+$, Cl$^-$, and K$^+$ concentrations in the roots and leaves of *Metasequoia glyptostroboides* seedlings amended with synthetic (Stockosorb, Luquasorb) and natural (Konjac glucomannan, KGM) hydrophilic polymers along with those not amended with polymers (Non-Polymer). (**A**): Na$^+$ concentrations. (**B**): Cl$^-$ concentrations. (**C**): K$^+$ concentrations. Each column represents the mean of three to five individual plants and bars represent the standard error of the mean. Columns labelled with different letters (a–e) are significantly different at the $p < 0.05$ level.

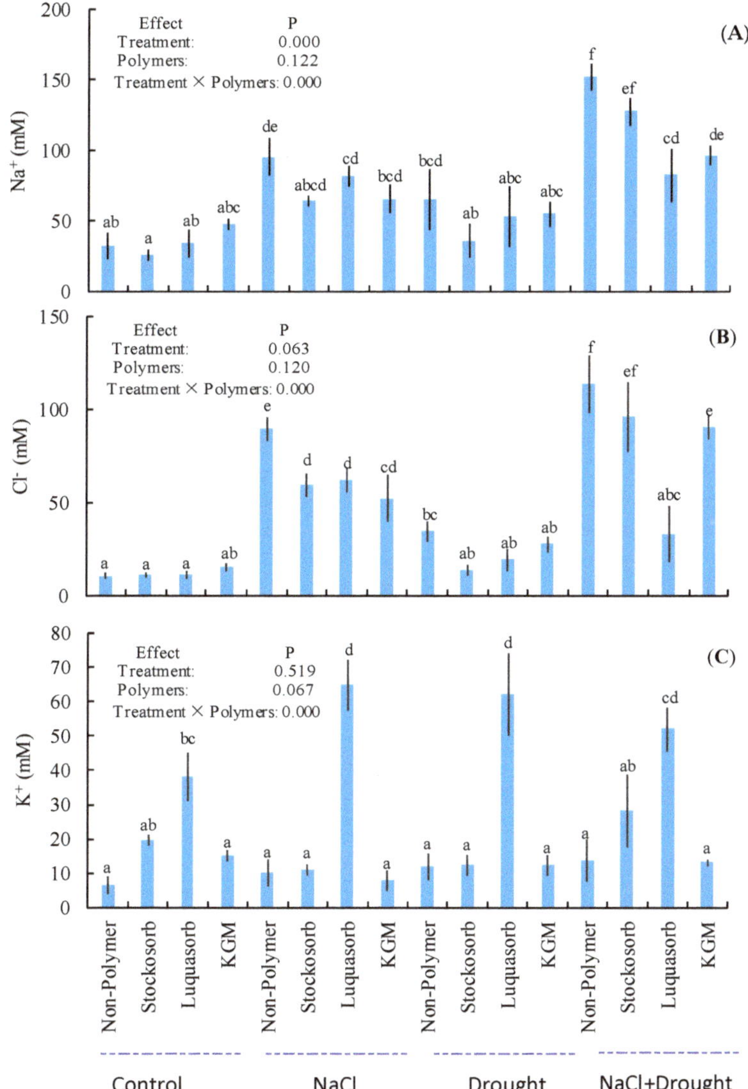

Figure 8. Effects of drought and salt stress on Na^+, Cl^-, and K^+ concentrations in soils supplemented without synthetic (Stockosorb, Luquasorb) and natural (Konjac glucomannan, KGM) hydrophilic polymers along with soils not supplemented with polymers (Non-Polymer). (**A**): Na^+ concentration. (**B**): Cl^- concentration. (**C**): K^+ concentration. Each column represents the mean of three to five soil samples and bars represent the standard error of the mean. Columns labeled with different letters (a–f) are significantly different at the $p < 0.05$ level.

4. Discussion

4.1. Hydrophilic Polymers Increase Water Supply under Drought Stress

Synthetic hydrophilic hydrogels (Stockosorb and Luquasorb) and natural KGM polymer amendments caused relative increases in growth and photosynthesis (Figures 3, 5 and 6) in plants subjected to drought conditions. These results support those reported in Syvertsen and Dunlop (2004), who found that polymer amendments improved the growth of drought-stressed citrus [36]. Similarly, Stockosorb polymers were shown to prolong the survival of *Pinus halepensis* [37], *Citrus* spp. [38], and *Populus Popularis* seedlings subjected to drought [28]. In this study, leaf injury caused by drought stress occurred due to increased leaf membrane permeability (Table 1; Figures 1 and 2). The application of Stockosorb, Luquasorb, and KGM decreased MP in drought-treated *M. glyptostroboides* leaves (Figure 2), thus delaying the occurrence of leaf injury (Table 1; Figure 1). Similarly, Shi et al. (2010) demonstrated that Stockosorb and Luquasorb polymers reduce the growth inhibition of *P. popularis* cuttings and delay leaf injury during drought stress [28]. Our results show that plants amended with the biopolymer exhibited a more pronounced delay of leaf injury relative to those treated with synthetic polymers (Table 1).

Growth and photosynthesis enhancement exhibited by plants treated with polymers occurred due to improved soil water retention under drought conditions [39]. Hydrogel amendments remarkably increased the available water capacity of the soil [28,39,40]. This water could then be absorbed by adjacent roots with up to 95% available to plants [22]. Increased root water supply in polymer-amended plants resulted in improved whole-plant water consumption and leaf transpiration per unit of surface area (Figures 4 and 5). Our data indicate that Stockosorb amendments provided plant roots with a larger amount of water than Luquasorb and KGM polymers (Figures 4 and 5). The larger Stockosorb granules were able to absorb and store larger amounts of water in the soil environment and can thus function as water sources during drought conditions [28]. *M. glyptostroboides* roots can grow interspersed with Stockosorb aggregates, which improves the water distribution [27,28]. The powder type polymers, KGM in particular, displayed an enhanced ability to moderate water supply to plant roots (Figures 4 and 5). This improved long-term water use efficiency under drought conditions [23,28]. The glucomannan structure of the KGM polymer confers excellent water retention capabilities [31,41]. This allows the KGM-amended soil to maintain the water supply to *M. glyptostroboides* roots over prolonged periods of water stress and thereby delays the occurrence of leaf injury (Table 1; Figure 1). These results support those of previous studies on hydrogel-treated *P. popularis* [28]. The powder polymers also more effectively slowed the rate of water supply to plants (and thus prolonged the duration of water supply) relative to the granular-type polymer [28].

4.2. Hydrophilic Polymers Alleviate Salt Stress

Under saline stress, plants amended with hydrophilic polymers exhibited improved shoot height increment, leaf gas exchange, and biomass relative to plants not amended with polymers (Figures 3 and 5). Plants subjected to saline conditions but amended with Stockosorb exhibited a more pronounced effect to lower the salt inhibition of transpiration (Figures 4 and 5) and photosynthesis (Figures 5 and 6) than those amended with Luquasorb and KGM. Enhanced tolerance of saline conditions mainly resulted from improved ion relations in polymer-amended plants. In this study, Stockosorb and KGM polymers markedly reduced Na^+ and Cl^- concentrations in roots and leaves under salt stress (Figure 7). Excessive salt accumulation poses a significant threat to woody plants subjected to saline conditions [3,8,9,32,35,42–46]. Previous studies have shown that hydrophilic polymer amendments inhibited salt accumulation in poplar species' roots and shoots regardless of salt-resistant and salt-sensitive genotypes [26–28]. Reduced salt ion accumulation in roots and shoots resulted from the salt-retention and dilution capacity of the polymer amendments [28] (Figure 8). Water retention by Stockosorb, Luquasorb, and KGM polymers specifically causes the dilution of soil Na^+ and Cl^- concentrations [28,47] and limits salt absorption by the surrounding roots.

In addition to reducing the uptake of salt ions, the two synthetic hydrogels, and especially Luquasorb, enhanced K^+ uptake in salt-stressed roots (Figure 7). Maintaining K^+ homeostasis allows plants to withstand toxicity from salt ions [45,48,49]. Plants amended with the two hydrophilic polymers exhibited improved K^+ retention due to increased exchangeable K^+ concentrations (Figure 8). These results were consistent with our previous studies in which *P. popularis* plants subjected to saline conditions but treated with Stockosorb and Luquasorb exhibited increased K^+ concentrations in leaves and roots [27,28].

4.3. Hydrophilic Polymers Enhance Plant Tolerance of Combined Drought and Salt Stresses

When *M. glyptostroboides* plants were exposed to combined drought and salt stress, roots could not absorb enough water to compensate for the water loss in shoots (Figures 4 and 5). Excessive accumulation of Na^+ and Cl^- also caused ion toxicity and membrane injury (Figures 1 and 7; Table 1). The hydrophilic polymers increased the soil available water for plants, which in turn increased water-consumption and leaf gas-exchange (Figures 4 and 5). In addition to improving root water supply, hydrophilic polymers lowered soil salt concentrations in the dryer, saline soil by increasing the water-holding capacity (Figure 8). As a consequence, the uptake and transport of salt ions were restricted in polymer-amended plants (Figure 7). Shi et al. (2010) similarly showed that hydrophilic polymers increased the ability of *P. popularis* to tolerate combined drought and salt stresses, primarily due to their water- and salt-retention capacities [28]. Plants amended with Stockosorb and Luquasorb exhibited improved K^+ nutritional status in roots. This enabled *M. glyptostroboides* plants to maintain K^+/Na^+ homeostasis under water and salt stress (Figures 7 and 8). Amendments with hydrophilic polymers improved water supply and K^+/Na^+ homeostasis under drought and salt stresses, which enhanced the growth and survival of *M. glyptostroboides*.

Relative to drought- and salt-stressed plants amended with Stockosorb or Luquasorb polymers, drought- and salt-stressed plants amended with the KGM biopolymer exhibited less pronounced growth/biomass improvements (Figure 3). This arose due to (i) fewer K^+ ions and (ii) poor aeration in KGM-amended soil. KGM can absorb water and swell to form a sticky gel. This gel, however, can reduce the air supply to roots and thereby limit root cell respiration. Combined drought and salt stress therefore lowered root and shoot growth for KGM-amended plants relative to plants amended with Stockosorb and Luquasorb granules.

5. Conclusions

The study showed that natural and synthetic hydrophilic polymers increased the growth and survival of *M. glyptostroboides* plants under drought, saline, and combined drought and saline stresses. We found that plants subjected to drought stress but amended with Stockosorb received more water delivery to roots relative to drought-stressed plants amended with Luquasorb and KGM polymers. KGM, a powder polymer, apparently limited the water supply to *M. glyptostroboides* roots and maintained a continuous water supply under long-term water stress. Under salt stress, the water retained in the hydrophilic polymers diluted salt concentrations in the soil, which in turn limited salt uptake by roots. Plants subjected to saline stress but amended with the two synthetic hydrogels, especially Luquasorb, exhibited enhanced K^+ absorption in roots and resisted Na^+ and Cl^- toxicity. The high water- and salt-retention capacities of the hydrophilic polymers enhanced *M. glyptostroboides* tolerance of combined drought and saline stress.

Author Contributions: J.L., X.M., G.S., D.Z., and X.Z. (Xiaojiang Zheng) performed experiments, collected data, and carried out all analyses; X.Z. (Xiaoyang Zhou), C.L., and S.L. provided technical assistance to J.L., X.M., and G.S.; J.L. wrote this manuscript; R.Z. and S.C. supervised and revised the writing.

Funding: The research was supported jointly by the National Natural Science Foundation of China (Grant Nos. 31770643, 31570587), Beijing Natural Science Foundation (Grant Nos. 6182030, 6172024), the Research Project of the Chinese Ministry of Education (Grant No. 113013A), Fundamental Research Funds for the Central Universities (Grant No. 2017ZY07), the Program of Introducing Talents of Discipline to Universities (111 Project, Grant No. B13007), the Research Project from the Science and Technology Department of Henan Province (Grant No. 172102310677), and the Doctoral Fund of Henan Polytechnic University (Grant No. B2012-028).

Acknowledgments: We thank BASF Corporation (Germany) for providing the Luquasorb product, potassium polyacrylate (powder type).

Conflicts of Interest: The authors declare no conflict of interest.

References

1. Kumar, M. Crop plants and abiotic stresses. *J. Biomol. Res. Ther.* **2013**, *3*, e125. [CrossRef]
2. Harfouche, A.; Meilan, R.; Altman, A. Molecular and physiological responses to abiotic stress in forest trees and their relevance to tree improvement. *Tree Physiol.* **2014**, *34*, 1181–1198. [CrossRef] [PubMed]
3. Polle, A.; Chen, S. On the salty side of life: Molecular, physiological and anatomical adaptation and acclimation of trees to extreme habitats. *Plant Cell Environ.* **2015**, *38*, 1794–1816. [CrossRef] [PubMed]
4. Rennenberg, H.; Loreto, F.; Polle, A.; Brilli, F.; Fares, S.; Beniwal, R.S.; Gessler, A. Physiological responses of forest trees to heat and drought. *Plant Biol.* **2006**, *8*, 556–571. [CrossRef] [PubMed]
5. Xie, R.; Zhang, J.; Ma, Y.; Pan, X.; Dong, C.; Pang, S.; He, S.; Deng, L.; Yi, S.; Zheng, Y.; et al. Combined analysis of mRNA and miRNA identifies dehydration and salinity responsive key molecular players in citrus roots. *Sci. Rep.* **2017**, *7*, 42094. [CrossRef] [PubMed]
6. Kumar, M.; Choi, J.; An, G.; Kim, S.R. Ectopic expression of *OsSta2* enhances salt stress tolerance in rice. *Front. Plant Sci.* **2017**, *8*, 316. [CrossRef] [PubMed]
7. Fu, R.; Zhang, M.; Zhao, Y.; He, X.; Ding, C.; Wang, S.; Feng, Y.; Song, X.; Li, P.; Wang, B. Identification of salt tolerance-related microRNAs and their targets in Maize (*Zea mays* L.) using high-throughput sequencing and degradome analysis. *Front. Plant Sci.* **2017**, *8*, 864. [CrossRef] [PubMed]
8. Chen, S.; Polle, A. Salinity tolerance of *Populus*. *Plant Biol.* **2010**, *12*, 317–333. [CrossRef] [PubMed]
9. Chen, S.; Hawighorst, P.; Sun, J.; Polle, A. Salt tolerance in *Populus*: Significance of stress signaling networks, mycorrhization, and soil amendments for cellular and whole-plant nutrition. *Environ. Exp. Bot.* **2014**, *107*, 113–124. [CrossRef]
10. Kumar, M.; Lee, S.C.; Kim, J.Y.; Kim, S.J.; Aye, S.S.; Kim, S.R. Over-expression of dehydrin gene, *OsDhn1*, improves drought and salt stress tolerance through scavenging of reactive oxygen species in rice (*Oryza sativa* L.). *J. Plant Biol.* **2014**, *57*, 383–393. [CrossRef]
11. Azzam, R.A. Polymeric conditioner gels for desert soils. *Commun. Soil Sci. Plant Anal.* **1983**, *14*, 739–760. [CrossRef]
12. Woodhouse, J.; Johnson, M.S. Effect of superabsorbent polymers on survival and growth of crop seedlings. *Agric. Water Manag.* **1991**, *20*, 63–70. [CrossRef]
13. Allahdadi, I.; Moazzen-Ghamsari, B.; Akbari, G.A.; Zohorianfar, M.J. *Investigating the Effect of Different Rates of Superabsorbent Polymer (Superab A200) and Irrigation on the Growth and Yield of Zea Mays, 3rd Specialized Training Course and Seminar on the Application of Superabsorbent Hydrogels in Agriculture*; IPPI: Tehran, Iran, 2005; pp. 52–56.
14. Al-Humaid, A.I.; Moftah, A.E. Effects of hydrophilic polymers on the survival of buttonwood seedlings grown under drought stress. *J. Plant Nutr.* **2007**, *30*, 53–66. [CrossRef]
15. Guiwei, Q.; Varennes, A.D.; Cunha-Queda, C. Remediation of a mine soil with insoluble polyacrylate polymers enhances soil quality and plant growth. *Soil Use Manag.* **2008**, *24*, 350–356. [CrossRef]
16. El-Hady, O.A.; El-Kader, A.A.; Shafi, A.M. Physico-bio-chemical properties of sandy soil conditioned with acrylamide hydrogels after cucumber plantation. *Aust. J. Basic Appl. Sci.* **2009**, *3*, 3145–3151.
17. Jobin, P.; Caron, J.; Bernier, P.Y.; Dansereau, B. Impact of two hydrophilic acrylic-based polymers on the physical properties of three substrates and the growth of *Petunia hybrida* 'Brilliant Pink'. *J. Am. Soc. Hort. Sci.* **2004**, *129*, 449–457.
18. Hüttermann, A.; Orikiriza, L.J.B.; Agaba, H. Application of superabsorbent polymers for improving the ecological chemistry of degraded or polluted lands. *Clean-Soil Air Water* **2009**, *37*, 517–526. [CrossRef]
19. Apostol, K.G.; Jacobs, D.F.; Dumroese, R.K. Root desiccation and drought stress responses of bare root *Quercus rubra* seedlings treated with a hydrophilic polymer root dip. *Plant Soil* **2009**, *315*, 229–240. [CrossRef]
20. Abedi-Koupai, J.; Sohrab, F. Evaluating the application of superabsorbent polymers on soil water capacity and potential on three soil textures. *Iran. J. Polym. Sci. Technol.* **2004**, *17*, 163–173.
21. Guterres, J.; Rossato, L.; Pudmenzky, A.; Doley, D.; Whittaker, M.; Schmidt, S. Micron-size metal-binding hydrogel particles improve germination and radicle elongation of Australian metallophyte grasses in mine waste rock and tailings. *J. Hazard. Mater.* **2013**, *248–249*, 442–450. [CrossRef] [PubMed]
22. Bhardwaj, A.K.; Shainberg, I.; Goldstein, D.; Warrington, D.N.; Levy, G.J. Water retention and hydraulic of cross-linked polyacrylamides in sandy soils. *Soil Sci. Soc. Am. J.* **2007**, *71*, 406–412. [CrossRef]

23. Agaba, H.; Orikiriza, L.J.B.; Obua, J.; Kabasa, J.D.; Worbes, M.; Hüttermann, A. Hydrogel amendment to sandy soil reduces irrigation frequency and improves the biomass of *Agrostis stolonifera*. *Agric. Sci.* **2011**, *2*, 544–550. [CrossRef]

24. El-Sayed, H.; Kirkwood, R.C.; Graham, N.B. The effects of a hydrogel polymer on the growth of certain horticultural crops under saline conditions. *J. Exp. Bot.* **1991**, *42*, 891–899. [CrossRef]

25. Szmidt, R.A.K.; Graham, N.B. The effect of poly (ethylene oxide) hydrogel on crop growth under saline conditions. *Acta Hortic.* **1991**, *287*, 211–218. [CrossRef]

26. Chen, S.; Zommorodi, M.; Fritz, E.; Wang, S.; Hüttermann, A. Hydrogel modified uptake of salt ions and calcium in *Populus euphratica* under saline conditions. *Trees* **2004**, *18*, 175–183.

27. Shao, J.; Chen, S.; Wang, R.; Zhang, X. Enhancement of hydrogel on salt resistance of *Populus popularis* '35–44' and its mechanism. *J. Beijing For. Univ.* **2007**, *29*, 79–84. (In Chinese)

28. Shi, Y.; Li, J.; Shao, J.; Deng, S.; Wang, R.; Li, N.; Sun, J.; Zhang, H.; Zhu, H.; Zhang, Y.; et al. Effects of Stockosorb and Luquasorb polymers on salt and drought tolerance of *Populus popularis*. *Sci. Hortic.* **2010**, *124*, 268–273. [CrossRef]

29. Dorraji, S.S.; Golchin, A.; Ahmadi, S. The effects of hydrophilic polymer and soil salinity on corn growth in sandy and loamy soils. *Clean-Soil Air Water* **2010**, *38*, 584–591. [CrossRef]

30. Koroskenyi, B.; Mccarthy, S.P. Synthesis of acetylated konjac glucomannan and effect of degree of acetylation on water absorbency. *Biomacromolecules* **2001**, *2*, 824–826. [CrossRef] [PubMed]

31. Zhang, C.; Chen, J.; Yang, F. Konjac glucomannan, a promising polysaccharide for OCDDS. *Carbohyd. Polym.* **2014**, *104*, 175–181. [CrossRef] [PubMed]

32. Chen, S.; Li, J.; Wang, S.; Fritz, E.; Hüttermann, A.; Altman, A. Effects of NaCl on shoot growth, transpiration, ion compartmentation, and transport in regenerated plants of *Populus euphratica* and *Populus tomentosa*. *Can. J. For. Res.* **2003**, *33*, 967–975. [CrossRef]

33. Wang, R.; Chen, S.; Deng, L.; Fritz, E.; Hüttermann, A.; Polle, A. Leaf photosynthesis, fluorescence response to salinity and the relevance to chloroplast salt compartmentation and anti-oxidative stress in two poplars. *Trees* **2007**, *21*, 581–591. [CrossRef]

34. Storey, R. Salt tolerance, ion relations and the effects of root medium on the response of citrus to salinity. *Aust. J. Plant Physiol.* **1995**, *22*, 101–114. [CrossRef]

35. Chen, S.; Li, J.; Wang, S.; Hüttermann, A.; Altman, A. Salt, nutrient uptake and transport and ABA of *Populus euphratica*; a hybrid in response to increasing soil NaCl. *Trees* **2001**, *15*, 186–194. [CrossRef]

36. Syvertsen, J.P.; Dunlop, J.M. Hydrophilic gel amendments to sand soil can increase growth and nitrogen uptake efficiency of citrus seedlings. *Hortic. Sci.* **2004**, *39*, 267–271.

37. Hüttermann, A.; Zommorodi, M.; Reise, K. Addition of hydrogels to soil for prolonging the survival of *Pinus halepensis* seedlings subjected to drought. *Soil Till. Res.* **1999**, *50*, 295–304. [CrossRef]

38. Arbona, V.; Iglesias, D.J.; Jacas, J.; Primo-Millo, E.; Talon, M.; Gómez-Cadenas, A. Hydrogel substrate amendment alleviates drought effects on young citrus plants. *Plant Soil* **2005**, *270*, 73–82. [CrossRef]

39. Chirino, E.; Vilagrosa, A.; Vallejo, V.R. Using hydrogel and clay to improve the water status of seedlings for dryland restoration. *Plant Soil* **2011**, *344*, 99–110. [CrossRef]

40. Narjary, B.; Aggarwal, P.; Singh, A.; Chakraborty, D.; Singh, R. Water availability in different soils in relation to hydrogel application. *Geoderma* **2012**, *187–188*, 94–101. [CrossRef]

41. Tatirat, O.; Charoenrein, S. Physicochemical properties of konjac glucomannan extracted from konjac flour by a simple centrifugation process. *LWT-Food Sci. Technol.* **2011**, *44*, 2059–2063. [CrossRef]

42. Chen, S.; Li, J.; Fritz, E.; Wang, S.; Hüttermann, A. Sodium and chloride distribution in roots and transport in three poplar genotypes under increasing NaCl stress. *For. Ecol. Manag.* **2002**, *168*, 217–230. [CrossRef]

43. Chen, S.; Li, J.; Wang, T.; Wang, S.; Polle, A.; Hüttermann, A. Osmotic stress and ion-specific effects on xylem abscisic acid and the relevance to salinity tolerance in poplar. *J. Plant Growth Regul.* **2002**, *21*, 224–233. [CrossRef]

44. Sun, J.; Chen, S.; Dai, S.; Wang, R.; Li, N.; Shen, X.; Zhou, X.; Lu, C.; Zheng, X.; Hu, Z.; et al. NaCl-induced alternations of cellular and tissue ion fluxes in roots of salt-resistant and salt-sensitive poplar species. *Plant Physiol.* **2009**, *149*, 1141–1153. [CrossRef] [PubMed]

45. Sun, J.; Dai, S.; Wang, R.; Chen, S.; Li, N.; Zhou, X.; Lu, C.; Shen, X.; Zheng, X.; Hu, Z.; et al. Calcium mediates root K$^+$/Na$^+$ homeostasis in poplar species differing in salt tolerance. *Tree Physiol.* **2009**, *29*, 1175–1186. [CrossRef] [PubMed]

46. Sun, J.; Wang, M.; Ding, M.; Deng, S.; Liu, M.; Lu, C.; Zhou, X.; Shen, X.; Zheng, X.; Zhang, Z.; et al. H_2O_2 and cytosolic Ca^{2+} signals triggered by the PM H^+-coupled transport system mediate K^+/Na^+ homeostasis in NaCl-stressed *Populus euphratica* cells. *Plant Cell Environ.* **2010**, *33*, 943–958. [CrossRef] [PubMed]

47. Liu, Z.; Han, G.; Jiang, Y.; Liu, W. Research on keep water properties of Konjac Powder. *Acad. Period. Farm Prod. Process* **2005**, *42*, 16–18. (In Chinese)

48. Shabala, S.; Cuin, T.A. Potassium transport and plant salt tolerance. *Physiol. Plant.* **2008**, *133*, 651–669. [CrossRef] [PubMed]

49. Munns, R.; Tester, M. Mechanisms of salinity tolerance. *Ann. Rev. Plant Biol.* **2008**, *59*, 651–681. [CrossRef] [PubMed]

Article

Characteristics and Expression Analysis of *FmTCP15* under Abiotic Stresses and Hormones and Interact with DELLA Protein in *Fraxinus mandshurica* Rupr.

Nansong Liang [1,2], Yaguang Zhan [1,2,*], Lei Yu [1,2], Ziqing Wang [1,2] and Fansuo Zeng [1,2]

1   State Key Laboratory of Tree Genetics and Breeding, Northeast Forestry University, Harbin 150040, China; liangnansong@nefu.edu.cn (N.L.); leiyunefu@126.com (L.Y.); wangziqing1001@126.com (Z.W.); youpractise@126.com (F.Z.)
2   Department of Forest Bioengineering, Northeast Forestry University, Harbin 150040, China
*   Correspondence: zhanyaguang2014@126.com; Tel.: +86-0451-8219-1752

Received: 10 March 2019; Accepted: 14 April 2019; Published: 17 April 2019

Abstract: The TEOSINTE BRANCHED1, CYCLOIDEA, and PROLIFERATION CELL FACTOR (TCP) transcription factor is a plant-specific gene family and acts on multiple functional genes in controlling growth, development, stress response, and the circadian clock. In this study, a class I member of the TCP family from *Fraxinus mandshurica* Rupr. was isolated and named *FmTCP15*, which encoded a protein of 362 amino acids. Protein structures were analyzed and five ligand binding sites were predicted. The phylogenetic relationship showed that FmTCP15 was most closely related to Solanaceae and Plantaginaceae. FmTCP15 was localized in the nuclei of *F. mandshurica* protoplast cells and highly expressed in cotyledons. The expression pattern revealed the *FmTCP15* response to multiple abiotic stresses and hormone signals. Downstream genes for transient overexpression of *FmTCP15* in seedlings were also investigated. A yeast two-hybrid assay confirmed that FmTCP15 could interact with DELLA proteins. *FmTCP15* participated in the GA-signaling pathway, responded to abiotic stresses and hormone signals, and regulated multiple genes in these biological processes. Our study revealed the potential value of *FmTCP15* for understanding the molecular mechanisms of stress and hormone signal responses.

Keywords: molecular cloning; functional analysis; TCP; DELLA; GA-signaling pathway; *Fraxinus mandshurica* Rupr.

1. Introduction

The survival of plants requires balancing the regulation of growth, development, and stress response. Plants need to utilize many mechanisms to respond to stress, while environmental changes may affect development. A large number of complex transcription factors (TFs) and genes are required and involved, and there is a novel TF family that participates in these processes.

The TEOSINTE BRANCHED1, CYCLOIDEA, and PROLIFERATION CELL FACTOR (TCP) gene family is a plant-specific TF family that was first identified in 1999. TCP is named after the first three identified members: *TEOSINTE BRANCHED1* (*tb1*) in maize (*Zea mays*), *CYCLOIDEA* (*CYC*) in snapdragon (*Antirrhinum majus*), and the *PROLIFERATING CELL FACTORS 1* and *2* (PCF1 and PCF2) in rice (*Oryza sativa*) [1–3]. The TCP gene family encodes a 59-amino-acid residue, noncanonical basic helix-loop-helix (bHLH) motif called the TCP domain, which allows DNA binding and protein–protein interactions [3–5]. Based on the differences between the TCP domains, the TCP family was divided into two major classes: class I (also known as PCF or TCP-P) and class II (also known as TCP-C) [6–8]. Class II was further divided into two clades, named CINCINNATA (CIN) [9] and CYCLOIDEA/TEOSINTE BRANCHED1(CYC/TB1) [5,10].

Class I TCP genes have been reported to promote plant growth and proliferation. In meristematic tissues, *PCF1/PCF2* from rice and *AtTCP20* from *Arabidopsis thaliana* act as transcriptional activators of *PCNA* and *CYCB1;1*, respectively [3,11]. However, the latest research indicates that class I TCP genes also participate in stress adaptation. In rice, *OsPCF2* positively regulates the *OsNHX1* gene by binding to its promoter and responds to salt and drought stress tolerance [12]. *OsTCP19* responds to water deficit and salt stress and interacts with *OsABI4* and *OsULT1* to function in abiotic stress response and abscisic acid (ABA) signaling [13]. The CIN clade of class II is mainly involved in regulating organ development, such as floral organ, leaf, and lateral organ development [14–17]. In *Arabidopsis*, CIN is required for the arrest of cell division in the peripheral regions of the leaf. *jaw-D* mutants, *cin* loss-of-function mutants, in which *TCP2*, *TCP3*, *TCP4*, *TCP10*, and *TCP24* were all strongly reduced, achieved highly crinkled leaves [14]. The CYC/TB1 clade of class II is mainly involved in regulating shoot branching, floral transition, organ identity, and development. In *Arabidopsis*, one of TCP family members, *BRANCHED1* (*BRC1*), is expressed in axillary buds and responses to endogenous and environmental signals and leads to branch suppression [18,19]. In rice, *OsTB1* interacts with *OsMADS57* and targets *Dwarf14* (*D14*) to control the outgrowth of axillary buds [20]. It is noteworthy that the two different classes of TCPs are believed to share common targets [21]. In *Arabidopsis*, *LIPOXYGENASE 2* (*LOX2*) was identified as a common target of *TCP20* (from class I) and *TCP4* (from class II); additionally, *TCP20* could inhibit but *TCP4* could induce the expression of *LOX2* [22]. These results show a proposed model by which classes I and II TCP proteins may act antagonistically. In tomato (*Solanum lycopersicum*), classes I and II SlTCP proteins can form homo- and heterodimers [23]. These results show a proposed model by which classes I and II TCP may act antagonistically and form functional protein complexes to regulate biological processes.

Although TCP gene functions have been shown to be responsive to both development and various stresses in model plants, their roles in forestry trees are less known. *Fraxinus mandshurica* Rupr., a member of the Oleaceae family, is a broad-leaved tree and is well known as the most valuable hardwood tree wildly distributed in the conifer and hardwood mixed forest in northeastern China [24–26]. As an important economic and timber species, studies of *F. mandshurica* have mainly focused on its seed germination [27], nutritional growth [28], ecological characteristics [29], and disease control [30]. However, there have been few reports on the genes that contribute to resistance to abiotic stresses and development responses in *F. mandshurica*. In this study, *FmTCP15*, a class I TCP transcription factor, was isolated. The gene structure, phylogenetic relationship, subcellular localization, transcript levels in different tissues, expression under abiotic stresses and hormone signaling, as well as the expression of downstream genes of *FmTCP15* were analyzed. We found that *FmTCP15* was mainly induced by cold, salt and drought stress, and gibberellic acid (GA3). Overexpressing *FmTCP15* caused a significant change in the expression of a series of key genes involved in stress response and the GA-signaling pathway. Moreover, FmTCP15 could interact with DELLA proteins (FmRGA and FmGAI), which are key proteins of the GA-signaling pathway. The interaction relationship between FmTCP15 and DELLA proteins may enhance the ability of the plant to resist stresses. Therefore, we postulate that *FmTCP15* could regulate DELLA proteins at both transcriptional and post-transcriptional levels. *FmTCP15* may regulate stress genes and balance plant growth and development through the GA-signaling pathway.

2. Materials and Methods

2.1. Plant Material and Growth Conditions

F. mandshurica seeds from the Northeast Forestry University experimental forest farm were used. The seeds were surface sterilized and grown at 25 °C under long-day conditions (16 h light/8 h dark) on a standard field. For expression analysis, main root, lateral root, xylem, phloem, cotyledon, function leaves, and petiole were harvested from 30-day-old seedlings.

2.2. Cloning and Identification of FmTCP15 Gene

The MiniBEST Plant RNA Extraction Kit (Takara Bio, Inc., Shiga, Japan) was used for total RNA extraction. The cDNA synthesized was created using the PrimeScript First Strand cDNA Synthesis Kit (Takara Bio Inc., Shiga, Japan). The full-length cDNA of *FmTCP15* was obtained by PCR using the primers *FmTCP15-F* (5′-ACCCATTCTTGAACCAACCTATC-3′) and *FmTCP15-R* (5′-CCAAACCCTAAATCCTCCACAT-3′). The sequence of the *FmTCP15* gene was submitted to GenBank with the accession number KX905157.

2.3. Sequence Features, Protein Modeling, and Phylogenetic Analysis of FmTCP15

Protparam was used for analysis of the physical and chemical properties (molecular mass and isoelectric point) (http://web.expasy.org/protparam/) [31]. TMPred was used for the prediction of transmembrane regions and orientation (https://embnet.vital-it.ch/software/TMPRED_form.html) [32]. The I-TASSER server was used to produce the protein model (http://zhanglab.ccmb.med.umich.edu/ITASSER/) [33,34]. ModRefiner online software was further used to refine the protein model (http://zhanglab.ccmb.med.umich.edu/ModRefiner/) [35]. The NCBI database BLAST method (https://blast.ncbi.nlm.nih.gov/Blast.cgi) was used to search for homologous sequences of *FmTCP15*. Multiple sequence alignment was performed using CLC Genomics Workbench 12. The conserved domains of FmTCP15 were analyzed by the NCBI Conserved Domain Database (http://www.ncbi.nlm.nih.gov/Structure/cdd/cdd.shtml). The phylogenetic tree was constructed by MEGA 5.0 with the neighbor-joining method and 1000 replicates of bootstrap analysis.

2.4. Subcellular Localization of FmTCP15 Proteins

The FmTCP15 coding region was introduced into the pROKII-GFP expression vector driven by the *CaMV-35S* promoter and fused to the 5′ green fluorescence protein (GFP) gene to generate 35S::FmTCP15-GFP using the specific primers *pRTCP15-GFP-F* (5′-GGTACCGATAC TCGAGATGGAAGGATTAGGTGATGA-3′) and *pRTCP15-GFP-R* (5′-CACGGGTCATCTCGAGTGA ATGGTGGTTCGTAGTC-3′). The constructed vector was transformed into *F. mandshurica* xylem protoplasts, which were isolated from xylem, as described for poplar protoplast constructs, with some modifications [36]. Fluorescence signals of the 35S::FmTCP15-GFP fusion protein were examined using a confocal microscope (Zeiss Confocal Microscopy, model LSM410, Zeiss, Jena, Germany).

2.5. Abiotic Stresses and Hormone Signal Treatments

Thirty-day-old *F. mandshurica* seedlings were selected and subjected to different abiotic stresses and hormone signaling treatments. For cold treatment, seedlings were transferred into liquid Murashige and Skoog (MS) medium and placed at 4 °C. For salt and drought stress treatments, seedlings were transferred into liquid MS medium, which contained 200 mm/L NaCl and 20% w/v PEG6000, respectively. For hormone treatment, seedlings were transferred into liquid MS medium containing 100 µmol/L ABA (abscisic acid) and 100 µmol/L GA3 (gibberellic acid), respectively. For control, untreated seedlings were transferred into liquid MS medium at 25 °C. The stresses and hormones were treated at 0, 6, 12, 24, 48, and 72 h.

2.6. Transient Overexpression of FmTCP15 Gene

The constructed pROKII-GFP expression vector 35S::FmTCP15-GFP and empty vector pROKII-GFP were used in overexpression of *FmTCP15* and negative control plants, respectively, and the two vectors were transformed into *Agrobacterium tumefaciens*. Twenty-day-old *F. mandshurica* wild-type (WT) seedlings were used for transient transformation, which followed the method described for Birch (*Betula platyphylla* Suk.) [37]. After coculture with *Agrobacterium*, transient overexpression of the *FmTCP15* whole seedlings was collected.

2.7. Analysis of Gene Expression of FmTCP15 and Downstream Genes

The quantified RNA was reverse-transcribed into cDNA using the PrimeScript™ RT reagent kit with gDNA Eraser (Perfect Real-Time) (Takara Bio Inc., Shiga, Japan). Quantitative RT-PCR (qRT-PCR) was conducted in a 7500 Real-Time PCR system (Applied Biosystems, Forster City, CA, USA) using the Takara SYBR® Premix Ex Taq™ II (Perfect Real Time) (Takara Bio, Inc., Shiga, Japan). All reactions were performed in triplicate to ensure technical and biological reproducibility, and the relative abundance of the transcripts was calculated using 7500 Software v 2.0.6 (Applied Biosystems, Forster City, CA, USA) using the comparative $2^{-\Delta\Delta CT}$ method [38]. The qRT-PCR primer pairs are shown in Table S1. Tubulin was used as an internal control to determine the expression levels of the target genes.

2.8. Yeast Two-Hybrid Protein–Protein Interaction Assays

The yeast two-hybrid (Y2H) assay was carried out using the Matchmaker™ Gold Yeast Two-Hybrid System (Clontech Laboratories Inc., Mountain View, CA, USA). The full-length coding sequence of *FmTCP15* was recombined into the pGBKT7 (BD) bait vector and pGADT7 (AD) prey vector, respectively. The candidate downstream genes of *FmTCP15* were recombined into the pGADT7 (AD) vector. The gene-specific primers are listed in Table S2. The pGBKT7-p53×pGADT7-T and pGBKT7-lam×pGADT7-T were used as positive and negative controls, respectively. The plasmids of bait and prey vectors were co-transformed into the yeast strain Y2HGOLD by using the lithium acetate method. The self-activation ability of the bait vector was tested. The transformed strains were further serially cultured on selective media, including SD/-Leu/-Trp (DDO), SD/-Ade/-Leu/-Trp (TDO), SD/-Ade/-His/-Leu/-Trp/Aureobasidin A (AbA^r) (QDO/A), and SD/-Ade/-His/-Leu/-Trp/AbAr/X-α-Gal (QDO/A/X) with 125 µM AbA^r and 4 mg mL^{-1} X-α-Gal and incubated at 30 °C for 3–5 days.

3. Results

3.1. Nucleotide Sequence Cloning and Protein Modeling of FmTCP15 Gene

The *Arabidopsis* AtTCP15 amino acid sequence was used to blast against the *F. mandshurica* TSA database. Then, we obtained the predicted cDNA sequence of the *AtTCP15* homologous gene in *F. mandshurica*. According to the sequence, specific primers were designed and the *FmTCP15* full-length cDNA sequence was obtained (GenBank: KX905157). The open reading frame (ORF) was 1089 bp and encoded a protein of 362 amino acids with a predicted molecular mass of 39.0 kDa and a theoretical isoelectric point (pI) of 6.67. The transmembrane prediction showed that the FmTCP15 protein had four possible transmembrane helices, located at 29–53, 129–149, 255–279, and 294–310 aa, which indicated that the FmTCP15 protein may have transmembrane capabilities. Secondary structure analysis of the FmTCP15 protein revealed that FmTCP15 consisted of α-helix (65.54%), extended strand (5.78%), and random coil (28.67%). The conserved domains of the FmTCP15 protein indicated that FmTCP15 contained a TCP domain ranging from 75 to 136 aa and belonged to the class I TCP superfamily.

For three-dimensional (3D) structure modeling, the FmTCP15 protein sequence was submitted to the I-TASSER server. The PDB template 2nbiA was used for homology modeling (identity 86.1%, coverage 93.9%). The FmTCP15 protein model was achieved (Figure 1A,B). Then, using the COACH method, five ligand binding sites (THR117, GLU119, TRP120, LEU121, and LEU122) were predicted (Figure 1C,D). Next, based on homologous Gene Ontology (GO) templates in PDB, we predicted GO terms for the FmTCP15 protein. The FmTCP15 protein had molecular functions GO:0032559 and GO:0035639 (adenyl ribonucleotide binding and purine ribonucleotide triphosphate binding); biological processes GO:0044255, GO0032787, and GO:0046394 (cellular lipid metabolic process, monocarboxylic acid metabolic process, and carboxylic acid biosynthetic process); and cellular components GO:0032991 and GO:0044445 (macromolecular complex).

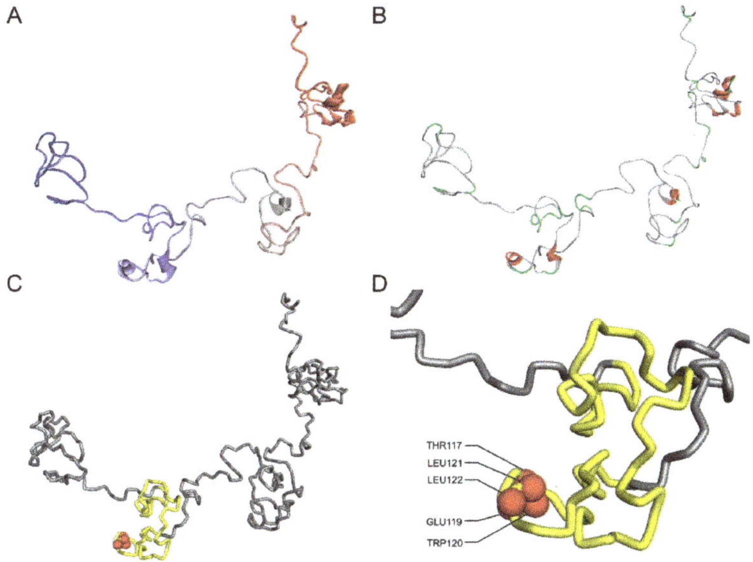

Figure 1. Three-dimensional protein model of FmTCP15 protein. (**A**) FmTCP15 3D structure displayed in solid ribbon style. N-to-C terminals: color residues in a continuous gradient from blue at the N-terminus through white to red at the C-terminus. (**B**) FmTCP15 3D structure displayed in solid ribbon style, secondary type: colors according to secondary structure. Helices are red, turns are green, and coils are white. (**C**) FmTCP15 3D structure displayed in tube style: the TCP domain is yellow, and the five predicted ligand binding sites (THR117, GLU119, TRP120, LEU121, and LEU122) are red balls. (**D**) The zoom-in view of the TCP domain of FmTCP15.

3.2. Homology Analysis and Phylogenetic Relationship of FmTCP15

For homologous alignment, FmTCP15 and 13 other amino acid sequences were aligned using CLC Genomics Workbench 12 (Table S3). The results revealed that the FmTCP15 shared high similarities with homologous genes from other species (Figure 2A). FmTCP15 and the 13 other genes shared a TCP conserved domain, and five ligand binding sites (THR117, GLU119, TRP120, LEU121, and LEU122) were also conserved in these species (Figure 2A).

To analyze the phylogenetic relationships between FmTCP15 and the homologous sequences of TCP15s in other plants, we constructed a phylogenetic tree. FmTCP15 and 42 other amino acid sequences were used for tree construction, including 34 dicotyledons, 7 monocotyledons, and 1 bryophyte (Table S4). The phylogenetic tree revealed a clear boundary between the TCP proteins of dicotyledons, monocotyledons, and bryophytes (Figure 2B). FmTCP15 was most closely related to the Solanaceae and Plantaginaceae families, such as CaTCP14 (*Capsicum annuum*, XP_016562806.1), NtTCP14-like (*Nicotiana tabacum*, XP_016464955.1), SlTCP17 (*Solanum lycopersicum*, NP_001233815.1), SpTCP14-like (*Solanum pennellii*, XP_015077184.1), StTCP14-like (*Solanum tuberosum*, XP_006354786.1), and AmTCP (*Antirrhinum majus*, CAE45599.1). Together with OeTCP14-like (*Olea europaea* var. sylvestris, XP_022844971.1), also an Oleaceae family protein, these nine sequences were grouped into one clade.

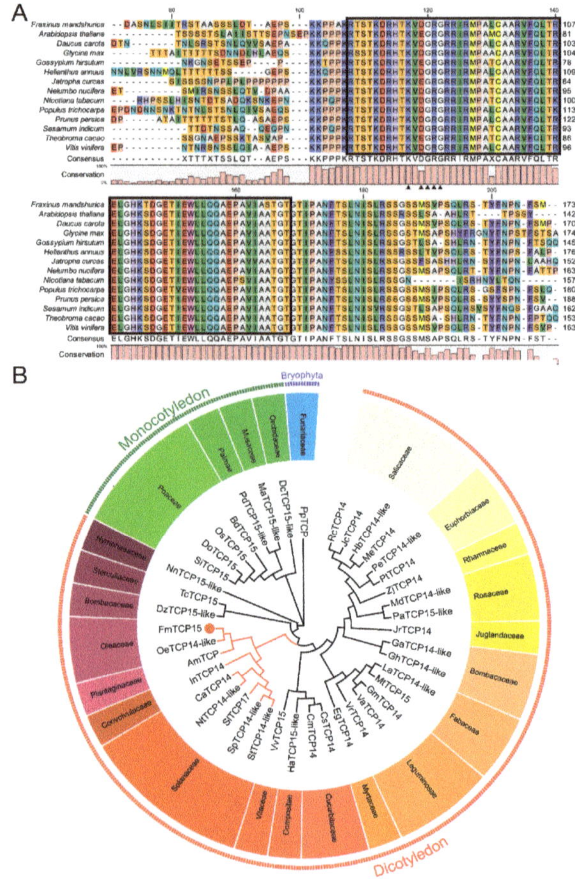

Figure 2. Multiple sequence alignment and phylogenetic relationship of FmTCP15 protein. (**A**) Homolog alignment of FmTCP15 protein. The colored bars at the bottom represent the conservation percentage. The black box and black triangle represent the TCP conserved domain and five ligand binding sites, respectively. (**B**) Phylogenetic relationship of FmTCP15 proteins. The phylogram of FmTCP15 proteins is presented in circle mode (displays only topology). The phylogenetic tree was constructed using the neighbor-joining method with MEGA 5.0 software (Oxford University Press: New York, NY, USA). FmTCP15 is denoted by a red dot. FmTCP15 (*Fraxinus mandshurica*), DzTCP15-like (*Durio zibethinus*), HaTCP15-like (*Helianthus annuus*), InTCP14 (*Ipomoea nil*), CmTCP14 (*Cucumis melo*), CsTCP14 (*Cucumis sativus*), RcTCP14 (*Ricinus communis*), HbTCP14-like (*Hevea brasiliensis*), JcTCP14 (*Jatropha curcas*), MeTCP14 (*Manihot esculenta*), LaTCP14-like (*Lupinus angustifolius*), MtTCP15 (*Medicago truncatula*), PpTCP (*Physcomitrella patens*), JrTCP14 (*Juglans regia*), GmTCP14 (*Glycine max*), VaTCP14 (*Vigna angularis*), VrTCP14 (*Vigna radiata* var. *radiata*), GaTCP14-like (*Gossypium arboreum*), GhTCP14-like (*Gossypium hirsutum*), MaTCP15-like (*Musa acuminata* subsp. *malaccensis*), EgTCP14 (*Eucalyptus grandis*), NnTCP15-like (*Nelumbo nucifera*), OeTCP14-like (*Olea europaea* var. *sylvestris*), DcTCP15-like (*Dendrobium catenatum*), PdTCP15-like (*Phoenix dactylifera*), AmTCP (*Antirrhinum majus* subsp. *majus*), BdTCP15 (*Brachypodium distachyon*), DoTCP15 (*Dichanthelium oligosanthes*), OsTCP15 (*Oryza sativa Japonica Group*), SiTCP15 (*Setaria italica*), ZjTCP14 (*Ziziphus jujuba*), MdTCP14-like (*Malus domestica*), PaTCP15-like (*Prunus avium*), PeTCP14-like (*Populus euphratica*), PtTCP14 (*Populus trichocarpa*), CaTCP14 (*Capsicum annuum*), NtTCP14-like (*Nicotiana tabacum*), SlTCP17 (*Solanum lycopersicum*), SpTCP14-like (*Solanum pennellii*), StTCP14-like (*Solanum tuberosum*), TcTCP15 (*Theobroma cacao*), and VvTCP15 (*Vitis vinifera*). Different colors represent different families; different phylum marked outside the circle.

3.3. Subcellular Localization of FmTCP15 Proteins

Subcellular localization is crucial for understanding protein function. To further investigate the subcellular localization of FmTCP15, we constructed *F. mandshurica* xylem protoplasts, which we observed with a confocal microscope after protoplast transformation with pROK2-FmTCP15-GFP vector (35S::TCP15-GFP). The results showed that the fluorescence signals were mainly concentrated in the nucleus, which demonstrated that FmTCP15 was located in the nucleus (Figure 3).

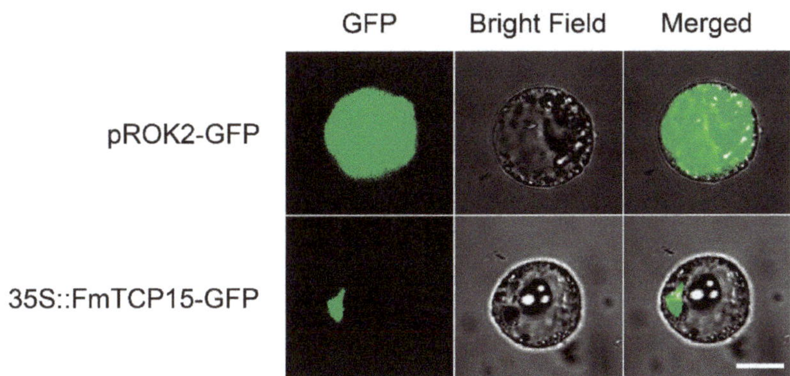

Figure 3. Subcellular localization of FmTCP15. The photographs were taken under darkfield illumination for green fluorescence localization (GFP), bright-field illumination to examine cell morphology (Bright field), and merged-field illumination (Merged). The bar represents 10 μm.

3.4. Expression Pattern of FmTCP15 in Different Tissues and Treatments

To determine the tissue-specific expression pattern of *FmTCP15*, we performed qRT-PCR in different tissues, including the main root, lateral root, xylem, phloem, cotyledon, function leaves, petiole, and seed. The expression level of *FmTCP15* was expressed highest in the cotyledons, followed by the xylem, and expressed lowest in the seeds (Figure 4A).

To investigate the expression pattern of *FmTCP15* under abiotic stresses and hormone signals, *F. mandshurica* seedlings were exposed to cold, salt, and drought stress treatments, as well as ABA and GA3 hormone signal. The results showed that, in response to abiotic stress treatment, *FmTCP15* gene expression was induced under cold, salt, and drought conditions (Figure 4B–F). However, the induction pattern was different. Cold stress induced *FmTCP15* to a high value from 6 to 12 h after initiation of the treatment, with a peak value at 12 h (Figure 4B). Salt stress induced *FmTCP15* to a peak value at 6 h after initiation of treatment (Figure 4C). In contrast, drought stress induced *FmTCP15* to a double peak expression pattern: the first peak value was at 6 h, and the second peak value was at 24 h (Figure 4D). For the hormone signal treatments, *FmTCP15* was downregulated after initiation of ABA treatment, with a double valley pattern: the first valley was at 24 h and the second valley was at 72 h (Figure 4E). GA3 induced *FmTCP15* expression, with a peak at 6 h (Figure 4F). These results indicate that *FmTCP15* responded to cold, salt, and drought abiotic stresses and ABA and GA3 treatments and imply that *FmTCP15* may participate in growth and development, as well as stress responses.

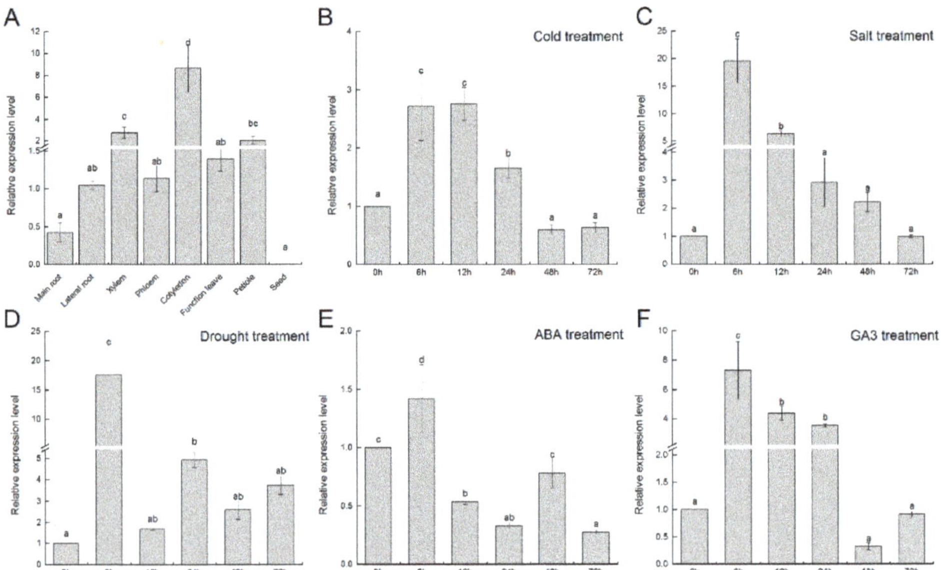

Figure 4. Expression patterns of *FmTCP15*. (**A**) Expression patterns of *FmTCP15* gene in different tissues. (**B–F**) Expression patterns of *FmTCP15* under different stress and hormone signal treatments: (**B**) cold (4 °C), (**C**) salt (200 mm/L NaCl), (**D**) drought (20% w/v PEG6000), (**E**) abscisic acid (100 μmol/L ABA), and (**F**) gibberellic acid (100 μmol/L GA3). Different letters above bars within statistically significant differences between different times of the treatments at the $p < 0.05$ level according to Duncan's multiple range test.

3.5. Functional Assay of Transient Overexpression of FmTCP15 in F. mandshurica

In plants, *TCP15* has multiple functions and is involved in a variety of growth, development, and stress responses, such as seed germination; leaf development; stem elongation; and the response to dehydration, salinity, and cold. In addition, hormones such as GA3 and JA acid, as well as auxin pathways, are also involved [39–41]. We detected the expression of key genes of these biological processes and TCP family downstream genes, as reported in a previous study. To elucidate the function of *FmTCP15* in *F. mandshurica*, an *Agrobacterium*-mediated transient expression system was used [39]. After coculture with *Agrobacterium*, transient overexpression of the *FmTCP15* plant was obtained. The transgenic expression levels of *FmTCP15* and *GFP* were determined by using qRT-PCR, which verified that the transformation was successful (Figure S1). qRT-PCR was used to examine the expression level of downstream genes (Figure 5). The results showed that the development-related genes *FmRGA*, *FmIAA3*, and *FmDAR1* were upregulated (Figure 5). Stress-response genes *FmAOS*, *FmABI5*, and *FmRAP2.1* and the cold regulation gene *FmCBF1* were also significantly upregulated (Figure 5). In addition, the circadian clock gene *FmCCA1* was also upregulated (Figure 5). However, one of the stress-response genes, *FmRAP2.12*, was suppressed (Figure 5). It is noteworthy that *FmTCP2* from class II and *FmTCP14* from class I showed significant upregulated or downregulated expression, respectively (Figure 5). Other genes did not show significant changes. These results indicated that *FmTCP15* directly regulated a series of development-related and abiotic stress-response genes and may act as a key node for developmental and stress responses.

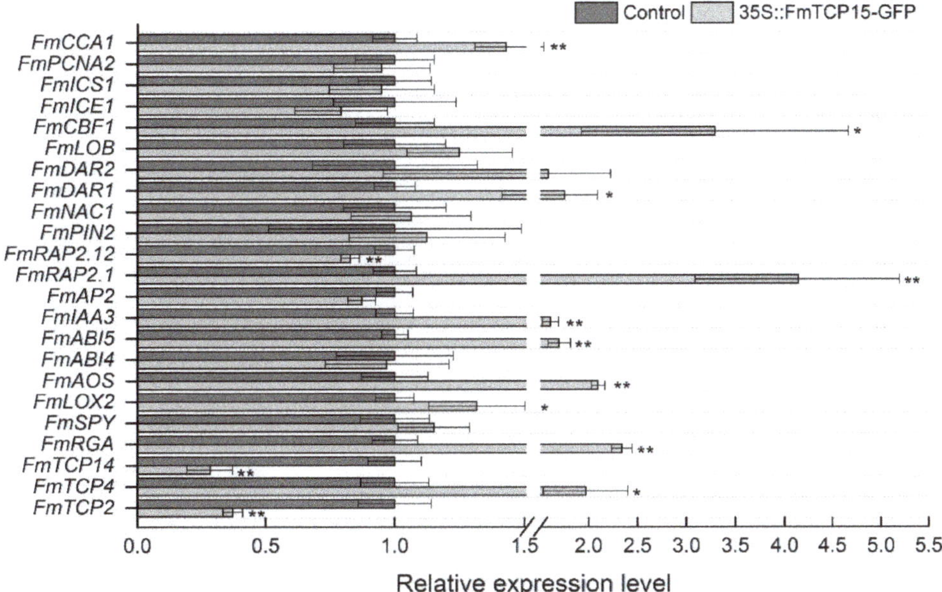

Figure 5. Quantitative assay results for downstream genes of transient overexpression of *FmTCP15* seedlings. The expression profile data of downstream genes of FmTCP15 were obtained through qRT-PCR. * Indicates a significant difference between control and 35S::FmTCP15-GFP ($p < 0.05$). ** Indicates a highly significant difference between control and 35S::FmTCP15-GFP ($p < 0.01$) according to Duncan's multiple range test.

3.6. Protein–Protein Interactions between FmTCP15 and DELLA Proteins

It is well known that the TCP domain provides TCPs the ability to form homo- and/or heterodimer protein complexes that are involved in the transcriptional activation of a series biological processes [22,42]. Previous studies showed that TCP proteins could interact with DELLA proteins and participate in the GA-regulated signaling regulatory pathway [24]. To determine whether TCP15 directly targets the downstream genes, particularly RGA1 and GAI (i.e., the two DELLA proteins), we conducted a Y2H assays. The full-length coding sequence of FmTCP15 was recombined into the pGBKT7 (BD) vector and pGADT7 (AD) vector, respectively. The candidate downstream genes of FmTCP15, such as *FmRGA*, *FmGAI*, *FmIAA3*, *FmDAR1*, *FmAOS*, *FmABI5*, *FmRAP2.1*, *FmCBF1*, *FmCCA1*, and *FmRAP2.12*, and the other TCP family members, such as *FmTCP2* and *FmTCP14*, were recombined into the pGADT7 (AD) vector. Y2H assays revealed that with the candidate downstream genes, FmTCP15 could interact with FmRGA and FmGAI (Figure 6). In addition, FmTCP15 could also interact with FmTCP2 from class II and FmTCP14 from class I (Figure 6). These results indicated that FmTCP15 proteins interact with DELLA-family proteins and may indirectly respond to stresses throughout the GA-signaling pathway.

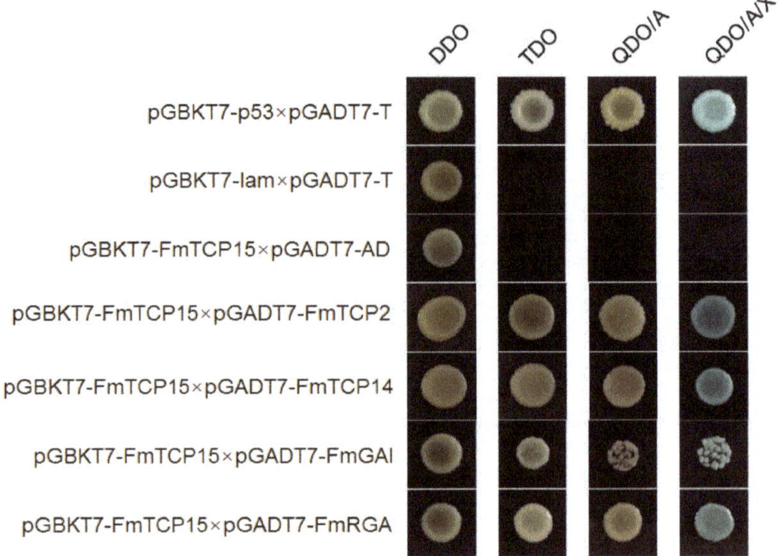

Figure 6. Yeast two-hybrid protein–protein interaction assays between FmTCP15 and DELLA proteins. The coding sequences of *FmTCPs*, *FmRGA*, and *FmGAI* were cloned intopGADT7 (AD) and pGBKT7 (BK) vectors. Transformants were assayed for growth on DDO, TDO, QDO/A, and QDO/A/X nutritional selection medium, and turning blue in the presence X-α-Gal was scored as a positive interaction.

4. Discussion

TCP transcription factors are a class of plant-specific transcription factors that play a very important role in many growth processes by directly or indirectly influencing plant hormonal signaling, the cell cycle, and the circadian clock. Research on TCPs has been conducted on various species, but the function of TCP family members in *F. mandshurica* has not been found. In addition, the molecular mechanisms of responses to abiotic stresses, as well as growth and development, in *F. mandshurica* are still scarcely understood. In this study, *FmTCP15* was isolated from *F. mandshurica* and subjected to a detailed bioinformatics analysis. The GO analysis indicated that FmTCP15 had protein-binding ability and participated multiple biological processes. Homology analysis and a phylogenetic tree of FmTCP15 revealed that FmTCP15 was most closely related to the Solanaceae and Plantaginaceae families (Figure 2B).

The tissue-specific expression pattern showed that *FmTCP15* was expressed highest in cotyledons, indicating that *FmTCP15* may play a role in seed germination and seedling development. Subcellular localization showed FmTCP15 was located in the nucleus. Furthermore, we observed that *FmTCP15* responded to both abiotic stresses and hormone signals. *FmTCP15* was mainly induced by cold, salt, and drought stress treatments and GA3 treatment but downregulated under ABA treatment (Figure 4B–F). Moreover, overexpressing *FmTCP15* caused a significant change in the expression of a series of key genes involved in the GA3, JA acid, and auxin pathways, as well as stress response and the circadian clock (Figure 5). In cotton, overexpression of *GhTCP14* altered the distribution of auxin and upregulated the expression levels of auxin-related genes such as *AUX1*, *PIN2*, and *IAA3* [43]. *IAA3* belongs to the Aux/IAA family and plays very important roles in regulating root growth and lateral root development [44]. A recent study showed that under cold and drought conditions, a series of *MeTCP* genes were significantly upregulated and functioned in resistance to abiotic stresses in cassava [41]. In this study, we found that overexpression of *FmTCP15* increased the expression level of *FmCBF* (Figure 5). The homologous gene of DREB1/CBF in rice has been found to be involved in cold tolerance and chilling acclimation [45]. In response to stress, we found that *FmRAP2.1* and

FmRAP2.12 were upregulated and downregulated by overexpressed *FmTCP15*, respectively. *RAP2.1* and *RAP2.12* are known as ERF/AP2 transcription factor genes from the ETHYLENE RESPONSE TRANSCRIPTION FACTOR (ERF) family and are involved in oxygen sensing and stress response. In tomato, *SlTCP12*, *SlTCP15*, and *SlTCP18* could bind to AP2/ERF proteins and may be indirectly involved in ethylene-dependent ripening [23]. In a study of *OsTCP19* in rice, a homologous gene of *FmTCP15*, *OsTCP19*, influenced lipid droplet (LD) synthesis and metabolism. In *Arabidopsis*, Rice *OsTCP19* overexpression transgenics upregulated *ABI4* and then promoted the expression of *diacylglycerolacetyl transferase (DGAT1)*, which is a triacylglycerol (TAG) biosynthesis gene that leads to the accumulation of LDs in vegetative tissue [14]. It was reported that TCPs may regulate *CIRCADIAN CLOCK ASSOCIATED1 (CCA1)* and participate in the regulation of the circadian clock. *AtTCP20* and *AtTCP22* acted as coactivators with *LIGHT-REGULATED WD1 (LWD1)* and regulated the expression of *CCA1* in *Arabidopsis* [46]. However, fold changes of FmTCP15 downstream genes were not very high, which may be related to the transient transformation efficiency not being compared with stable transformation.

Notably, overexpression of FmTCP15 significantly increased the expression level of *FmRGA* (Figure 6), and the Y2H assay also revealed the protein level interaction between FmTCP15 and DELLA proteins (FmRGA and FmGAI) (Figure 6). Our results showed that the regulatory relationship between FmTCP15 and DELLAs are at the transcriptional and post-transcriptional level. It is well known that the central function of GA3 is to regulate growth and development. However, increasing evidence has shown that GA3 responds to abiotic stresses. Reducing the levels of GA3 or signal transduction will lead to the restriction of plant growth, which is conducive for plant response to cold, salt, and osmotic stresses [47,48]. DELLA proteins are the key regulatory components of the GA-signaling pathway that repress the transcription of GA-responsive genes to restrain plant growth [49–51]. For the stress response, DELLA proteins could mediate the crosstalk between GA and ABA and regulate the balance between seed dormancy and germination [52]. A recent work has shown that during cold temperatures, the transcription of CBF3 will activate, subsequently decrease the bioactive level of GA3, lead to the accumulation of DELLAs, and thus enhance plant resistance to low temperatures [53]. In *Arabidopsis*, the DELLA protein RGA interacted with the TCP DNA-binding motif of TCP14 and TCP15, negatively controlled the expression of cell-cycle progression genes, and restricted plant height [24]. However, we suspected that the interaction between TCPs and DELLA may enhance plant resistance to stress. We surmised that, on the one hand, FmTCP15 interacts with DELLA and regulates plant growth, development, and responses to stress by participating in the GA-signaling pathway. On the other hand, FmTCP15 indirectly regulates downstream abiotic response genes to respond to abiotic stresses. Interestingly, we observed that *FmTCP2* from class II and *FmTCP14* from class I showed significant upregulated or downregulated expression in overexpressing *FmTCP15* plants. In addition, *FmTCP4* from class II was significantly upregulated. Y2H assay also demonstrated the interaction between FmTCP15, FmTCP2, and FmTCP14. These results indicated that there were interactions between classes I and II TCPs, and that the molecular mechanism of TCPs responding to hormone signaling pathways, growth, development, and abiotic stresses is flexible, complex, and multifunctional.

In summary, we revealed that *FmTCP15* was significantly induced by cold, salt, and drought stress and GA3. Overexpressing *FmTCP15* caused a change in the expression of a series of key genes involved in stress response, including *FmRAP2.1*, *FmRAP2.12*, and *FmCBF*; plant growth regulation, including *FmIAA3* and *FmDAR1*; and the GA-signaling pathway genes, including DELLA-family members (*FmRGA* and *FmGAI*), which are key proteins of the GA-signaling pathway. These results showed the FmTCP15 directly responded to the stresses. Moreover, FmTCP15 could interact with FmRGA and FmGAI at the protein level. As previously shown, the DELLA proteins are the key regulatory components of the GA-signaling pathway and restrain plant growth but enhance plant resistance to stress responses. The interactive relationship between FmTCP15 and DELLA proteins may enhance the ability of plants to resist stress responses, which suggests that its crucial function is indirectly responsive to stresses throughout the GA-signaling pathway. Regarding the gene diversity functions of the TCP family of genes, in the future, research on the molecular mechanisms of TCPs still needs to be enriched.

5. Conclusions

FmTCP15 encoded a protein of 362 amino acids with a TCP domain. The phylogenetic relationship showed that FmTCP15 was most closely related to the Solanaceae and Plantaginaceae families. FmTCP15 was localized in the nuclei and highly expressed in cotyledons. *FmTCP15* was mainly induced by cold, salt, and drought stress and GA3 but downregulated by ABA, revealing the *FmTCP15* response to multiple abiotic stresses and hormone signals. Overexpressing *FmTCP15* caused the expression change of the key genes in the stress and hormone responses and circadian clock. FmTCP15 interacted with DELLA proteins (FmGAI and FmRGA). Thus, on the one hand, *FmTCP15* directly regulated the abiotic stress-response genes and responded to stresses. On the other hand, *FmTCP15* interacted with the GA-signaling pathway genes, such as DELLA-family members, and indirectly responded to stresses throughout the GA-signaling pathway and enhanced the ability of the plant to resist stress responses. *FmTCP15* presents a variety of functions, and the molecular mechanism of *FmTCP15* still requires in-depth research.

Supplementary Materials: The following are available online at http://www.mdpi.com/1999-4907/10/4/343/s1, Figure S1: Quantitative assay results for *GFP* and *FmTCP15* genes in transient overexpression of *FmTCP15* seedlings, Table S1: Primers for quantitative RT-PCR (qRT-PCR), Table S2: Primers for yeast two-hybrid assay (Y2H), Table S3: Amino acid sequences for multiple sequence alignment, Table S4: Amino acid sequences for phylogenetic tree construction.

Author Contributions: The experiments were designed by N.L. and Y.Z. The manuscript was written by N.L. The experiments were carried out by N.L., L.Y. and Z.W. The experimental materials were collected by N.L. and L.Y. The manuscript was reviewed and edited by N.L., Y.Z. and F.Z.

Funding: This research was funded by a grant from the National Key Research and Development Project of China (No. 2017YFD0600605-01) and the National Natural Science Foundation of China (NSFC) (No. 31270697).

Conflicts of Interest: The authors declare no conflict of interest.

References

1. Luo, D.; Carpenter, R.; Vincent, C.; Copsey, L.; Coen, E. Origin of floral asymmetry in *Antirrhinum*. *Nature* **1996**, *383*, 794–799. [CrossRef] [PubMed]
2. Doebley, J.; Stec, A.; Hubbard, L. The evolution of apical dominance in maize. *Nature* **1997**, *386*, 485–488. [CrossRef] [PubMed]
3. Cubas, P.; Lauter, N.; Doebley, J.; Coen, E. The TCP domain: A motif found in proteins regulating plant growth and development. *Plant J.* **1999**, *18*, 215–222. [CrossRef] [PubMed]
4. Kosugi, S.; Ohashi, Y. PCF1 and PCF2 specifically bind to cis elements in the rice proliferating cell nuclear antigen gene. *Plant Cell.* **1997**, *9*, 1607–1619. [CrossRef] [PubMed]
5. Dhaka, N.; Bhardwaj, V.; Sharma, M.K.; Sharma, R. Evolving Tale of TCPs: New Paradigms and Old Lacunae. *Front Plant Sci.* **2017**, *8*, 479. [CrossRef] [PubMed]
6. Kosugi, S.; Ohashi, Y. DNA binding and dimerization specificity and potential targets for the TCP protein family. *Plant J.* **2002**, *30*, 337–348. [CrossRef]
7. Martín-Trillo, M.; Cubas, P. TCP genes: A family snapshot ten years later. *Trends Plant Sci.* **2010**, *15*, 31–39. [CrossRef] [PubMed]
8. Navaud, O.; Dabos, P.; Carnus, E.; Tremousaygue, D.; Hervé, C. TCP transcription factors predate the emergence of land plants. *J. Mol. Evol.* **2007**, *65*, 23–33. [CrossRef]
9. Girardi, C.L.; Rombaldi, C.V.; Dal Cero, J.; Nobile, P.M.; Laurens, F.; Bouzayen, M.; Quecini, V. Genome-wide analysis of the AP2/ERF superfamily in apple and transcriptional evidence of ERF involvement in scab pathogenesis. *Sci. Hortic.* **2013**, *151*, 112–121. [CrossRef]
10. Howarth, D.G.; Donoghue, M.J. Phylogenetic analysis of the "ECE"(CYC/TB1) clade reveals duplications predating the core eudicots. *Proc. Natl. Acad. Sci. USA* **2006**, *103*, 9101–9106. [CrossRef] [PubMed]

11. Li, C.; Potuschak, T.; Colón-Carmona, A.; Gutiérrez, R.A.; Doerner, P. *Arabidopsis* TCP20 links regulation of growth and cell division control pathways. *Proc. Natl. Acad. Sci. USA* **2005**, *102*, 12978–12983. [CrossRef] [PubMed]

12. Almeida, D.M.; Gregorio, G.B.; Oliveira, M.M.; Saibo, N.J. Five novel transcription factors as potential regulators of OsNHX1 gene expression in a salt tolerant rice genotype. *Plant Mol. Biol.* **2017**, *93*, 61–77. [CrossRef] [PubMed]

13. Mukhopadhyay, P.; Tyagi, A.K.; Tyagi, A.K. OsTCP19 influences developmental and abiotic stress signaling by modulating ABI4-mediated pathways. *Sci. Rep.* **2015**, *5*, 9998. [CrossRef]

14. Palatnik, J.F.; Allen, E.; Wu, X.; Schommer, C.; Schwab, R.; Carrington, J.C.; Weigel, D. Control of leaf morphogenesis by microRNAs. *Nature* **2003**, *425*, 257–263. [CrossRef]

15. Koyama, T.; Furutani, M.; Tasaka, M.; Ohme-Takagi, M. TCP transcription factors control the morphology of shoot lateral organs via negative regulation of the expression of boundary-specific genes in *Arabidopsis*. *Plant Cell.* **2007**, *19*, 473–484. [CrossRef]

16. Schommer, C.; Palatnik, J.F.; Aggarwal, P.; Chételat, A.; Cubas, P.; Farmer, E.E.; Nath, U.; Weigel, D. Control of jasmonate biosynthesis and senescence by miR319 targets. *PLoS Biol.* **2008**, *6*, e230. [CrossRef]

17. Yang, X.; Zhao, X.; Li, C.; Liu, J.; Qiu, Z.; Dong, Y.; Wang, Y. Distinct regulatory changes underlying differential expression of TEOSINTE BRANCHED1-CYCLOIDEA-PROLIFERATING CELL FACTOR genes associated with petal variations in zygomorphic flowers of *Petrocosmea spp.* of the family Gesneriaceae. *Plant Physiol.* **2015**, *169*, 2138–2151. [PubMed]

18. Gonzalezgrandio, E.; Pozacarrion, C.; Oscar, C.; Sorzano, S.; Cubas, P. BRANCHED1 Promotes Axillary Bud Dormancy in Response to Shade in *Arabidopsis*. *Plant Cell.* **2013**, *25*, 834–850. [CrossRef]

19. Reddy, S.K.; Holalu, S.V.; Casal, J.J.; Finlayson, S.A. Abscisic acid regulates axillary bud outgrowth responses to the ratio of red to far-red light. *Plant Physiol.* **2013**, *163*, 1047–1058. [CrossRef] [PubMed]

20. Guo, S.; Xu, Y.; Liu, H.; Mao, Z.; Zhang, C.; Ma, Y.; Zhang, Q.; Meng, Z.; Chong, K. The interaction between OsMADS57 and OsTB1 modulates rice tillering via DWARF14. *Nat. Commun.* **2013**, *4*, 1566. [CrossRef]

21. Costa, M.M.R.; Fox, S.; Hanna, A.I.; Baxter, C.; Coen, E. Evolution of regulatory interactions controlling floral asymmetry. *Development* **2005**, *132*, 5093–5101. [CrossRef] [PubMed]

22. Danisman, S.; Van der Wal, F.; Dhondt, S.; Waites, R.; de Folter, S.; Bimbo, A.; van Dijk, A.D.; Muino, J.M.; Cutri, L.; Dornelas, M.C. *Arabidopsis* class I and class II TCP transcription factors regulate jasmonic acid metabolism and leaf development antagonistically. *Plant Physiol.* **2012**, *159*, 1511–1523. [CrossRef]

23. Parapunova, V.; Busscher, M.; Busscher-Lange, J.; Lammers, M.; Karlova, R.; Bovy, A.G.; Angenent, G.C.; de Maagd, R.A. Identification, cloning and characterization of the tomato TCP transcription factor family. *BMC Plant Biol.* **2014**, *14*, 157. [CrossRef] [PubMed]

24. Davière, J.M.; Wild, M.; Regnault, T.; Baumberger, N.; Eisler, H.; Genschik, P.; Achard, P. Class I TCP-DELLA interactions in inflorescence shoot apex determine plant height. *Curr. Biol.* **2014**, *24*, 1923–1928. [CrossRef] [PubMed]

25. Torres, M.L.D.L.Y.; Palomares, O.; Quiralte, J.; Pauli, G.; Rodriguez, R.; Villalba, M. An Enzymatically Active β-1,3-Glucanase from Ash Pollen with Allergenic Properties: A Particular Member in the Oleaceae Family. *PLoS ONE* **2015**, *10*, e0133066. [CrossRef] [PubMed]

26. Song, J.; Yang, D.; Niu, C.; Zhang, W.; Wang, M.; Hao, G. Correlation between leaf size and hydraulic architecture in five compound-leaved tree species of a temperate forest in NE China. *For. Ecol. Manage.* **2017**, *418*, 63–72. [CrossRef]

27. Gang, Q.; Yan, Q.; Zhu, J. Effects of thinning on early seed regeneration of two broadleaved tree species in larch plantations: Implication for converting pure larch plantations into larch-broadleaved mixed forests. *Forestry* **2015**, *88*, 573–585. [CrossRef]

28. Huang, S.; Sun, X.; Zhang, Y.; Sun, H.; Wang, Z. Nutrient retranslocation from the fine roots of *Fraxinus mandshurica* and *Larix olgensis* in northeastern China. *J. For. Res.* **2016**, *27*, 1305–1312. [CrossRef]

29. Li, T.; Wu, J.; Chen, H.; Ji, L.; Yu, D.; Zhou, L.; Zhou, W.; Tong, Y.; Li, Y.; Dai, L. Intraspecific functional trait variability across different spatial scales: A case study of two dominant trees in Korean pine broadleaved forest. *Plant Ecol.* **2018**, *219*, 875–886. [CrossRef]

30. Drenkhan, R.; Solheim, H.; Bogacheva, A.M.; Riit, T.; Adamson, K.; Drenkhan, T.; Maaten, T.; Hietala, A.M. *Hymenoscyphus fraxineus* is a leaf pathogen of local Fraxinus species in the Russian Far East. *Plant Pathol.* **2017**, *66*, 490–500. [CrossRef]

31. Gasteiger, E.; Hoogland, C.; Gattiker, A.; Duvaud, S.e.; Wilkins, M.R.; Appel, R.D.; Bairoch, A. Protein identification and analysis tools on the ExPASy server. In *The Proteomics Protocols Handbook*, 1st ed.; John, M.W., Ed.; Humana Press: Clifton, NJ, USA, 2005; pp. 571–607.

32. Hofmann, K. TMbase-A database of membrane spanning proteins segments. *Biol. Chem. Hoppe-Seyler.* **1993**, *374*, 166.

33. Wang, Y.; Virtanen, J.; Xue, Z.; Tesmer, J.J.; Zhang, Y. Using iterative fragment assembly and progressive sequence truncation to facilitate phasing and crystal structure determination of distantly related proteins. *Acta Crystallogr. D* **2016**, *72*, 616–628. [CrossRef] [PubMed]

34. Wang, Y.; Virtanen, J.; Xue, Z.; Zhang, Y. I-TASSER-MR: Automated molecular replacement for distant-homology proteins using iterative fragment assembly and progressive sequence truncation. *Nucleic Acids Res.* **2017**, *45*, 429–434. [CrossRef]

35. Xu, D.; Zhang, Y. Improving the physical realism and structural accuracy of protein models by a two-step atomic-level energy minimization. *Biophys. J.* **2011**, *101*, 2525–2534. [CrossRef]

36. Lin, Y.; Li, W.; Chen, H.; Li, Q.; Sun, Y.; Shi, R.; Lin, C.; Wang, J.P.; Chen, H.; Chuang, L. A simple improved-throughput xylem protoplast system for studying wood formation. *Nat. Protoc.* **2014**, *9*, 2194–2205. [CrossRef]

37. Zhang, Y.; Wang, Y.; Wang, C. Gene overexpression and gene silencing in Birch using an *Agrobacterium*-mediated transient expression system. *Mol. Biol. Rep.* **2012**, *39*, 5537–5541. [CrossRef] [PubMed]

38. Livak, K.J.; Schmittgen, T.D. Analysis of relative gene expression data using real-time quantitative PCR and the $2^{-\Delta\Delta CT}$ method. *Methods* **2001**, *25*, 402–408. [CrossRef]

39. Kieffer, M.; Master, V.; Waites, R.; Davies, B. TCP14 and TCP15 affect internode length and leaf shape in *Arabidopsis*. *Plant J.* **2011**, *68*, 147–158. [CrossRef] [PubMed]

40. Resentini, F.; Felipo-Benavent, A.; Colombo, L.; Blázquez, M.A.; Alabadí, D.; Masiero, S. TCP14 and TCP15 mediate the promotion of seed germination by gibberellins in *Arabidopsis thaliana*. *Mol. Plant.* **2015**, *8*, 482–485. [CrossRef]

41. Lei, N.; Yu, X.; Li, S.; Zeng, C.; Zou, L.; Liao, W.; Peng, M. Phylogeny and expression pattern analysis of TCP transcription factors in cassava seedlings exposed to cold and/or drought stress. *Sci. Rep.* **2017**, *7*, 10016. [CrossRef] [PubMed]

42. Zhao, J.; Zhai, Z.; Li, Y.; Geng, S.; Song, G.; Guan, J.; Jia, M.; Wang, F.; Sun, G.; Feng, N. Genome-Wide Identification and Expression Profiling of the TCP Family Genes in Spike and Grain Development of Wheat (*Triticum aestivum* L.). *Front Plant Sci.* **2018**, *9*, 1282. [CrossRef] [PubMed]

43. Wang, M.; Zhao, P.; Cheng, H.; Han, L.; Wu, X.; Gao, P.; Wang, H.; Yang, C.; Zhong, N.; Zuo, J. The cotton transcription factor TCP14 functions in auxin-mediated epidermal cell differentiation and elongation. *Plant Physiol.* **2013**, *162*, 1669–1680. [CrossRef]

44. Tian, Q.; Uhlir, N.J.; Reed, J.W. *Arabidopsis* SHY2/IAA3 inhibits auxin-regulated gene expression. *Plant Cell.* **2002**, *14*, 301–319. [CrossRef] [PubMed]

45. Mao, D.; Chen, C. Colinearity and similar expression pattern of rice DREB1s reveal their functional conservation in the cold-responsive pathway. *PLoS ONE* **2012**, *7*, e47275. [CrossRef]

46. Wu, J.; Tsai, H.; Joanito, I.; Wu, Y.; Chang, C.; Li, Y.; Wang, Y.; Hong, J.C.; Chu, J.; Hsu, C.; et al. LWD–TCP complex activates the morning gene CCA1 in *Arabidopsis*. *Nat. Commun.* **2016**, *7*, 13181. [CrossRef]

47. Colebrook, E.H.; Thomas, S.G.; Phillips, A.; Hedden, P. The role of gibberellin signalling in plant responses to abiotic stress. *J. Exp. Biol.* **2014**, *217*, 67–75. [CrossRef] [PubMed]

48. Liu, B.; De Storme, N.; Geelen, D. Cold-Induced Male Meiotic Restitution in *Arabidopsis thaliana* Is Not Mediated by GA-DELLA Signaling. *Front Plant Sci.* **2018**, *9*, 91. [CrossRef] [PubMed]

49. Achard, P.; Gong, F.; Cheminant, S.; Alioua, M.; Hedden, P.; Genschik, P. The Cold-Inducible CBF1 Factor–Dependent Signaling Pathway Modulates the Accumulation of the Growth-Repressing DELLA Proteins via Its Effect on Gibberellin Metabolism. *Plant Cell.* **2008**, *20*, 2117–2129. [CrossRef]

50. King, K.E.; Moritz, T.; Harberd, N.P. Gibberellins Are Not Required for Normal Stem Growth in *Arabidopsis thaliana* in the Absence of GAI and RGA. *Genetics* **2001**, *159*, 767–776. [PubMed]

51. Yang, D.L.; Yao, J.; Mei, C.S.; Tong, X.H.; Zeng, L.J.; Li, Q.; Xiao, L.T.; Sun, T.; Li, J.; Deng, X.W. Plant hormone jasmonate prioritizes defense over growth by interfering with gibberellin signaling cascade. *Proc. Natl. Acad. Sci. USA* **2012**, *109*, 7152–7153. [CrossRef]

52. Verma, V.; Ravindran, P.; Kumar, P.P. Plant hormone-mediated regulation of stress responses. *BMC Plant Biol.* **2016**, *16*, 86. [CrossRef] [PubMed]
53. Zhou, M.; Chen, H.; Wei, D.; Ma, H.; Lin, J. *Arabidopsis* CBF3 and DELLAs positively regulate each other in response to low temperature. *Sci. Rep.* **2017**, *7*, 39819. [CrossRef] [PubMed]

Article

Non-Structural Carbohydrate Dynamics in Leaves and Branches of *Pinus massoniana* (Lamb.) Following 3-Year Rainfall Exclusion

Tian Lin [1,2], Huaizhou Zheng [3,4], Zhihong Huang [5], Jian Wang [3,4,*] and Jinmao Zhu [1,3]

1   College of Life Sciences, Fujian Normal University, Fuzhou 350117, China; tlin1984@fjut.edu.cn (T.L.); jmzhu@fjnu.edu.cn (J.Z.)
2   College of Ecological Environment and Urban Construction, Fujian University of Technology, Fuzhou 350118, China
3   Fujian Provincial Key Laboratory for Plant Eco-physiology, Fujian Normal University, Fuzhou 350007, China; zhz@fjnu.edu.cn
4   School of Geographical Sciences, Fujian Normal University, Fuzhou 350007, China
5   Department of Basic Medical Science, Fujian Medical University, Fuzhou 350004, China; hzh8512@fjmu.edu.cn
*   Correspondence: jwang@fjnu.edu.cn; Tel: +86-591-8348-3731

Received: 17 May 2018; Accepted: 31 May 2018; Published: 1 June 2018

Abstract: Drought-induced tree mortality is an increasing and global ecological problem. Stored non-structural carbohydrates (NSCs) may be a key determinant of drought resistance, but most existing studies are temporally limited. In this study, a 3-year 100% rainfall exclusion manipulation experiment was conducted to evaluate the response of NSC dynamics to drought stress in 25-year-old *Pinus massoniana* leaves and branches. The results showed: (1) compared with the control condition, leaf NSC concentration in the drought treatment increased 90% in the early stage (days 115–542) ($p < 0.05$), and then decreased 15% in the late stage (days 542–1032), which was attributed to water limitation instead of phenology; (2) the response of leaf NSCs to drought was more significant than branch NSCs, demonstrating a time lag effect; and (3) the response of *P. massoniana* to mild drought stress was to increase the soluble sugars and starch in the early stage, followed by an increase in soluble sugars caused by decreasing starch in the later stress period. Considering these results, mid-term drought stress had no significant effect on the total NSC concentration in *P. massoniana*, removing carbon storage as a potential adaptation to drought stress.

Keywords: drought; mid-term; non-structural carbohydrate; soluble sugar; starch; *Pinus massoniana*

1. Introduction

The frequency and intensity of tree mortality associated with drought is increasing globally [1–3]. As an important part of forest ecosystems, increased tree mortality could lead to worldwide large-scale forest die-off [4,5]. Such widespread events would have long-term impacts on ecosystem structure and functioning [2,6], as well as forest carbon storage capacity [7]. Severe and recurrent drought events have been identified as a major contributing factor to forest decline and mortality, resulting in feedback to atmospheric carbon dioxide (CO_2) and climate [8,9]. Despite a growing research interest surrounding the physiology of drought-induced tree mortality [10], our current understanding of the adaptive mechanisms remains poor, limiting our ability to predict widespread mortality events, their feedback in the future climate system, and the impact on ecosystem services provided to humans [11–13].

In existing studies, McDowell [14] formalized two non-exclusive hypotheses of hydraulic failure and carbon starvation to explain the impact of drought stress on water and carbon cycling. The hydraulic-failure hypothesis predicts that intense drought quickly reduces soil moisture and increases evaporative demand, causing the plant's water potential to fall below a critical level. This results in the cavitation of xylem conduits and rhizosphere, stopping the flow of water and desiccating plant tissues, eventually causing cellular death [15]. When droughts are less intense but prolonged, carbon starvation occurs. The carbon starvation hypothesis predicts that preventing desiccation via stomatal closure causes a decrease in the photosynthetic uptake of carbon, but the continued demand for carbohydrates to maintain metabolism depletes carbohydrate reserves [14,16]. Studies have attempted to verify this hypothesis, but the results were inconsistent due to differences in tree species, drought intensity and duration, as well as differences in measurement methods [1,17,18]. However, a growing number of studies are demonstrating an association between carbon reserve depletion and drought-induced mortality in different tree species [1,17,19–22].

Accumulated and stored non-structural carbohydrates (NSCs), as the primary products of photosynthesis, support growth and normal metabolism [23]. NSCs mainly consist of soluble sugars (e.g., sucrose, glucose, and fructose) and immobile starch. The dynamics of NSC storage are related to the balance between carbon source and carbon sink. After NSC compounds are assimilated during photosynthesis, the products are passively allocated first to metabolism, then to new tissue growth. When all demands for carbon have been met, the remaining carbon is stored. Stored NSCs provide an essential carbon pool buffer when demand for growth and maintenance exceeds the supply provided by photosynthesis. For example, this pool is essential in the dormant season, in response to environmental stress, as well as during other carbon deficit conditions to help to maintain hydraulic conductivity [16,24]. The role of NSCs as a buffer has been supported by previous studies [14,16,25]. However, due to differences in drought properties as well as tree size, age, tissue, and species, the lack of agreement among these studies about the effects of drought on NSC dynamics supports the need for further investigation [1,25] to better understand and predict forest ecosystem responses to global climate scenarios.

Different levels of drought intensity may induce different NSC dynamics in trees [14]. The following studies emphasized the influence of drought progression on NSC dynamics [26]. In the early stages of drought, NSCs may increase because cell turgor pressure reduces plant growth ahead of photosynthetic decline, as well as when a carbohydrate imbalance occurs between supply and demand [26–28]. Previous research found that photosynthesis was more sensitive to drought stress than respiration [29,30]. Thus, as drought conditions developed, carbohydrates produced by photosynthesis could no longer meet the need of respiration, and stored NSC compounds began to be consumed, leading to a decrease in NSCs. When the concentration of NSCs dropped to a certain level and failed to meet the demands of plant physiological metabolism, trees faced the risk of mortality [26]. However, most of the previous studies tended to measure NSC dynamics of seedlings in growing seasons under varying water supplies by carrying out short-term experiments [9,25,27,31–33]. Either no reduction or even increases in carbohydrate reserves were found under short-term drought stress [8,19,33]. A model used to estimate carbohydrate reserves determined that reserves should decline under exceptionally long-term drought stress [26,34]. The limited timeframe in the short-term experiments restricted the analysis of the carbohydrates dynamics caused by climate change. So, there is a lack of mid- or long-term drought-induced NSC dynamics experiments in adult trees, which are particularly critical for the prediction of tree and forest responses to future climate conditions.

Different tree tissues have different NSC concentrations as a result of drought stress. Carbon storage in different tissues depends on the role each of these parameters plays in response to drought stress [31,35,36]. Galiano et al. [37] found that trees subjected to severe drought could recover rapidly depending on their storage organs (e.g., lignotubers). As the main site of photosynthesis, leaves have stronger carbon assimilation ability, and became one of the main carbon sources to meet the carbon demand. The distribution in NSC concentration was related to the distance to the leaves (carbon source). The tissues near the carbon source were preferable for NSC to those far from the carbon source [38]. So, the distribution of NSCs in leaves and branches gradually became a research focus. Hartman et al. [1] examined NSC concentrations under drought conditions and found that leaves and branches showed little difference. Other studies that focused on different species or different tissues reported variable conclusions [39,40]. A study showed that mortality caused by drought may not be defined by the organism, but by the tree tissue [1]. As such, the concentration of NSCs stored in different tissues may provide a comprehensive understanding of the response of trees to drought stress, helping to predict the impact of future global climate change on forest ecosystem carbon cycling.

To better understand the responses to precipitation reduction of plant-stored NSC concentration dynamics, and to reveal the effects of drought stress on plant carbon supply status, we conducted a 3-year rainfall exclusion experiment from April 2013 to January 2016 in Chang Ting county, Fujian province, China. Here, we measured the concentration of stored NSCs in leaves and branches of adult *Pinus massoniana* (Lamb.), which is widely used for afforestation in Chang Ting country, grown in both control (natural) conditions and drought (100% rainfall exclusion) treatments. We aimed to (1) determine the effect of mid-term drought on the concentration and composition of NSCs; (2) characterize the seasonal dynamics in NSC concentration over the study period; and (3) study the relationship between NSCs, their composition, and soil moisture content.

2. Materials and Methods

2.1. Study Site

The study site was located in the town of HeTian, in Southeast Chang Ting, FuJian province, China (116°18′–116°31′ E, 25°33′–25°48′ N; 310 m asl), which is one of the regions in Fujian province experiencing serious soil erosion. The climate is characterized as subtropical monsoon with distinguishing wet (March to September) and dry (October to February) seasons. The study site had a mean annual rainfall of 1737 mm, which mainly fell from May to July, and a mean annual temperature between 17.5 and 18.8 °C. The lowest recorded temperature was −7.8 °C, and the highest recorded temperature was 39.8 °C from April 2013 to January 2016. Climate data were collected every 15 min from an automatic meteorological station installed at the study site to monitor temperature, precipitation (Figure 1), and soil moisture content (Figure 2). According to Figure 1, we used the linear combination of the two most important climatic factors (precipitation and temperature) to define the aridity condition, called the Bagnouls–Gaussen diagram [41]. The climate was defined as arid when the value of precipitation was lower than the double value of the mean temperature, "P-2T". As illustrated by Figure 1, the hydric balance inverted when the precipitation point was below the temperature. Thus, the results estimated that only October 2013, January 2014, April 2014, September 2014, October 2014, and October 2015 experienced soil water deficiency between April 2013 and January 2016. Therefore, we concluded that there was no aridity at the study site, which could be considered as a control group. The main soil type of this study site was red soil, derived from medium-to-coarse crystalline granite. Most of the soil exposed the soil core (B layer), the parent material layer (C layer), and the mother rock (D layer). Vegetation cover in the study area was low. The dominant species were *P. massoniana*, *Dicranopteris dichotoma (Thunb.)* Bernh., and less shrub and grass vegetation occurred under the forest.

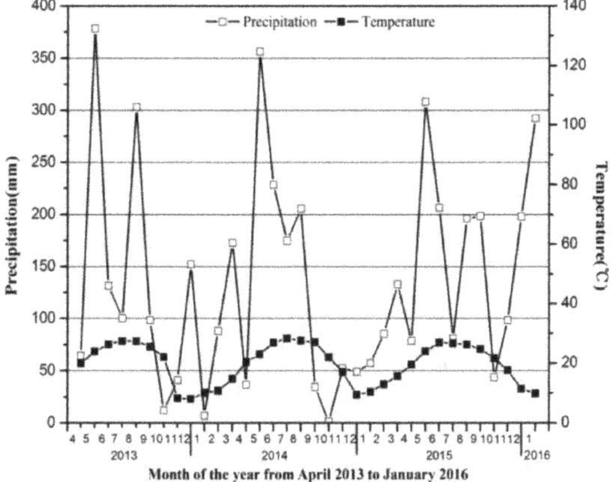

Figure 1. Bagnouls–Gaussen bio-climatic diagram.

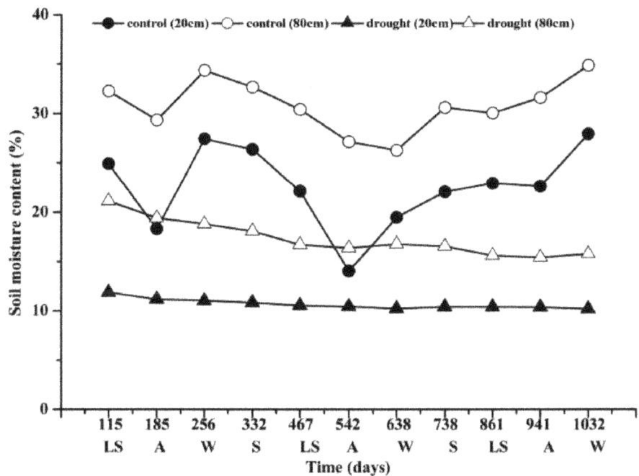

Figure 2. Relative change in daily average soil moisture content across the study period from August 2013 to January 2016 in the control and drought plots. Samples were collected in late summer (LS), autumn (A), winter (W), and late spring (S).

2.2. Experimental Design

A 3-year rainfall exclusion experiment under natural setting were running in the study area since April 2013. We set up four 20 × 20 m sample plots facing southwest on a 30° slope at 333 m above sea level. Inside these four sample plots, we had 42 adult *P. massoniana* individuals, nearly 25 years old, with 10 or 11 *P. massoniana* individuals in each sample, which were seeded by aircraft in the 1990s. On average, the *P. massoniana* individuals were 4.2 cm in diameter at breast height (DBH), 2.4 m in height, with a leaf length of 8.5 cm. Two assigned plots received a drought treatment. To avoid the influence of slope on sampling, these two plots included one from up-slope and the other from down-slope. The drought treatment, which was achieved by fixing 4 m of transparent ultraviolet (UV) paint wave tile (light transmittance 90%), installed parallel to the terrain so that rainfall could flow down the slope. PVC strips did not produce changes in environment temperature or humidity. In addition, an 80-cm deep ditch lined with aluminum was dug along the entire top edge of the drought treatment plots to prevent surface runoff inflow. The drought treatment resulted in decreased soil moisture over the 3-year period.

2.3. Leaf Water Potential

To test the effectiveness of the rainfall exclusion on plant water status, leaf water potentials were measured at dawn (Ψ_d) nearly once per month from April 2013 to March 2014 with a WP4 dew-point potential meter (Decagon Device, Pullman, WA, USA). The six target individuals, which included three from the up-slope plot and three from the down-slope plot were chosen from each treatment. Two south-facing braches from the upper and middle parts of each tree crown with healthy leaves were collected from each individual, immediately placed in a cold closet (0–4 °C), then taken to the laboratory to determine the leaf water potentials.

2.4. Sampling Methods

After 115 days of continuous 100% rainfall exclusion, three individuals 4.2 cm in diameter and 2.4 m in height within the center of each plot, to avoid the border effect (n = 12), were randomly selected. The six individuals from each treatment included three from the up-slope plot and three from the down-slope plot. Plant sampling dates were chosen according to the main climatic drivers at the study site. The sampling days were as follows: day 115 (18 August 2013, late summer), day 185 (27 October 2013 autumn), day 256 (6 January 2014, winter), day 332 (23 March 2014, spring), day 467 (1 August 2014, late summer), day 542 (15 October 2014, autumn), day 638 (30 December 2014, winter), day 738 (9 April 2015, spring), day 861 (10 August 2015, late summer), day 941 (29 October 2015, autumn), and day 1032 (28 January 2016, winter). On 28 January 2016, we found that two sampled individuals in the drought treatment had died, and decided to terminate the experiment. In all cases, leaves and branches were collected from each sampled individual to measure the concentrations of stored NSCs. To minimize diurnal variability in NSCs, four small branches exposed to the sun were always collected between 9:00 and 11:00 a.m., from the southeast and northwest direction of the upper and middle parts of each tree crown, above which we selected well-developed leaves. Leaf samples were 8.3 to 8.6 cm in length and branch samples were 0.3 to 0.5 cm in diameter (with bark removed). Samples were mixed to create one leaf sample and one branch sample. All the samples were immediately stored in a cold closet (0–4 °C) prior to performing laboratory analyses. To minimize continued enzymatic activity, all samples were microwaved at 800 W for 5 min, then dried at 65 °C for 48 h until a constant weight was reached. Leaf and branch samples (with bark removed) were ball-milled to a fine powder (Tissuelyser-24, Shanghai, China) for analysis.

2.5. NSC Analysis

We measured NSC concentration using the sum of soluble sugar and starch concentrations using the anthrone method [42] with some minor modifications. A 0.05 g subsample of ground tissue was placed in a 10 mL centrifuge tube, and 5 mL of distilled water was added. The sample was then subjected to 80 W ultrasonic disruption at normal atmospheric temperature for 30 min. The mixture was incubated at 100 °C in a water bath for 10 min and then cooled. The sample was then centrifuged at 4000 r/min for 20 min. The supernatant, after three extractions, was used for soluble sugar determination.

Starch was extracted from the residue at the top of the centrifuge tube, then 1.5 mL of 9.2 M $HClO_4$ and 3.5 mL of distilled water were added to the mixture and left to sit overnight. The following day, the sample was subjected to a boiling water bath for 10 min. After cooling to room temperature, the sample was centrifuged at 4000 r/min for 20 min. After three extractions, the supernatant was used for starch determination.

Both the soluble sugar and starch concentrations were determined based on the absorbance at 625 nm using the same anthrone reagent in a spectrophotometer (TU1901, Persee, Beijing, China) [42]. Finally, the corresponding content was deduced using the formula ($y = \frac{M \times V1 \times V2}{W \times C}$), where y is the soluble sugar or starch concentration expressed as mg/g, M is the glucose concentration determined from the standard curve, $V1$ is the sample extraction volume, $V2$ is the extract volume at color development, W is the sample weight, and C is the dilution ratio at color development.

2.6. Statistical Analyses

All results are reported as mean $\pm$ standard deviation (SD) for the six replicates. We used repeated measures of analysis of variance (ANOVA) to study the effects of treatment and time on the amount of stored NSCs and their composition. One-way ANOVA was used to detect the effects of different treatments on the above indices. Single linear regression models were used to compare the relationships among NSCs, soluble sugar, starch concentration, and the ratio of soluble sugar to starch for the soil moisture content. Significance and high significance levels for all tests were set at $p < 0.05$ and $p < 0.01$, respectively. All statistical analyses were performed using SPSS 19.0 (SPSS Inc., Chicago, IL, USA). Diagrams were drawn using Origin 8.0 software.

3. Results

3.1. Soil Moisture Content and Leaf Water Potential During Mid-Term Drought

Soil moisture content differed between drought and control treatments. During the 1032 days of continuous monitoring, the soil moisture content at 20 and 80 cm depths of the control treatment was significantly higher than the corresponding soil moisture in the drought group, by 11.87% and 13.54%, respectively (Figure 2). Nevertheless, the average soil moisture content in the drought group, 10.69% at 20 cm depth and 17.31% at 80 cm depth, appeared to be able to meet the demands of a normal metabolism. We proposed two mechanisms to explain this phenomenon: (1) moderate slope limited the degree of drying due to below-ground flow; and (2) complete stomatal closure occurred under drought conditions at the beginning of the experiment. Overall, with 1737 mm annual rainfall at the study site, the experiment might be defined as mild drought [43]. Mild drought conditions are more representative of the real scenarios that are expected under natural conditions. In order to check the effects of rainfall exclusion on *P. massoniana*, leaf water potentials were analyzed. The effect of increasing drought on Ψ_d is seen in Figure 3. From April 2013 to March 2014, Ψ_d in the drought group decreased over the course of the experiment. The mean value of Ψ_d (-3.73 MPa) in the drought group was significantly lower than that in the control group (Ψ_d: -2.80 MPa) ($p < 0.05$; Figure 3), suggesting that soil water availability was different between different treatments.

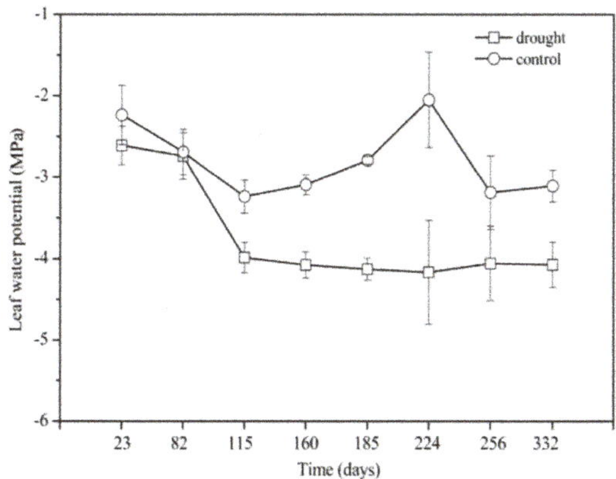

Figure 3. Monthly patterns of dawn leaf water potentials from April 2013 to March 2014.

Then in the control treatment, soil moisture content at depths of both 20 and 80 cm showed significant seasonal fluctuations ($p < 0.01$; Figure 2). In the drought group, soil moisture content at depths of both 20 and 80 cm were significantly lower than that of the control group at day 115 ($p < 0.01$; Figure 2). The remaining study period (days 185–1032) showed very little fluctuation in soil moisture content (Figure 2).

3.2. NSC Concentration and Composition in Different Treatments

Between treatments, the mean NSC concentration in the leaves in the drought group (183.91 mg/g) was somewhat higher than the control group (178.87 mg/g). The mean NSC concentration in the branches in the drought group (184.32 mg/g) was slightly lower than the control group (194.11 mg/g). No significant differences were found between the control and drought groups for either tissue type over the course of the study period. However, different treatments had different NSC dynamics: in the control treatment, total leaf NSC concentration showed substantial seasonal variability from 2013 to 2016 (Figure 4). In both 2013 and 2014 from spring through autumn, a gradual upward trend in total leaf NSC concentration was observed. The peak leaf NSC concentration value was observed on days 185 (27 October 2013, autumn) and 542 (15 October 2014, autumn) (Figure 4). The peak leaf NSC concentration in 2015 was observed in summer, and the seasonal dynamics differed from 2013 and 2014 (Figure 4). In the drought group, total leaf NSC concentration increased significantly in the early stages (days 115–542) ($p < 0.05$), then decreased in the late stages (days 542–1032) of the experimental period (Figure 4).

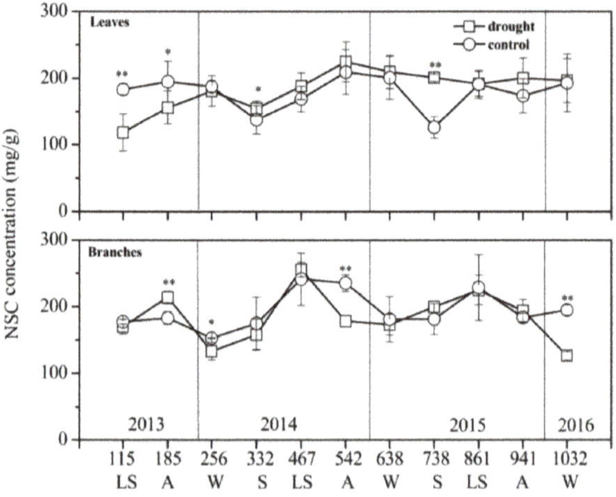

Figure 4. Non-structural carbohydrates (NSCs) in the leaves and branches of *Pinus massoniana* under control and drought treatments (mean ± SD). Samples were collected in late summer (LS), autumn (A), winter (W), and late spring (S) from 2013 to 2016. Variables marked with asterisks indicate a significant difference between the control and drought groups (** $p < 0.01$, * $p < 0.05$).

As for the branches, the control group also showed seasonal dynamics, with an annual peak in summer or autumn, and a valley in spring or winter. The branch NSC concentration dynamics of different treatments across three years differed from those of the leaves (Figure 4). From day 115 to day 467, the branches of the drought group displayed the same trend as the control group. From day 467 to day 542, NSC concentration decreased significantly ($p < 0.05$; Figure 4), then generally increased after declining in the drought group.

The effects of drought, sampling time, and their interaction on the soluble sugar concentration, starch concentration, NSC concentration, and the ratio of soluble sugar to starch were obvious (Table 1). The average concentration of soluble sugar in the leaves was 97.21 mg/g in the drought group and 89.98 mg/g in the control group, which is an increase of about 8.02%. Soluble sugar concentration in the leaves of the drought group was significantly lower than in the control group in the early stages of rainfall exclusion ($p < 0.05$; days 115–256). The gradual upward trend in the soluble sugar concentration of the drought group between days 256 and 1032 was greater than observed in the control group (Figure 5). However, the average leaf starch concentration decreased by 2.46% in the drought group compared to the control group (Figure 6). Leaf starch concentration in the drought group showed a trend of increasing first, then decreasing rapidly (Figure 6). The ratio of soluble sugar to starch in the drought group increased significantly with the increase in isolation time ($p < 0.05$; Figure 7), which had significant seasonal dynamics in the control group ($p < 0.05$; Figure 7). The effect of season on branch starch in the control group was not significant, but treatment effects were significant for soluble sugar concentration ($p < 0.01$; Figures 5 and 6; Table 1).

Table 1. Repeated measures analysis of variance (ANOVA) of nonstructural carbohydrate (NSC) concentration and composition by treatment and sampling date.

NSC Component	Source of Variations	Leaves			Branches		
		df	F	p	df	F	p
NSC	Treatment	1	0.006	0.941	1	3.636	0.063
	Time	10	4.991	0.000	10	13.877	0.000
	Treatment × time	10	1.702	0.100	10	3.607	0.001
Soluble sugar	Treatment	1	0.044	0.834	1	2.896	0.093
	Time	10	3.644	0.000	10	7.026	0.000
	Treatment × time	10	1.083	0.383	10	1.472	0.165
Starch	Treatment	1	0.870	0.355	1	0.054	0.817
	Time	10	12.738	0.000	10	22.253	0.000
	Treatment × time	10	2.070	0.041	10	5.191	0.000
Soluble sugar/starch	Treatment	1	0.842	0.361	1	0.248	0.620
	Time	10	7.668	0.000	10	6.872	0.000
	Treatment × time	10	0.634	0.780	10	0.714	0.708

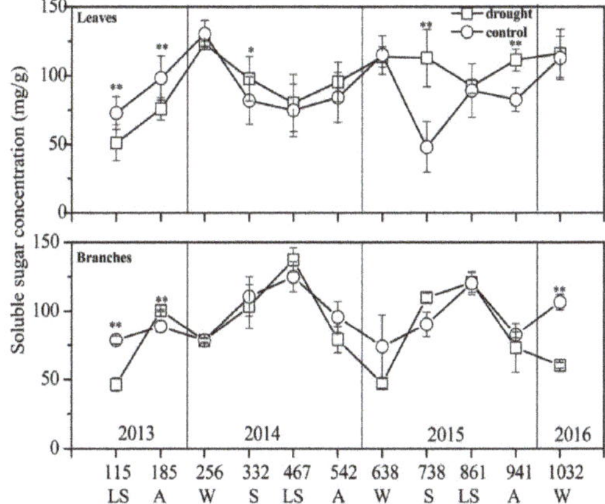

Figure 5. Soluble sugar concentrations in the leaves and branches of *Pinus massoniana* under control and drought treatments (mean ± SD). Samples were collected in late summer (LS), autumn (A), winter (W), and spring (S) from 2013 to 2016. Variables marked with asterisks indicate a significant difference between the control and drought groups (** $p < 0.01$, * $p < 0.05$).

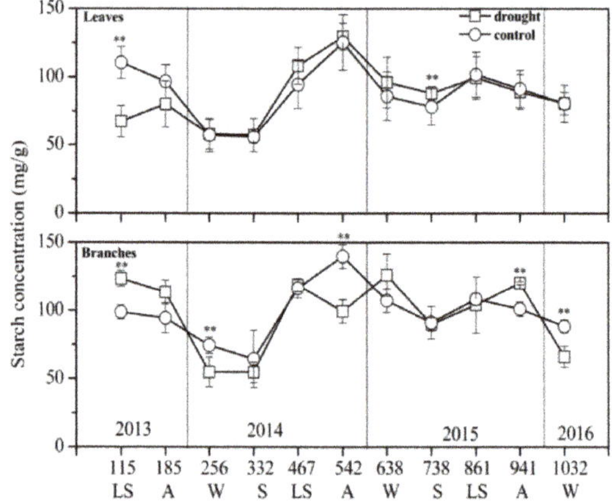

Figure 6. Starch concentrations in the leaves and branches of *Pinus massoniana* under control and drought treatments (mean ± SD). Samples were collected in late summer (LS), autumn (A), winter (W), and late spring (S) from 2013 to 2016. Variables marked with asterisks indicate a significant difference between the control and drought groups (** $p < 0.01$, * $p < 0.05$).

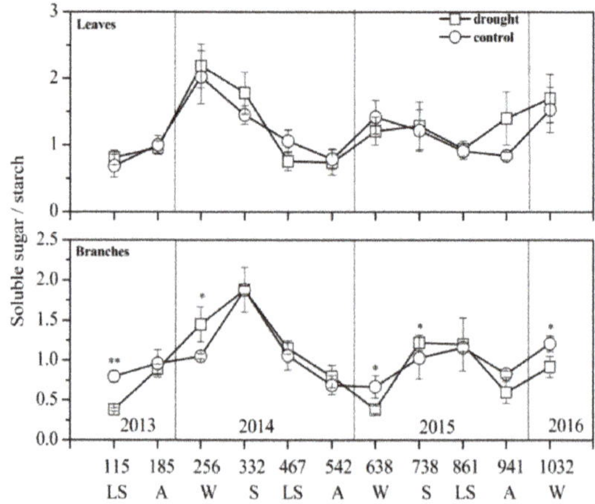

Figure 7. The ratio of soluble sugar to starch in leaves and branches of *Pinus massoniana* under control and drought treatments (mean ± SD). Samples were collected in late summer (LS), autumn (A), winter (W), and late spring (S) from 2013 to 2016. Variables marked with asterisks indicate a significant difference between the control and drought groups (** $p < 0.01$, * $p < 0.05$).

3.3. Relationship between NSC and Soil Moisture Content

In this study, negative linear correlations were found between deep soil (80 cm) moisture content and leaf NSC concentrations, and soluble sugar concentration in the drought treatment, respectively ($p < 0.05$; Figure 8). Leaf starch concentrations from days 115–542 showed negative linear correlations with deep soil (80 cm) ($p < 0.05$; Figure 9). While leaf starch concentration from days 638–1032 showed positive linear correlations with deep soil (80 cm) moisture content, the relationships were not significant, which is different from leaf starch concentration for days 115–1032.

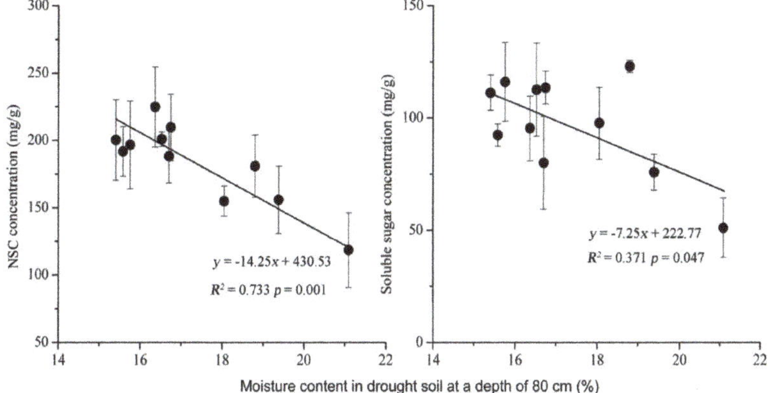

Figure 8. Relationships between NSC concentrations and soluble sugar in leaves and deep soil (80 cm) moisture content in the drought treatment.

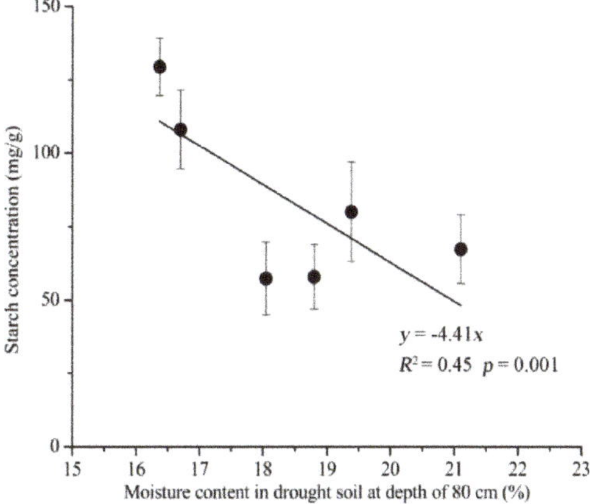

Figure 9. Relationships between concentrations of starch in leaves and deep soil (80 cm) moisture content in the drought treatment for days 115–542.

4. Discussion

4.1. Drought Effect on NSC Concentration

NSCs are considered one of the most critical aspects for tree survival under drought conditions [44]. Our study analyzed the time dynamics of NSCs in *P. massoniana* leaves and branches over a 3-year period, comparing six trees subjected to a drought treatment with six control trees. We found a common phenomenon: mild drought did not significantly alter carbon allocation to leaf and branch NSCs. Mild drought stress may lead to a slowdown in growth before any effects are observed on a carbohydrate level [31,33]. An important result of this drought simulation on leaf NSC dynamics was that NSC concentration significantly increased 89.73% in the early stages of drought (days 115–542) and decreased 14.7% in the late stages (days 542–1032), which was consistent with the conclusions provided by McDowell [26]. The reasons for this phenomenon are that in the early stages of drought stress, cellular turgidity restricted plant growth before photosynthesis decreased [45], and an imbalance in carbohydrate supply and demand caused an NSC surplus [27,28]. With continued drought and having a higher sensitivity to drought, photosynthesis decreased prior to respiration and carbohydrates produced by photosynthesis could no longer meet the normal physiological metabolic demands of trees, resulting in the consumption of stored NSCs [11,26,46].

From days 115 to 467, the *P. massoniana* in-branch NSC concentration in the drought treatment was similar to the control group, which decreased significantly. Compared with leaves, the NSC difference between treatments was observed from the beginning of the experiment, and the branches demonstrated a time lag phenomenon. This phenomenon suggests a reduced sensitivity of the branches to drought compared to leaves [47]. Branches may represent a carbon pool, promoting plant survival in a stressed environment.

When analyzing the leaf NSC seasonal dynamics, we found that phenology influenced NSCs, as indicated by the seasonal patterns of the control treatment [48]. The leaf NSC concentration in the control treatment gradually increased during the growing season, peaking when soil moisture content was lowest. This trend aligns with the standard conceptual model for NSCs, which states that the pool was depleted when demand exceeded supply and increased when the supply exceeded demand [49]. We propose three aspects to explain these phenomena: (1) NSCs increased during the growing season because the production of photoassimilates exceeded metabolic and growth demand for carbon; (2) carbohydrates draw down during the dormant season, when photosynthesis decreases and reserves are used to provide the energy required for normal metabolism [23,27]; and (3) during the growing season when water is deficient, the secondary metabolic processes that synthesize structural material were weakened and NSCs increased, resulting in peak NSC concentration during the dry season [36,50,51]. In terms of branches, the control group demonstrated a seasonal trend, but the consistency was not as obvious as for the leaves. Our sampling of leaf and branch NSCs provided an incomplete picture of the whole-tree NSC budget, but as the important tissues assimilating carbon, the leaves and branches could serve as an indicator of the NSC status of the entire tree.

4.2. Sugar Transfer and Transformation under Drought Stress

Transfer and transformation of NSCs in plants is an important adaptation mechanism to drought stress [24,52]. In our study, continued drought stress led to the increase in soluble sugar concentrations in leaves, and there was a significant negative linear correlation with soil moisture content. Many studies have agreed with this conclusion and stated that soluble sugar is an important aspect that helps adjust the osmotic potential of plant cells according to the available soil moisture [53]. The integrity of the conduit and a certain turgidity must be maintained to ensure the normal functioning of plant physiological processes under drought conditions [24]. When the soluble sugar concentration was low, starch was converted into sugar in order to resist stress [54]. However, a significant negative linear correlation between starch concentration and soil moisture content was observed from days 115–542, which is contrary to the results of many other studies [55–57]. Therefore, we further analyzed our results and found a positive linear correlation, although non-significant between days 638 and 1032. We have two reasons for analyzing this phenomenon: (1) In Chang Ting county, which is experiencing extensive soil erosion and is considered a severely degraded ecosystem, *P. massoniana* was a pioneer species for regional reforestation in the region. After nearly 25 years of cultivation, a unique ecological adaptation mechanism has developed: at the beginning of the mild drought, plant growth slows down, but the water retention in leaves can still maintain normal photosynthesis. Leaves act as carbon sources by assimilating carbon, and a normal metabolism is maintained, then many of the remaining carbohydrates are stored to resist stress enhancement at a later time. Owing to this, *P. massoniana* could maintain normal functioning [55–57] in the later period by decomposing starch to increase soluble sugar and the concentration of osmotic substances. (2) In the period between days 638 and 1032, the decrease of starch content was not significant. This may have occurred because this experimental period was during a mid-term drought, and as the isolation time increased, the decrease in starch became significant. However, up to now, the mid-term and long-term responses of stored NSC reserves to drought remained unclear.

Soluble sugar was not only used to resist stress, but also to adapt to changes influenced by seasonality. Analysis of the ratio of soluble sugar to starch in leaves and branches between different treatments showed that seasons affected the NSC storage dynamics. The ratio for *P. massoniana* leaves was similar to that reported for evergreen plants in China [58]. From summer to autumn, more starch was accumulated and converted into soluble sugar in winter and early spring to compensate for the loss of carbon due to decreasing photosynthetic capacity. However, few studies have been conducted on the branches; the ratio of soluble sugar to starch of branches was lower than found for *Larix gmelinii* [47]. Because of differences in terms of the length of the studies, no comparable data was available.

5. Conclusions

Drought events were associated a minimal NSC concentration change, potentially affecting the capacity of individuals to recover during mid-term mild drought [37]. In the sampled adult trees, a phenomenon where starch was transferred to soluble sugar to cope with drought stress was observed. During the drought process, leaves were the first to react, followed by branches, which showed a time lag effect. Our results highlight the importance of mid-term experiments to fully understand drought response, as our results differed from short-term studies [59]. Future global climate changes will directly affect NSC concentrations in plants, which in turn will affect physiological and biochemical processes [35,60]. Therefore, a better understanding of the controls and mechanisms of drought resistance is crucial for accurate predictions and mitigation strategies.

Author Contributions: T.L., H.Z. and J.Z. designed the experiment; T.L. and H.Z. carried out the field experiment; T.L. and J.W. performed the experiments in the lab; T.L. analyzed data; T.L. wrote the manuscript. T.L., Z.H. and J.W. revised and improved the manuscript.

Acknowledgments: This work was supported by National Natural Science Foundation of China (General Program: 31270659, 31200460) and the Natural Science Foundation of Fujian Province of China (Grant NO. 2016J01142). We thank Xiao-hui Zhong for her assistance in the field and lab. Then, we would also like to thank Alison Beamish at the University of British Columbia for her assistance with English language and grammatical editing of the manuscript.

Conflicts of Interest: The authors declare no conflict of interest.

References

1. Hartmann, H.; Ziegler, W.; Trumbore, S.; Knapp, A. Lethal drought leads to reduction in nonstructural carbohydrates in Norway spruce tree roots but not in the canopy. *Funct. Ecol.* **2013**, *27*, 413–427. [CrossRef]
2. Choat, B.; Jansen, S.; Brodribb, T.J.; Cochard, H.; Delzon, S.; Bhaskar, R.; Bucci, S.J.; Feild, T.S.; Gleason, S.M.; Hacke, U.G.; et al. Global convergence in the vulnerability of forests to drought. *Nature* **2012**, *491*, 752–755. [CrossRef] [PubMed]
3. Malone, S. Monitoring Changes in Water Use Efficiency to Understand Drought Induced Tree Mortality. *Forests* **2017**, *8*, 365. [CrossRef]
4. Peng, C.; Ma, Z.; Lei, X.; Zhu, Q.; Chen, H.; Wang, W.; Liu, S.; Li, W.; Fang, X.; Zhou, X. A drought-induced pervasive increase in tree mortality across Canada's boreal forests. *Nat. Clim. Chang.* **2011**, *1*, 467–471. [CrossRef]
5. Allen, C.D.; Macalady, A.K.; Chenchouni, H.; Bachelet, D.; McDowell, N.; Vennetier, M.; Kitzberger, T.; Rigling, A.; Breshears, D.D.; Hogg, E.H.; et al. A global overview of drought and heat-induced tree mortality reveals emerging climate change risks for forests. *For. Ecol. Manag.* **2010**, *259*, 660–684. [CrossRef]
6. Breshears, D.D.; Allen, C.D. The importance of rapid, disturbance-induced losses in carbon management and sequestration. *Glob. Ecol. Biogeogr.* **2002**, *11*, 1–5. [CrossRef]
7. Kuptz, D.; Fleischmann, F.; Matyssek, R.; Grams, T.E.E. Seasonal patterns of carbon allocation to respiratory pools in 60-yr-old deciduous (*Fagus sylvatica*) and evergreen (*Picea abies*) trees assessed via whole-tree stable carbon isotope labeling. *New Phytol.* **2011**, *191*, 160–172. [CrossRef] [PubMed]

8. Anderegg, W.R.L.; Kane, J.M.; Anderegg, L.D.L. Consequences of widespread tree mortality triggered by drought and temperature stress. *Nat. Clim. Chang.* **2012**, *3*, 30–36. [CrossRef]

9. Anderegg, W.R.L.; Anderegg, L.D.L. Hydraulic and carbohydrate changes in experimental drought-induced mortality of saplings in two conifer species. *Tree Physiol.* **2013**, *33*, 252–260. [CrossRef] [PubMed]

10. Sloan, J.L.; Jacobs, D.F. Leaf physiology and sugar concentrations of transplanted Quercus rubra seedlings in relation to nutrient and water availability. *New For.* **2012**, *43*, 779–790. [CrossRef]

11. Sevanto, S.; McDowell, N.G.; Dickman, L.T.; Pangle, R.; Pockman, W.T. How do trees die? A test of the hydraulic failure and carbon starvation hypotheses. *Plant Cell Environ.* **2014**, *37*, 153–161. [CrossRef] [PubMed]

12. McDowell, N.G.; Ryan, M.G.; Zeppel, M.J.B.; Tissue, D.T. Feature: Improving our knowledge of drought-induced forest mortality through experiments, observations, and modeling. *New Phytol.* **2013**, *200*, 289–293. [CrossRef] [PubMed]

13. Dai, G.; Yang, J.; Huang, C.; Sun, C.; Jia, L.; Ma, L. The Effects of Climate Change on the Development of Tree Plantations for Biodiesel Production in China. *Forests* **2017**, *8*, 207. [CrossRef]

14. McDowell, N.; Pockman, W.T.; Allen, C.D.; Breshears, D.D.; Cobb, N.; Kolb, T.; Plaut, J.; Sperry, J.; West, A.; Williams, D.G.; et al. Mechanisms of plant survival and mortality during drought: Why do some plants survive while others succumb to drought? *New Phytol.* **2008**, *178*, 719–739. [CrossRef] [PubMed]

15. Sperry, J.S. Hydraulic constraints on plant gas exchange. *Agric. For. Meteorol.* **2000**, *104*, 13–23. [CrossRef]

16. Adams, H.D.; Guardiola-Claramonte, M.; Barron-Gafford, G.A.; Villegas, J.C.; Breshears, D.D.; Zou, C.B.; Troch, P.A.; Huxman, T.E. Temperature sensitivity of drought-induced tree mortality portends increased regional die-off under global-change-type drought. *Proc. Natl. Acad. Sci. USA* **2009**, *106*, 7063–7066. [CrossRef] [PubMed]

17. Mitchell, P.J.; O'Grady, A.P.; Tissue, D.T.; White, D.A.; Ottenschlaeger, M.L.; Pinkard, E.A. Drought response strategies define the relative contributions of hydraulic dysfunction and carbohydrate depletion during tree mortality. *New Phytol.* **2013**, *197*, 862–872. [CrossRef] [PubMed]

18. Nardini, A.; Battistuzzo, M.; Savi, T. Shoot desiccation and hydraulic failure in temperate woody angiosperms during an extreme summer drought. *New Phytol.* **2013**, *200*, 322–329. [CrossRef] [PubMed]

19. Galvez, D.A.; Landhäusser, S.M.; Tyree, M.T. Low root reserve accumulation during drought may lead to winter mortality in poplar seedlings. *New Phytol.* **2013**, *198*, 139–148. [CrossRef] [PubMed]

20. Piper, F.I. Drought induces opposite changes in the concentration of non-structural carbohydrates of two evergreen Nothofagus species of differential drought resistance. *Ann. For. Sci.* **2011**, *68*, 415–424. [CrossRef]

21. Galiano, L.; Martínez-Vilalta, J.; Lloret, F. Carbon reserves and canopy defoliation determine the recovery of Scots pine 4 yr after a drought episode. *New Phytol.* **2011**, *190*, 750–759. [CrossRef] [PubMed]

22. Adams, H.D.; Zeppel, M.J.; Anderegg, W.R.; Hartmann, H.; Landhäusser, S.M.; Tissue, D.T.; Huxman, T.E.; Hudson, P.J.; Franz, T.E.; Allen, C.D.; et al. A multi-species synthesis of physiological mechanisms in drought-induced tree mortality. *Nat. Ecol. Evol.* **2017**, *1*, 1285–1291. [CrossRef] [PubMed]

23. Chapin, F.S.; Schulze, E.; Mooney, H.A. The Ecology and Economics of Storage in Plants. *Annu. Rev. Ecol. Syst.* **1990**, *21*, 423–447. [CrossRef]

24. Sala, A.; Woodruff, D.R.; Meinzer, F.C. Carbon dynamics in trees: Feast or famine? *Tree Physiol.* **2012**, *32*, 764–775. [CrossRef] [PubMed]

25. Sala, A.; Piper, F.; Hoch, G. Physiological mechanisms of drought-induced tree mortality are far from being resolved. *New Phytol.* **2010**, *186*, 274–281. [CrossRef] [PubMed]

26. McDowell, N.G.; Beerling, D.J.; Breshears, D.D.; Fisher, R.A.; Raffa, K.F.; Stitt, M. The interdependence of mechanisms underlying climate-driven vegetation mortality. *Trends Ecol. Evol.* **2011**, *26*, 523–532. [CrossRef] [PubMed]

27. Korner, C. Carbon limitation in trees. *J. Ecol.* **2003**, *91*, 4–17. [CrossRef]

28. Ayub, G.; Smith, R.A.; Tissue, D.T.; Atkin, O.K. Impacts of drought on leaf respiration in darkness and light in *Eucalyptus saligna* exposed to industrial-age atmospheric CO_2 and growth temperature. *New Phytol.* **2011**, *190*, 1003–1018. [CrossRef] [PubMed]

29. McCree, K.J.; Kallsen, C.E.; Richardson, S.G. Carbon Balance of Sorghum Plants during Osmotic Adjustment to Water Stress. *Plant Physiol.* **1984**, *76*, 898–902. [CrossRef] [PubMed]

30. McDowell, N.G. Mechanisms Linking Drought, Hydraulics, Carbon Metabolism, and Vegetation Mortality. *Plant Physiol.* **2011**, *155*, 1051–1059. [CrossRef] [PubMed]

31. Millard, P.; Sommerkorn, M.; Grelet, G.-A. Environmental change and carbon limitation in trees: A biochemical, ecophysiological and ecosystem appraisal. *New Phytol.* **2007**, *175*, 11–28. [CrossRef] [PubMed]

32. Regier, N.; Streb, S.; Cocozza, C.; Schaub, M.; Cherubini, P.; Zeeman, S.C.; Frey, B. Drought tolerance of two black poplar (*Populus nigra* L.) clones: Contribution of carbohydrates and oxidative stress defence. *Plant Cell Environ.* **2009**, *32*, 1724–1736. [CrossRef] [PubMed]

33. Woodruff, D.R.; Meinzer, F.C. Water stress, shoot growth and storage of non-structural carbohydrates along a tree height gradient in a tall conifer. *Plant Cell Environ.* **2011**, *34*, 1920–1930. [CrossRef] [PubMed]

34. Quentin, A.G.; Pinkard, E.A.; Ryan, M.G.; Tissue, D.T.; Baggett, L.S.; Adams, H.D.; Maillard, P.; Marchand, J.; Landhäusser, S.M.; Lacointe, A.; et al. Non-structural carbohydrates in woody plants compared among laboratories. *Tree Physiol.* **2015**, *35*, 1146–1165. [CrossRef] [PubMed]

35. Hoch, G.; Richter, A.; Korner, C. Non-structural carbon compounds in temperate forest trees. *Plant Cell Environ.* **2003**, *26*, 1067–1081. [CrossRef]

36. Würth, M.K.R.; Peláez-Riedl, S.; Wright, S.J.; Körner, C. Non-structural carbohydrate pools in a tropical forest. *Oecologia* **2004**, *143*, 11–24. [CrossRef] [PubMed]

37. Galiano, L.; Martinez-Vilalta, J.; Sabate, S.; Lloret, F. Determinants of drought effects on crown condition and their relationship with depletion of carbon reserves in a *Mediterranean holm* oak forest. *Tree Physiol.* **2012**, *32*, 478–489. [CrossRef] [PubMed]

38. Lacointe, A.; Kajji, A.; Daudet, F.-A.; Archer, P.; Frossard, J.-S.; Saint-Joanis, B.; Vandame, M. Mobilization of carbon reserves in young walnut trees. *Acta Bot. Gallica* **1993**, *140*, 435–441. [CrossRef]

39. Ryan, M.G. Tree responses to drought. *Tree Physiol.* **2011**, *31*, 237–239. [CrossRef] [PubMed]

40. Dietze, M.C.; Sala, A.; Carbone, M.S.; Czimczik, C.I.; Mantooth, J.A.; Richardson, A.D.; Vargas, R. Nonstructural Carbon in Woody Plants. *Annu. Rev. Plant Biol.* **2014**, *65*, 667–687. [CrossRef] [PubMed]

41. Boschetto, R.G.; Mohamed, R.M.; Arrigotti, J. Vulnerability to Desertification in a Sub-Saharan. Region: A First Local Assessment in Five Villages, of Southern Region of Malawi. *Ital. J. Agron.* **2010**, *3*, 91–101. [CrossRef]

42. Yemm, E.W.; Willis, A.J. The estimation of carbohydrates in plant extracts by anthrone. *Biochem. J.* **1954**, *57*, 508–514. [CrossRef] [PubMed]

43. Zhang, T.; Cao, Y.; Chen, Y.; Liu, G. Non-structural carbohydrate dynamics in Robinia pseudoacacia saplings under three levels of continuous drought stress. *Trees* **2015**, *29*, 1837–1849. [CrossRef]

44. Rosas, T.; Galiano, L.; Ogaya, R.; Peñuelas, J.; Martínez-Vilalta, J. Dynamics of non-structural carbohydrates in three Mediterranean woody species following long-term experimental drought. *Front. Plant Sci.* **2013**, *4*, 400. [CrossRef] [PubMed]

45. Bogeat-Triboulot, M.B.; Brosche, M.; Renaut, J.; Jouve, L.; Le Thiec, D.; Fayyaz, P.; Vinocur, B.; Witters, E.; Laukens, K.; Teichmann, T.; et al. Gradual Soil Water Depletion Results in Reversible Changes of Gene Expression, Protein Profiles, Ecophysiology, and Growth Performance in Populus euphratica, a Poplar Growing in Arid Regions. *Plant Physiol.* **2006**, *143*, 876–892. [CrossRef] [PubMed]

46. Klein, T.; Rotenberg, E.; Cohen-Hilaleh, E.; Raz-Yaseef, N.; Tatarinov, F.; Preisler, Y.; Ogée, J.; Cohen, S.; Yakir, D. Quantifying transpirable soil water and its relations to tree water use dynamics in a water-limited pine forest. *Ecohydrology* **2014**, *7*, 409–419. [CrossRef]

47. Du, Y.; Han, Y.; Wang, C.K. The influence of drought on non-structural carbohydrates in the needles and twigs of *Larix gmelinii*. *Acta Ecol. Sin.* **2014**, *34*, 6090–6099. (In Chinese)

48. Martínez-Vilalta, J.; Sala, A.; Asensio, D.; Galiano, L.; Hoch, G.; Palacio, S.; Piper, F.I.; Lloret, F. Dynamics of non-structural carbohydrates in terrestrial plants: A global synthesis. *Ecol. Monogr.* **2016**, *86*, 495–516. [CrossRef]

49. Richardson, A.D.; Carbone, M.S.; Keenan, T.F.; Czimczik, C.I.; Hollinger, D.Y.; Murakami, P.; Schaberg, P.G.; Xu, X. Seasonal dynamics and age of stemwood nonstructural carbohydrates in temperate forest trees. *New Phytol.* **2013**, *197*, 850–861. [CrossRef] [PubMed]

50. Tissue, D.T.; Wright, S.J. Effect of Seasonal Water Availability on Phenology and the Annual Shoot Carbohydrate Cycle of Tropical Forest Shrubs. *Funct. Ecol.* **1995**, *9*, 518–527. [CrossRef]

51. Latt, C.R.; Nair, P.K.R.; Kang, B.T. Reserve carbohydrate levels in the boles and structural roots of five multipurpose tree species in a seasonally dry tropical climate. *For. Ecol. Manag.* **2001**, *146*, 145–158. [CrossRef]

52. Zeppel, M.J.B.; Adams, H.D.; Anderegg, W.R.L. Mechanistic causes of tree drought mortality: Recent results, unresolved questions and future research needs. *New Phytol.* **2011**, *192*, 800–803. [CrossRef] [PubMed]

53. Iannucci, A.; Russo, M.; Arena, L.; Di Fonzo, N.; Martiniello, P. Water deficit effects on osmotic adjustment and solute accumulation in leaves of annual clovers. *Eur. J. Agron.* **2002**, *16*, 111–122. [CrossRef]

54. Nguyen, P.V.; Dickmann, D.I.; Pregitzer, K.S.; Hendrick, R. Late-season changes in allocation of starch and sugar to shoots, coarse roots, and fine roots in two hybrid poplar clones. *Tree Physiol.* **1990**, *7*, 95–105. [CrossRef] [PubMed]

55. Fischer, C.; Höll, W. Food reserves of Scots pine (*Pinus sylvestris* L.). *Trees* **1991**, *5*, 187–195. [CrossRef]

56. Zwiazek, J.J. Cell wall changes in white spruce (*Picea glauca*) needles subjected to repeated drought stress. *Physiol. Plant.* **1991**, *82*, 513–518. [CrossRef]

57. Amundson, R.G.; Kohut, R.J.; Laurence, J.A.; Fellows, S.; Colavito, L.J. Moderate water stress alters carbohydrate content and cold tolerance of red spruce foliage. *Environ. Exp. Bot.* **1993**, *33*, 383–390. [CrossRef]

58. Li, N.; He, N.; Yu, G. Non-structural carbohydrates in leaves of tree species from four typical forest in China. *Acta Bot. Boreal-Occident. Sin.* **2015**, *35*, 1846–1854. (In Chinese)

59. Hollister, R.D.; Webber, P.J.; Tweedie, C.E. The response of Alaskan arctic tundra to experimental warming: Differences between short- and long-term responses. *Glob. Chang. Biol.* **2005**, *11*, 525–536. [CrossRef]

60. Ericsson, T.; Rytter, L.; Vapaavuori, E. Physiology of carbon allocation in trees. *Biomass Bioenergy* **1996**, *11*, 115–127. [CrossRef]

 *forests*

Article

Nitrogen Nutrition of European Beech Is Maintained at Sufficient Water Supply in Mixed Beech-Fir Stands

Ruth-Kristina Magh [1,*], Fengli Yang [1,2], Stephanie Rehschuh [3], Martin Burger [1], Michael Dannenmann [3], Rodica Pena [4], Tim Burzlaff [1], Mladen Ivanković [5] and Heinz Rennenberg [1,6]

[1] Institute of Forest Sciences, Chair of Tree Physiology, University of Freiburg, Georges-Koehler-Allee 54/54, 79110 Freiburg, Germany; fengli.yang@ctp.uni-freiburg.de (F.Y.); burger@gmx.de (M.B.); tim.burzlaff@ctp.uni-freiburg.de (T.B.); heinz.rennenberg@ctp.uni-freiburg.de (H.R.)

[2] College of Urban and Rural Development and Planning, Mianyang Normal University, Xianren Road 30, Mianyang 621000, China

[3] Institute of Meteorology and Climate Research, Atmospheric Environmental Research (IMK-IFU), Karlsruhe Institute of Technology (KIT), Kreuzeckbahnstrasse 19, 82467 Garmisch-Partenkirchen, Germany; stephanie.rehschuh@kit.edu (S.R.); michael.dannenmann@kit.edu (M.D.)

[4] Büsgen-Institute, Department of Forest Botany and Tree Physiology, University of Göttingen, Büsgenweg 2, 37077 Göttingen, Germany; Rodica.Pena@forst.uni-goettingen.de

[5] Hrvatski Šumarski Institute, Zavod za Genetiku, Oplemenjivanje Šumskog Drveća i Sjemenarstvo Cvjetno naselje 41, 10450 Jastrebarsko, Croatia; mladeni@sumins.hr

[6] College of Sciences, King Saud University, P.O. Box 2455, Riyadh 11451, Saudi Arabia

* Correspondence: ruth.magh@ctp.uni-freiburg.de; Tel.: +49-761-203-96824

Received: 17 October 2018; Accepted: 21 November 2018; Published: 23 November 2018

Abstract: Research highlights: Interaction effects of coniferous on deciduous species have been investigated before the background of climate change. Background and objectives: The cultivation of European beech (*Fagus sylvatica* L.) in mixed stands has currently received attention, since the future performance of beech in mid-European forest monocultures in a changing climate is under debate. We investigated water relations and nitrogen (N) nutrition of beech in monocultures and mixed with silver-fir (*Abies alba* Mill.) in the Black Forest at different environmental conditions, and in the Croatian Velebit at the southern distribution limit of beech, over a seasonal course at sufficient water availability. Material and methods: Water relations were analyzed via $\delta^{13}C$ signatures, as integrative measures of water supply assuming that photosynthesis processes were not impaired. N nutrition was characterized by N partitioning between soluble N fractions and structural N. Results: In the relatively wet year 2016, water relations of beech leaves, fir needles and roots differed by season, but generally not between beech monocultures and mixed cultivation. At all sites, previous and current year fir needles revealed significantly lower total N contents over the entire season than beech leaves. Fir fine roots exhibited higher or similar amounts of total N compared to needles. Correlation analysis revealed a strong relationship of leaf and root $\delta^{13}C$ signatures with soil parameters at the mixed beech stands, but not at pure beech stands. While glutamine (Gln) uptake capacity of beech roots was strongly related to soil N in the monoculture beech stands, arginine (Arg) uptake capacities of beech roots were strongly related to soil N in mixed stands. Conclusions: Leaf N contents indicated a facilitative effect of silver-fir on beech on sites where soil total N concentrations where low, but an indication of competition effect where it was high. This improvement could be partially attributed to protein contents, but not to differences in uptake capacity of an individual N source. From these results it is concluded that despite similar performance of beech trees at the three field sites investigated, the association with silver-fir mediated interactive effects between species association, climate and soil parameters even at sufficient water supply.

Keywords: *Fagus sylvatica* L.; *Abies alba* Mill.; N nutrition; mixed stands; pure stands; soil N; water relations

1. Introduction

European beech (*Fagus sylvatica* L.) and silver-fir (*Abies alba* MILL.) possess a similar migration history after the last glaciation into Central Europe. Both species found refuge areas in the south, southwest, and southeast of Europe in order to survive the last glacial period [1]. Silver-fir migrated into Central Europe from three of its five refugia, one located in the Balkans, one in Central Italy, and the third in Central/Eastern France [2]. The locations of beech refugia are still a matter of debate, but presumably also beech migrated from several refuge areas into Central Europe, mainly from a population in the Alpine-Slovenia-Istria region [3]. The populations originating from the Calabrian refugium seem to be restricted to Italy [4]. Today, European beech is a key species in Central European forests, where it dominates under many climatic conditions [5], and is the most spread of all beech species [6]. Its climate optimum allowed the colonization of large areas all over Europe, only limited by excessive water supply in very moist habitats, or restricted water supply in the south, where its populations decrease in favor of conifers and oaks under dry and warm conditions [7]. Silver-fir occurs mainly in mountainous regions scattered all over Europe, and is thought to restrain its distribution range to cool and moist climate conditions [2,7–9], though this belief is questioned to date [10]. Mixed beech-fir forests ('*Abieti-Fagetum*' [5]) constituted the typical association in the Black Forest in Southern Germany before the 18th century [11]. Today, these mixed forests are found mainly at the southern distribution limit of beech, i.e., in the mountainous regions of the Dinaric Alps or the Iberian Peninsula [12].

The quasi-monoculture beech forests found in Germany today are the result of a combination of anthropogenic and natural processes. To counteract the promotion of pure plantations of Norway spruce (*Picea abies* (L.) KARST), beech plantations have been favored by forestry practices due to its economic benefits. For example, beech forests are promoted in the state of Baden-Württemberg, Germany, towards a desired extend of 30% of the total forested area [13]. Since beech trees are very competitive, they can occupy between 80 and 100% of the canopy area in forests [14]. Additionally, beech seedlings can grow in the understory of a closed canopy by adapting their photosynthetic performance [15,16] and by increasing their specific leaf area (SLA) to increase light capture and carbon gain [17–19], an indication of high shade tolerance. Thus, adult beeches favor their offspring by enabling its growth under low light conditions for a prolonged period of time [20–22]. However, if a "forest gap" opens, juvenile beech trees can react in the subsequent growing season by the development of leaves adapted to high light conditions and, hence, increase growth [19,23]. These features led to a coverage of large areas of Central Europe with beech forests, where moderate site conditions apply [3].

Despite the dominance of beech in Central European forests, numerous studies showed that beech trees are relatively drought-sensitive [7,24,25]. The distribution limit of beech is found in regions where long severe winters and summer drought arise [26]. As a consequence, the Balkan Peninsula constitutes the southernmost distribution area of beech, where it is likely to perform at its physiological limit. Under the auspices of climate change, we can expect forest responses towards heat, drought, and extreme events especially in this region [27–29], with implications for nutrient acquisition [30], forest productivity [31,32], and forest dieback [33–35].

In a changing climate with increased frequencies of "drying-wetting-cycles", the future performance of beech forests in Central Europe is questioned [36–38]. To cope with a changing climate, the development of adaption approaches is thought to be required fostering compositional, functional and/or structural complexity of forests [39,40]. To meet this aim, in the Black Forest in southwest Germany quasi-monoculture stands of beech are currently diversified by re-introducing naturally occurring species such as silver-fir (*Abies alba* Mill.) [39,41]. This species combination is frequently found in mountain forests of Europe forming natural associations with varying beech:fir ratios also at the southern distribution limit of beech [42]. Under these conditions, silver-fir was found to lose its sensitivity to summer droughts under nutrient limitation [43] and is thought to benefit from enhanced water supply provided by increased stem-flow and concentrated infiltration of water at the stem base of beech [44,45]. Vice versa, relatively shallow-rooting beech trees may benefit from the association with deep-rooting fir under drought conditions from the access of fir to additional water sources [46]. However, facilitative effects such as improved water supply by silver-fir's taproot via hydraulic lift [47] during drought might be superimposed by competition effects during years of sufficient water supply, like aboveground competition for light [48,49], or belowground competition for nutrients [50,51].

To address the extend of the interaction effect of beech and fir, we analyzed water relations and N nutrition of beech at three forest sites, naturally differing in nutrient supply and climate in a wet year that provided sufficient water to the stands at all sites. Two forest sites were located in the "Black Forest" in southwest Germany and one at the southern distribution limit of beech in the Dinaric Alps in Croatia. We investigated the tree vegetation over a seasonal course, trying to link the water and nutrient availability in the soil to the performance of the investigated stands. We hypothesized that even at sufficient water supply (1) the performance of beech and silver-fir differed between the forest site at its southern distribution limit and the two stands in the "Black Forest"; (2) an interactive effect of firs on the performance of beech can be identified by comparing N partitioning and acquisition of beeches in mixed versus pure beech stands; (3) soil N contents, climate, and species associations interactively affect the performance of beech also at sufficient water supply.

2. Materials and Methods

2.1. Field Sites and Experimental Approaches

For the present experiments, three field sites were chosen differing in N supply and water holding capacity, i.e., the Freiamt (EM) and Conventwald (CO) sites in Germany and a field site in Croatia (CR) (Table S5). The EM field site is located in the Black Forest area at approx. (approximately) 400 m a.s.l. (above sea level). The mean annual air temperature is 9.6 °C and the annual precipitation amounts to 1100 mm. The soil parent material is sandstone and is characterized as Dystric Cambisol with a mean field capacity of approx. 18 vol%. The average soil depth amounts to 80–100 cm. The vegetation consists of 70% of beech, 15% silver-fir, and 15% larch.

The CO site is located 15 km from the EM site at ca. 700 m a.s.l., revealing a mean annual air temperature of 7.3 °C and an annual precipitation of 1777 mm. The soil is classified as hyperdystric skeletic folic Cambisol developed on paragneiss. The mean field capacity of approx. 19 vol% is similar to the EM site. Soil depth at the CO site is very heterogenic and varies between 50 and 100 cm, with a high amount of skeleton. Vegetation composition differed between the two sub-sites used for the present study. In the "pure beech" sub-site it consists of approx. 55% beech, 10% silver-fir, and 35% spruce. In the "mixed forest" sub-site we found approx. 35% beech, 35% silver-fir, 5% spruce, 10% douglas fir, and 5% larch.

The CR field site is located near the city of Gospić in the Velebit Mountains at an elevation of approx. 900 m a.s.l. (Table S5). The mean annual air temperature amounts to 7.5 °C and the annual precipitation to ca. 2230 mm (577 mm during the vegetation period, climate station Baške Oštarije, 924 m; 1987–2010). The soil developed on limestone parent bedrock and is classified as Chromic Cambisol, exhibiting a field capacity of ~33 vol%. The average soil depth amounts to ca. 45 cm. As for the CO site, two sub-sites were included in the studies at the CR site. The vegetation composition surrounding the "pure beech" sub-site consisted of 57.6% beech, 40% silver-fir and 3% others; the "mixed forest" sub-site was composed of 38% beech, 50% silver-fir and 12% other deciduous species (personal communication, forest direction of Gospić).

For the experiments at each site, beeches were chosen growing either surrounded by beeches as neighboring trees (i.e., "pure stands"), or with silver-firs as the neighboring species (i.e., mixed stands).

2.2. Soil Sampling and Analyses

For the CO site, soil characterization and other soil data were provided by the University of Freiburg, Chair of Soil Ecology (Lang and Krüger, personal communication). For soil characterization, soil of the field sites EM and CR was sampled in July 2016, to a depth of approx. 100 cm and 45 cm, respectively, corresponding to the beginning of soil parent rock material. For the EM site three pits were dug and three replicas were taken from each horizon, with ten further drillings around every pit to account for spatial variability in topsoil. For the CR site a total of 10 pits—including at least three pits under pure beech, three under pure fir as well as three under a mix of beech and fir, each—were dug and three replicated samples were taken from each horizon. All samples were combined to one sample per horizon and pit and were consequently air-dried. Soil texture was determined by wet sieving and sedimentation (DIN ISO 11277) in a commercial laboratory (Agrolab Group, Sarstedt, Germany).

To quantify SOC and TN stocks, soil samples of 100 cm^3 were taken with a soil corer in five to eight replicates per soil pit and horizon and used for the determination of soil bulk density via drying at 105 °C for 24 h. For total carbon analysis, as well as for carbonate and total nitrogen (TN) investigation, soil was sieved to 2 mm, grinded and the contents were analyzed in a commercial laboratory (Dr. Janssen's laboratory, Gillersheim, Germany), using VDLUFA (Verband Deutscher Landwirtschaftlicher Untersuchungs- und Forschungsanstalten) Method IA 5.3.1 for carbonate, DIN ISO 10694 for total carbon, and DIN ISO 13878 for TN. Soil organic carbon (SOC) was calculated as the difference between total carbon and carbonate-C content. For the analysis of extractable ammonium and nitrate concentrations in EM and CO soil, three replicated samples (30 g) of the Ah horizon and the Bv horizon as well as leaf litter samples (10 g) were extracted with 0.5 M K_2SO_4 (Merck KGaA, Darmstadt, Germany) [52]. For CR, 30 replicated samples of the Ah and Bv horizon as well as leaf litter were extracted with 1 M KCl (Merck KGaA, Darmstadt, Germany). Tests showed that for Chromic Cambisol, extraction with 1 M KCl revealed the same inorganic N concentrations compared to extraction with 0.5 M K_2SO_4. For the extractions, samples were shaken with the extraction solution for one hour in 250 mL plastic bottles (Carl Roth GmbH, Karlsruhe, Germany) at 170 rpm. Subsequently, extracts were filtered using vacuum pumps, glass fiber filters (Whatman GF/A, Springfield, UK), and finally syringe filters (0.45 µm) (Schleicher and Schuell, Dassel, Germany) and frozen immediately. Concentrations of NH_4^+ and NO_3^- were analyzed colorimetrically by a commercial laboratory (Dr. Janssen's laboratory, Gillersheim, Germany) using the VDLUFA method A 6141 [53].

Field capacity was estimated as the soil volumetric water content (VWC) two days after rain events in spring. Volumetric water content was gained from continuous measurements using TDR (Time Domain Reflectometry, 5TM, and GS1 sonsors, Decagon Devices, Inc., Pullman, WA, USA) probes—calibrated by gravimetric measurements—at a depth of 5 and 40 cm for the CR site. The average VWC for the CR site was calculated from seven sensors in 5 cm depth and five sensors in 40 cm depth. For the EM site, field capacity was calculated as the water held against gravity [54]. Further description of the latter method can be found elsewhere [46].

2.3. Sampling of Plant Material

At each field site, European beech trees (*Fagus sylvatica* L.) and silver-firs (*Abies alba* MILL.) of similar size and age were selected. Sampling was conducted three times over the growing season, the first time after beech leaves were just fully developed (i.e., "Spring"), a second time during a hotter and drier period ca. six to eight weeks later (i.e., "Summer"), and a third time just after some beech leaves started to change color (though only still intact and green leaves were sampled; i.e., "Fall"). At each sampling campaign, branches facing the adjacent trees in the upper surface of the crown, of approximately 30–45 cm in length, were collected by professional tree climbers equipped with a throw line and pole saws. To limit diurnal variation and variation by light availability, branches were harvested between 10 a.m. to 2 p.m. at approximately the same tree height.

From each beech twig, three fully developed, intact leaves were collected; from twigs of silver firs, previous-year needles were excised. To collect root samples, fine roots (<2 mm diameter) were carefully removed from the soil near the tree trunk, rinsed with demineralized water, and dried with paper tissues. Subsamples of leaves/needles and roots were immediately shock-frozen in liquid nitrogen in the field. For further analysis, samples were homogenized with mortar and pestle under liquid nitrogen and stored at $-20\ °C$. The rest of the samples were dried in the oven at $60\ °C$ for approx. 3 days until weight constancy. Hydration state of leaf and root material was calculated as the difference between fresh weight (FW) and dry weight (DW) divided by the DW [55].

2.4. Element and Stable Isotope Analyses of C and N

Total N and total C contents as well as $\delta^{15}N$ and $\delta^{13}C$ signatures were measured as previously described elsewhere [56]. For this purpose, oven dried samples were ground with a ball mill (Retsch MM 400, Retsch GmbH, Haan, Germany), aliquots (1.0–1.5 mg) were loaded into tin capsules (IVA Analysentechnik, Meerbusch, Germany) and measured in an isotope ratio mass spectrometer (Delta V Plus, Thermo Finnigan MAT, GmbH, Bremen, Germany) coupled via a Conflo III interface with an element analyser (NA2500, CE Instruments, Milan, Italy). Working standards (glutamic acid) were calibrated against the primary standards of the U.S. Geological Survey 40 (USGS 40, glutamic acid $\delta^{13}C_{PDB} = -26.39‰$) and USGS 41 (glutamic acid $\delta^{13}C_{PDB} = 37.63‰$) for $\delta^{13}C$, and USGS 25 (ammonium sulphate, $\delta^{15}N_{Air} = -30.4‰$) and USGS 41 (glutamic acid $\delta^{15}N_{Air} = 47.600‰$) for $\delta^{15}N$ and analyzed after every tenth sample to account for a potential instrument drift over time as described by Simon et al. [21].

2.5. Protein and Amino Acid Analyses

Soluble protein in plant material was determined using a modification of the method described by Du et al. [57]. Approximately 50 mg frozen homogenized leaf/needle or 80 mg root material was mixed with double the amount of polyvinylpolypyrrolidone (PVPP, Sigma-Aldrich Inc., St. Louis, MO, USA) and extracted with 1 mL Tris/HCl-Buffer (50 mM Tris-Cl (Sigma-Aldrich Inc., St. Louis, MO, USA), 1 mM EDTA (Sigma-Aldrich Inc., St. Louis, MO, USA), 1 mM DTT (Sigma-Aldrich Inc., St. Louis, MO, USA), 100 μM PMSF (Sigma-Aldrich Inc., St. Louis, MO, USA), 15% Glycerol (Sigma-Aldrich Inc., St. Louis, MO, USA) (v/v), 0.1% Triton-100 (Sigma-Aldrich Inc., St. Louis, MO, USA) (v/v), pH 8.0) by vortexing and incubation for 30 min at $4\ °C$. After centrifugation for 10 min at $12,000\times g$ and $4\ °C$, 500 μL aliquots of the supernatant were transferred into new tubes, mixed with 500 μL 10% trichloroacetic acid (TCA, Merck KGaA, Darmstadt, Germany), vortexed and incubated for 10 min at room temperature. After 10 min centrifugation at $12,000\times g$ and $4\ °C$, the pellet was dissolved in 500 μL 1 M KOH (Merck KGaA, Darmstadt, Germany) by shaking for 30 min at $4\ °C$. For quantification of total soluble protein of leaf material (i.e., leaves and needles), 5 μL aliquots of the extract were pipetted in triplicate into a disposable polystyrene micro cuvette plate and mixed with 200 μL Bradford reagent (Amresco, Solon, OH, USA). The mixtures were incubated for 10 min at room temperature and the optical density was measured at 595 nm using a microplate reader (Sunrise-basic Tecan, Grödlg, Austria). For root material 25 μL aliquots were mixed with 500 μL Bradford reagent. Optical density was measured in a spectrophotometer (Type DU 650, Beckman Coulter, Brea, CA, USA) at 595 nm. In both cases bovine serum albumin (BSA, Sigma-Aldrich, Taufkirchen, Germany) was used as a standard. Protein-N was calculated from the mean N content in BSA (15.9%).

The extraction of amino acids from plant material was conducted after the method of Winter et al. [58] and Hu et al. [59] with slight modifications. Approximately 50 mg of frozen leaf or 80 mg root powder were mixed with 1 mL methanol/chloroform (Merck KGaA, Darmstadt, Germany) (3.5/1.5, v/v) and 0.2 mL Hepes buffer (containing 20 mM Hepes (Merck KGaA, Darmstadt, Germany), 5 mM EGTA (Sigma-Aldrich Inc., St. Louis, MO, USA), 10 mM NaF (Sigma-Aldrich Inc., St. Louis, MO, USA)). After incubation on ice for 30 min, 0.6 mL distilled water was added and the mixture was centrifuged for 10 min at $14,000\times g$ at $4\ °C$. Aliquots of 1 mL of the supernatant were transferred into new tubes and stored on ice. Subsequently, the extraction of the plant material was repeated and 1 mL aliquots of the supernatants were combined. Aliquots of 0.1 mL of the combined supernatants and 0.1 mL ninhydrin solution (1:1 mixture of a solution containing 4.2 g monohydrate citric acid (Merck KGaA, Darmstadt, Germany) and 0.16g anhydrous stannous chloride (Sigma-Aldrich Inc., St. Louis, MO, USA) in 40 mL 1 M NaOH (Merck KGaA, Darmstadt, Germany), made up to 100 mL with distilled water (pH 5.0) and a solution containing 4 g ninhydrin (Sigma-Aldrich Inc., St. Louis, MO, USA) in 100 mL ethylene glycolmonomethyl ether (Merck KGaA, Darmstadt, Germany) were boiled for 30 min. Subsequently, 1.25 mL isopropanol (Merck KGaA, Darmstadt, Germany) (50%) were added and the samples were incubated for 15 min at room temperature in the dark. The optical density was determined with a UV-DU650 spectrophotometer (DU 650, Beckman Coulter, Brea, CA, USA) at 570 nm. L-glutamine (Sigma-Aldrich Inc., St. Louis, MO, USA) was used as standard.

2.6. Calculation of Structural N

Structural N in leaves/needles and roots was calculated by subtracting the N fractions of total amino acids and soluble proteins from total N. Calculations were based on the dry weight of the samples. Dannenmann et al. [60] as well as Simon et al. [21] found that inorganic N did not contribute significantly to total N in temperate forest trees and, therefore, inorganic N was neglected in this calculation.

2.7. N Uptake Experiments

Net N uptake experiments were conducted using the N-enrichment technique described by Gessler et al. [61]. Fine roots still attached to the tree were dug out, rinsed with distilled water to remove adhering soil particles, dried with cellulose paper and submersed into 4 mL of an artificial soil solution for 2 h between 10 a.m. and 2 p.m. to avoid diurnal variation in N uptake capacity [62]. The artificial soil solution resembled the mineral soil onsite in concentration of anions and cations as well as in pH; it contained 100 µM KNO_3 (Merck KGaA, Darmstadt, Germany), 90 µM $CaCl_2$ $2H_2O$ (Merck KGaA, Darmstadt, Germany), 70 µM $MgCl_2$ $6H_2O$ (Merck KGaA, Darmstadt, Germany), 50 µM KCl (Merck KGaA, Darmstadt, Germany), 24 µM $MnCl_2$ $4H_2O$ (Merck KGaA, Darmstadt, Germany), 20 µM NaCl (Merck KGaA, Darmstadt, Germany), 10 µM $AlCl_3$ (Merck KGaA, Darmstadt, Germany), 7 µM $FeSo_4$ $7H_2O$ (Merck KGaA, Darmstadt, Germany), 6 µM K_2HPO_4 (Merck KGaA, Darmstadt, Germany), 1 µM NH_4Cl (Merck KGaA, Darmstadt, Germany), as well as the amino acids glutamine (Sigma-Aldrich Inc., St. Louis, MO, USA) (25 µM) and arginine (Sigma-Aldrich Inc., St. Louis, MO, USA) (25 µM). Thus, the artificial soil solution consisted of four different N sources, ammonium (NH_4^+) and nitrate (NO_3^-) as inorganic and the amino acids glutamine (Gln) and arginine (Arg) as organic N sources. The latter constitute the most abundant amino compounds in beech roots [63]. Uptake experiments were carried out with only one of the four N sources being offered as ^{15}N-labeled compound ($K^{15}NO_3$, $^{15}NH_4Cl$ and double labeled $^{15}N^{13}C$-Gln/$^{15}N^{13}C$-Arg, Cambridge Isotope Laboratories, Inc., Andover, MA, USA) with three replicates per N-labeled source and tree. In addition, control solutions without label were applied to account for the natural abundance of ^{15}N in the roots. After being submersed for 2 h, the roots were cut off, washed twice with 0.5 M $CaCl_2$ (Merck KGaA, Darmstadt, Germany), dried with cellulose paper, and dried in an oven at 60 °C for 48 h. Subsequently, the samples were ground and 1.2–2 mg powder was weighed into tin capsules (IVA Analysetechnik, Meerbusch, Germany). Samples were analyzed with an element analyzer (NA 1108, Fisons, Rodano, Italy, or NA 1110, CE Instruments, Milan, Italy) coupled via a Conflo III interface to an isotope ratio mass spectrometer (Delta Plus, Thermo Finnigan MAT GmbH, Bremen, Germany). The working standard glutamic acid, calibrated against the international standard USGS 41 ($\delta^{15}N_{air}$ = 47.600) for $\delta^{15}N$, was measured after each tenth sample. For calculation of the net N uptake capacity (NUC), Equation (1) was applied [64]:

$$NUC = \sum_{n=1}^{2} \frac{\Delta^{15}N_n * cN_n * DW * 10^5}{\Delta t * FW * M(N)} \tag{1}$$

where *NUC* is the net uptake capacity (nmol g^{-1} FW h^{-1}), $\Delta^{15}N_n$ the difference in ^{15}N abundance (% total N) of the sample and the ^{15}N abundance of a control sample (natural ^{15}N abundance), *cN* the total N (% of root dry weight), *DW* the dry weight of the root segment (g), Δt the incubation time, *FW* the fresh weight of the sample incubated in the artificial soil solution (g), and *M(N)* the molecular weight of N (15 g mol^{-1}).

2.8. Statistical Analysis

All statistical analyses were performed using reference [65]. A detailed list of all R packages used can be found in Table S9. A linear mixed model was fitted on the data in order to analyze differences between identification (ID; i.e., BB = pure beech; BF = beech in mixed stands or FB = fir in mixed stands, respectively), time of season (season; i.e., spring, summer, and fall, respectively), and field sites (CO; EM; CR). We fitted the model twice on the data, once including all three possible IDs (i.e., BB, BF, and FB), and again with only BB and BF, to correctly meet the requirement of hypothesis 2, if there were differences between BB and BF stands. Since the results were identical for both models, we only documented the "full" model including all three IDs.

ID, season and site were treated as fixed factors, whereas the tree number served as random factor due to the repeated measurements on the same trees. δ^{13}C, total N, structural N, protein N, and amino acid N were the dependent variables of each model. Bulk leaf material (leaves and needles) and root material were analyzed separately. Normal distribution of residuals and homoscedasticity were checked visually (residual-fitted values–plots [66,67] and supported by the Shapiro-Wilk test when checking for normal distribution of residuals, and by Levene's test for homoscedasticity. If these assumptions were not met, raw data were transformed by either cube root transformation, or by "Tukey's ladder of Powers". Subsequently, transformed data were fitted again and the assumptions were checked visually as above. Backward elimination of non-significant effects by the principle of marginality was conducted to find the best suitable model. To assess significant differences "ls-means" was used as a wrapper, comparing the means pairwise based on the model output. The same approach was used on the N uptake data, distinguishing between N sources (arginine A, glutamine G, nitrate NO$_3^-$, and ammonium NH$_4^+$), site, season, and ID.

Correlation analyses were conducted using Spearman's rank correlation coefficient (ρ) to identify relations between leaf and root bulk material and soil parameters, as well as for analyzing relationships between the N uptake capacity of the four different N sources. Soil parameters were categorized into three depths (litter, Ah-horizon, and Bv-horizon) to ensure comparability between the three investigated sites. Consequently, these three depths were correlated separately with each plant parameter. Correlation was tested for its significance and considered a "true" correlation, when $\rho \geq \pm 0.5$ was met.

3. Results

For our first hypothesis we analyzed silver-fir and beech performance at three different sites and the first part of the results is dedicated to this hypothesis, thus including silver-fir and the two associations of beech (BB and BF). The second part, consisting mainly of bivariate correlation analysis, is dedicated to investigate our second and third hypothesis and, thus, excludes samples obtained from silver-firs (FB) for soil and plant parameter correlations. Since adult beech mainly takes up nitrogen in fall [21], we correlated soil and plant data for the fall measurements exclusively.

3.1. Inorganic Soil Nitrogen Contents in Fall

Ammonium-N contributes most to mineral soil N and was significantly higher than nitrate-N (Table S5). Mineral soil N concentrations differed significantly between the litter layer and the mineral soil (i.e., Ah and Bv). Due to high variation of inorganic nitrogen contents in the litter layer, the differences between the EM and CR sites (revealing twofold higher values compared to the CO site) and the CO site were not significant (Table S5). We observed similar patterns at the three sites, with mineral N concentrations being highest in the litter layer and decreasing significantly to the Ah- and Bv-horizon (without differences between the latter) (Table S5).

3.2. $\delta^{13}C$ Signatures in Leaves/Needles and Roots

$\delta^{13}C$ signatures of beech leaves and fir needles were significantly more enriched in spring, became more depleted in summer and did not differ in fall, regardless if associated in pure or mixed stands (Figure 1, Table S1). Previous as well as current year fir needles at the CR site were significantly more enriched compared to the EM site (Table S1 and S2). Compared to previous year needles, current year needles tended to be more depleted at every sampling point of the season (Table S2). We found that roots were significantly more enriched in summer and fall compared to spring at all three sites, again regardless of ID. At the CR site, fine roots signatures of pure as well as mixed beech stands were significantly more enriched compared to the EM site. We did not observe significant differences in $\delta^{13}C$ signatures between pure and mixed beech stands at any site and at any time of the season (Table S3).

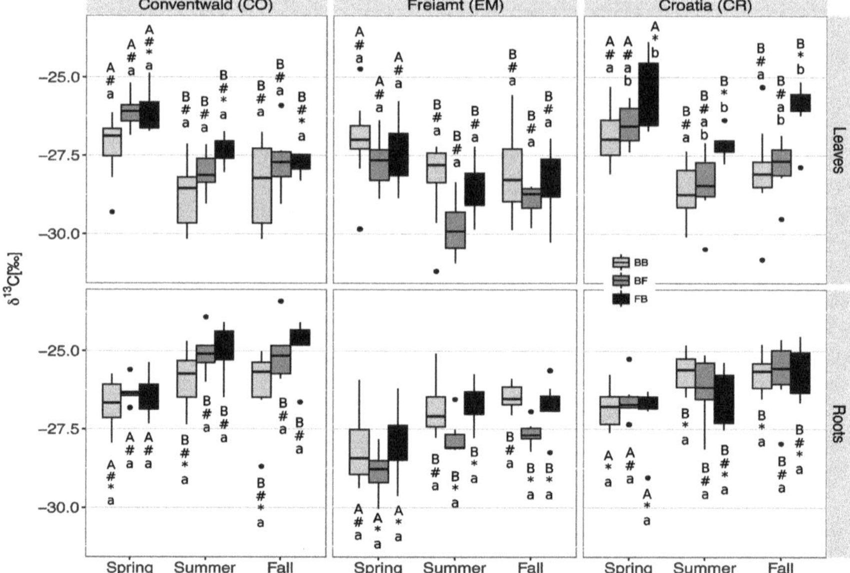

Figure 1. Seasonal changes in $\delta^{13}C$ signatures of leaves/previous year needles and roots. Top row reveals bulk leaf material, bottom row bulk root material, obtained from three field sites, i.e., "Conventwald" (CO), "Freiamt" (EM) and "Croatia" (CR), respectively. Three shades of grey depict the three different species associations, namely "BB" light grey—beeches in pure beech stands, "BF" grey—beeches in mixed beech-fir-stands, and "FB" dark grey—firs in mixed beech-fir-stands at three sampling time points within the season of 2016, i.e., "Spring" (early June), "Summer" (early September), and "Fall" (mid-September to early October). Box plots include median (solid black line), standard deviation (black error bars), and outliers (black dots) which were generated with $n = 10$ (CO)/11(EM)/12(CR) for "BB", $n = 6$ (CO)/8(EM)/6(CR) for "BF", and $n = 6$(CO)/9(EM)/6(CR) at each given time point. Capital letters indicate differences between the same ID and site, but at different seasons; lower case letters indicate differences between different IDs, but at the same season and site; hashtags and asterisks indicate differences between different sites, but same ID and season. Sharing the same letter/symbol indicates no significant difference at $p \leq 0.05$.

3.3. N Contents and N Partitioning in Leaves/Needles and Roots

Previous and current year fir needles revealed significantly lower total N contents over the entire season than beech leaves disregarding sites (Figure 2, Table S1). At the same time, fir fine roots exhibited more or similar amounts of total N compared to the needles (Figure 2, Tables S1 and S3). At the EM and CR site, beech leaves in mixed stands contained significantly higher amounts of total N in summer and fall compared to the CO site. We did not observe significant differences between pure and mixed beech stands due to high natural variations in beech leaf N content, except in fall at the CR site, where beeches in mixed stands revealed significantly higher amounts of total N compared to pure beeches.

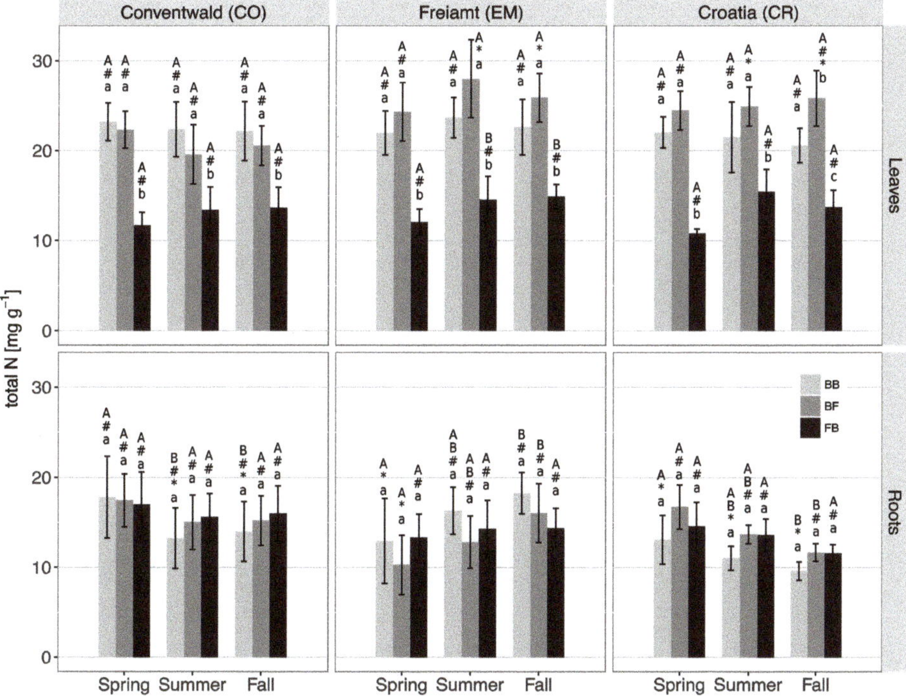

Figure 2. Seasonal changes of total nitrogen contents of leaves/previous year needles and roots. Top row reveals bulk leaf material, bottom row bulk root material, obtained from three field sites, i.e., "Conventwald" (CO), "Freiamt" (EM) and "Croatia" (CR), respectively. Three shades of grey depict the three different species associations, namely "BB" light grey—beeches in pure beech stands, "BF" grey—beeches in mixed beech-fir-stands, and "FB" dark grey—firs in mixed beech-fir-stands at three sampling time points within the season of 2016, i.e., "Spring" (early June), "Summer" (early September), and "Fall" (mid-September to early October). Bar plots include standard deviation (black error bars), which were generated with $n = 10$ (CO)/11(EM)/12(CR) for "BB", $n = 6$ (CO)/8(EM)/6(CR) for "BF", and $n = 6$(CO)/9(EM)/6(CR) for "FB" at each given time point. Capital letters indicate differences between the same ID and site, but at different seasons; lower case letters indicate differences between different IDs, but at the same season and site; hashtags and asterisks indicate differences between different sites, but same ID and season. Sharing the same letter/symbol indicates no significant difference at $p \leq 0.05$.

While previous and current year fir needles revealed low protein N contents throughout the whole season, protein N contents of beech leaves were significantly lower in spring compared to summer and fall (Figure 3 Table S1). Differences in protein N contents were not observed between pure and mixed stands, neither in leaves nor in fine roots of beech, except at the CR site in fall where, as already observed for total N contents, protein N amounts in the mixed beeches were significantly higher compared to the pure beech stands. In fine roots, the differences between the two species were negligible throughout the seasons, but as previously observed for leaves, the protein N content in roots of both species became significantly higher from spring to summer, and even ascended more in fall at the CR site (Figure 3, Tables S1 and S3). Protein N contents of fir roots were much higher (up to five-fold) compared to needles, but similar to beech roots (Figure 3).

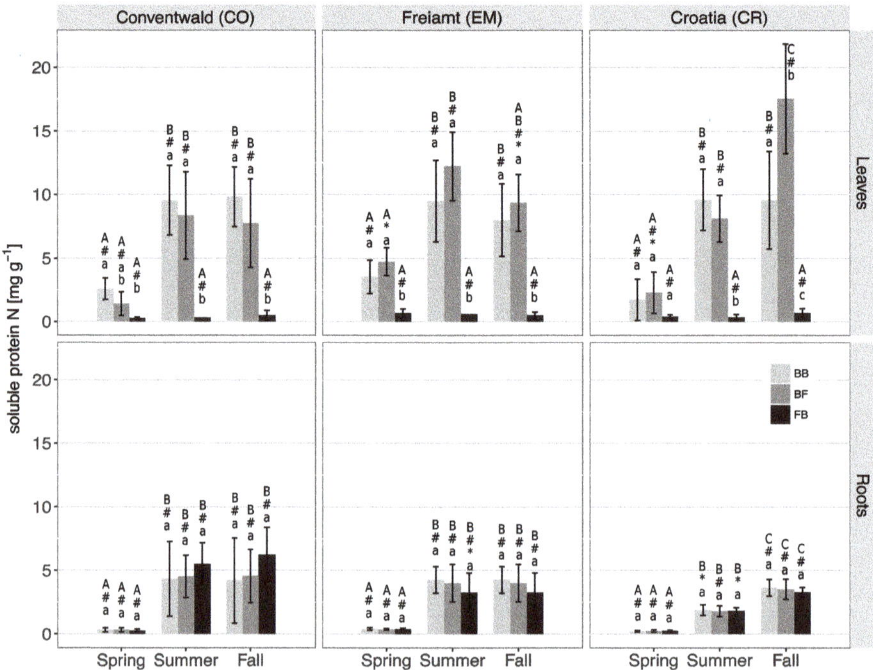

Figure 3. Seasonal changes in soluble protein N contents of leaves/previous year needles and roots. Top row reveals bulk leaf material, bottom row bulk root material, obtained from three field sites, i.e., "Conventwald" (CO), "Freiamt" (EM) and "Croatia" (CR), respectively. Three shades of grey depict the three different species associations, namely "BB" light grey—beeches in pure beech stands, "BF" grey—beeches in mixed beech-fir-stands, and "FB" dark grey—firs in mixed beech-fir-stands at three sampling time points within the season of 2016, i.e., "Spring" (early June), "Summer" (early September), and "Fall" (mid-September to early October). Bar plots include standard deviation (black error bars), which were generated with $n = 10$ (CO)/11(EM)/12(CR) for "BB", $n = 6$ (CO)/8(EM)/6(CR) for "BF", and $n = 6$(CO)/9(EM)/6(CR) at each given time point. Capital letters indicate differences between the same ID and site, but at different seasons; lower case letters indicate differences between different IDs, but at the same season and site; hashtags and asterisks indicate differences between different sites, but same ID and season. Sharing the same letter/symbol indicates no significant difference at $p \leq 0.05$.

Total amino acid N concentrations of leaves/needles did not reveal general differences over the seasons regardless of affiliation. At the CO site, amino acid N contents of beech leaves (i.e., BB and BF) significantly increased from spring to summer and fall (Figure 4, Table S1). In fine roots, a similar increase of amino acid N concentration was significant. Beech roots from monoculture revealed the highest amino acid N concentrations at the EM, and the lowest at the CR site (Table S3). Amino acid N contents in roots of both species tended to be higher than in leaves, except for spring when similar contents were observed (Figure 4, Tables S1 and S3). Again, we did not discover differences between pure and mixed beech stands at any time or site.

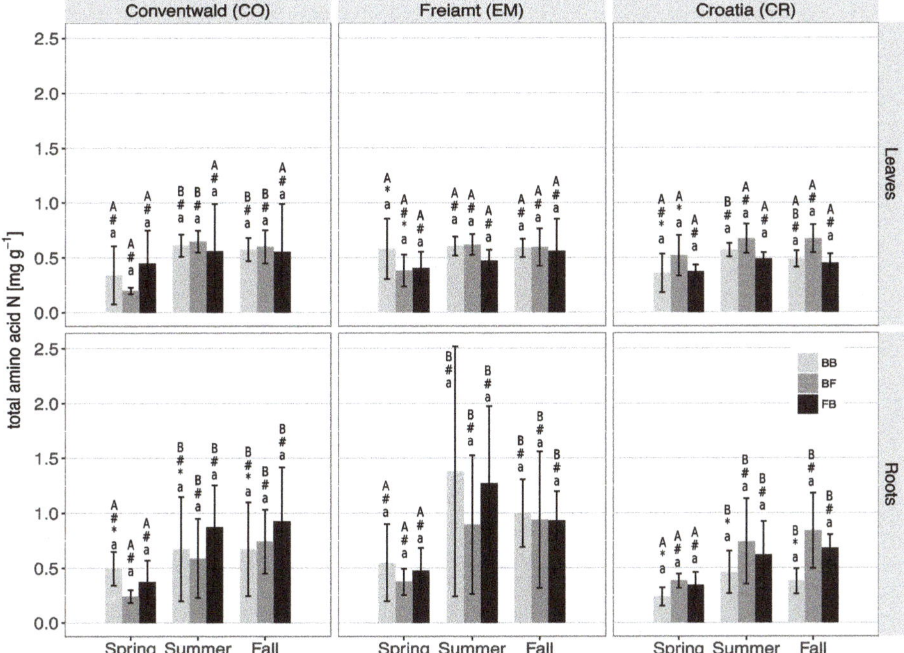

Figure 4. Seasonal changes of total amino acid N contents of leaves/previous year needles and roots. Top row reveals bulk leaf material, bottom row bulk root material, obtained from three field sites, i.e., "Conventwald" (CO), "Freiamt" (EM) and "Croatia" (CR), respectively. Three shades of grey depict the three different species associations, namely "BB" light grey—beeches in pure beech stands, "BF" grey—beeches in mixed beech-fir-stands, and "FB" dark grey—firs in mixed beech-fir-stands at three sampling time points within the season of 2016, i.e., "Spring" (early June), "Summer" (early September), and "Fall" (mid-September to early October). Bar plots include standard deviation (black error bars), which were generated with $n = 10$ (CO)/11(EM)/12(CR) for "BB", $n = 6$ (CO)/8(EM)/6(CR) for "BF", and $n = 6$(CO)/9(EM)/6(CR) at each given time point. Capital letters indicate differences between the same ID and site, but at different seasons; lower case letters indicate differences between different IDs, but at the same season and site; hashtags and asterisks indicate differences between different sites, but same ID and season. Sharing the same letter/symbol indicates no significant difference at $p \leq 0.05$.

For structural N a different seasonal pattern emerged. In spring, the structural N contents of beech leaves were significantly higher (CO and CR (BB)) or tended to be higher (EM and CR (BF)) compared to summer and fall (Figure 5, Table S1). In previous year fir needles, structural N was significantly lower in spring compared to beech leaves at every site. These differences vanished in summer and fall, when beech leaves became depleted in structural N. Current year needles contained more structural N than previous year needles (Table S2). In fine roots, at the CO and CR site, structural N contents dropped significantly from spring to summer and fall (Figure 5, Tables S1 and S3). Also, the structural N contents of roots in spring at the CO site were significantly higher than at the EM, and for mixed beech also at the CR site (Table S3).

Overall, we found a tendency towards improved N nutrition of beech leaves in mixed stands compared to monocultures though this trend was only rarely significant. Compared to beech leaves, silver-fir needles exhibited different N partitioning at all sites and times. Regardless of species, root N contents and partitioning deviated strongly from leaves in a season dependent manner. Differences in N partitioning between the sites were small in both leaves/needles and roots.

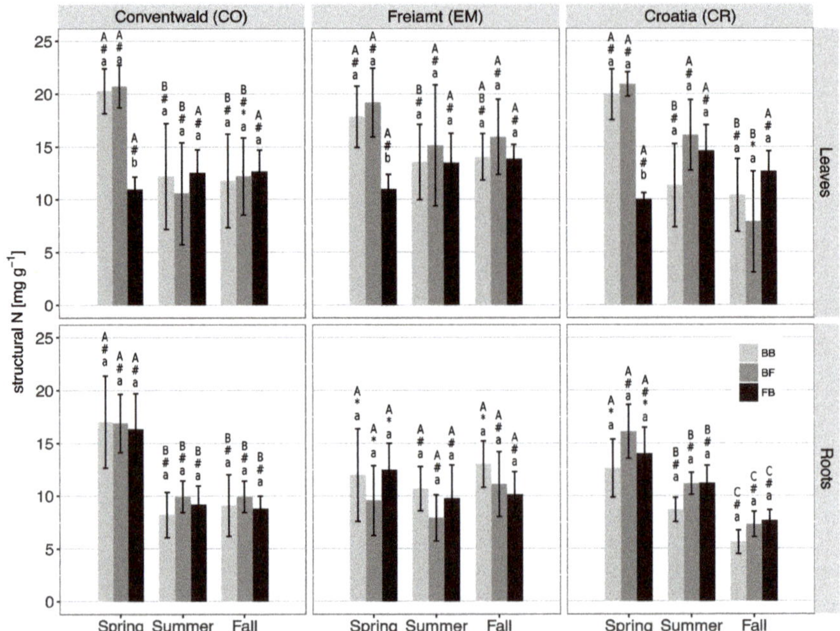

Figure 5. Seasonal changes of structural nitrogen contents of leaves/previous year needles and roots. Top row reveals bulk leaf material, bottom row bulk root material, obtained from three field sites, i.e., "Conventwald" (CO), "Freiamt" (EM) and "Croatia" (CR), respectively. Three shades of grey depict the three different species associations, namely "BB" light grey—beeches in pure beech stands, "BF" grey—beeches in mixed beech-fir-stands, and "FB" dark grey—firs in mixed beech-fir-stands at three sampling time points within the season of 2016, i.e., "Spring" (early June), "Summer" (early September), and "Fall" (mid-September to early October). Bar plots include standard deviation (black error bars), which were generated with n = 10 (CO)/11(EM)/12(CR) for "BB", n = 6 (CO)/8(EM)/6(CR) for "BF", and n = 6(CO)/9(EM)/6(CR) at each given time point. Capital letters indicate differences between the same ID and site, but at different seasons; lower case letters indicate differences between different IDs, but at the same season and site; hashtags and asterisks indicate differences between different sites, but same ID and season. Sharing the same letter/symbol indicates no significant difference at $p \leq 0.05$.

3.4. N Uptake Capacities of the Roots

Species or association effects within the same N source between beech trees in pure stands (BB), beech trees in mixed stands (BF) and firs in mixed stands (FB) were not observed, irrespective of the site or season (Figure 6, Table S3). In spring, Arg uptake capacities did not differ between the sites for both species. Gln uptake capacities of beech monoculture at the CR site revealed significantly higher values compared to the EM site. The same pattern was observed for NH_4^+ and NO_3^- uptake capacity of beech in monoculture (Table S4). Uptake capacities between each inorganic N source were similar for CO and CR, whereas the uptake capacities between organic and inorganic sources differed significantly from each other (Table S4). At the EM site, uptake capacities of pure beech and fir in mixed stands for Arg were higher compared to Gln uptake capacities, which both were higher than NH_4^+ and NO_3^- uptake capacities. Similar results were obtained in mixed stands for beech.

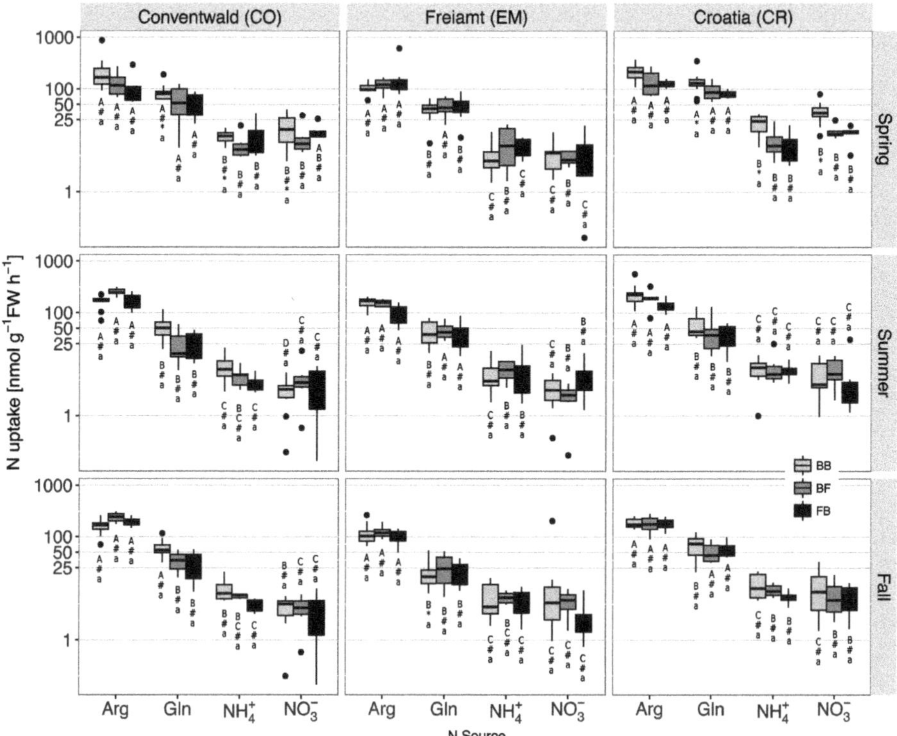

Figure 6. Seasonal changes in N uptake capacities of fine roots for inorganic and organic N sources. Top row indicates the three field sites, i.e., "Conventwald" (CO), "Freiamt" (EM) and "Croatia" (CR), respectively. Three shades of grey depict the three different species associations, namely "BB" light grey—beeches in pure beech stands, "BF" grey—beeches in mixed beech-fir-stands, and "FB" dark grey—firs in mixed beech-fir-stands at three sampling time points within the season of 2016, i.e., "Spring" (early June), "Summer" (early September), and "fall" (mid-September to early October). Box plots include median (solid black line), standard deviation (black error bars), and outliers (black dots) which were generated with $n = 10$ (CO)/11(EM)/12(CR) for "BB", $n = 6$ (CO)/8(EM)/6(CR) for "BF", and $n = 6$(CO)/9(EM)/6(CR) at each given time point. Please note that the y-axis is logarithmic. Significant differences in N uptake capacity are to be read as follows: Capital letters indicate differences between different N sources, but the same ID and site; lower case letters indicate differences between IDs at the same site and N source; hashtags and asterisks indicate differences between sites, but the same IDs and N sources. Sharing the same letter/symbol indicates no significant difference at $p \leq 0.05$.

In summer, across the three investigated sites (CO, EM, CR), uptake capacities showed a similar pattern for every N source (Arg, Gln, NH_4^+, NO_3^-). At the CO site, uptake capacities of beech in monoculture differed significantly from each other for every N source (Table S4). For BF and FB the uptake capacity for Arg was significantly higher than for Gln, which in turn was higher than the uptake capacities of both inorganic N sources. At the EM site for BF and FB neither the organic nor the inorganic N sources differed from each other, but differed from another. In beech monocultures, highest uptake capacity was observed for Arg, followed by Gln and the two inorganic sources, which did not differ from each other (Table S4). A similar pattern was found at the CR site for every affiliation (BB, BF and FB).

In fall, across the three investigated sites (CO, EM, CR), the uptake capacity pattern was similar for every N source except for pure beech at EM where Gln uptake capacity was significantly lower compared to the same association at CO and CR. When comparing the N uptake capacities for each source at each site, patterns were different from spring and summer. For beech monocultures at the CO site, neither uptake capacities of organic sources nor of inorganic sources differed amongst themselves, but differed from each other. For BF and FB Arg uptake capacity was highest, followed by Gln and the inorganic sources, which did not differ from each other. At the EM site, the latter was true for all three associations and also for BB at the CR site. For BF and FB at the CR site, neither the organic nor the inorganic N sources differed amongst themselves, but from each other (Table S4).

Overall we found inorganic N uptake capacities to be much lower compared to organic N uptake capacities, regardless of IDs, season, or site. Differences in uptake capacities between the two inorganic N sources were small and preferences were not found (Figure 6, Table S4). However, between the organic N sources, uptake capacities of Arg were often significantly higher than Gln uptake capacities. Over the seasonal course, uptake patterns differed inconsistently, though with Arg uptake capacities always being the highest.

3.5. Correlations between Environmental-, Soil-, and Plant Parameters

3.5.1. Environmental and Plant Parameters Relations

We investigated the relationship between the environmental parameters (temperature, precipitation, and altitude) on one side and the $\delta^{13}C_{plant}$ in bulk leaf/needle/root material on the other side (Table 1). Correlation analyses revealed that fir (current and previous year) needle ^{13}C correlated highly significant with altitude, indicating an enrichment in ^{13}C with increasing altitude. This correlation was weaker for beech leaves at the mixed stand, and no correlation was found for beech trees in monoculture. Beech root $\delta^{13}C$ in mixed stands revealed a positive correlation with altitude and precipitation, suggesting enrichment in ^{13}C with increasing altitude and precipitation, but a strongly negative correlation with temperature, indicating a depletion in root ^{13}C signatures with increasing air temperature.

Table 1. Correlation between $\delta^{13}C_{plnat}$ and environmental parameters. Leaf (L13C)/previous year needle bulk $\delta^{13}C$ (LY13C), current year needle bulk $\delta^{13}C$ (CY13C), and root bulk $\delta^{13}C$ (R13C) were correlated with elevation (above sea level), mean annual average temperature (Temp), and annual precipitation (Prec). Correlation coefficient is Spearman's rho (ρ) followed by asterisks depicting corresponding *p*-values: $p < 0.0001$ "****"; $p < 0.001$ "***"; $p < 0.01$ "**"; $p < 0.05$ "*". Correlations with $\rho \geq \pm 0.5$ were considered as "true" relation.

Pure Beech	LC13	R13C	Elevation	Temp	
L13C					
R13C	−0.32 **				
Elevation	−0.04	0.42 ****			
Temp	0.08	−0.39 ****	−0.58 ****		
Prec	−0.04	0.42 ****	1.00 ****	−0.58 ****	

Mixed Beech	LC13	R13C	Elevation	Temp	
L13					
R13C	0.2				
Elevation	0.40 **	0.59 ****			
Temp	−0.46 ***	−0.77 ****	−0.53 ****		
Prec	0.40 **	0.59 ****	1.00 ****	−0.53 ****	

Mixed Fir	LY13C	CY13C	R13C	Elevation	Temp
LY13C					
CY13C	0.74 ****				
R13C	0.05	0.17			
Elevation	0.66 ****	0.71 ****	0.25		
Temp	−0.46 ***	−0.40 **	−0.47 ***	−0.67 ****	
Prec	0.66 ****	0.71 ****	0.25	1.00 ****	−0.67 ****

3.5.2. Soil and Plant Parameter Relations in Fall

Inorganic soil N concentrations decreased significantly with increasing soil TN concentrations in the upper two soil layers (i.e., litter and Ah-horizon) for both IDs (i.e., BB and BF), but this effect was more pronounced at the BF stands (Table S7). At the Bv-horizon, we observed a significant increase in inorganic soil N concentrations with increasing TN, which was more pronounced at the BB stands (Table S6).

Bulk leaf parameters at the BB stands did not correlate with soil TN, SOC or inorganic N concentrations. For bulk root material, total N, amino acid N, and structural N contents decreased with increasing soil TN, SOC and inorganic N concentrations, though the correlation coefficients only indicated a weak relation, except for the Bv-horizon, where it was very pronounced. Bulk leaf δ^{13}C signatures did not correlate with any soil parameter, whereas for bulk root δ^{13}C signatures the correlation revealed a soft enrichment in roots with increasing TN, SOC and at the Bv-horizon also with inorganic N concentrations (Table S6).

At the BF stands, leaf total N decreased with increasing soil TN and SOC concentrations, and also with increasing inorganic N at the Bv-horizon, but increased with inorganic N in the litter layer. The same relationship applied to the protein N contents of bulk leaf material, again only for the litter layer. Leaf protein N levels decreased with increasing soil inorganic N levels measured in Ah- and Bv-horizons. Total N of bulk roots increased with increasing inorganic N contents of the Ah-horizon, and the same applied for root structural N. Bulk leaf δ^{13}C signatures became enriched with increased levels of soil TN and SOC at the litter layer, but were depleted when inorganic N contents of the Ah-horizon increased. These observations were more pronounced for bulk root δ^{13}C signatures, with an enrichment of ^{13}C in the roots with increasing amounts of TN and SOC in the soil at all depths. As observed for leaf bulk δ^{13}C signatures, root bulk δ^{13}C became depleted when inorganic soil N increased at the Ah-horizon (Table S7).

Overall, we observed differing correlation patterns for pure and mixed beech stands in fall. Correlation analysis of plant with soil parameters revealed that for pure beech stands the relation between root and soil parameters were more pronounced and no such observation was made for bulk leaf material, whereas for mixed beech stands leaf total N exhibited a strong relation with soil TN, SOC and inorganic N concentrations. Furthermore, correlations between leaf and root δ^{13}C on one site and the soil on the other site were more pronounced in mixed compared to pure beech stands.

3.5.3. Relations of N Uptake and Soil Parameters in Fall

For the BB stands, we observed only weak relations between increasing Arg uptake capacities and increasing soil TN, SOC and inorganic N concentrations in Ah- and Bv-horizons. For Gln uptake capacities at the Bv-horizon, the correlations were much stronger with soil TN and SOC at all three soil depths (Table S8). For NH_4^+ uptake capacities, the relations were again weak and only applicable at the Bv-horizon for TN, SOC, and inorganic N.

At the BF stands, increases in Arg uptake capacities were associated with increased soil TN and SOC, but not with inorganic N. The other N sources used in this experiment were not found to be related to soil parameters (Table S8).

However, we observed distinctly different uptake patterns between pure and mixed beech stands. At the BB stands, NH_4^+ uptake capacities were associated with increased Arg- and Gln uptake capacities, and NO_3^- with increased NH_4^+ uptake capacities (Table S8). This pattern could not be observed at the BF stands, where only a strong increase in Gln uptake capacities correlated with increased NH_4^+ uptake capacities (Table S8).

4. Discussion

4.1. Beech and Silver-Fir Performances are Similar at the Croatian and Black Forest Field Sites

Compared to standard nutrient levels of beech leaves, the N levels found for mixed beech at the EM and CR sites were in the lower to upper normal range, or even enhanced [68,69]. This result was surprising, since these sites were initially thought to reveal restrictive tree performance due to higher drought sensitivity (EM: higher MAAT (mean annual air temperature), less precipitation, smaller water holding capacity; CR: shallow soil profile (45 cm depth)). The foliar N concentrations of silver-fir were below the values thought to indicate a good nutrient status [70,71]. Previous year needles exhibited the lowest N amounts in spring indicating re-translocation of N into current year needles, which in turn revealed highest amounts of N in spring likely due to relocation [72]. Both, previous and current year needles highly vary in their amounts of N and range between insufficiency (<12.3 g kg^{-1}) in spring (previous) and summer/fall (current), and the normal range (12.3–15.5 g kg^{-1}) in fall [69]. N uptake capacities revealed high variations within the same ID (Figure 6, Table S4), but similar uptake capacities at the Black Forest and Croatian field sites.

Though all $\delta^{13}C_{plant}$ signatures did not indicate water limitation (Table S1), they varied up to 5‰ between the EM and CR site. For fir needles this observation was significant, whereas beech leaves revealed no significant differences between the sites, though signatures at the CR site tended to be more enriched compared to EM. These observations might be a consequence of different site conditions. Even though the CR site revealed the highest yearly precipitation rate, precipitation in summer only amounts to a quarter of that. Also it had the most shallow soil (only up to 45 cm; Table S5), though revealing the highest water holding capacity. When investigating plant bulk material of beech along a climate gradient across Europe, a positive relation between altitude and $\delta^{13}C$ signatures has been reported by [73]. Our field sites are located on altitudes ranging from 400 to 900 m a.s.l. and correlation analysis confirmed this positive relation, where $\delta^{13}C$ signatures become more enriched with increasing height. This observation also has been made in previous studies with other species [74–76] and was attributed to the reduction of O_2 partial pressure combined with an increase in assimilation rates [74].

From the present results, we were not able to confirm Hypothesis 1 regarding N partitioning or N acquisition strategies, and only partially confirmed this hypothesis for water supply and photosynthesis indicated by $\delta^{13}C$ signatures. However, studies during a wet year subsequent to extended periods of drought are likely to show different results.

Water scarcity in beeches not only result in leaf stomatal closure and, thereby, reduced discrimination against $^{13}CO_2$ in leaves, but also in a reduction of root water retrieval from the soil and reduced nutrient acquisition resulting in reduced competitiveness [77–79]. Concomitant with the lack of precipitation, soluble proteins in leaves and roots are degraded mediating an increase of free amino acids [80]. Ultimately, beeches react towards drought with decreasing root growth and root exudates, thereby deteriorating their nutrient uptake capacity [81]. This observation was confirmed by observations in a greenhouse experiment, where N uptake and consequent allocation of N into the fine roots was reduced [80]. Projecting these findings into years with repetitive drought periods and into consecutive dry years, the effects on forest ecosystems will be severe. Thus, the number of drought periods in dry years and also the number of dry years preceding a wet year will play a major role when investigating N nutrition during a wet year. Therefore, long-term studies are required even for evaluating N nutrition during a wet year in areas subject to climate change. Hence, the present study performed in such areas only provides "baseline" information of mixed and pure beech stands.

4.2. The Performance of Beech Tends to Be Influenced by the Presence of Silver-Fir

We did not observe any differences in N uptake capacities regardless of site or association (pure vs. mixed). Significant differences in N contents and partitioning were only rarely observed and, in these cases, indicate improved N nutrition upon fir admixture rather than competition between beech and fir for N. A possible explanation for this observation might be the similar mineral N contents at the three investigation sites (Table S5), even though the three sites were characterized by major differences in soil total nitrogen concentrations (Table S5). The similar soil mineral N concentrations may have mediated similar tree N uptake capacities (Figure 6, Table S4), and subsequently, mostly similar total N contents and N partitioning (Figures 2–5, Tables S1–S3) at the field sites studied.

Beeches grown in mixture at the field sites more susceptible towards drought (EM and CR, see above) revealed a trend towards higher foliar N contents (Figure 2) indicating a facilitative effect of silver-fir on beech on sites where soil total N concentrations were low. It supports the "stress-gradient hypothesis" (SGH), stating that facilitation was more "common in communities developing under high physical stress (...)" [82]. According to this hypothesis, such an effect would be observable at sites providing a benign environment and at these sites competition should dominate tree performance. Therefore, the tendency towards lower total foliar N at the CO site in our study is consistent with SGH (Figure 2, Tables S1 and S5). Many studies report that species richness results in enhanced forest productivity [83]. When mixing of evergreen and deciduous species, this effect is attributed to a reduction of light competition [84,85], a greater absorption for photosynthetically active radiation [86], and improved nitrogen nutrition [87]. However, the assessment of mixture effects on the trees' nutritional status is still a matter of debate [51,88].

From the present results we conclude that the presence of silver-fir does not cause a general change in N acquisition and partitioning of beech trees, thereby rejecting hypothesis 2. However, we also did not observe a competition effect of fir admixture on beech, indicating that beech trees experience largely the same inter- and intra-specific competition for nutrients with or without growing in mixture with silver firs at sufficient water availability.

4.3. What Is the Driving Force of the N Nutrition of Beech?

In order to assess our third hypothesis stating that soil N contents, climate and species associations interactively affect the performance of beech, we correlated environmental, soil derived and plant derived parameters with each other. We did not find a consistent general pattern that could enable us to relate any specific parameter to performance, but we rather revealed the complexity of the ecosystems investigated. Well-known and established in literature, we still could not relate soil total N and foliar N concentrations [88,89] at the pure beech stands, indicating that soil N was not a main influencing factor at these sites. This might be explained by the fact that soil total N stocks are largely bound in polymeric organic compounds, thereby not being available to plants. It is only microbial processes, such as depolymerization and ammonification, that produce the small plant-available soil monomeric organic N or mineral N pools [90]. In mixed stands, foliar N concentrations even decreased with higher soil TN concentrations [88]. In contrast, the mineral N concentrations of the litter layer revealed a positive correlation with foliar N at the mixed stands, but not mineral N concentrations at the other soil depths, indicating that at the mixed stands the litter layer can determine the N concentrations of the foliage, as previously reported [90]. Additionally, when including N uptake capacities in the equation, we found that in the pure beech stands the relationship between Gln uptake capacities and soil total N was strong, while at the mixed stand (both, beeches and firs) at the same time, Arg uptake capacities increased with increasing total N in the soil, but neither site revealed a strong relation between mineral soil N and N uptake capacities. This difference between pure to mixed stands in plant-soil interaction poses the question, how fir admixture alters soil parameters. In the mixed stands, the soil respiration was found to be decreased in comparison to the pure stands (Rehschuh and Dannenmann; personal communication). Therefore, reduced mineralization rates of soil organic matter in mixed stands due to more recalcitrant fir litter may have mediated lower N acquisition by beech roots in the mixed stands.

Thus, the present results support our third hypothesis that soil N contents, climate and species associations interactively affect the performance of beech, but it remains unclear which factor influences most/least and in which direction.

4.4. Silver-Fir Total N Contents Indicate the Roots as Storage Location

As previously observed, fir needle N contents were lower compared to beech leaves and root N contents were similar for both species [46]. Furthermore, the fir root N contents were the same as current/previous year needles or even exceeded their N contents (Tables S2 and S3). Needle N contents were comparable to the ones of Scots pine in Sweden, though revealing a higher variability [91], to the ones in Radiata Pine [92], though the ones from this study again indicated a wider range. In comparison to spruce stands in Bavaria, we found that total N contents in previous year needles were lower or similar, but similar or higher compared to values obtained from Douglas fir in the same study [88]. Evergreen species are thought to store N mainly in the youngest age class of needles [93–95], where N is mostly stored in the form of photosynthetic proteins (i.e., RuBisCo) [96,97] and provided for initial growth and development to the next needle generation [98]. For silver-fir this view cannot be confirmed from the present observations, since we (a) found similar/exceeding amounts of total N in the roots of fir, and (b) could neither find decreased amounts of protein N in previous year needles in spring nor increased amounts in fall. Therefore, we propose that N storage in roots of silver firs plays a major role, as previously observed for the evergreen *Nothofagus fusca* [99].

5. Conclusions

At sufficient water availability, beneficial effects of silver-fir on the N nutrition of beeches might be restricted to sites with low N supply, where an increase in soil available N would increase the nutritional status of the trees. Additionally, in future studies a closer interrelation of soil and tree analyses would improve the assessment of dependencies of the alteration of plant and soil parameters. In the present study, the analyses of soil parameters remained only at a descriptive level.

Supplementary Materials: The following are available online at http://www.mdpi.com/1999-4907/9/12/733/s1, Table S1: δ^{13}C signatures and N partitioning of the previous year leaves/needles sorted by site, season and ID, Table S2: δ^{13}C signatures and N partitioning of the current year leaves/needles sorted by site, season and ID, Table S3: δ^{13}C signatures and N partitioning of the roots sorted by site, season and ID, Table S4: Net N uptake capacities of beech and fir roots sorted by site, N source, season and ID, Table S5: Basic climate, vegetation, and soil properties of the investigated field sites, Table S6: Correlation between pure beech plant parameters and soil parameters in fall, Table S7: Correlation between mixed beech plant parameters and soil parameters in fall, Table S8: Correlation between N uptake capacities of pure and mixed beech stands in fall, Table S9: Packages used in statistical analysis in R.

Author Contributions: R.-K.M. and F.Y. contributed equally to this work. Conceptualization: M.D. and H.R.; data curation: R.-K.M., F.Y., S.R., M.B. and R.P.; formal analysis: R.-K.M. and S.R.; funding acquisition: M.D. and H.R.; investigation: R.-K.M., F.Y., S.R. and M.B.; project administration: T.B., M.I. and H.R.; resources: T.B. and M.I.; supervision: M.D. and H.R.; visualization: R.-K.M.; writing—original draft: R.-K.M., F.Y. and S.R.; writing—review and editing: R.P., M.D. and H.R.

Funding: The present study is part of the project "Buchen-Tannen-Mischwälder zur Anpassung von Wirtschaftswäldern an Extremereignisse des Klimawandels (BuTaKli)" within the program "Waldklimafonds" (No. 28W-C-1-069-01) which was financially supported via the Bundesanstalt für Landwirtschaft und Ernährung (BLE), Germany, by the Bundesministerium für Ernährung und Landwirtschaft (BMEL) and the Bundesministerium für Umwelt, Naturschutz, Bau und Reaktorsicherheit (BMUB) based on the decision of the German Federal Parliament.

Acknowledgments: We gratefully acknowledge the help of Cornelia Herschbach, Monika Eiblmeier, Marcus Zistl-Schlingmann, Jan Holweg, Lukas Krentzel, Martin Fuchs, and Jannik Menz during sample collection and lab work. Data on tree composition of the forest stands were kindly provided by the forestry districts "Breisgau-Hochschwarzwald" and "Emmendingen". We sincerely thank the forest administration in Gospic (Croatia) for information on the local forest sites and support whenever needed. In addition, we express our gratitude to Friederike Lang and Jaane Krüger for providing soil analysis data from the Conventwald site.

Conflicts of Interest: The authors declare no conflict of interest.

References

1. Taberlet, P.; Fumagalli, L.; Wust-Saucy, A.G.; Cosson, J.F. Comparative Phylogeography and Postglacial Colonization Routes in Europe. *Mol. Ecol.* **1998**, *7*, 453–464. [CrossRef] [PubMed]

2. Konnert, M.; Bergmann, F. The Geographical Distribution of Genetic Variation of Silver Fir (*Abies alba, Pinaceae*) in Relation to Its Migration History. *Plant Syst. Evol.* **1995**, *196*, 19–30. [CrossRef]

3. Magri, D.; Vendramin, G.G.; Comps, B.; Dupanloup, I.; Geburek, T.; Gömöry, D.; Latałowa, M.; Litt, T.; Paule, L.; Roure, J.M.; et al. A New Scenario for the Quaternary History of European Beech Populations: Palaeobotanical Evidence and Genetic Consequences. *New Phytol.* **2006**, *171*, 199–221. [CrossRef] [PubMed]

4. Demesure, B.; Comps, B.; Petit, R.J. Chloroplast DNA Phylogeography of the Common Beech (*Fagus sylvatica* L.) in Europe. *Evolution* **1996**, *50*, 2515–2520. [CrossRef] [PubMed]

5. Ellenberg, H. *Vegetation Ecology of Central Europe*; Cambridge University Press: Cambridge, UK, 1988.

6. Fang, J.; Lechowicz, M.J. Climatic Limits for the Present Distribution of Beech (*Fagus* L.) Species in the World. *J. Biogeogr.* **2006**, *33*, 1804–1819. [CrossRef]

7. Ellenberg, H.; Leuschner, C. *Vegetation Mitteleuropas Mit Den Alpen: In Ökologischer, Dynamischer Und Historischer Sicht*; UTB Ulmer: Stuttgart, Germany, 2010; p. 1333.

8. Frey, W.; Lösch, R. *Geobotanik*; Spektrum Akademischer Verlag: Heidelberg, Germany, 2010; p. 600.

9. Ellenberg, H. *Vegetation Ecology of Central Europe*, 4th ed.; Cambridge University Press: Cambridge, UK, 2009.

10. Tinner, W.; Colombaroli, D.; Heiri, O.; Henne, P.D.; Steinacher, M.; Untenecker, J.; Vescovi, E.; Allen, J.R.M.; Carraro, G.; Conedera, M.; et al. The Past Ecology of *Abies alba* Provides New Perspectives on Future Responses of Silver Fir Forests to Global Warming. *Ecol. Monogr.* **2013**, *83*, 419–439. [CrossRef]

11. Oberdorfer, E. *Süddeutsche Pflanzengesellschaften*; Gustav Fischer Verlag: Jena, Germany, 1957; p. 564.

12. Blanco, E.; Casado, M.A.; Costa, M.; Escribano, R.; García, M.; Génova, M.; Gómez, A.; Gómez, F.; Moreno, J.C.; Morla, C. *Los Bosques Ibéricos. Una Interpretación Geobotánica*, 4th ed.; Planeta: Barcelona, Spain, 1997; p. 572.

13. Schraml, U.; Volz, K. Conversion of Coniferous Forests in Social and Political Perspectives. Findings from Selected Countries with Special Respect to Germany. *Norway Spruce Convers.—Options Conseq. EFI Res. Rep.* **2004**, *18*, 97–119.

14. Guckland, A.; Jacob, M.; Flessa, H.; Thomas, F.M.; Leuschne, C. Acidity, Nutrient Stocks, and Organic-Matter Content in Soils of a Temperate Deciduous Forest with Different Abundance of European Beech (*Fagus sylvatica* L.). *J. Plant Nutr. Soil Sci.* **2009**, *172*, 500–511. [CrossRef]

15. Lichtenthaler, H.K. The Stress Concept in Plants: An Introduction. *Ann. N. Y. Acad. Sci.* **1998**, *851*, 187–198. [CrossRef] [PubMed]

16. Kunstler, G.; Curt, T.; Bouchaud, M.; Lepart, J. Growth, Mortality, and Morphological Response of European Beech and Downy Oak along a Light Gradient in Sub-Mediterranean Forest. *Can. J. For. Res.* **2005**, *35*, 1657–1668. [CrossRef]

17. Curt, T.; Coll, L.; Prévosto, B.; Balandier, P.; Kunstler, G.; Truţă, E.; Căpraru, G.; Surdu, Ş.; Zamfirache, M.M.; Olteanu, Z.; et al. Plasticity in Growth, Biomass Allocation and Root Morphology in Beech Seedlings as Induced by Irradiance and Herbaceous Competition. *Ann. For. Sci.* **2005**, *59*, 51–60. [CrossRef]

18. Bonosi, L. *The Influence of Light and Size on Photosynthetic Performance, Light, Interception, Biomass Partitioning and Tree Architecture in Open Grown Acer pseudoplatanus, Fraxinus excelsior and Fagus sylvatica Seedlings*; Albert-Ludwigs-Univ., Waldbau-Inst.: Freiburg, Germany, 2006.

19. Petritan, A.M.; von Lüpke, B.; Petritan, I.C. Influence of Light Availability on Growth, Leaf Morphology and Plant Architecture of Beech (*Fagus sylvatica* L.), Maple (*Acer pseudoplatanus* L.) and Ash (*Fraxinus excelsior* L.) Saplings. *Eur. J. For. Res.* **2009**, *128*, 61–74. [CrossRef]

20. Johnson, J.D.; Tognetti, R.; Michelozzi, M.; Pinzauti, S.; Minotta, G.; Borghetti, M. Ecophysiological Responses of *Fagus sylvatica* Seedlings to Changing Light Conditions—The Interaction of Light Environment and Soil Fertility on Seedling Physiology. *Physiol. Plant* **1997**, *101*, 124–134. [CrossRef]

21. Simon, J.; Dannenmann, M.; Gasche, R.; Holst, J.; Mayer, H.; Papen, H.; Rennenberg, H. Competition for Nitrogen between Adult European Beech and Its Offspring Is Reduced by Avoidance Strategy. *For. Ecol. Manag.* **2011**, *262*, 105–114. [CrossRef]

22. Petritan, A.M.; Von Lüpke, B.; Petritan, I.C. Effects of Shade on Growth and Mortality of Maple (*Acer pseudoplatanus*), Ash (*Fraxinus excelsior*) and Beech (*Fagus sylvatica*) Saplings. *Forestry* **2007**, *80*, 397–412. [CrossRef]

23. Annighöfer, P. Stress Relief through Gap Creation? Growth Response of a Shade Tolerant Species (*Fagus sylvatica* L.) to a Changed Light Environment. *For. Ecol. Manag.* **2018**, *415*, 139–147.

24. Fotelli, M.N.; Rennenberg, H.; Holst, T.; Mayer, H.; Geßler, A. Carbon Isotope Composition of Various Tissues of Beech (*Fagus sylvatica*) Regeneration Is Indicative of Recent Environmental Conditions within the Forest Understorey. *New Phytol.* **2003**, *159*, 229–244. [CrossRef]

25. Peuke, A.D.; Gessler, A.; Rennenberg, H. The Effect of Drought on C and N Stable Isotopes in Different Fractions of Leaves, Stems and Roots of Sensitive and Tolerant Beech Ecotypes. *Plant Cell Environ.* **2006**, *29*, 823–835. [CrossRef]

26. Röhrig, E.; Bartsch, N.; von Lüpke, B.; Dengler, A. *Waldbau Auf Ökologischer Grundlage*; Ulmer: Stuttgart, Germany, 2006; p. 479.

27. Schlesinger, W.H.; Dietze, M.C.; Jackson, R.B.; Phillips, R.P.; Rhoades, C.C.; Rustad, L.E.; Vose, J.M. Forest Biogeochemistry in Response to Drought. *Glob. Chang. Biol.* **2016**, *22*, 2318–2328. [CrossRef] [PubMed]

28. Allen, C.D.; Macalady, A.K.; Chenchouni, H.; Bachelet, D.; McDowell, N.; Vennetier, M.; Kitzberger, T.; Rigling, A.; Breshears, D.D.; Hogg, E.H. (Ted); et al. A Global Overview of Drought and Heat-Induced Tree Mortality Reveals Emerging Climate Change Risks for Forests. *For. Ecol. Manag.* **2010**, *259*, 660–684. [CrossRef]

29. Vitali, V.; Forrester, D.I.; Bauhus, J. Know Your Neighbours: Drought Response of Norway Spruce, Silver Fir and Douglas Fir in Mixed Forests Depends on Species Identity and Diversity of Tree Neighbourhoods. *Ecosystems* **2018**, *21*, 1215–1229. [CrossRef]

30. Kreuzwieser, J.; Gessler, A. Global Climate Change and Tree Nutrition: Influence of Water Availability. *Tree Physiol.* **2010**, *30*, 1221–1234. [CrossRef] [PubMed]

31. Choat, B.; Jansen, S.; Brodribb, T.J.; Cochard, H.; Delzon, S.; Bhaskar, R.; Bucci, S.J.; Feild, T.S.; Gleason, S.M.; Hacke, U.G.; et al. Global Convergence in the Vulnerability of Forests to Drought. *Nature* **2012**, *491*, 752–755. [CrossRef] [PubMed]

32. Fuhrer, J.; Beniston, M.; Fischlin, A.; Frei, C.; Goyette, S.; Jasper, K.; Pfister, C. Climate Risks and Their Impact on Agriculture and Forests in Switzerland. *Clim. Chang.* **2006**, *79*, 79–102. [CrossRef]

33. McDowell, N.; Pockman, W.T.; Allen, C.D.; Breshears, D.D.; Cobb, N.; Kolb, T.; Plaut, J.; Sperry, J.; West, A.; Williams, D.G.; et al. Mechanisms of Plant Survival and Mortality during Drought: Why Do Some Plants Survive While Others Succumb to Drought? *New Phytol.* **2008**, *178*, 719–739. [CrossRef] [PubMed]

34. Breshears, D.D.; Cobb, N.S.; Rich, P.M.; Price, K.P.; Allen, C.D.; Balice, R.G.; Romme, W.H.; Kastens, J.H.; Floyd, M.L.; Belnap, J.; et al. Regional Vegetation Die-off in Response to Global-Change-Type Drought. *Proc. Natl. Acad. Sci. USA* **2005**, *102*, 15144–15148. [CrossRef] [PubMed]

35. Bigler, C.; Bräker, O.U.; Bugmann, H.; Dobbertin, M.; Rigling, A. Drought as an Inciting Mortality Factor in Scots Pine Stands of the Valais, Switzerland. *Ecosystems* **2006**, *9*, 330–343. [CrossRef]

36. Rennenberg, H.; Seiler, W.; Matyssek, R.; Gessler, A.; Kreuzwieser, J. European Beech (*Fagus sylvatica* L.)—A Forest Tree without Future in the South of Central Europe. *Allg. Forst Und Jagdzeitung* **2004**, *175*, 210–224.

37. IPCC. *Climate Change 2013: The Physical Science Basis. Contribution of Working Group I to the Fifth Assessment Report of the Intergovernmental Panel on Climate Change*; Cambridge University Press: Cambridge, UK; New York, NY, USA, 2013; p. 1535.

38. Lindner, M.; Maroschek, M.; Netherer, S.; Kremer, A.; Barbati, A.; Garcia-Gonzalo, J.; Seidl, R.; Delzon, S.; Corona, P.; Kolström, M.; et al. Climate Change Impacts, Adaptive Capacity, and Vulnerability of European Forest Ecosystems. *For. Ecol. Manag.* **2010**, *259*, 698–709. [CrossRef]

39. BMEL. *Der Wald in Deutschland: Ausgewählte Ergebnisse Der Dritten Bundeswaldinventur*; Bundesministerium Für Ernährung Und Landwirtschaft: Berlin, Germany, 2014.

40. Puettmann, K.J.; Coates, K.D.; Messier, C. *A Critique of Silviculture. Managing for Complexity*; Island Press: Washington DC, USA, 2008; p. 189.

41. Fritz, P.; Jenssen, M. *Ökologischer Waldumbau in Deutschland–Fragen Antworten, Perspektiven*; Oekom: München, Germany, 2006; p. 351.

42. Robakowski, P.; Wyka, T.; Samardakiewicz, S.; Kierzkowski, D. Growth, Photosynthesis, and Needle Structure of Silver Fir (*Abies alba* Mill.) Seedlings under Different Canopies. *For. Ecol. Manag.* **2004**, *201*, 211–227. [CrossRef]

43. Lebourgeois, F.; Gomez, N.; Pinto, P.; Mérian, P. Mixed Stands Reduce *Abies alba* Tree-Ring Sensitivity to Summer Drought in the Vosges Mountains, Western Europe. *For. Ecol. Manag.* **2013**, *303*, 61–71. [CrossRef]

44. PALUCH, J.G.; GRUBA, P. Effect of Local Species Composition on Topsoil Properties in Mixed Stands with Silver Fir (*Abies alba* Mill.). *Forestry* **2012**, *85*, 413–426. [CrossRef]

45. Koch, A.S.; Matzner, E. Heterogeneity of Soil and Soil Solution Chemistry under Norway Spruce (*Picea abies* Karst.) and European Beech (*Fagus sylvatica* L.) as Influenced by Distance from the Stem Basis. *Plant Soil* **1993**, *151*, 227–237. [CrossRef]

46. Magh, R.-K.; Grün, M.; Knothe, V.E.; Stubenazy, T.; Tejedor, J.; Dannenmann, M.; Rennenberg, H. Silver-Fir (*Abies alba* MILL.) Neighbors Improve Water Relations of European Beech (*Fagus sylvatica* L.), but Do Not Affect, N. Nutrition. *Trees* **2018**, *32*, 337–348. [CrossRef]

47. Caldwell, M.M.; Dawson, T.E.; Richards, J.H. Hydraulic Lift: Consequences of Water Efflux from the Roots of Plants. *Oecologia* **1998**, *113*, 151–161. [CrossRef] [PubMed]

48. Simon, J.; Li, X.; Rennenberg, H.; Näsholm, T. Competition for Nitrogen between European Beech and Sycamore Maple Shifts in Favour of Beech with Decreasing Light Availability. *Tree Physiol.* **2014**, *34*, 49–60. [CrossRef] [PubMed]

49. Chen, H.Y.; Klinka, K.; Mathey, A.-H.; Wang, X.; Varga, P.; Chourmouzis, C. Are Mixed-Species Stands More Productive than Single-Species Stands: An Empirical Test of Three Forest Types in British Columbia and Alberta. *Can. J. For. Res.* **2003**, *33*, 1227–1237. [CrossRef]

50. Casper, B.B.; Jackson, R.B. Plant Competition Underground. *Annu. Rev. Ecol. Syst.* **1997**, *28*, 545–570. [CrossRef]

51. Rothe, A.; Binkley, D. Nutritional Interactions in Mixed Species Forests: A. Synthesis. *Can. J. For. Res.* **2001**, *31*, 1855–1870. [CrossRef]

52. Dannenmann, M.; Bimüller, C.; Gschwendtner, S.; Leberecht, M.; Tejedor, J.; Bilela, S.; Gasche, R.; Hanewinkel, M.; Baltensweiler, A.; Kögel-Knabner, I.; et al. Climate Change Impairs Nitrogen Cycling in European Beech Forests. *PLoS ONE* **2016**, *11*, e0158823. [CrossRef] [PubMed]

53. Hoffmann, G. *Die Untersuchung von Böden—Methodenbuch Band 1*, 4th ed.; VDLUFA-Verlag: Darmstadt, Germany, 1991.

54. Arbeitskreis Standortskartierung. *Forstliche Standortsaufnahme: Begriffe, Definitionen, Einteilungen, Kennzeichnungen, Erläuterungen*, 7th ed.; Arbeitskreis Standortskartierung in der Arbeitsgemeinschaft Forsteinrichtung: Eching, München, 2016.

55. Contin, D.R.; Soriani, H.H.; Hernández, I.; Furriel, R.P.M.; Munné-Bosch, S.; Martinez, C.A. Antioxidant and Photoprotective Defenses in Response to Gradual Water Stress under Low and High Irradiance in Two Malvaceae Tree Species Used for Tropical Forest Restoration. *Trees* **2014**, *28*, 1705–1722. [CrossRef]

56. Simon, J.; Waldhecker, P.; Brüggemann, N.; Rennenberg, H. Competition for Nitrogen Sources between European Beech (*Fagus sylvatica*) and Sycamore Maple (*Acer pseudoplatanus*) Seedlings. *Plant Biol.* **2010**, *12*, 453–458. [CrossRef] [PubMed]

57. Du, B.; Jansen, K.; Junker, L.V.; Eiblmeier, M.; Kreuzwieser, J.; Gessler, A.; Ensminger, I.; Rennenberg, H. Elevated Temperature Differently Affects Foliar Nitrogen Partitioning in Seedlings of Diverse Douglas Fir Provenances. *Tree Physiol.* **2014**, *34*, 1090–1101. [CrossRef] [PubMed]

58. Winter, H.; Lohaus, G.; Heldt, H.W. Phloem Transport of Amino Acids in Relation to Their Cytosolic Levels in Barley Leaves. *Plant Physiol.* **1992**, *99*, 996–1004. [CrossRef] [PubMed]

59. Hu, B.; Simon, J.; Kuster, T.M.; Arend, M.; Siegwolf, R.; Rennenberg, H. Nitrogen Partitioning in Oak Leaves Depends on Species, Provenance, Climate Conditions and Soil Type. *Plant Biol.* **2013**, *15*, 198–209. [CrossRef] [PubMed]

60. Dannenmann, M.; Simon, J.; Gasche, R.; Holst, J.; Naumann, P.S.; Kögel-Knabner, I.; Knicker, H.; Mayer, H.; Schloter, M.; Pena, R.; et al. Tree Girdling Provides Insight on the Role of Labile Carbon in Nitrogen Partitioning between Soil Microorganisms and Adult European Beech. *Soil Biol. Biochem.* **2009**, *41*, 1622–1631. [CrossRef]

61. Gessler, A.; Schneider, S.; Sengbusch v., D.; Weber, P.; Hanemann, U.; Huber, C.; Rothe, A.; Kreutzer, K.; Rennenberg, H. Field and Laboratory Experiments on Net Uptake of Nitrate and Ammonium by the Roots of Spruce (*Picea abies*) and Beech (*Fagus sylvatica*) Trees. *New Phytol.* **1998**, *138*, 275–285. [CrossRef]

62. Geßler, A.; Kreuzwieser, J.; Dopatka, T.; Rennenberg, H. Diurnal Courses of Ammonium Net Uptake by the Roots of Adult Beech (*Fagus sylvatica*) and Spruce (*Picea abies*) Trees. *Plant Soil* **2002**, *240*, 23–32. [CrossRef]

63. Stoelken, G.; Simon, J.; Ehlting, B.; Rennenberg, H. The Presence of Amino Acids Affects Inorganic N Uptake in Non-Mycorrhizal Seedlings of European Beech (*Fagus sylvatica* L.). *Tree Physiol.* **2010**, *30*, 1118–1128. [CrossRef] [PubMed]

64. Kreuzwieser, J.; Fürniss, S.; Rennenberg, H. Impact of Waterlogging on the N-Metabolism of Flood Tolerant and Non-Tolerant Tree Species. *Plant Cell Environ.* **2002**, *25*, 1039–1049.

65. Team, R.C. *R: A Language and Environment for Statistical Computing*; R Foundation for Statistical Computing: Vienna, Austria, 2017.

66. Dorman, C. *Parametrische Statistik. Verteilungen, Maximum Likelihood Und GLM in R*; Springer: Heidelberg, Germany, 2013; Volume 53.

67. Crawley, M.J. *Statistik Mit R*; Wiley-VCH Verlag GmbH & Co. KGaA: Weinheim, Germany, 2012; p. 424.

68. Mellert, K.H.; Göttlein, A. Comparison of New Foliar Nutrient Thresholds Derived from van Den Burg's Literature Compilation with Established Central European References. *Eur. J. For. Res.* **2012**, *131*, 1461–1472. [CrossRef]

69. Göttlein, A.; Baier, R.; Mellert, K.H. Neue Ernährungskennwerte Für Die Forstlichen Hauptbaumarten in Mitteleuropa—Eine Statistische Herleitung Aus VAN DEN BURG'S Literaturzusammenstellung. *Allg. Forst-und Jagdzeitung* **2011**, *182*, 173–186.

70. Bergman, W. *Ernährungsstörungen Bei Kulturpflanzen, Entstehung Und Diagnose*; Gustav Fischer Verlag: Jena, Germany, 1983; p. 614.

71. Peguero-Pina, J.J.; Camarero, J.J.; Abadía, A.; Martín, E.; González-Cascón, R.; Morales, F.; Gil-Pelegrín, E. Physiological Performance of Silver-Fir (*Abies alba* Mill.) Populations under Contrasting Climates near the South-Western Distribution Limit of the Species. *Flora* **2007**, *202*, 226–236. [CrossRef]

72. Millard, P.; Grelet, G.A. Nitrogen Storage and Remobilization by Trees: Ecophysiological Relevance in a Changing World. *Tree Physiol.* **2010**, *30*, 1083–1095. [CrossRef] [PubMed]

73. Keitel, C.; Matzarakis, A.; Rennenberg, H.; Gessler, A. Carbon Isotopic Composition and Oxygen Isotopic Enrichment in Phloem and Total Leaf Organic Matter of European Beech (*Fagus sylvatica* L.) along a Climate Gradient. *Plant Cell Environ.* **2006**, *29*, 1492–1507. [CrossRef]

74. Körner, C.; Farquhar, G.D.; Wong, S.C. Carbon Isotope Discrimination by Plants Follows Latitudinal and Altitudinal Trends. *Oecologia* **1991**, *88*, 30–40. [CrossRef] [PubMed]

75. Hultine, K.R.; Marshall, J.D. Altitude Trends in Conifer Leaf Morphology and Stable Carbon Isotope Composition. *Oecologia* **2000**, *123*, 32–40. [CrossRef] [PubMed]

76. Warren, C.R.; McGrath, J.F.; Adams, M.A. Water Availability and Carbon Isotope Discrimination in Conifers. *Oecologia* **2001**, *127*, 476–486. [CrossRef] [PubMed]

77. Fotelli, M.N.; Rienks, M.; Rennenberg, H.; Geßler, A. Climate and Forest Management Affect ^{15}N-Uptake, N Balance and Biomass of European Beech Seedlings. *Trees—Struct. Funct.* **2004**, *18*, 157–166. [CrossRef]

78. Peuke, A.D.; Rennenberg, H. Carbon, Nitrogen, Phosphorus, and Sulphur Concentration and Partitioning in Beech Ecotypes (*Fagus sylvatica* L.): Phosphorus Most Affected by Drought. *Trees* **2004**, *18*, 639–648. [CrossRef]

79. Rennenberg, H.; Loreto, F.; Polle, A.; Brilli, F.; Fares, S.; Beniwal, R.S.; Gessler, A. Physiological Responses of Forest Trees to Heat and Drought. *Plant Biol.* **2006**, *8*, 556–571. [CrossRef] [PubMed]

80. Fotelli, M.N.; Rennenberg, H.; Geßler, A. Effects of Drought on the Competitive Interference of an Early Successional Species (*Rubus fruticosus*) on *Fagus sylvatica* L. Seedlings: 15N Uptake and Partitioning, Responses of Amino Acids and Other, N. Compounds. *Plant Biol.* **2002**, *4*, 311–320. [CrossRef]

81. Meier, I.C.; Leuschner, C. Belowground Drought Response of European Beech: Fine Root Biomass and Carbon Partitioning in 14 Mature Stands across a Precipitation Gradient. *Glob. Chang. Biol.* **2008**, *14*, 2081–2095. [CrossRef]

82. Bertness, M.D.; Callaway, R. Positive Interactions in Communities. *Trends Ecol. Evol.* **1994**, *9*, 187–191. [CrossRef]

83. Morin, X.; Fahse, L.; Scherer-Lorenzen, M.; Bugmann, H. Tree Species Richness Promotes Productivity in Temperate Forests through Strong Complementarity between Species. *Ecol. Lett.* **2011**, *14*, 1211–1219. [CrossRef] [PubMed]

84. Ishii, H.; Asano, S. The Role of Crown Architecture, Leaf Phenology and Photosynthetic Activity in Promoting Complementary Use of Light among Coexisting Species in Temperate Forests. *Ecol. Res.* **2010**, *25*, 715–722. [CrossRef]

85. Sapijanskas, J.; Paquette, A.; Potvin, C.; Kunert, N.; Loreau, M. Tropical Tree Diversity Enhances Light Capture through Plastic Architectural Changes and Spatial and Temporal Niche Differences. *Ecology* **2014**, *95*, 2479–2492. [CrossRef]

86. Forrester, D.I.; Ammer, C.; Annighöfer, P.J.; Barbeito, I.; Bielak, K.; Bravo-Oviedo, A.; Coll, L.; del Río, M.; Drössler, L.; Heym, M.; et al. Effects of Crown Architecture and Stand Structure on Light Absorption in Mixed and Monospecific *Fagus sylvatica* and *Pinus sylvestris* Forests along a Productivity and Climate Gradient through Europe. *J. Ecol.* **2018**, *106*, 746–760. [CrossRef]

87. Berger, T.W.; Untersteiner, H.; Toplitzer, M.; Neubauer, C. Nutrient Fluxes in Pure and Mixed Stands of Spruce (*Picea abies*) and Beech (*Fagus sylvatica*). *Plant Soil* **2009**, *322*, 317–342. [CrossRef]

88. Rothe, A.; Ewald, J.; Hibbs, D.E. Do Admixed Broadleaves Improve Foliar Nutrient Status of Conifer Tree Crops? *For. Ecol. Manag.* **2003**, *172*, 327–338. [CrossRef]

89. Maire, V.; Wright, I.J.; Prentice, I.C.; Batjes, N.H.; Bhaskar, R.; van Bodegom, P.M.; Cornwell, W.K.; Ellsworth, D.; Niinemets, Ü.; Ordonez, A.; et al. Global Effects of Soil and Climate on Leaf Photosynthetic Traits and Rates. *Glob. Ecol. Biogeogr.* **2015**, *24*, 706–717. [CrossRef]

90. Rennenberg, H.; Dannenmann, M. Nitrogen Nutrition of Trees in Temperate Forests—The Significance of Nitrogen Availability in the Pedosphere and Atmosphere. *Forests* **2015**, *6*, 2820–2835. [CrossRef]

91. Tarvainen, L.; Lutz, M.; Räntfors, M.; Näsholm, T.; Wallin, G. Increased Needle Nitrogen Contents Did Not Improve Shoot Photosynthetic Performance of Mature Nitrogen-Poor Scots Pine Trees. *Front. Plant Sci.* **2016**, *7*, 1051. [CrossRef] [PubMed]

92. Nambiar, E.K.S.; Fife, D.N. Growth and Nutrient Retranslocation in Needles of Radiata Pine in Relation to Nitrogen Supply. *Ann. Bot.* **1987**, *60*, 147–156. [CrossRef]

93. Millard, P.; Proe, M.F. Storage and Internal Cycling of Nitrogen in Relation To Seasonal Growth of Sitka Spruce. *Tree Physiol.* **1992**, *10*, 33–43. [CrossRef] [PubMed]

94. Millard, P.; Hester, A.; Wendler, R.; Baillie, G. Interspecific Defoliation Responses of Trees Depend on Sites of Winter Nitrogen Storage. *Funct. Ecol.* **2001**, *15*, 535–543. [CrossRef]

95. Sadanandan Nambiar, E.K.; Bowen, G.D. Uptake, Distribution and Retranslocation of Nitrogen by *Pinus radiata* from 15N-Labelled Fertilizer Applied to Podzolized Sandy Soil. *For. Ecol. Manag.* **1986**, *15*, 269–284. [CrossRef]

96. Camm, E. Photosynthetic Responses in Developing and Year-Old Douglas-Fir Needles during New Shoot Development. *Trees* **1993**, *8*, 61–66. [CrossRef]

97. Millard, P.; Sommerkorn, M.; Grelet, G.A. Environmental Change and Carbon Limitation in Trees: A Biochemical, Ecophysiological and Ecosystem Appraisal. *New Phytol.* **2007**, *175*, 11–28. [CrossRef] [PubMed]

98. Chapin, E.S.; Kedrowski, R.A. Seasonal Changes in Nitrogen and Phosphorus Fractions and Autumn Retranslocation in Evergreen and Deciduous Taiga Trees. *Ecology* **1983**, *64*, 376–391. [CrossRef]

99. Stephens, D.W.; Millard, P.; Turnbull, M.H.; Whitehead, D. The Influence of Nitrogen Supply on Growth and Internal Recycling of Nitrogen in Young *Nothofagus fusca* Trees. *Funct. Plant Biol.* **2001**, *28*, 249–255. [CrossRef]

 *forests*

Article

Soil Silicon Amendment Increases *Phyllostachys praecox* Cold Tolerance in a Pot Experiment

Zhuang Zhuang Qian [1,2], Shun Yao Zhuang [1,*], Qiang Li [1] and Ren Yi Gui [3]

[1] State Key Lab of Soil and Sustainable Agriculture, Institute of Soil Science, Chinese Academy of Sciences, Nanjing 210008, China; zzqian@njfu.edu.cn (Z.Z.Q.); qli1074@gmail.com (Q.L.)
[2] College of Forestry, Nanjing Forestry University, Nanjing 210037, China
[3] State Key Lab of Sub-Tropical Sivlculture, Zhejiang A & F University, Hangzhou 311300, China; gry@zafu.edu.cn
* Correspondence: syzhuang@issas.ac.cn

Received: 15 March 2019; Accepted: 7 May 2019; Published: 10 May 2019

Abstract: Cultivated bamboos are occasionally subjected to cold stress in winter, and silicon could improve their cold tolerance. However, evidence of the effect of Si on bamboos is still limited. Therefore, a batch and pot experiment was conducted for six months to investigate the effects of different Si fertilizer application rates (0, 0.5, 1.0, 2.0, 4.0, and 8.0 g kg^{-1} of soil weight) on the physiological responses and photosynthesis parameters of *Phyllostachys praecox* under a simulated cold stress condition. The cold temperature was set to 5 °C, 0 °C, and −5 °C, successively. The bamboo biomass increased significantly when the Si amendment rate was at least 2.0 g kg^{-1} ($P = 0.002$), and the highest biomass increase and root-to-canopy ratio were obtained with the 4.0 g kg^{-1} Si amendment. Furthermore, the Si contents in all organs of the bamboos increased with the increase of the Si amendment rate. The highest content of Si among the other organs was observed in the leaf, and the content was 68.95 mg kg^{-1} with the treatment of 4.0 g kg^{-1}. With the application of Si, the photosynthesis rate of bamboo leaves was significantly increased ($P = 0.008$). The Si-amended bamboo exhibited a cold tolerance that was associated with stimulating antioxidant systems, and the enzyme activities of superoxide dismutase, peroxidase, and catalase increased with the increase of the Si amendment rate, whereas the malondialdehyde content and cell membrane permeability decreased with all Si treatments. A low temperature of −5 °C exerted effects on the bamboo leaf chloroplasts, but the ultrastructures of the chloroplasts remained intact after Si treatment. These findings suggest that Si fertilizer enhances bamboo growth and the tolerance of bamboo plants to cold stress. However, a high application rate (8.0 g kg^{-1}) caused a decline in the bamboo biomass, compared to T4. Thus, a Si fertilization rate of 2.0~8.0 g kg^{-1} is recommended for bamboos under cold conditions.

Keywords: bamboo forest; cold stress; physiological response; silicon fertilization

---

1. Introduction

Silicon is not a necessary element for higher plants, but it is essential for obtaining a high and sustainable yield for *Poaceae* crops. Si promotes the growth of various plants, especially those under abiotic and biotic stress conditions [1]. It is efficient in alleviating abiotic stresses, including salt stress, metal toxicity, drought stress, radiation damage, nutrient imbalance, high temperature, and freezing [2–9]. Previous studies have suggested that the following possible mechanisms underlie the Si-enhanced resistance of plants to abiotic stress: Stimulating antioxidant systems [10–13], reducing the transpirational bypass flow [14], reducing malondialdehyde (MDA), and improving root traits and the photosynthetic rate [15].

Cold stress is an abiotic stress that causes severe damage to membranes [16], and an increase in reactive oxygen species (ROS) in plants caused by freezing as well as increased lipid peroxidation, arising from the accumulation of ROS, are the major causes of membrane damage. MDA is a harmful lipid peroxidation product, which could reflect the extent of oxidative damage [17]. Under an ROS burst, superoxide dismutase (SOD) constitutes the first line of defense against ROS [18]. The activity of SOD has marked positive effects on the antioxidant capacity of plants. Peroxidase (POD) is generally considered to be a merely ROS-detoxifying enzyme [19]. Catalase (CAT) is a specific enzyme that catalyzes the dismutation of H_2O_2 into O_2 and H_2O, preventing the damaging effects of H_2O_2 accumulation and protecting cells from oxidative stress [20]. Silicon has been shown to ameliorate the damage of cold stress on plants. Liu et al. [21] found that Si addition increased antioxidant activities and decreased the MDA of *Cucumis sativus cv.* under chilling stress. Zhan et al. [22] also showed that Si amendment alleviates chilling stress in *D. brandisii* plantlets, exhibiting increased CAT and SOD activities and decreased MDA. Moreover, Si forms deposits and undergoes polymerization, forming "phytoliths" in cells and intercellular spaces [23,24]. The strength and rigidity of the tissues improves [25], and thus cold tolerance is enhanced. Several researchers concluded that Si not only acts as a physical or mechanical barrier in plants but is also involved in metabolic and physiological activities [26].

Despite the wealth of information on the beneficial effects of Si on plants, the mechanisms underlying Si-mediated alleviation under freeze stress remain poorly understood. *Phyllostachys praecox* is among the bamboos with edible shoots in the region of Southeast China. This bamboo species has a high yield and plays an important role in the local economy. Accordingly, its sustainable production is essential for farmers. However, it is subjected to low temperatures and freeze stress in winter and often shows poor growth. Zhou et al. [27] reported that the ice damage rate of typical subtropical forests varied between 25% and 81%. A previous study found that the shoot output decreased by 32% per hectare in 2008 [28], and the bamboo freeze injury index of Lin-an City was 51% in 2016 [29]. Thus, measures to mitigate low-temperature stress are beneficial for bamboo production. Graminaceous plants absorb much more Si than other species [30]. Similar to many *Poaceae* species, bamboos accumulate Si, which may alleviate stress at low temperatures and freeze stress. However, Si uptake and accumulation in *P. praecox*, and the effect of Si on the resistance of *P. praecox* to low temperatures, have not been studied. Therefore, this study aimed to (1) investigate how Si amendment to *P. praecox* affects Si accumulation, (2) elucidate the relationship between Si content and *P. praecox* growth, and (3) explore the mechanism underlying the enhanced resistance of plants to low temperatures, after Si addition, by performing a plant physiological indicator analysis.

2. Materials and Methods

2.1. Bamboo and Soil

P. praecox is a bamboo that is mainly cultivated in Lin-an City (30°16′24.58″ N, 119°35′14.59″ E), with intensive management. Bamboos usually grow from seeds, underground rhizomes, or *P. praecox* flowers in the study site, but they do not form seeds, and they extend through the underground rhizome. Therefore, we selected bamboo rhizome for our pot experiment. The length of rhizomes was approximately 200 mm, their diameter was approximately 10 mm, and their weight was approximately 300 g. The rhizome sprouts of *P. praecox* were excavated in 2015 from the bamboo field and cultured in a greenhouse. Before the pot experiment, rhizome sprouts with similar sizes and weights were incubated.

Soil (0–25 cm) in the pot experiment was collected from the bamboo garden in Zhejiang Agriculture and Forestry, Lin-an City, ZheJiang Province, China. The soil type was classified as Ultisol. The soil basic physicochemical properties were as follows: pH, 5.31; available Si, 43.51 mg kg^{-1}; soil organic matter, 16.8 g kg^{-1}; soil total nitrogen, 769 mg kg^{-1}; available potassium, 25 mg kg^{-1}; hydrolyzed nitrogen, 81.62 mg kg^{-1}; and available phosphorus, 68.5 mg kg^{-1}. The soil sample was air-dried, passed through a 2 mm mesh sieve, and, for the improvement of the soil structure, sieved soils were mixed with perlite (bamboo rhizomes prefer soil that is loose and permeable and are easily damaged under hypoxia conditions, so perlite can increase the permeability of soil, which is beneficial for the growth of bamboo) in a 3:1 ratio (volume volume^{-1}) for the pot experiment.

2.2. Bamboo Pot Experiment

The bamboo pot experiment was conducted in 2016 in the greenhouse in Zhejiang Agriculture and Forestry, China. The light transmittance rate in the greenhouse was 88%, and the temperature ranged from 20 to 28 °C, which was suitable for bamboo growth. The height of the experimental plastic pots was 250 mm, and the diameter of the pots was 300 mm. Each pot had five holes for drainage in the bottom and was filled with 3 kg of soil. The experiment involved six treatments, with Si fertilizer application rates of 0, 0.5, 1.0, 2.0, 4.0, or 8.0 g kg^{-1}, marked as T0, T0.5, T1, T2, T4, and T8, respectively. Each treatment had 20 pots, and four bamboos were planted in each pot. The photosynthesis parameters (photosynthesis rate, water use efficiency, CO_2 of intercellular space, and stomatal conductance) of the bamboo leaves were measured every month. After six months, five bamboo pots from each treatment were randomly selected for biomass measurement. The root, rhizome, leaf, and stem of the bamboo plant were collected for further analysis. The root-to-canopy ratio was calculated by the underground biomass and overground biomass, and the equation was:

$$\text{root} - \text{to} - \text{canopy ratio} = \frac{\text{underground biomass}}{\text{overground biomass}}. \tag{1}$$

Low Temperature Incubation

Three bamboo pots (bamboos in pots were basic, with the same height and ground diameter) were chosen randomly from every treatment for further incubation. The three bamboo pots from each treatment were divided into three culture boxes, and the temperature of the three culture boxes was set to 5, 0, and −5 °C. A total of 18 bamboo pots were cultured for three days under the specified temperatures. After the three types of temperature treatments (5, 0 and −5 °C), the functional leaves of bamboo were collected for plant physiological analysis.

2.3. Si in Plant Measurement

The excavated bamboo was washed and divided into root, leaf, stem, and rhizome. It was then soaked in 0.5 M HCl for 20 s, followed by three to four rinses in distilled water, and then dried to stable weight at 65 °C. After being dried, the bamboo samples were ground and passed through a 0.25 mm mesh sieve for Si measurement. The samples were then microwave digested in a mixture of 3 mL of 62% (w w^{-1}) HNO_3, 3 mL of 30% (w w^{-1}) hydrogen peroxide, and 2 mL of 46% (w w^{-1}) hydrofluoric acid (HF), and the digested sample was diluted to 100 mL with 4% (w v^{-1}) boric acid. The Si concentration in the digested solution was determined by the colorimetric molybdenum blue method at 600 nm [31].

2.4. Photosynthesis Parameter Measurement and Physiological Indicator Analysis

For the photosynthesis parameter measurement, three sunlight-exposed leaves of each bamboo plant were measured, two bamboo plants were measured in a single pot, and three pots were measured per treatment. The photosynthesis rate (Pn, μmol $CO_2 \bullet$ m^{-2} s^{-1}), transpiration rate (Tr, mmol $H_2O \bullet$m^{-2} s^{-1}), water use efficiency (WUE, μmol $CO_2 \bullet$mmol^{-1} H_2O), CO_2 of intercellular space (Ci, μL$\bullet$L^{-1}), and stomatal conductance (Gs, mmol$\bullet$m^{-2} s^{-1}) of the bamboo functional leaves were measured using a GFS-3000 (WALZ, Effeltrich, Germany). During the measurement, the environmental condition was set at a sunlight intensity of 1300 μmol$\bullet$m$^{-2}\bullet$s^{-1}, ambient CO_2 concentration of 400 ± 20 μmol$\bullet$m$^{-2}\bullet$s^{-1}, and leaf temperature of 25 ± 1 °C [32].

After three days of low-temperature incubations at 5, 0, and −5 °C, the SOD, POD, CAT, MDA, and cell membrane permeability (CMP) activities of the leaves were measured by the nitroblue tetrazolium reduction [33], guaiacol colorimetric [34], ultraviolet absorption [35], thiobarbituric acid [36], and electric conductivity methods, respectively [37].

2.5. Observation of Bamboo Leaf Chloroplast Ultrastructure

Mature bamboo leaves of the T0 and T4 treatments were collected from the $-5\,^{\circ}$C culture boxes and cut into small pieces (1 mm $\times$ 1 mm). The T0 and T4 treatments were chosen, because T0 had no Si amendment, and T4 showed the highest leaf biomass among all treatments. We hypothesized that T4 could exhibit a better visual expression than the other treatments. The chosen section was located in the middle of the leaf, at 1 cm from the main vein. Small pieces of these leaf materials were immersed in 4% (m m^{-1}) glutaraldehyde overnight at 4 $^{\circ}$C, thoroughly rinsed with 0.1 M phosphate buffer (pH 7.0), and fixated with 1% (m m^{-1}) osmium acid for 14 h at $-4\,^{\circ}$C. Then, the samples were sequentially immersed in 50% alcohol for 15 min, 70% alcohol for 15 min, 80% alcohol for 15 min (2 times), 90% alcohol for 15 min (2 times), 95% alcohol for 15 min (2 times), 100% alcohol for 10 min (2 times), and 100% acetone for 10 min (2 times). Finally, the samples were embedded in molds with epoxy resin (Epson 812) at 60 $^{\circ}$C for 24 h. Ultrathin sections were stained by uranyl acetate and lead citrate and observed under a microscope (XSP-8CA, Shanghai, China). Cell photomicrographs were taken by transmission electron microscopy (JEM-1200EX, Tokyo, Japan) [38].

2.6. Data Analysis

The study was carried out in a completely randomized design. The SOD, POD, and CAT activity, CMP and MDA concentration, photosynthesis parameters (photosynthesis rate, water use efficiency, CO$_2$ of intercellular space, and stomatal conductance) in the plant leaves, biomass, and root-to-canopy ratio were examined statistically by an analysis of the variance and means of three replicates were subjected to Duncan's test at a 5% probability level using IBM SPSS Statistics 20.0 (SPSS Inc., Chicago, IL, USA).

Furthermore, a multiple regression analysis was adopted. The relationship between the Si content in the plant and the amended rate can be described by the following equation:

$$Y = ax^2 + bx + c \tag{2}$$

where Y is the Si content in bamboo, x is the Si content amended in the pot, and a, b, and c are the parameters for the equation.

The relationship between the Si amendment and enzyme content under $-5\,^{\circ}$C treatment could also be described by Equation (2), where Y is the SOD, POD, and CAT activity and MDA concentration in the bamboo leaves, and x is the Si content amended in the pot. Correlation coefficients and multiple regression coefficients were calculated using Microsoft Excel.

3. Results

3.1. Bamboo Biomass Change with Si Amendment

The bamboo biomass increased significantly when the Si amendment rate was above T2, after six months of incubation (Figure 1) ($P = 0.002$). However, no significant difference was found in the T2, T4, and T8 treatments ($P > 0.05$). The high rate of Si amendment (T8) led to a slight decline in biomass, compared to T4. Similarly, the root-to-canopy ratio showed the same trend as the biomass ratio increased. T4 had the highest biomass increase and root-to-canopy ratio among the treatments. These results indicated that Si amendment improves bamboo growth under low temperatures.

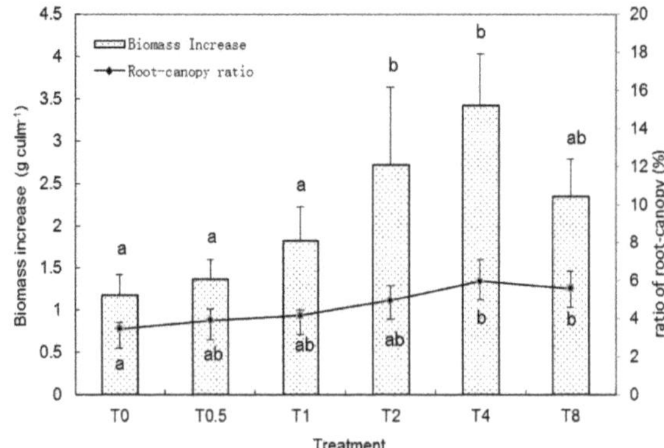

Figure 1. Bamboo biomass change after Si amendment. Values followed by the same letter(s) are not significantly different at $P < 0.05$, according to Duncan's multiple range tests. The measurements were taken after six months of Si application at room temperature.

3.2. Si Content in Bamboo Plants

The Si contents in the leaf, stem, rhizome, and root of the bamboos increased with the increase of the amendment rate (Table 1). In the leaves, the highest Si content was 68.95 mg kg⁻¹, which occurred in T4. In the stem, rhizome, and root, the highest Si contents were all observed in T8. Relatively, the Si content in the leaf was higher than in the stem, rhizome, or root. The Si content in the bamboo plant body can be arranged as follows: leaf > stem > rhizome > root. The results also showed that the total Si content in the bamboo was closely related to the Si amendment rate. Table 2 shows that the Si content in all bamboo organs could be well fitted by the equation, and the equations for the leaf, stem, rhizome, and root were $y = -1.2901x^2 + 15.311x + 26.607$, $y = -0.3743x^2 + 4.5539x + 10.923$, $y = -0.1309x^2 + 1.5841x + 3.2716$, and $-0.1003x^2 + 1.5574x + 3.1215$, respectively. All R^2 values were above 0.95. The maximum values of the leaf, stem, rhizome, and root biomass were obtained at 5.93, 6.08, 6.05, and 7.76 mg kg⁻¹, respectively. This result suggests that bamboo had a maximal Si uptake, regardless of the Si amendment rate.

Table 1. Si content in bamboo parts with various Si treatments (mg kg⁻¹). Data are the means ± SD of three replicates. Different letters in the same column indicate significant differences, based on Duncan's multiple range tests at the 0.05 level. The measurements were taken after six months of Si application at room temperature.

Treatment	Leaf	Stem	Rhizome	Root
T0	27.77 ± 4.70 a[1]	11.18 ± 1.53 a	3.50 ± 0.72 a	3.07 ± 0.99 a
T0.5	31.93 ± 3.68 a	11.35 ± 1.40 a	3.31 ± 0.81 a	3.69 ± 1.18 a
T1	43.19 ± 3.25 b	16.71 ± 4.37 ab	4.82 ± 0.95 a	5.33 ± 0.71 b
T2	48.88 ± 2.13 b	18.88 ± 3.69 b	5.64 ± 1.10 ab	6.14 ± 0.74 bc
T4	68.95 ± 10.11 c	22.60 ± 3.71 c	7.92 ± 2.51 bc	7.16 ± 1.11 c
T8	66.26 ± 9.20 c	23.50 ± 5.57 c	9.13 ± 3.25 c	7.63 ± 1.19 c

Table 2. Parameters for the equation for the Si content in the bamboo plant and the Si-amended plant. Significant differences were based on Duncan's multiple range tests at the 0.05 level.

	a	*b*	*c*	R^2	*P*
Leaf	−1.290	15.31	26.61	0.9828	0.003
Stem	−0.3743	4.554	10.92	0.9566	0.009
Rhizome	−0.1003	1.557	3.122	0.9789	0.003
Root	−0.1309	1.584	3.272	0.9585	0.008

3.3. Photosynthesis Parameters

As shown in Table 3, with Si amendment, the photosynthesis rate (Pn) of the bamboo leaves increased linearly, from 7.03 µmol $CO_2 \bullet$ m^{-2} s^{-1} to 11.89 µmol $CO_2 \bullet$ m^{-2} s^{-1}, and Pn increased significantly when the Si amendment rate was above 4.0 g kg^{-1} ($P = 0.008$). However, the bamboo transpiration rate (Tr) was significantly reduced, from 3.74 mmol $H_2O \bullet$m^{-2} s^{-1} to 1.93 mmol $H_2O \bullet$m^{-2} s^{-1} ($P = 0.001$). The WUE increased from 2.80 µmol $CO_2 \bullet$mmol^{-1} H_2O to 7.39 µmol $CO_2 \bullet$mmol^{-1} H_2O, and it increased significantly when the Si amendment rate was above 4.0 g kg^{-1} ($P = 0.01$). The CO_2 concentration of the intercellular space and stomatal conductance showed no significant difference with all the Si amending treatments ($P > 0.05$).

Table 3. Photosynthesis parameters of the bamboo leaf with various treatments. Data are the means ± SD of three replicates. Different letters in the same column indicate significant differences, based on Duncan's multiple range tests at the 0.05 level. The measurements were taken after six months of Si application at approximately 25 °C.

Treatment	Photosynthesis Rate (Pn) (µmol $CO_2 \bullet$ m^{-2} s^{-1})	Transpiration Rate (Tr) (mmol $H_2O \bullet$m^{-2} s^{-1})	Water Use Efficiency (WUE) (µmol $CO_2 \bullet$mmol^{-1} H_2O)	CO_2 of Intercellular Space (Ci) (µL L^{-1})	Stomatal Conductance (G_s) (mmol$\bullet$m^{-2} s^{-1})
T0	7.03 ± 2.69 a	3.74 ± 1.26 a	2.80 ± 1.83 a	224.33 ± 45.89 a	109.0 ± 14.91 a
T0.5	7.01 ± 0.65 a	2.64 ± 0.89 b	2.38 ± 0.58 ab	240.62 ± 40.19 a	99.4 ± 28.6 a
T1	8.11 ± 3.97 ab	2.5 ± 0.32 b	3.84 ± 1.58 ab	241.12 ± 45.21 a	97.63 ± 12.83 a
T2	9.57 ± 2.6 abc	2.58 ± 0.89 b	3.94 ± 1.17 ab	240.19 ± 47.57 a	111.07 ± 30.65 a
T4	10.11 ± 2.55 bc	2.27 ± 0.6 b	4.55 ± 1.12 b	228.47 ± 19.73 a	103.96 ± 17.01 a
T8	11.89 ± 1.63 c	1.96 ± 0.75 b	7.39 ± 3.79 c	230.99 ± 37.23 a	118.7 ± 38.06 a

3.4. Physiological Indicators Treated with Low Temperature

Figure 2 shows that the SOD activity was significantly higher at 0 °C than at 5 °C and −5 °C ($P = 0.001$). With Si amendment, the SOD activity increased with all temperature treatments and was in the order of T8 > T4 > T2 > T1 > T0.5. As shown in Figure 3, the low temperature of −5 °C significantly reduced the POD activity of the bamboo leaves ($P = 0.001$). The Si amendment increased the POD activity of leaves at various temperatures, and no significant difference was found at 5 °C ($P > 0.05$). At 0 °C and −5 °C, the POD significantly increased with the increasing rate ($P = 0.007$, $P = 0.002$). Similarly, with no more than 8 g kg^{-1} application rates, the SOD and CAT activity of the bamboo leaves increased with the increasing Si amendment (Table 4). The highest CAT activity was recorded at −5 °C with all Si treatments (Figure 4). Unlike the SOD and CAT, the Si amendment decreased the MDA concentration (Figure 5). The highest MDA concentration was recorded at −5 °C. At 0 °C and −5 °C, the MDA significantly decreased with the increasing rate ($P = 0.01$, $P = 0.002$) (Table 4). The CMP of the bamboo leaves increased with the decreasing temperature (Figure 6). The Si amendment could reduce the CMP at various temperatures, and the CMP decreased significantly when the Si amendment rate was above 2.0 g kg^{-1} at −5 °C (Figure 1) ($P = 0.001$).

Table 4. Parameters for the equation for the Si amendment and enzyme content at −5 °C. Significant differences were based on Duncan's multiple range tests at the 0.05 level. SOD, superoxide dismutase; POD, peroxidase; CAT, catalase; MDA, malondialdehyde.

	a	b	c	R^2	P
SOD	−1.783	23.266	633.6	0.89389	0.016
POD	−0.7196	9.935	112.1	0.96943	0.002
CAT	−0.3301	6.390	27.49	0.98354	9.82×10^{-4}
MDA	0.05001	−1.202	27.95	0.97371	0.002

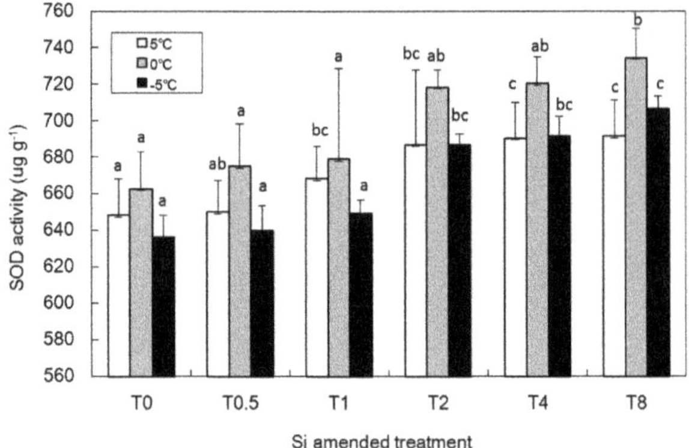

Figure 2. Effect of Si amendment on the SOD activity of leaves at various temperatures. Data are the means ± SD of three replicates. Different letters at the same temperature indicate significant differences, based on Duncan's multiple range tests at the 0.05 level (the same hereinafter). SOD, superoxide dismutase. SOD activities were measured after three days of low-temperature incubations at 5 °C, 0 °C, and −5 °C (the same hereinafter).

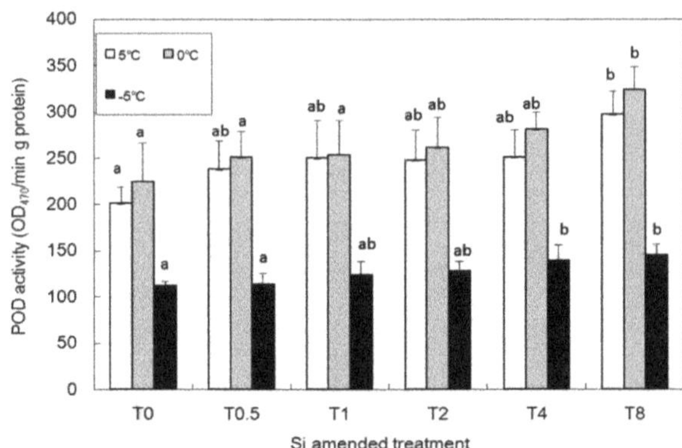

Figure 3. Effect of Si amendment on the POD activity of leaves at various temperatures. POD, peroxidase.

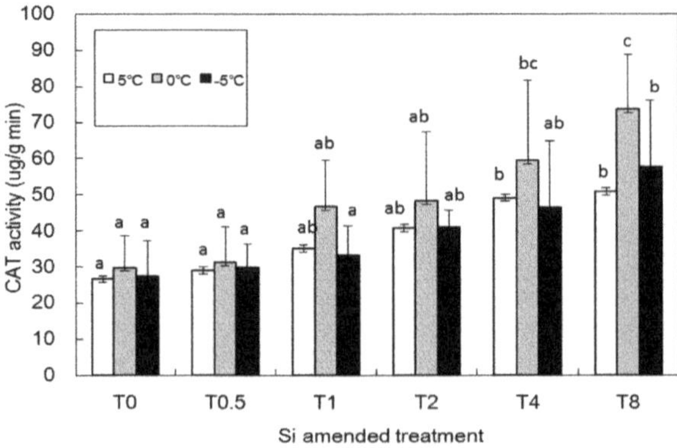

Figure 4. Effect of Si amendment on the CAT activity of leaves at various temperatures. CAT, catalase.

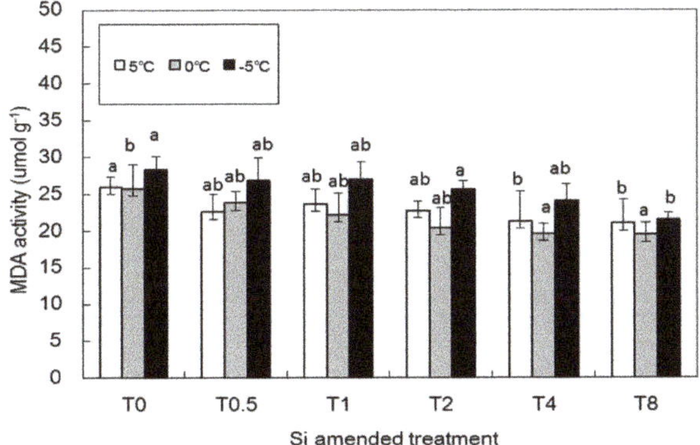

Figure 5. Effect of Si amendment on the MDA concentration of leaves at various temperatures. MDA, malondialdehyde.

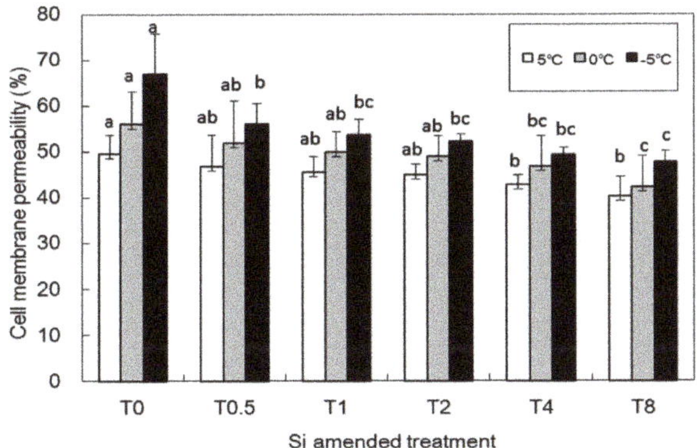

Figure 6. Effect of Si amendment on the cell membrane permeability of leaves at various temperatures.

3.5. Effect of Si Amendment on Bamboo Leaf Chloroplast Ultrastructure

Figure 7 shows that the low temperature of −5 °C exerted a significant effect on the bamboo leaf chloroplast. The chloroplast swelled to a circular shape and separated with the cell membrane. Even worse, the chloroplast membrane ruptured, and the grana disintegrated, while some dissolved. The osmiophillic number increased, and some small vesicles were found in the cell matrix (Figure 7A,B). The chloroplast showed an intact ultrastructure with Si treatment (Figure 7C,D) and came in close contact with the cell membrane. The membrane of the chloroplast was distinct and full. The application of Si prevented low-temperature stress in the bamboo leaf chloroplast and decreased the degree of damage affecting bamboo growth.

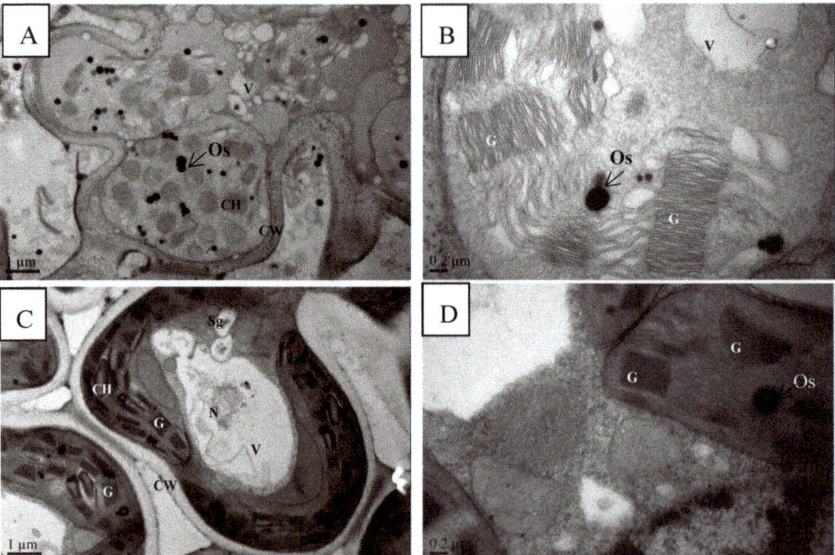

Figure 7. Effect of Si amendment on the chloroplast ultrastructure of bamboo leaves at −5 °C (**A**) 15,000×, chloroplast of control; (**B**) 40,000×, chloroplast of control; (**C**) 15,000×, chloroplast of Si treatment; (**D**) 40,000×, chloroplast of Si treatment). Os: Osmiophile globule; Cw: cell wall; V: vacuole; Ch: chloroplast; G: Granum; N: nucleus; Sg: starch grain. The ultrastructure of bamboos leaves (T0 and T4) were observed after three days of −5 °C temperature incubations.

4. Discussion

Temperature determines the distribution of bamboo [28,39,40], and *P. praecox* is usually subjected to low temperatures. Thus, Si amendment has been considered one of the best methods to mitigate low-temperature stress in *P. praecox*, because Si is beneficial for bamboo growth [41,42].

The results of this study indicate that Si amendment significantly improves the bamboo root-to-canopy ratio ($P = 5 \times 10^{-4}$), which indicate that the growth of underground organs increased, such that bamboo roots easily absorbed nutrients and water from the soil. Moreover, Si increased the total biomass of bamboo and especially accumulated in bamboo leaves, which was beneficial for the bamboo growth and enhanced the cold tolerance of bamboo. In our findings, Si significantly increased the photosynthesis rate of bamboo leaves ($P = 0.008$) and the ultrastructures of the chloroplasts, which was consistent with Detmann et al. [43]. Moreover, Si improved the leaf thickness and surface area of chloroplasts. In addition, Si improved the erectness of leaves, so as to indirectly improve the whole-plant photosynthesis [44]. The WUE increased significantly with Si amendment ($P = 0.01$) due to increased photosynthesis and reduced water Tr. The stomatal conductance and mesophyll conductance were limitations to the photosynthetic capacity of leaves [45], but there was no significant difference of stomatal conductance in all the Si amending treatments ($P > 0.05$), which suggested that stomatal produce posed a less significant limitation on photosynthesis. However, the high application rate of Si fertilization may have adverse effects on bamboo growth. Active and passive Si uptake both existed in plants, and in the passive process, the supply of silicon in the bamboo depends on the availability of $Si(OH)_4$ in the soil of their growth area and the rates of water uptake and evaporation [42]. When too much silicon accumulated in the bamboo, the transpiration rate significantly decreased ($P = 0.001$), and the passive process decreased, which caused the decrease in biomass. Thus, Si fertilization is an effective measure for improving the tolerance of bamboos to cold stress, and a Si fertilization rate of 2.0~8.0 g kg^{-1} is recommended for bamboo growth.

The increased growth was closely related to the stimulation in the enzymatic antioxidant system under cold stress. ROS, including superoxide ($O_2 \bullet-$), H_2O_2, and hydroxyl radicals ($OH\bullet$), were generated in the plants due to environmental stresses [46,47]. Chloroplasts are the major sites for ROS generation, and oxygen generated in the chloroplasts during photosynthesis can accept electrons passing through the photosystems, thus forming $O_2\bullet-$ [48]. When plants were exposed to cold stress factors, an excessive accumulation of ROS could result in oxidative damage to plant tissues [49–51]. Moradtalab et al. [52] suggested that the major effect of Si on improving the cold tolerance of plants is the mitigation of oxidative stress. Therefore, cold stress resistance is associated with an enhanced antioxidative defense system, which includes antioxidant compounds and several antioxidative enzymes. In our study, Si application stimulated the SOD, POD, and CAT activity of bamboo leaves at various temperatures. Moreover, the content of SOD, POD, and CAT activity increased with the increase of the Si application rate. SOD is the major $O_2\bullet-$ scavenger, and its enzymatic action results in H_2O_2 and O_2 formation. Since H_2O_2 can diffuse directly across the membrane, H_2O_2 produced in chloroplasts, mitochondria, and other organelles can also diffuse into the peroxisomes and be scavenged by CAT [53]. The lowest POD content was observed at $-5\,°C$. A low content of H_2O_2 in plant cells is removed by POD. When the H_2O_2 content is very high, CAT is mainly responsible for its removal [54]. Temperature stress results in an increase in the MDA content [55]. Therefore, a low temperature of $-5\,°C$ increased the MDA content of bamboo leaves, and the decreased MDA and CMP content indicated that the stability of the cell membrane was improved, the CMP was decreased, and the membrane lipid peroxidation was alleviated, induced by the low temperature, which enhanced the cold tolerance of the bamboos. Plants have developed an antioxidant system, including CAT, guaiacol peroxidase (GPX), glutathione peroxidase (GSH-Px), and the ascorbate–glutathione cycle, to scavenge ROS. The ascorbate–glutathione cycle, including glutathione (GSH), ascorbate (AsA), and related enzymes, such as glutathione reductase (GR) and dehydroascorbate reductase (DHAR), is an important way to scavenge the toxic products in chloroplasts and other non-photosynthetic tissues [56]. Our study did not examine the dynamic changes of GR and DHAR, but showed that the antioxidant enzymes of SOD, POD, and CAT were stimulated by Si addition under cold conditions. In addition, under a low temperature, the morphological structures of organelles change, along with the damaged chloroplasts and starch granules [57]. In our study, the shape of chloroplasts was improved by Si amendment, and the chloroplast membranes were stabilized. Our results supported previous studies that found that, under cold stress, Si amendment enhanced oxidative defense systems and the chlorophyll content, and reduced the absolute electrolyte leakage quantity of corn plantlets [21,22,58,59]. These results indicate that Si enhances the resistance of plants to cold stress. In addition, further field studies should be conducted to test the effects of silicon in improving bamboo cold tolerance. Our low temperature treatment was not conducted for as long as the field situation required, and our study of the mechanism of Si-mediated cold stress needs further development.

The mechanism of Si in improving bamboo cold tolerance was as follows: (1) Si improves the bamboo biomass, the Si content in bamboo leaves, the chlorophyll content, and the ultrastructure of chloroplast, which was beneficial in increasing the photosynthesis rate. (2) Antioxidant systems are stimulated by Si, and play an important role in alleviating the peroxidation damage induced by cold stress.

5. Conclusions

Our results show that Si application improved the growth of *P. praecox* and increased the biomass, root-to-canopy ratio, and *Pn* of bamboo leaves. The method of soil Si amendment was effective in alleviating cold stress in *P. praecox*. It resulted in an improved chloroplast and cell structure, increased CAT, POD, and SOD activities, and decreased MDA and CMP levels. These findings suggest that Si is beneficial for bamboo growth, although a high rate of application ($8\,g\,kg^{-1}$) may have adverse effects on bamboo growth, so we recommended a Si fertilization rate range of $2.0{\sim}8.0\,g\,kg^{-1}$, and Si fertilizer enhances the tolerance of bamboo plants to cold stress. Accordingly, Si fertilization is recommended for bamboos under cold conditions. Furthermore, more field studies need to be conducted to test the practical application of these findings.

Author Contributions: Conceptualization, Z.Z.Q., S.Y.Z., Q.L., and R.Y.G.; Methodology, Z.Z.Q., S.Y.Z., Q.L., and R.Y.G.; Validation, Z.Z.Q., S.Y.Z., Q.L., and R.Y.G.; Formal Analysis, S.Y.Z.; Resources, S.Y.Z. and R.Y.G.; Visualization, Z.Z.Q., S.Y.Z., Q.L., and R.Y.G.; Investigation, Z.Z.Q. and Q.L.; Resources, S.Y.Z. and R.Y.G.; Software, Z.Z.Q. and Q.L.; Data Curation, Z.Z.Q. and Q.L.; Writing—Original Draft Preparation, Z.Z.Q. and S.Y.Z.; Writing—Review and Editing, Z.Z.Q. and S.Y.Z.; Supervision, S.Y.Z. and R.Y.G.; Project Administration, S.Y.Z.; Funding Acquisition, S.Y.Z. and R.Y.G.

Funding: The authors are grateful for financial support from the National Natural Science Foundation of China (41671296), National Key R & D Project of China (2016FYE0112700), and Science and Technology Department of Zhejiang Province of China (2017C02016).

Acknowledgments: We are grateful for experimental support from Yuhe Zhang and Xin Wang.

Conflicts of Interest: The authors declare no conflict of interest.

References

1. Epstein, E. Silicon. *Annu. Rev. Plant Phys.* **1999**, *50*, 641–664. [CrossRef] [PubMed]
2. Goto, M.; Ehara, H.; Karita, S.; Takabe, K.; Ogawa, N.; Yamada, Y.; Ogawa, S.; Yahaya, M.S.; Morita, O. Protective effect of silicon on phenolic biosynthesis and ultraviolet spectral stress in rice crop. *Plant Sci.* **2003**, *164*, 349–356. [CrossRef]
3. Agarie, S.; Hanaoka, N.; Ueno, O.; Miyazaki, A.; Kubota, F.; Agata, W.; Kaufman, P.B. Effects of silicon on tolerance to water deficit and heat stress in rice plants (*Oryza sativa* L.), monitored by electrolyte leakage. *Plant Prod. Sci.* **1998**, *1*, 96–103. [CrossRef]
4. Liang, Y.C.; Zhang, W.H.; Chen, Q.; Liu, Y.L.; Ding, R.X. Effect of exogenous silicon (Si) on H+- ATPase activity, phospholipids and fluidity of plasma membrane in leaves of salt-stressed barley (*Hordeum vulgare* L.). *Environ. Exp. Bot.* **2006**, *57*, 212–219. [CrossRef]
5. Liang, Y.C.; Shen, Q.R.; Shen, Z.G.; Ma, T. Effects of silicon on salinity tolerance of two barley cultivars. *J. Plant Nutr.* **1996**, *19*, 173–183. [CrossRef]
6. Liang, Y.C. Effects of silicon on enzyme activity, and sodium, potassium and calcium concentration in barley under salt stress. *Plant Soil.* **1999**, *209*, 217–224. [CrossRef]
7. Liang, Y.C.; Zhu, J.; Li, Z.J.; Chu, G.X.; Ding, Y.F.; Zhang, J.; Sun, W.C. Role of silicon in enhancing resistance to freezing stress in two contrasting winter wheat cultivars. *Environ. Exp. Bot.* **2008**, *64*, 286–294. [CrossRef]
8. Liang, Y.C.; Yang, C.G.; Shi, H.H. Effects of silicon on growth and mineral composition of barley grown under toxic levels of aluminum. *J. Plant Nutr.* **2001**, *24*, 229–243. [CrossRef]
9. Iwasaki, K.; Matsumura, A. Effect of silicon on alleviation of manganese toxicity in pumpkin (*Cucurbita moschata* Duch cv. Shintosa). *Soil Sci. Plant Nutr.* **1999**, *45*, 909–920. [CrossRef]
10. Shi, Q.H.; Bao, Z.Y.; Zhu, Z.J.; He, Y.; Qian, Q.Q.; Yu, J.Q. Silicon-mediated alleviation of Mn toxicity in Cucumis sativus in relation to activities of superoxide dismutase and ascorbate peroxidase. *Phytochemistry* **2005**, *66*, 1551–1559. [CrossRef] [PubMed]
11. Zhu, Z.J.; Wei, G.Q.; Li, J.; Qian, Q.Q.; Yu, J.Q. Silicon alleviates salt stress and increases antioxidant enzymes activity in leaves of salt-stressed cucumber (*Cucumis sativus* L.). *Plant Sci.* **2004**, *167*, 527–533. [CrossRef]
12. Liang, Y.C.; Zhang, W.; Chen, Q.; Ding, R. Effect of silicon on H+- ATPase and H+- PPase activity, fatty acid composition and fluidity of tonoplast vesiciles from roots of salt-stressed barley (*Hordeum vulagare* L.). *Environ. Exp. Bot.* **2004**, *102*, 34–39.
13. Ding, Y.; Liang, Y.C.; Zhu, J.; Li, Z. Effects of silicon on plant growth, photosynthetic parameters and soluble sugar content in leaves of wheat under drought stress. *Plant Nutr. Fert. Sci.* **2007**, *13*, 471–478. (In Chinese)
14. Yeo, A.R.; Flowers, S.A.; Rao, G.; Welfare, K.; Senanayake, N.; Flowers, T.J. Silicon reduces sodium uptake in rice (*Oryza sativa* L.) in saline conditions and this is accounted for by a reduction in the transpirational bypass flow. *Plant Cell Environ.* **1999**, *22*, 559–565. [CrossRef]
15. Chen, W.; Yao, X.Q.; Cai, K.Z.; Chen, J.N. Silicon alleviates drought stress of rice plants by improving plant water status, photosynthesis and mineral nutrient absorption. *Biol. Trace Elem. Res.* **2011**, *142*, 67–76. [CrossRef]

16. Steponkus, P.L. Role of the plasma membrane in freezing injury and cold acclimation. *Annu. Rev. Plant. Physiol.* **1984**, *35*, 543–584. [CrossRef]

17. Zhang, Q.; Zhang, J.Z.; Chow, W.S.; Sun, L.L.; Chen, J.W.; Chen, Y.J.; Peng, C.L. The influence of low temperature on photosynthesis and antioxidant enzymes in sensitive banana and tolerant plantain (*Musa* sp.) cultivars. *Photosynthetica* **2011**, *49*, 201–208. [CrossRef]

18. Alscher, R.G.; Erturk, N.; Heath, L.S. Role of superoxide dismutases (SODs) in controlling oxidative stress in plants. *J. Exp. Bot.* **2002**, *53*, 1331–1341. [CrossRef]

19. Kawano, T. Roles of the reactive oxygen species-generating peroxidase reactions in plant defense and growth induction. *Plant Cell Rep.* **2003**, *21*, 829–837. [PubMed]

20. Polidoros, A.N.; Scandalios, J.G. Role of hydrogen peroxide and different classes of antioxidants in the regulation of catalase and glutathione S-transferase gene expression in maize (*Zea mays* L.). *Physiol. Plantarum.* **1999**, *106*, 112–120. [CrossRef]

21. Liu, J.J.; Lin, S.H.; Xu, P.L.; Wang, X.J.; Ba, J.G. Effects of exogenous silicon on the activities of antioxidant enzymes and lipid peroxidation in chilling-stressed cucumber leaves. *Agric. Sci. China* **2009**, *8*, 1075–1086. [CrossRef]

22. Zhan, H.; Zhang, L.; Deng, L.; Niu, Z.; Li, M.; Wang, C.; Wang, S. Physiological and anatomical response of foliar silicon application to *Dendrocalamus brandisii* plantlet leaves under chilling. *Acta Physiol. Plant.* **2018**, *40*, 208. [CrossRef]

23. Xu, C.X.; Liu, Y.L. Si absorption, transport and accumulation in plants. *Acta Bot. Boreali Occid. Sin.* **2006**, *26*, 1071–1078. (In Chinese)

24. Ma, J.F.; Tamai, K.; Yamaji, N.; Mitani, N.; Konishi, S.; Katsuhara, M.; Ishiguro, M.; Murata, Y.; Yano, M. A silicon transporter in rice. *Nature* **2006**, *440*, 688–691. [CrossRef] [PubMed]

25. Ma, J.F.; Yamaji, N. Si uptake and accumulation in higher plants. *Trends Plant Sci.* **2006**, *11*, 392–397. [CrossRef]

26. Ma, J.F. Role of silicon in enhancing the resistance of plants to biotic and abiotic stresses. *Soil Sci. Plant Nutr.* **2004**, *50*, 11–18. [CrossRef]

27. Zhou, B.; Wang, X.; Cao, Y.; Ge, X.; Gu, L.; Meng, J. Damage assessment to subtropical forests following the 2008 Chinese ice storm. *iForest* **2017**, *10*, 406–415. [CrossRef]

28. Zhong, Y.; Zhuang, S.Y.; He, J.C.; Gui, R.Y. Intensive management measures: Effects on freeze injury of *Phyllostachys praecox*. *Chin. Agric. Sci. Bull.* **2018**, *34*, 95–100. (In Chinese)

29. He, D.; Song, H.; Qiu, Z.; Shi, G.; Chen, J. Analysis on effect of snow disaster on *Phyllostachys praecox* cv. *Prevernalis* stand and countermeasures. *J. Zhejiang For. Sci. Tech.* **2008**, *5*, 54–56. (In Chinese)

30. Liang, Y.C.; Sun, W.; Zhu, Y.G.; Christie, P. Mechanisms of silicon-mediated alleviation of abiotic stresses in higher plants: A review. *Environ. Pollut.* **2007**, *147*, 42. [CrossRef] [PubMed]

31. Ma, J.F.; Tamai, K.; Ichii, M.; Wu, G.F. A rice mutant defective in Si uptake. *Plant Physiol.* **2002**, *30*, 2111–2117. [CrossRef] [PubMed]

32. Yue, J.; Fu, Z.; Zhang, L.; Zhang, Z.; Zhang, J. The positive effect of different 24-epiBL pretreatments on salinity tolerance in *Robinia pseudoacacia* L. seedlings. *Forests* **2019**, *10*, 4. [CrossRef]

33. Giannopolitis, C.N.; Ries, S.K. Superoxide dismutases: I. Occurrence in higher plants. *Plant Physiol.* **1977**, *59*, 309–314. [CrossRef]

34. Maehly, A. The assay of catalases and peroxidases. *Methods Biochem. Anal.* **1954**, 357–424. [CrossRef]

35. Chance, B.; Maehly, A.C. Assay of catalase and peroxidases. *Methods Enzymol.* **1955**, *1*, 764–775.

36. Xu, P.L.; Guo, Y.K.; Bai, J.G.; Shang, L.; Wang, X.J. Effects of long-term chilling on ultrastructure and antioxidant activity in leaves of two cucumber cultivars under low light. *Physiol. Plant.* **2008**, *132*, 467–478. [CrossRef]

37. Zhang, Y.; Huang, X.; Chen, Y. *Experimental Course of Plant Physiology*, 2nd ed.; Higher Education Press: Beijing, China, 2009; pp. 129–139. (In Chinese)

38. Tian, H.Q.; Kuang, A.; Musgrave, M.E.; Russell, S.D. Calcium distribution in fertile and sterile anther of a photoperiod-sensitive genic male-sterile rice. *Planta* **1998**, *204*, 183–192. [CrossRef]

39. Jiang, P.; Ye, Z.; Xu, Q. Effect of mulching on soil chemical properties and enzyme activities in bamboo plantation of *phyllostachy praecox*. *Commun. Soil Sci. Plan.* **2002**, *33*, 11. [CrossRef]

40. Gu, D.X.; Chen, S.L.; Zheng, W.M.; Mao, X.Q. Review of the ecological adaptability of bamboo. *J. Bamboo Res.* **2010**, *29*, 17–23. (In Chinese)

41. Li, Z.; Lin, P.; He, J.; Yang, Z.; Lin, Y. Silicon's organic pool and biological cycle in moso bamboo community of Wuyishan Biosphere Reserve. *J. Zhejiang Univ.-SC B* **2006**, *7*, 849–857. [CrossRef]

42. Ding, T.P.; Zhou, J.X.; Wan, D.F.; Chen, Z.Y.; Wang, C.Y.; Zhang, F. Silicon isotope fractionation in bamboo and its significance to the biogeochemical cycle of silicon. *Geochim Cosmochim Ac.* **2008**, *72*, 1395. [CrossRef]

43. Detmann, K.C.; Araújo, W.L.; Martins, S.C.; Sanglard, L.M.; Reis, J.V.; Detmann, E.; Rodrigues, F.A.; Nunes-Nesi, A.; Fernie, A.R.; DaMatta, F.M. Silicon nutrition increases grain yield, which, in turn, exerts a feed-forward stimulation of photosynthetic rates via enhanced mesophyll conductance and alters primary metabolism in rice. *New Phytol.* **2012**, *196*, 752–762. [CrossRef] [PubMed]

44. Tamai, K.; Ma, J.F. Reexamination of silicon effects on rice growth and production under field conditions using a low silicon mutant. *Plant Soil.* **2008**, *307*, 21–27. [CrossRef]

45. Flexas, J.; Barbour, M.M.; Brendel, O.; Cabrera, H.M.; Carriqui, M.; Diaz-Espejo, A.; Douthe, C.; Dreyer, E.; Ferrio, J.P.; Gago, J.; et al. Mesophyll diffusion conductance to CO_2: An unappreciated central player in photosynthesis. *Plant Sci.* **2012**, *193*, 70–84. [CrossRef] [PubMed]

46. Alscher, R.G.; Donahue, J.L.; Cramer, C.L. Reactive oxygen species and antioxidants: Relationship in green cells. *Physiol. Plant.* **1997**, *100*, 224–233. [CrossRef]

47. Mittler, R. Oxidative stress, antioxidants and stress tolerance. *Trends Plant Sci.* **2002**, *7*, 405–410. [CrossRef]

48. Gill, S.S.; Tuteja, N. Reactive oxygen species and antioxidant machinery in abiotic stress tolerance in crop plants. *Plant Physiol. Bioch.* **2010**, *48*, 909–930. [CrossRef]

49. Baek, K.H.; Skinner, D.Z. Production of reactive oxygen species by freezing stress, and the protective roles of antioxidant enzymes in plants. *J. Agric. Chem. Environ.* **2012**, *1*, 34–40. [CrossRef]

50. Saeidnejad, A.H.; Pouramir, F.; Naghizedek, M. Improving chilling tolerance of maize seedlings under cold conditions by spermine application. *Not Sci. Biol.* **2012**, *4*, 110–117. [CrossRef]

51. Moradtalab, N.; Weinmann, M.; Walker, F.; Höglinger, B.; Ludewig, U.; Neumann, G. Silicon improves chilling tolerance during early growth of maize by effects on micronutrient homeostasis and hormonal balances. *Front Plant Sci.* **2018**, *9*, 420. [CrossRef]

52. Neto, A.D.D.; Prisco, J.T.; Eneas, J.; de Abreu, C.E.B.; Gomes, E. Effect of salt stress on antioxidative enzymes and lipid peroxidation in leaves and roots of salt-tolerant and salt-sensitive maize genotypes. *Environ. Exp. Bot.* **2006**, *56*, 87–94. [CrossRef]

53. Foyer, C.H.; Lopez-Delgado, H.; Dat, J.F.; Scott, I.M. Hydrogen peroxide and glutathione-associated mechanisms of acclamatory stress tolerance and signaling. *Physiol. Plant.* **1997**, *100*, 241–254. [CrossRef]

54. Zgallai, H.; Steppe, K.; Lemeur, R. Effects of different levels of water stress on leaf water potential, stomatal resistance, protein and chlorophyll content and certain anti-oxidative enzymes in tomato plants. *J. Integr. Plant Biol.* **2006**, *48*, 679–685. [CrossRef]

55. Sgherri, C.L.M.; Loggini, B.; Puliga, S.; Navari-Izzo, F. Antioxidant system in Sporobolus stapfianus: Changes in response to desiccation and rehydration. *Phytochemistry* **1994**, *35*, 561–565. [CrossRef]

56. Wang, L.J.; Huang, W.D.; Li, J.Y.; Liu, Y.F.; Shi, Y.L. Peroxidation of membrane lipid and Ca^{2+} homeostasis in grape mesophyll cells during the process of cross-adaptation to temperature stresses. *Plant Sci.* **2004**, *167*, 71–77. [CrossRef]

57. Ji, S.X.; Dai, S.J.; Liu, W. The advances of plants in response and adaption to low temperature stress. *Chin. Bull Life Sci.* **2010**, *22*, 1013–1019. (In Chinese)

58. He, Y.; Xiao, H.; Wang, H.; Chen, Y.; Yu, M. Effect of silicon on chilling-induced changes of solutes, antioxidants, and membrane stability in seashore paspalum turfgrass. *Acta Physiol. Plant.* **2010**, *32*, 487–494. [CrossRef]

59. Chen, B.; Shi, Y.; Chen, Z. Studies on the effect of different P, K, Si fertilizer of corn cold-resistant. *Chin. Agric. Sci. Bull.* **2011**, *27*, 85–89. (In Chinese)

 *forests*

Article

Salicylic Acid Alleviated Salt Damage of *Populus euphratica*: A Physiological and Transcriptomic Analysis

Shupei Rao [1,2], Chao Du [1], Aijia Li [1,2], Xinli Xia [1,2,3], Weilun Yin [1,2] and Jinhuan Chen [1,2,*]

[1]   College of Biological Sciences and technology, Beijing Forestry University, Beijing 100083, China; raoshupei@bjfu.edu.cn (S.R.); duchao@bjfu.edu.cn (C.D.); aijiali@bjfu.edu.cn (A.L.); xiaxl@bjfu.edu.cn (X.X.); yinwl@bjfu.edu.cn (W.Y.)
[2]   National Engineering Laboratory for Tree Breeding, Beijing Forestry University, Beijing 100083, China
[3]   Beijing Advanced Innovation Center for Tree Breeding by Molecular Design, Beijing Forestry University, Beijing 100083, China
*   Correspondence: chenjh@bjfu.edu.cn; Tel.: +86-10-6233-8371

Received: 2 April 2019; Accepted: 10 May 2019; Published: 16 May 2019

Abstract: *Populus euphratica* Oliv. is a model tree for studying abiotic stress, especially salt stress response. Salt stress is one of the most extensive abiotic stresses, which has an adverse effect on plant growth and development. Salicylic acid (SA) is an important signaling molecule that plays an important role in modulating the plant responses to abiotic stresses. To answer whether the endogenous SA can be induced by salt stress, and whether SA effectively alleviates the negative effects of salt on poplar growth is the main purpose of the study. To elucidate the effects of SA and salt stress on the growth of *P. euphratica*, we examined the morphological and physiological changes of *P. euphratica* under 300 mM NaCl after treatment with different concentrations of SA. A pretreatment of *P. euphratica* with 0.4 mM SA for 3 days effectively improved the growth status of plants under subsequent salt stress. These results indicate that appropriate concentrations of exogenous SA can effectively counteract the negative effect of salt stress on growth and development. Subsequently, transcripts involved in salt stress response via SA signaling were captured by RNA sequencing. The results indicated that numerous specific genes encoding mitogen-activated protein kinase, calcium-dependent protein kinase, and antioxidant enzymes were upregulated. Potassium transporters and Na^+/H^+ antiporters, which maintain K^+/Na^+ balance, were also upregulated after SA pretreatment. The transcriptome changes show that the ion transport and antioxidant enzymes were the early enhanced systems in response of *P. euphratica* to salt via SA, expanding our knowledge about SA function in salt stress defense in *P. euphratica*. This provides a solid foundation for future study of functional genes controlling effective components in metabolic pathways of trees.

Keywords: *Populus euphratica*; salt stress; salicylic acid; malondialdehyde; differentially expressed genes

---

1. Introduction

Plants are exposed to multiple environmental stresses such as salinity, drought, and extreme temperatures that are harmful to plants. In particular, salinity stress, which usually occurs in arid and semi-arid regions, severely affects plant growth and development throughout the world [1]. Many studies confirm the inhibitory effect of salt stress on physiological and biochemical processes, which inhibits photosynthesis and destroys cell membranes. The question of how stressed plants can diminish the deleterious effects of salt stress has drawn a lot of attention. Salicylic acid (SA) is a naturally synthesizing endogenous signaling molecule that activates plant growth and defense responses to systemic acquired resistance (SAR) and biotic resistance [2–5]. It was first discovered to be a plant immune signal, produced by a pathogen challenge to induce SAR. SA exhibits a very broad-spectrum defense against reinvasion of pathogens and has been widely investigated due to its functions in plant immunity against diseases or biotic resistance [6]. The accumulation of endogenous SA can induce the expression of pathogenesis-related genes, by imitating endogenous phenolic signaling molecules and activating the development of SAR, thereby increasing resistance of plants [7]. In addition to biotic resistance, an increasing number of studies indicate that SA may modulate abiotic tolerance and have potential functions in response to plant salinity resistance. SA could generate a wide range of metabolic and physiological responses in plants by affecting their growth and development. For example, foliar SA application has been reported to markedly promote growth and relative water content and reduce electrolyte leakage of cucumber plants grown under salt stress [8,9]. During salt stress, SA can also improve sunflower growth and photosynthesis [10]. However, the effective concentration of SA to alleviating salt damages depends on the plant species, genotypes, developmental stages, and tissue types. To date, the abiotic tolerance response mechanism of SA, especially in trees, still needs to be elucidated. Further studies are needed to determine the function of SA in abiotic tolerance, and the early response genes to SA and investigate whether abiotic stress can increase the levels of endogenous SA and how the SA signaling pathway is involved in abiotic stress response.

Populus euphratica Oliv. is a special poplar species found in saline and semi-arid desert regions extending from western China, through the Middle East and Central Asia, to North Africa [11,12]. Being the only arboreal species in the largest shifting sand desert in the world [13], *P. euphratica* has the ability to adapt to hostile and extreme environmental stresses, especially salt, which makes it a model species for elucidating mechanisms of abiotic stress resistance in trees [14]. Research on the mechanism of resistance in *P. euphratica* should greatly contribute to the cultivation of woody plants and the protection of *P. euphratica* germplasm resources [15–17]. Previously, scientists completed the whole-genome sequencing of *P. euphratica* and obtained dozens of salt stress responsive genes in this species [18]. However, whether SA influences or participates in the abiotic stress response of *P. euphratica* remains to be elucidated.

In view of the above description, the main purpose of this study was to determine whether endogenous SA could be induced under salt stress in *P. euphratica*, and whether exogenous SA could effectively alleviate the negative effects of salt stress on plant growth. By studying the physiological, biochemical parameters, and genomic transcriptional profile of SA-pretreated *P. euphratica* seedlings under salt stress, the theoretical foundation for the expansion of abiotic stress response mechanism in regions with high salinity was further established.

2. Materials and Methods

2.1. Plant Materials

The *P. euphratica* seeds were collected from the Inner Mongolia Autonomous Region of China. The seeds were first sterilized and cultured in tissue-cultured plantlets for growth and propagation. Then, they were transferred to plastic pots for 1 month in the greenhouse of Beijing Forestry University. After that, seedlings with similar stem length and number of expanded leaves were selected and grown hydroponically in half-strength Hoagland's solution in a hydroponic box for another month. The environment conditions were as follows: 16 h photoperiod, 25 ± 2 °C/18 ± 2 °C (day/night) temperature, 250 µmol m^{-2} s^{-1} light intensity, and 65% ± 5% relative humidity. Each hydroponic box dish contained 12 seedlings.

2.2. Quantification of SA after Salt Treatment

P. euphratica seedlings were divided into three groups, which were treated with 300 mM NaCl solution for 0, 1, and 6 h, respectively. The salt-treated leaves were snap-frozen in a liquid nitrogen environment and ground into a fine powder by vigorous vortexing. Approximately 50 mg of powder per tube was immediately weighed with a 1/10,000 balance weight. Afterward, 0.5 mL of extraction solvent (dichloromethane, 99.9%) was added, followed by 50 μL of 0.2 ng μL^{-1} IH4 SA and 10 μL of 1 ng μL^{-1} D6 abscisic acid (ABA) as internal standards. The tubes were then placed on a shaker at a speed of 100 rpm under 4 °C for 30 min. After centrifugation at 13,000 rpm for 5 min, two phases were formed, and approximately 900 μL of the solvent from the lower phase was transferred for later use. Finally, the samples were evaporated to dryness and dissolved in 0.1 mL methanol. The contents of SA and ABA were determined using ultra performance liquid chromatography/tandem mass spectrometry (UPLC-MS/MS) (Agilent 5500, Agilent Technologies Inc., Santa Clara, CA, USA), by injecting 50 μL of sample solution into the reverse-phase C18 Gemini UPLC column. The area of peaks in the chromatogram was quantified using MassHunter software (Agilent Technologies Inc., Santa Clara, CA, USA).

2.3. Effects of SA and NaCl Treatment on Morphology and Physiology

To test the effects of SA treatments on *P. euphratica* growth, SA was dissolved in ethanol and Tween-20 (0.1% dilute solution), and double-distilled water was added to obtain an SA mother liquor concentration. We further divided the plants into four groups and cultured them in half-strength Hoagland's solution containing SA at concentrations of 0, 0.4, 1.0, and 2.0 mM. After 3 days, all the seedlings were rinsed with distilled water to remove residual SA and then transferred to half-strength Hoagland's solution supplemented with 300 mM NaCl. The performances were recorded on the 3rd day after NaCl treatment. To clearly observe the effect of SA on alleviating salt stress from the phenotype, we performed the same experiment using salt treatment on the 7th day of lethality during phenotypic observation.

The chlorophyll contents of the leaves were measured using the Minolta Chlorophyll Meter (SPAD-502, Konica Minolta, Osaka, Japan) in vivo. The seedlings growing under normal growth condition were used as untreated control. Six random leaves of the upper, middle, and lower parts of each plant were selected in the same position, and each leaf was measured thrice. At the end of experiment, the leaves from three replications per treatment were harvested for fresh weight (FW) determination.

2.4. Determination of Malondialdehyde (MDA) Content and Antioxidant Enzyme

The MDA contents were determined following the previously reported method [19,20]. The leaves of seedlings (0.50 g) were homogenized in 10 mL of 10% (*w/v*) trichloroacetic acid (TCA) and then centrifuged at 10,000 × *g*. Approximately 2 mL of 0.6% (*w/v*) thiobarbituric acid containing 10% (*w/v*) TCA was added. The samples were heated for 30 min in a boiling water bath at 100 °C. The absorbance of the supernatant was recorded at 532, 450, and 600 nm after the samples cooled. The MDA content was calculated as follows: C (μmol g^{-1}) = 6.45 × (A532 − A600) − 0.56 × A450. Peroxidase (POD) activities were assayed according to Meloni et al. [21]. The activity of superoxide dismutase (SOD) was determined according to the method described by Becana et al. [22]. Three replicate measurements were made for each treatment, and the results were averaged.

2.5. Determination of K and Na Contents

After various treatments, the leaves of seedlings were dried in an oven at 70 °C to constant weight. The dried samples were weighed and pulverized, and then digested at 260 °C with 2 mL of 30% H_2O_2 and 5 mL of H_2SO_4 in a microwave oven (CEM mars240/50, CEM Inc., Matthews, NC, USA). Then, the samples were cooled and subjected to atomic absorption spectrometer (Varian, Spectraa-220) to determine the concentrations of K and Na.

2.6. RNA Extraction, Library Preparation, and RNA Sequencing (RNA-seq)

Leaves of *P. euphratica* seedlings pretreated with 0 and 0.4 mM of SA solution were frozen in liquid nitrogen immediately after harvest at 0 h and 6 h time points and stored at −80 °C for later use.

Total RNA was extracted using RNAprep Pure Plant Kit (TIANGEN BIOTECH Co. Ltd., Beijing, China). After extracting and treating the RNA with DNase I, we used Oligo (dT) to isolate mRNA (Messenger RNA), which was further fragmented to synthesize cDNA by reverse transcription. After using the Agilent 2100 Bioanalyzer (Agilent Technologies Inc., Santa Clara, CA, USA) and ABI StepOnePlus Real-Time PCR System (Applied Biosystems, Waltham, MA, USA) to quantify and qualify the sample library, the cDNA library was sequenced using Illumina HiSeq 2000™ (Illumina, San Diego, CA, USA).

2.7. Processing of Sequence Data

The raw data for sequencing included low quality, linker contamination, and reads with unknown base N content, which need to be removed prior to data analysis to ensure reliable results. We used SOAPnuke (https://github.com/BGI-flexlab/SOAPnuke, The Beijing Genomics Institute, Shenzhen, Guang Dong, China), a filter software developed by BGI company, to remove reads containing adaptor, reads with unknown base N content greater than 5%, and low-quality bases (>20% of the bases with a quality score ≤10). After filtering, the remaining reads were called "clean reads" and stored in FASTQ (A format that stores biological sequences and corresponding quality assessments) format [23].

2.8. Mapping Reads to the P. euphratica Genome and Function Annotation

We used HISAT2 (http://www.ccb.jhu.edu/software/hisat, The Center for Computational Biology at Johns Hopkins University, Baltimore, MD, USA) and Bowtie2 (http://bowtie-bio.sourceforge.net/Bowtie2/index.shtml, The Center for Computational Biology at Johns Hopkins University, Baltimore, MD, USA) to compare RNA-seq reads from control and treated samples with reference genomic sequences of *P. euphratica* [24,25]. The abundance of transcripts was calculated using the RSEM method of gene expression by mapping the transcript to RNA-seq reads and expressed as fragments per kilobase of transcript per million mapped reads (FPKM) [25,26].

2.9. Differential Expression Analysis

To identify differentially expressed genes (DEGs) in SA-pretreated and non-SA-pretreated salt-stressed *P. euphratica* leaves, we used the DEGseq method based on the Poisson distribution model for differential gene detection [27]. To improve the accuracy of DEGs, we defined the false discovery rate (FDR) ≤0.001 and $\log_2$ ratio ≥1 as thresholds to discriminate significant DEGs. The differential genes were functionally classified by Gene Ontology (GO), using the Web Gene Ontology Annotation Plot (WEGO) software package. The phyper function of the R software was used for enrichment analysis with the threshold of FDR ≤0.001. Transcription factors (TFs) were predicted using HMMsearch (http://hmmer.org, European Bioinformatics Institute, Cambridge, Cambs, England) [28].

2.10. Pathway Enrichment Analysis of DEG

Pathway enrichment analysis was based on the Kyoto Encyclopedia of Genes and Genomes (KEGG) database (http://www.genome.jp/kegg/, Kyoto University Bioinformatics Center, Sakyo ku, Kyoto, Japan) and performed as described previously [14]. Q value was used for determining the threshold of significance in multiple tests. Pathways with a Q value of ≤0.05 were considered significantly enriched for DEGs.

2.11. Validation of DEGs through qRT-PCR (Quantitative Real-Time PCR)

To validate the reliability of RNA-seq experiments, we randomly selected eight functional DEGs from RNA-seq results. Total RNA was extracted using *P. euphratica* leaves from the same three samples. Reverse transcription was performed using FastQuant RT Kit with DNase (TIANGEN BIOTECH Co., Ltd., Beijing, China) to synthesize the first-strand cDNA. qRT-PCR was performed in an optical 96-well reaction plate using the ABI PRISM 7500 real-time PCR system (Applied Biosystems). SuperReal PreMix Plus SYBR Green (TIANGEN BIOTECH Co., Ltd., Beijing, China) was used in the experiments.

Each reaction contained 12.5 µL of SYBR Premix ExTaq, 0.5 µL of ROX Reference Dye, 2.0 µL of cDNA samples, and 1.0 µL of gene-specific primers, with a final volume of 25 µL. The thermal cycle program used was as follows: 95 °C for 10 s, 45 cycles at 95 °C for 5 s, and 60 °C for 40 s. Real-time PCR data were analyzed using the $2^{-\Delta\Delta CT}$ method [29].

2.12. Statistical Analysis

Statistical tests were performed with the SPSS software (version 20.0, SPSS Inc., Chicago, IL, USA). Duncan' test was used to analyze the data unless otherwise indicated. The means were compared on the basis of $p < 0.05$.

3. Results

3.1. Increased SA Content in P. euphratica under Salt Stress

To illustrate the relationship between endogenous SA and salt stress, we evaluated the SA content by UPLC-MS/MS after short-term salt treatment. Given that ABA has been shown to be involved in the adaptation to environmental stresses, it was used as a reference phytohormone in the measurement. UPLC- MS/MS was used for the quantitative profiling of SA and ABA contents in *P. euphratica* leaves. Figure 1 shows the SA and ABA content results of a typical MRM chromatogram of internal standard solution and sample under control calculated by Analyst® 1.5 software (Agilent Technologies Inc., Santa Clara, CA, USA). Their internal standard is shown in Figure S1. For the quantitative analysis of SA and ABA, their contents under salt treatment at different time points were calculated by comparing with their internal standards. Our results showed that the levels of SA and ABA increased significantly at 1 and 6 h after application of 300 mM NaCl, respectively. The SA and ABA contents were approximately 200 and 100 ng/g FW$^-$ at 1 h, respectively, and both reached a concentration of about 350 ng/g FW^{-a} at 6 h. This result indicated that the content of SA and ABA increased significantly during the first 6 h of salt treatment, but the rate of SA increased was lower than that of ABA.

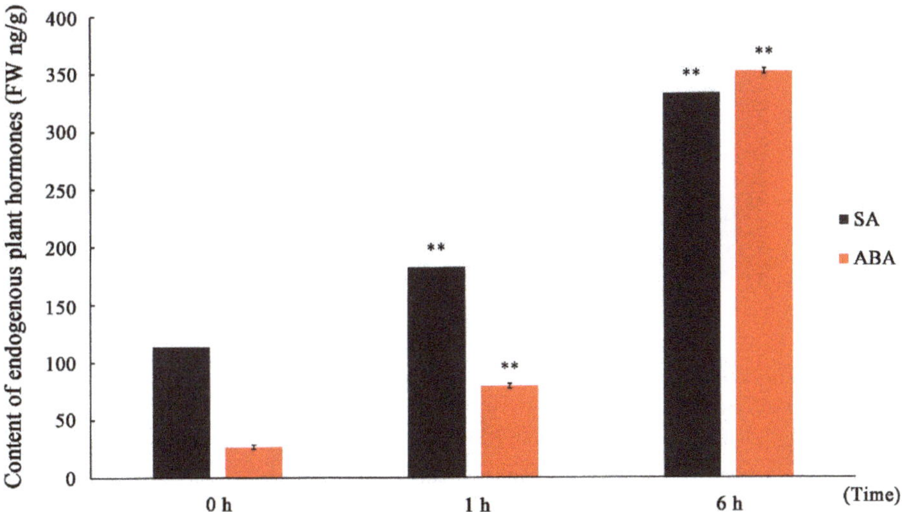

Figure 1. Content of endogenous plant hormones (ABA and SA) in *P. euphratica* leaf samples after treated with 300 mM NaCl. Data represent the mean ± SD of three independent experiments. X axis represents the treatment time of 300 mM NaCl. Y axis represents the content of endogenous plant hormones. "FW" denotes fresh weight of plant sample. Asterisks indicate a significant hormonal difference between control and salt-stressed plants ($n = 3$, ** $p < 0.01$,* $p < 0.05$, Duncan' test).

3.2. Morphological and Physiological Response of SA Treatment to Salt Tolerance of P. euphratica Seedlings

The general morphological traits of the plants at the end of the experiment are strongly dependent on the treatment. To demonstrate the relationship between SA pretreatment and salt stress response performance at the phenotypic and physiological level, we treated *P. euphratica* with SA concentration gradients of 0, 0.4, 1, and 2 mM, followed by a 300 mM NaCl treatment. When treated with 300 mM NaCl for 3 days, the leaves of *P. euphratica* showed obvious curl and chlorosis (Figure 2b). Most leaves of 0.4 mM SA-pretreated seedlings remained unfolded and green when treated with the same concentration of NaCl (Figure 2c). However, when the SA concentration was increased to 1 mM, the whole plant began to curl (Figure 2d). *P. euphratica* was treated continuously with 300 mM NaCl. After 7 days of 300 mM NaCl treatment, the leaves of *P. euphratica* seedlings without SA treatment withered (Figure 2e). Based on the phenotype, the 0.4 mM SA significantly alleviated the negative effect of salt stress (Figure 2f). Although the leaves became curly, they still remained green. The seedlings pretreated with 0.4 mM SA also displayed better performance than the seedlings treated with higher concentrations (1 mM and 2 mM) of SA (Figure 2g–h), indicating that moderate SA treatment on the phenotype could alleviate the damage caused by salt stress to *P. euphratica*.

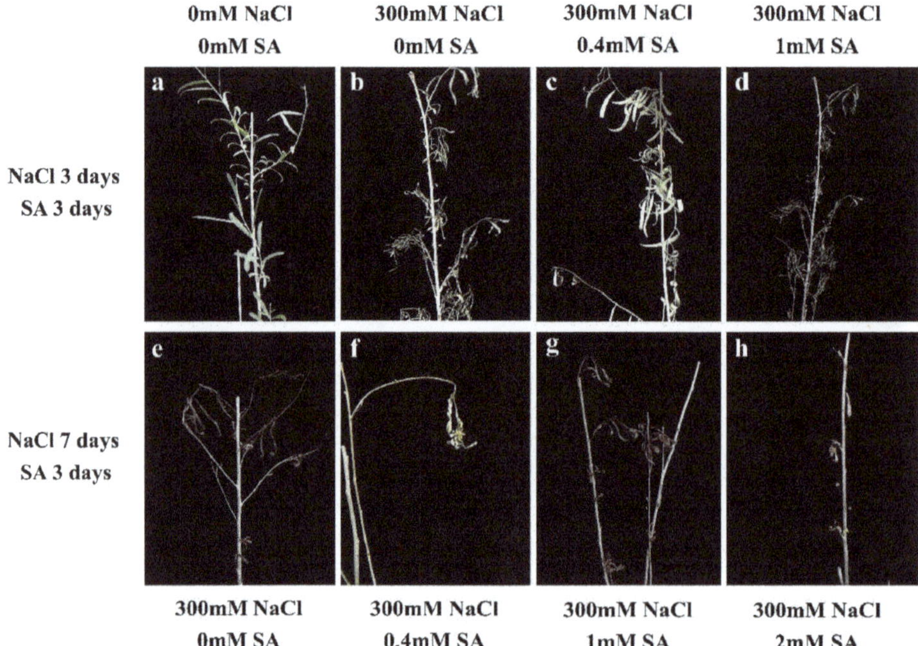

Figure 2. Phenotypes of *P. euphratica* seedlings treated with SA and salt stress. (**a**) Seedling grown under normal condition. (**b**) Seedling only treated with 300 mM NaCl for 3 d. (**c**) Seedling treated with 0.4 mM SA (3 d) + 300 mM NaCl (3 d). (**d**) Seedling treated with 1 mM SA (3 d) + 300 mM NaCl (3 d). (**e**) Seedling only treated with 300 mM NaCl for 7 d. (**f**) Seedling treated with 0.4 mM SA (3 d) + 300 mM NaCl (7 d). (**g**) Seedling treated with 1 mM SA (3 d) + 300 mM NaCl (7 d). (**h**) Seedling treated with 2 mM SA (3 d) + 300 mM NaCl (7 d).

We further measured physiological indices of *P. euphratica* seedlings before and after salt treatment with or without SA pre-application. Overall, 0.4 mM SA-pretreated seedlings exhibited less chlorosis and relatively increased total FW after 3 days of salt stress treatment compared to the salt only treated group. The total FW in the 0.4 mM SA + salt treatment group for 3 days was very close to the untreated control, and it was 48.3%, 28.3%, and 30.8% higher than those in the salt treatment—1 mM SA + salt pretreatment, and 2 mM SA + salt pretreatment groups, respectively (Figure 3a). The chlorophyll content in the 0.4 mM SA + salt treatment group was 29.9%, 28.3%, and 30.8% higher than those in the salt treatment, 1 mM SA + salt pretreatment, and 2 mM SA + salt pretreatment groups, respectively (Figure 3b). The phenotypic results were confirmed by MDA content measurement. As shown in Figure 3c, the MDA content was reduced significantly in 0.4 mM SA-treated samples compared with salt-treated and 1 or 2 mM SA-pretreated samples. Furthermore, a significant accumulation of MDA (over three times higher) was observed in the 2 mM pretreatment + 300 mM NaCl treatment samples compared with direct salt treatment, indicating that when the concentration of SA exceeded a certain range, it would cause damage to the plants. The results showed that salinity caused membrane damage to *P. euphratica* seedlings. The oxidative damage of plasma membrane was alleviated with the addition of 0.4 mM exogenous SA, but not with 1 or 2 mM SA. Therefore, the application of appropriate concentrations of exogenous SA could ameliorate membrane deterioration under salt stress and facilitate the maintenance of membrane functions in *P. euphratica*.

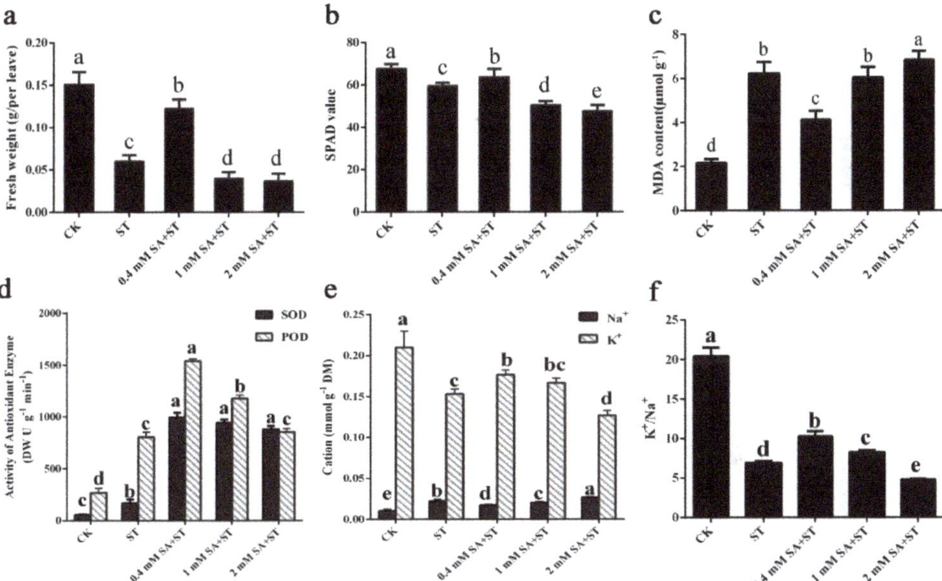

Figure 3. Effects of SA on the chlorophyll, fresh weight, MDA, SOD, and POD in *P. euphratica* grown under salt stress. (**a**) Chlorophyll content indicated by chlorophyll meter SPAD-502. (**b**) Fresh weight. (**c**) Effects of application different concentrations of SA on the malondialdehyde (MDA) content of *P. euphratica* grown under salt stress. (**d**) SOD (black) and POD (oblique line): activities of *P. euphratica* grown under salt stress with or without different concentrations of SA pretreatment. (**e**) Na$^+$ (black) and K$^+$ (oblique line): contents of leaves with different treatment. (**f**) K$^+$/Na$^+$ ratio with different treatment. ST means salt treated, 0.4 mM SA+ST, 1.0 mM SA+ST and 2.0 mM SA+ST indicate different concentrations of SA solution plus with subsequent 300 mM salt treatment, respectively. CK refers to *P. euphratica* grown under natural conditions without any treatments. ST means *P. euphratica* treated only with 300 mM NaCl. n mM SA+ST stands for *P. euphratica* which was pretreated with n mM SA and then treated with 300 mM NaCl (n = 0.4, 1, 2). To determine significant differences between treatments, Duncan's Test was applied. Data indicate mean ± SE (n = 3). The different letters present on the column indicate significant differences at $p < 0.05$.

3.3. Antioxidant Enzyme and Ion Contents of P. euphratica under Salt Stress and SA Treatments

POD and SOD are crucial antioxidants that can scavenge reactive oxygen species (ROS). Thus, we detected the activities of these two enzymes in *P. euphratica* seedlings before and after salt treatment with or without SA pre-application. The results indicated that the activity levels of POD and SOD were dramatically elevated in seedlings suffering salt stress in comparison to those growing under normal growth conditions (Figure 3d). The SOD and POD activities in *P. euphratica* were also significantly affected by SA pretreatment. SOD was increased more than threefold under the SA-pretreated compared with the only salt-treated conditions, but the difference between the 0.4 mM SA + salt treatment and the 1 or 2 mM SA + salt treatment groups was not significant (Figure 3d). In addition, POD was increased by 92.1% ($p < 0.05$) and 48.6% under 0.4 or 1 mM SA + NaCl conditions, respectively, compared with the only salt treated plants. However, when pretreated with 2 mM SA, the POD activity of seedlings decreased to a level similar to that of seedlings treated only with salt (Figure 3d). The SOD and POD activities of plants treated with 0.4 mM SA were both enhanced to a high level compared with those of non-SA-treated plants under salt stress.

After 3 days of salt treatment, K^+ content in leaves of *P. euphratica* seedlings was significantly lower, while Na^+ content was significantly higher than that of the control (Figure 3e). When pretreated with 0.4 and 1 mM SA, K^+ content increased significantly compared with only salt treated seedlings, but it was still lower than that of control plants (Figure 3e). As for 2 mM SA pretreated seedlings, K^+ content decreased to the lowest level while the Na^+ content reached the highest level. The lowest Na^+ content and the highest K^+/Na^+ were shown in 0.4 mM SA pretreated group (Figure 3 e–f). The results showed that moderate concentration of SA pretreatment play an important role in the balance of K^+ and Na^+ in *P. euphratica*.

3.4. Analysis and Mapping to Transcriptome Sequencing

Leaves from non-SA-pretreated salt control and 0.4 mM SA-pretreated plants were sampled. Their cDNAs were prepared and sequenced independently for three replicates. The cDNA libraries were sequenced using Illumina HiSeq 2000™. Raw data were deposited in the National Center for Biotechnology Information (NCBI) database under the accession number PRJNA503730. After trimming of the sequencing reads containing low-quality, adaptor-polluted and ambiguous reads, 13.70 Gb clean bases from all samples remained for transcriptome assembly. To elucidate the potential molecular events of the DEG profiles, all clean reads were aligned with the reference *P. euphratica* genome database (GCF_000495115.1_PopEup_1.0). All groups were able to map more than 90% sequences to the genome, and more than 70% of the distinct tags were uniquely mapped to the reference sequence. The length distribution of all transcripts is shown in Figure S2. The correlation of the three replicates of the gene expression in the control and SA-pretreated samples was higher than 0.99, indicating that the sequencing results were well reproducible (Figure S3).

3.5. Functional Analysis of DEGs

To investigate changes in transcription levels caused by SA pretreatment, we analyzed DEGs between SA-pretreated and untreated samples. According to the FPKM value of unigenes, we detected 5537 upregulated and 2231 downregulated unigenes (Table S1), as well as 552 unannotated novel DEGs between the only salt-treated and SA-pretreated groups (Table S2). The differential genes between the SA-pretreated and untreated groups were functionally classified by Gene Ontology (GO) and Kyoto Encyclopedia of Genes and Genomes (KEGG) analyses.

We used hypergeometric testing to determine the results of up- and downregulation of significantly enriched GO categories compared with the genomic background ($p \leq 0.05$, after Bonferroni correction) by three ontology categories, including molecular function, cellular components, and biological processes (Figure 4). The top five GO terms for biological processes were "cellular process," "metabolic process," "regulation of biological process," "response to stimulus process," and "biological regulation" after SA and salt treatment. Most of the cellular components and molecular functions were mapped to the "catalytic activity," "binding," "cell," and "cell part." To understand the functional classification of DEGs, GO class enrichment analysis was performed using FDR with an adjusted p value ≤ 0.05 as cutoff value. Among the enriched GO terms, "catalytic activity" and "protein kinase activity" have a high number of upregulated DEGs and may play a very important role in regulating signal transduction (Figure 5b). SA pretreatment to alleviate salt stress may be significantly upregulated by the expression of some protein kinase genes, which enhance the ability to catalyze protein phosphorylation, resulting in changes in ion channel proteins and channel gates.

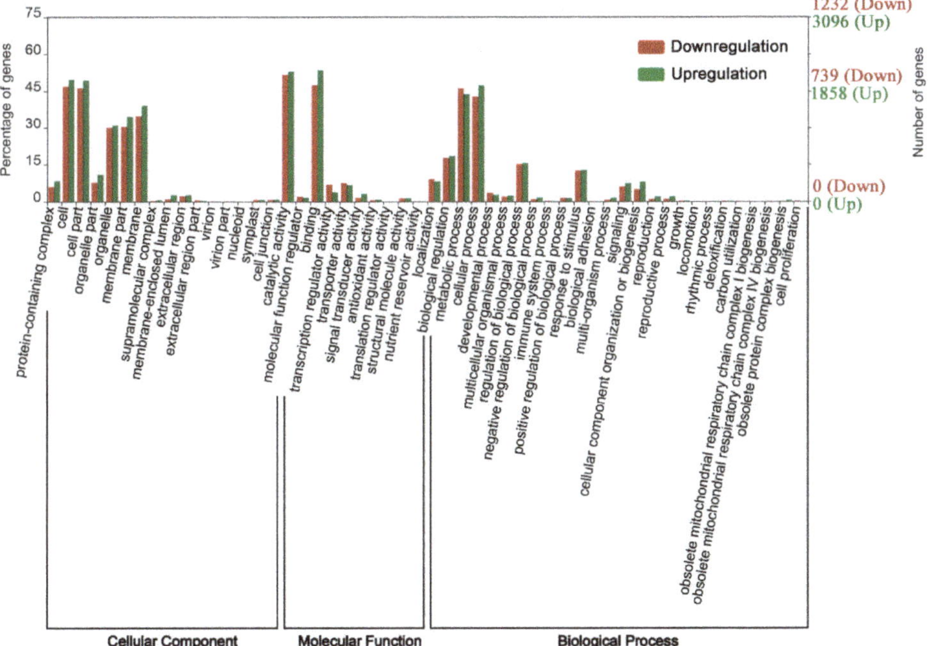

Figure 4. Gene Ontology classification of upregulated and downregulated differential genes. The X axis represents different GO terms. The left Y axis represents the percentage of genes and the right Y axis represents the number of genes. The upper number in the right Y-axis refers to Down-regulation, and the number below represents Up-regulation. Red columns refer to down-regulated genes, green columns refer to up-regulated genes.

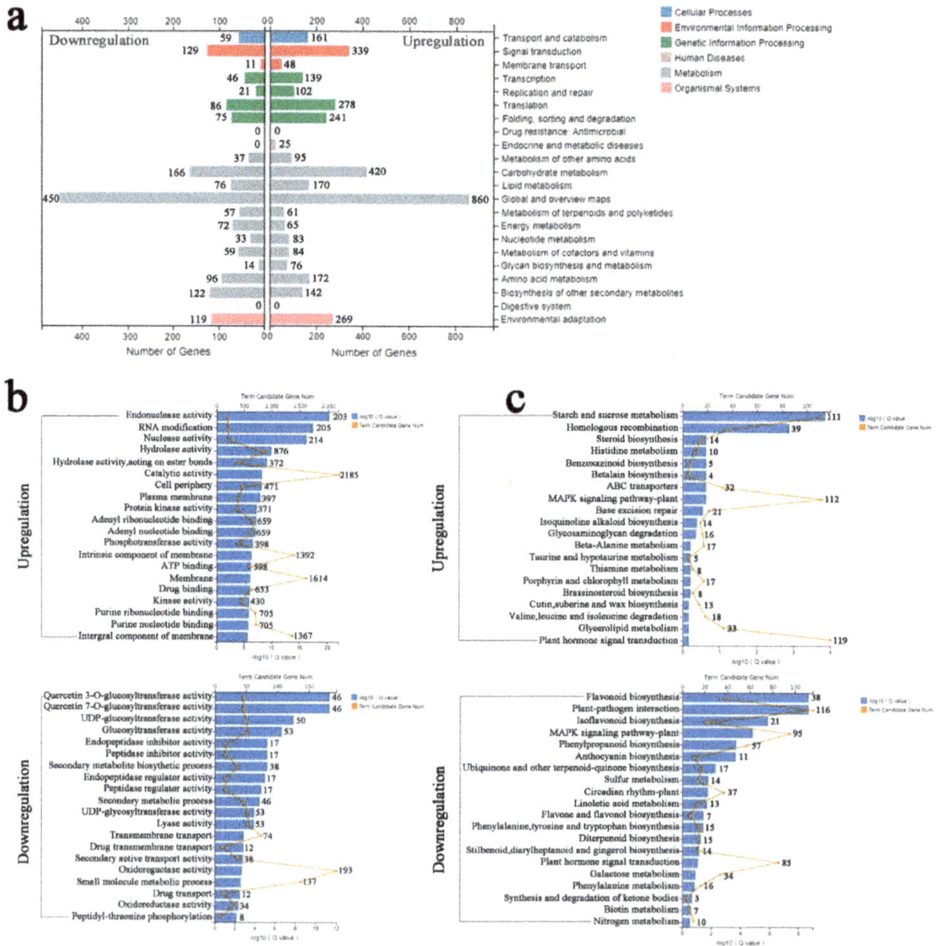

Figure 5. Upregulation and downregulation of GO enrichment and KEGG analysis of differential genes. (**a**) KEGG classification of upregulated and downregulated DEGs. The left side is the upregulated KEGG analysis, and the right side is the downregulated KEGG analysis. (**b**) Upregulation and downregulation of GO enrichment analysis of differential genes. (**c**) Upregulation and downregulation of KEGG enrichment of differential genes. The blue columns are expressed as $-\log_{10}$ (Q value), and the orange line graph represents the number of candidate genes. The gene numbers were indicated after the columns.

Furthermore, according to the KEGG pathway database, the upregulated and downregulated DEGs were separately classified. A total of 269 upregulated and 119 downregulated DEGs were involved in the "environmental adaptation" approach with 339 upregulated and 129 downregulated DEGs involved in the "signal transduction" pathway, which indicated that many SA-induced genes are involved in signal transduction and environmental stress response in *P. euphratica* (Figure 5a). When FDR ≤ 0.01, the KEGG term was regarded as significantly enriched. As a result, the upregulated DEGs produced by SA pretreatment were mainly enriched in "starch and sucrose metabolism," "mitogen-activated protein kinase (MAPK) signaling pathway-plant," and "plant hormone signal transduction," indicating that genes related to MAPK signaling and phytohormone signal transduction pathway were significantly upregulated after SA pretreatment. Application of SA may enhance the salt tolerance of plants by regulating the MAPK signaling pathway and plant hormone signaling pathways (Figure 5c).

3.6. Expression of TF Regulates SA to Confer Salt Tolerance

The TF is a trans-acting factor that interacts with *cis*-factors to enhance or inhibit gene expression and regulates plant growth and development and response to the external environment. Family prediction and classification of TFs revealed that MYB, AP2-EREBP, bHLH, NAC, and WRKY were the top five TF families, accounting for 50% of the total TFs that were differentially expressed between control and SA-pretreated samples (Figure 6). This result indicated that MYB, AP2-EREBP, bHLH, NAC, and WRKY TFs play a leading role in the alleviation of salt stress in *P. euphratica* by SA. Based on the results of GO and KEGG analysis combined with the analysis of TFs, DEGs under SA treatment were mainly located in pathways related to signal transduction. A heat map analysis was performed on the top 4 TFs, protein kinases, PODs, and ion transport-related genes (Figure 7).

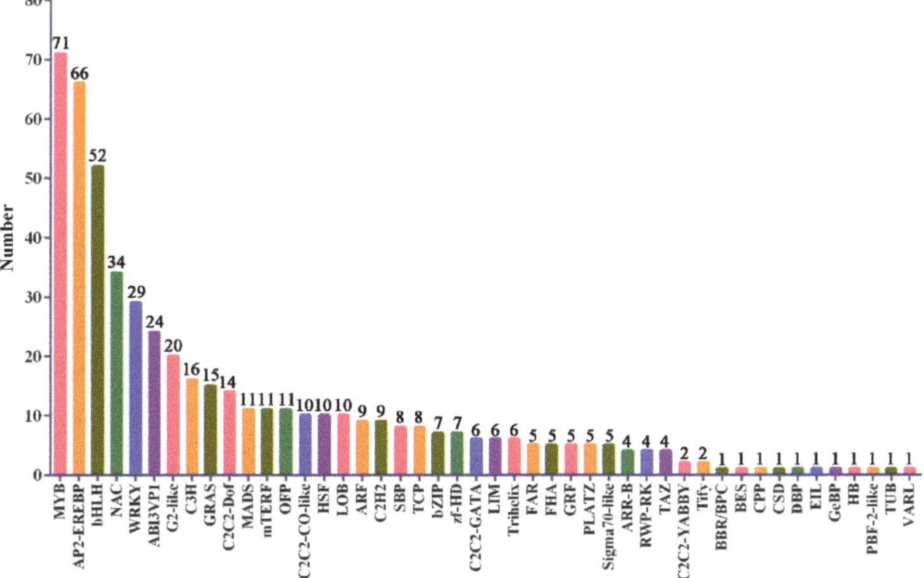

Figure 6. Family of transcription factors differentially expressed between control and SA pretreated salt stressed samples. The X axis represents a different family of transcription factors, and the Y axis represents the number of transcription factors contained in each transcription factor family. The number above the column indicates the number of transcription factors.

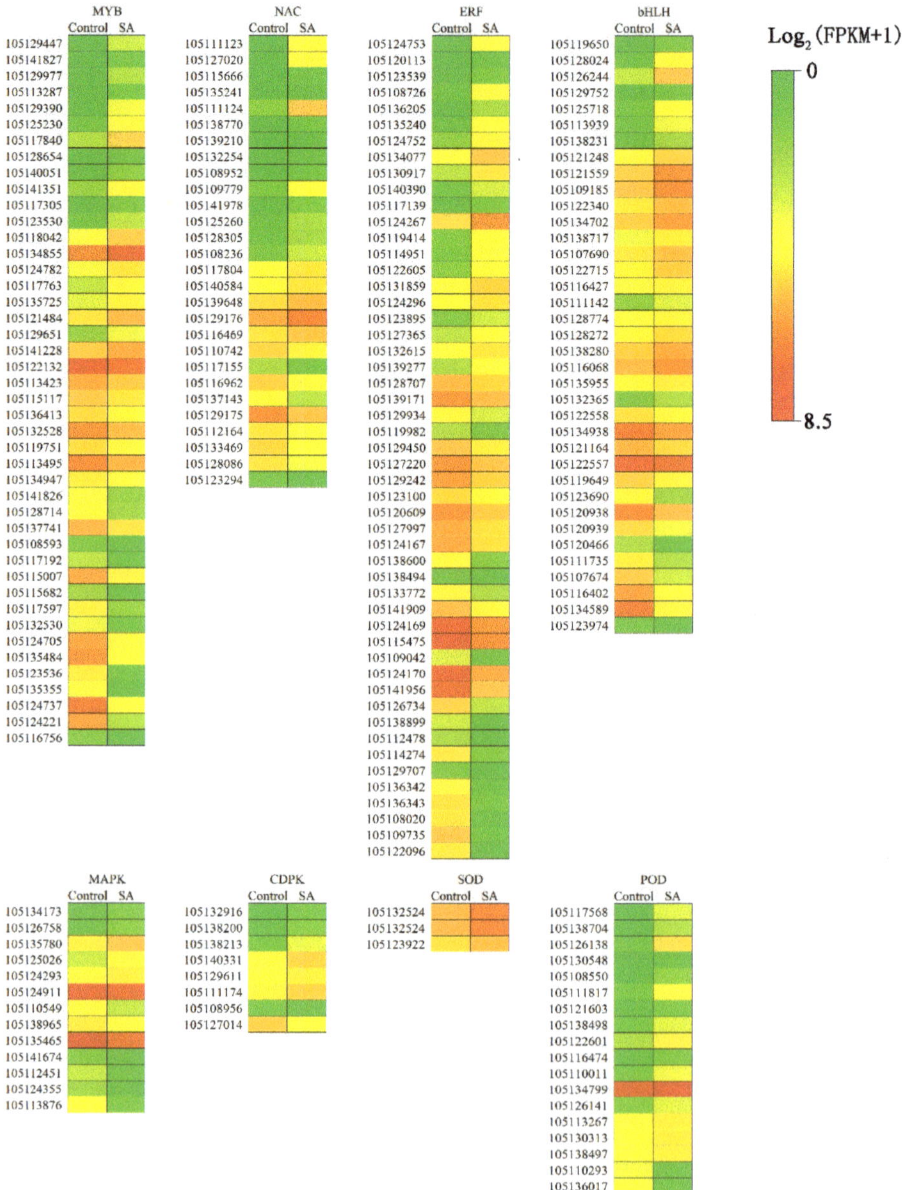

Figure 7. Expression profiles of differentially expressed genes regulated SA to alleviate salt tolerance. Data represent the mean ± SD (Standard Deviation) of three independent experiments. The heat map of each related protein and transcription factor family is represented by the $\log_2$ (FPKM + 1) of Control check and SA. Here, the FPKM means fragments per kilobase of transcript per million mapped reads. The range of all values varies from 0 to 8.5. The greener is closer to the minimum value, and the red is closer to the maximum value. The gene ID is on the left side.

3.7. Confirmation of Differentially Expressed Candidate Genes by qRT-PCR Analysis

To confirm the reliability of the transcriptomic data, we examined the expression level of eight unigenes by qPCR using the primers listed in Table S3. The qPCR results of the eight unigenes in the 0.4 mM SA pretreatment group were mostly consistent with the RNA-seq data, confirming the reliability of our transcriptome data (Figure 8).

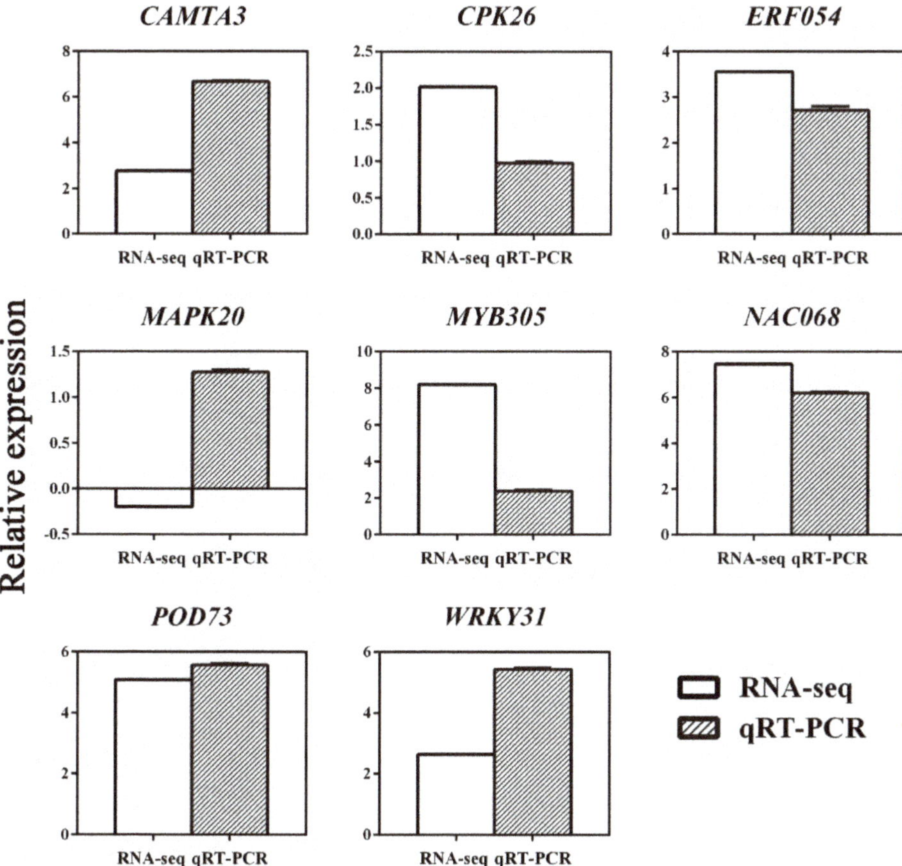

Figure 8. Comparison of expression patterns of 8 candidate genes between RNA-seq (white) and qRT-PCR (oblique line) for selected transcripts. Data represent the mean ± SD of three independent experiments. X axis is the selected gene ID, and Y axis is the $\log_2$ Ratio.

4. Discussion

SA stimulates a wide range of metabolic and physiological response in plants. These responses improve plant growth, transpiration rates, stomatal regulation, and photosynthesis [1]. Recently, an increasing number of researchers have studied the abiotic resistance mechanism of SA in different crops and flowers and found that SA enhances the salt tolerance of tomato and *Dianthus superbus* mainly by enhancing photosynthesis and alleviating membrane damage [1]. *P. euphratica* is a model tree widely used to study abiotic stress response mechanisms. However, the influence of SA on salt stress of trees is rarely studied, and the precise function mediated by SA in poplar remains to be elucidated. Therefore, *P. euphratica* should be used as a material to study the response mechanism of SA involved in abiotic stress. According to a previous study, SA applied beyond a certain range may be detrimental [30]. Thus, the exogenous SA concentrations and the mode of application must be determined as a prerequisite. Here, we evaluated the effects of SA on improving the salt tolerance of *P. euphratica* under laboratory conditions and further performed transcriptomic analysis at this concentration to gain new insights into the effects of SA in relation to salt resistance endpoints. The results suggested that 0.4 mM SA may be considered as a potential growth regulator to improve salinity stress resistance of poplar, because the application of exogenous SA before salt treatment reduced the adverse effects of salinity and increased the salt resistance of *P. euphratica*.

Important developmental plant processes, such as photosynthesis, protein synthesis, and carbohydrate, and lipid metabolism, are negatively affected when subjected to salt stress [31]. To deal with salt-induced ionic toxicity, osmotic stress, and secondary oxidative stress, salt-tolerant plant species often develop a variety of strategies to adapt to saline environments, such as detoxifying ROS and regulating ionic homeostasis [32]. ROS is the product of aerobic metabolism. It is a class of important signaling molecules in organism, including superoxide anion Radical O_2^-, hydroxyl (-OH), and hydrogen peroxide (H_2O_2). Plants may accumulate ROS in the metabolic process during drought and salinity stress conditions [33]. Too much accumulated ROS in plant cells have a widespread toxic effect, causing peroxidation of lipids, and damaging DNA and proteins, eventually leading to cell death [34]. Recent studies have found that salt stress may increase the concentration of H_2O_2 and O_2^- in cells, leading to decreased growth of chlorophyll, protein degeneration, cellular dysfunction, or even death [35,36]. Salt stress also leads to a decrease in the ratio of ascorbic acid/dehydroascorbic acid (ASA/DHA) and a reducing of glutathione/oxidized glutathione (GSH/GSSG), which causes redox imbalance [37]. ASA has antioxidant activity and could act as a free radical scavenger to remove ROS [38]. When ASA is oxidized, it is converted into an unstable monodehydroascorbic acid (MDA) and loses its function. MDA can reform ASA after the catalysis of MDA reductase or through recycling from DHA by DHA reductase with glutathione. This pathway of controlling hydrogen peroxide and recycling ASA is known as the Foyer–Halliwell–Asada cycle [39]. In the chloroplast, the Foyer–Halliwell–Asada cycle is directly related to the SOD catalytic reaction that converts O_2^- to oxygen and H_2O_2 to form a water–water cycle [40]. Low concentrations of SA may improve plants' ability to resist dehydration by promoting the activity of antioxidant enzymes, including POD, SOD, and catalase [41,42]. SOD can catalyze the conversion of O_2^- to H_2O_2, and POD can convert H_2O_2 to H_2O [43]. Ascorbate peroxidase (APX) is a key enzyme in the ascorbate–glutathione cycle, which uses ASA to reduce H_2O_2 to H_2O. According to our biochemical analysis results, the activity of these enzymes was correspondingly and significantly increased in the cells of *P. euphratica* after SA induction. Furthermore, our sequencing data indicated that two candidate genes encoding SOD, fifteen genes encoding POD and one gene encoding APX were obviously upregulated after the application of SA (Table S4). Considering that SOD, POD, and APX could function in the scavenging of ROS to improve salt tolerance [44], we speculated that moderate SA could mitigate the salt damage to cells by activating SOD, POD, and APX.

The compartmentalization of Na^+ in vacuoles is another important strategy employed by plants to increase salt resistance. However, there are some contrary results in terms of SA involvement in ionic homeostasis under salinity environment. Exogenous SA minimizes Na^+ uptake while increasing tissue concentrations of K^+ in maize under salinity stress [45]. The application of SA to tomato plants inhibited K^+ uptake and increased Na^+ uptake [46]. The concentration of Na^+ and Cl^- in salt-treated spinach was not affected by SA, which seemed to be a neutral result [47]. Thus, it is not easy to draw a conclusion on the role of SA in the maintenance of ionic homeostasis under salinity stress. Here, the regulatory role of SA in the induction of genes regarding membrane transporters controlling K^+ homeostasis, Na^+ uptake, and Na^+ redistribution during salt stress was analyzed. According to our data, SA-treated *P. euphratica* showed obvious alterations in the transcription of K^+ channel and K^+ transporter protein genes. Four K^+ transporter genes encoding K^+ transporter 2/10/11/13 and two K^+ channel genes encoding *AKT1* homologous genes accumulated after SA administration (Table S5). Further investigation on gene expression related to sodium transport revealed that the *HKT1* gene significantly accumulated their transcripts in SA-pretreated *P. euphratica* compared with the only salt treated control (Table S5). The HKT gene family is a type of Na$^+$ or K$^+$ transporter or Na$^+$–K$^+$ co-transporter associated with salt tolerance stress in plants and plays a decisive role in regulating intracellular Na$^+$/K$^+$ homeostasis [48]. Among them, *HKT1* is a high-affinity K$^+$ transporter encoded in roots and leaves mainly by controlling Na$^+$ input in roots [49]. Garriga et al. [50] conducted a functional study of *AtHKT1* in the model plant *Arabidopsis thaliana* L. Heynh and found that *AtHKT1*, which mediates Na$^+$ transport, increases plant tolerance to salt. The Na$^+$ efflux was dependent on Na$^+$/H$^+$ exchanger/antiporter localized to the plasma membrane. The Na$^+$/H$^+$ exchanger was highly induced in the SA-pretreated group, which indicated that SA pretreatment might help to extrude Na$^+$ of cytoplasm by a plasma membrane-localized Na$^+$/H$^+$ exchanger. In addition to the Na$^+$ efflux, the redistribution of Na$^+$ also plays an important role, mainly relying on the tonoplast-localized Na$^+$/H$^+$ exchanger, such as *NHX1*. The tonoplast-localized Na$^+$/H$^+$ exchanger could absorb Na$^+$ into the vacuole to regulate intracellular pH and maintain Na$^+$ levels in the cytoplasm [51,52]. Generally, we proposed that SA pretreatment triggered or enhanced the salt response, presumably by regulating the genes related to ionic homeostasis in *P. euphratica* by a working model indicated by Figure 9. The exogenous SA was supposed to convert into endogenous SA at first. When the subsequent salt treatment conducted, the ion transporting system might be evoked, resulting in an improved performance during salt stress. We could conclude that SA participates in ionic homeostasis in salt stress resistance in *P. euphratica* at a specific dose, for example, 0.4 mM for *P. euphratica*.

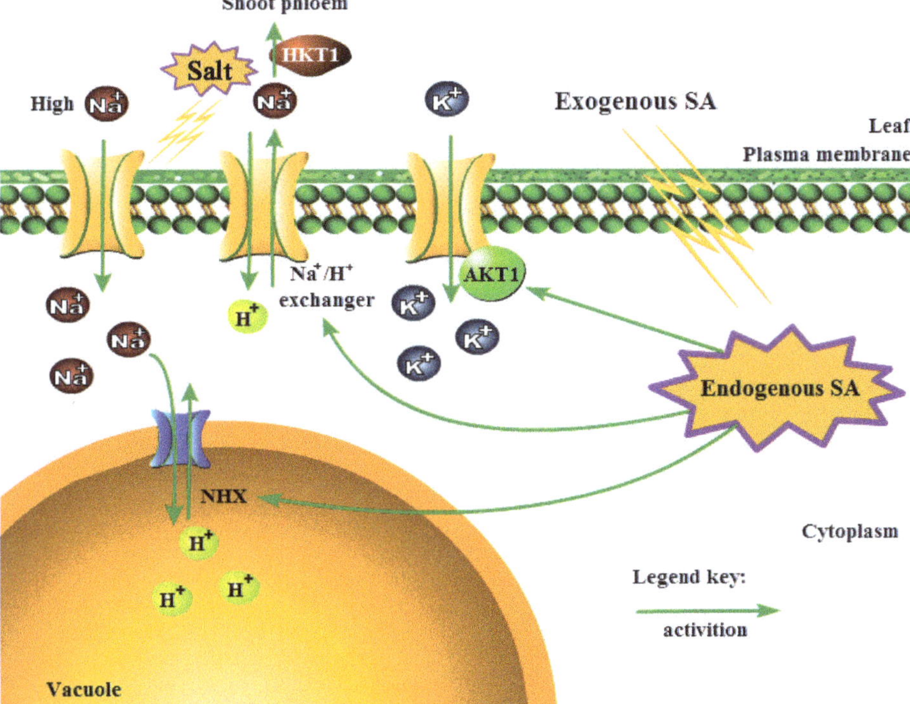

Figure 9. A proposed working model for sodium and potassium transporting in *P. euphratica* leaves in relation to SA treatment and subsequent salt treatment. The model is based on the cation accumulation result and RNA-seq result. The exogenous SA that sprayed initially was supposed to convert into endogenous SA. When the subsequent salt treatment conducted, the ion transporting system might be evoked as described.

Identifying the possible candidate genes and the signaling pathways related to abiotic stress resistance is an essential step toward understanding the abiotic stress resistance characteristics of various plants [46,53]. Here, DEG mining and pathway enrichment analysis based on RNA-seq data revealed certain genes were involved in SA-induced salt stress response of *P. euphratica*. Among the functional genes, TFs play an important role in growth, development, defense, and stress response of plants [54]. Stress-inducible TFs such as MYB, DREB, and WRKY can bind to elements in the promoter region of functional genes, such as *rd22*, *rd29A*, and *cor15A* [55,56]. These TF-regulating responsive genes are important for improving plant defense in salinity, drought, and other stresses in *Arabidopsis* [57].

In our study, the transcripts of NAC, ERF, MYB, bHLH, and WRKY accumulated in the 0.4 mM SA pretreatment group (Table S6). The NAC family genes are involved in response to salinity and drought stresses [58]. For example, *SlNAC1* and *SlNAM1* can be induced by salinity [59]. The overexpression of three *NAC* genes (*NAC019*, *NAC055*, *NAC072*) significantly enhances the tolerance of *Arabidopsis* to drought [60]. According to our RNA-seq data, 19 members in the *NAC* gene family were upregulated, which illustrates that the *NAC* gene family is involved in SA-mediated salt response signal and may be one of the important mechanisms of SA-induced plant resistance to abiotic stress in *P. euphratica*. AP2-EREBP is a super gene family, including five branches of AP2, EREBP, DREB, ERF, and others. The ERF gene family was highly activated in the SA pretreatment group. *ERF3*, which is associated with abiotic stresses such as high salinity, drought, and defense against pathogen infection, was significantly upregulated. Studies have shown that *GmERF3* is involved in the signaling pathways of SA [61]. We speculated that ERFs are also involved in the process of SA signal transduction for salt defense in *P. euphratica*. The MYB gene family is another important TF family in abiotic stress defense. The accumulation of flavonoid compounds is an important characteristic for plants to confront abiotic stress. The *MYB* is widely involved in the regulation of flavonoid metabolic pathways. *MYB4* belongs to the R2R3-MYB subclass and is one of the key TFs regulating abiotic stress response. Pasquali et al. [62] transformed *Osmyb4* into apple plants, improving the physiological and biochemical adaptability of plants in drought and low-temperature stress. Overexpression of *Arabidopsis MYB44* can reduce the expression level of ABA signaling negative regulator phosphatase 2C (PP2C), thereby enhancing the salt tolerance of plants [63]. The bHLH and WRKY gene families are both closely related to plant resistance to salt stress [61]. Considering that the genes in these families were significantly upregulated in the SA pretreatment group, we concluded that SA mediates salt stress defense by regulating specific TF gene families, such as NAC, MYB, ERF, WRKY, and bHLH.

A large number of genes encoding MAPK pathway components were identified in the way of MAPK cascades [64]. The basic MAPK cascade consists of MAPK, MAPK kinase (MAPKK), with the latter activating the former, and MAPK kinase kinase (MAPKKK) that activates MAPKK. This signaling pathway participates in the defense against stresses by transmitting stress signaling and activating the expression of resistance genes [65]. A previous study found that the MAPK cascade is involved in the SA signaling pathway of pathogen resistance, especially in defense responses against bacteria and oomycetes. The MAPK signaling cascade pathway also plays an essential role in abiotic stress response of various plants [59,66]. In *Arabidopsis*, studies have elucidated the function of MAPK signaling components in abiotic stress response. Both *AtMPK4* and *AtMPK6* are activated by cold, salt, and drought stresses, and *MPKK1–MPKK2–MPK4/MPK6* is activated by cold and salt stresses [67,68]. *MAPK9* is specifically expressed in the stomatal pores of *Arabidopsis*, and its transcriptional expression is upregulated under salt stress and is involved in a series of signal transductions, which significantly increases its salt tolerance [69]. The overexpression of *NPK1* can enhance the tolerance of tobacco to drought and salt by activating oxidative stress signaling [70]. *MPK4* in the *Nicotiana* species is involved in responses to ozone treatment through the regulation of stomatal movement [71]. SA rapidly induces the tobacco *SIPK*, which has high homology with MAPK family genes and contains the characteristic region of MAPK [72]. A recent study suggested that the constitutive expression of *MAPK3* would induce strong accumulation of SA in *Arabidopsis* [73], which indicates the relationship of

MAPK3 in participating in SA signaling. Ten putative poplar *MAPKK* genes (*PtMKKs*) and 21 putative poplar *MAPK* genes (*PtMPKs*) have been identified in the *P. trichocarpa* genome [74]. MPK4-silenced poplar exhibits high foliar content of free SA and elevated cellular ROS level and antioxidant enzyme activities [75], indicating the function of *MAPK* genes in cellular ROS/SA homeostasis and water management in woody plant species. However, in contrast to *Arabidopsis* and *Nicotiana*, few MAPKs have been functionally characterized, and their role in trees in the regulation of abiotic stresses and the SA pathway remains to be fully elucidated [75]. Our RNA-seq data indicated that many MAPK cascade candidate genes in *P. euphratica* were upregulated under SA treatment (Table S7). The results indicated that the SA signaling system interacted with the MAPK cascade pathway in *P. euphratica*. We therefore speculated that SA could trigger the MAPK cascade in response to abiotic stresses in trees.

CDPK is a kind of calcium sensor protein that plays an important role in intracellular Ca^{2+} signal transduction [76]. The CDPK mRNA often accumulates under various abiotic stresses including drought, salt, and cold and further causes the expression of related genes in plant cells that respond to these abiotic stresses [76,77]. A variety of plant hormones, such as gibberellin, ABA, cytokinin, indole acetic acid, and ethylene, can enhance the activation of CDPK [78]. SA can induce the expression of *CDPK* genes in a similar manner to mechanical wounding induction in tomato [79]. However, whether SA can enhance plant resistance to salt stress by inducing the differential expression of *CDPK* genes remains to be elucidated. According to our RNA-seq data, six candidate *CDPK* genes were upregulated after the application of 0.4 mM SA and salt treatment compared with the non-SA-treated control (Table S8). These genes may participate in processes of salt stress resistance. Thus, we conclude that *CDPK* genes are involved in a SA-mediated response to abiotic stress in *P. euphratica*.

5. Conclusions

We have initially provided an answer as to whether and how SA improves salt resistance in *P. euphratica*. SA minimized the deleterious effect of salt on the adaptation of *P. euphratica*, which was attributed to the improved growth, higher activity of the antioxidant enzymes and better potassium-sodium balance. Gene mining uncovered the genomic differentially expressed unigenes caused by SA pretreatment that contribute to salt stress adaptation. SA induced the expression level of antioxidant enzymes encoding genes, MAPK, CDPK, and ionic homeostasis genes, allowing *P. euphratica* to tolerate salt stress. This study is the first to provide insight into the SA function in response of *P. euphratica* to abiotic stress. A genome-level SA-induced gene expression for *P. euphratica* would be an important prerequisite for a deeper understanding of SA function in salt stress adaptation in *P. euphratica*. Furthermore, the specific target genes and their mediated signaling pathways may be involved in the mechanism by which SA mediates salt stress response in *P. euphratica*. Further research of these genes may unravel the mechanism of salt stress adaptation via SA pathways in trees.

Supplementary Materials: The following are available online at http://www.mdpi.com/1999-4907/10/5/423/s1, Figure S1: Typical MRM chromatogram of internal standards solution of *D4-SA* and *D6-ABA* and sample of ABA and SA. The retention times of D4-SA, SA, D6-ABA, ABA were 5.02, 4.99, 5.84, and 5.85 minutes, respectively. Figure S2: The length distribution of transcripts. Figure S3: Correlation analysis of transcriptome data between different treatments. Table S1: The list of up and down regulated DEGs. Table S2: The length distribution of transcripts. Table S3: Genes with regarding primer sequences used for qPCR. Table S4: Differentially expressed unigenes encoding SOD, POD, and APX. Table S5: Differentially expressed unigenes involved in putative sodium and potassium balance. Table S6: Differentially expressed unigenes encoding the top five transcription factors. Table S7: Differential expressed unigenes in MAPK cascade in *P. euphratica*. Table S8: Differential expressed unigenes in CDPK cascade in *P. euphratica*.

Author Contributions: J.C. conceived and designed research. C.D. and A.L. conducted experiments. S.R. analyzed data. S.R. and J.C. wrote the manuscript. X.X. and W.Y. revised the manuscript. All authors read and approved the manuscript.

Funding: This research was supported by the National Natural Science Foundation of China (31670610; 31370597).

Acknowledgments: We are grateful to Beijing Forestry University for providing greenhouses to maintain the growth of the seedlings.

Conflicts of Interest: The authors declare no conflict of interest.

References

1. Mimouni, H.; Wasti, S.; Manaa, A.; Gharbi, E.; Chalh, A.; Vandoorne, B.; Lutts, S.; Ben Ahmed, H. Does Salicylic Acid (SA) Improve Tolerance to Salt Stress in Plants? A Study of SA Effects On Tomato Plant Growth, Water Dynamics, Photosynthesis, and Biochemical Parameters. *Omics-a J. Integr. Biol.* **2016**, *20*, 180–190. [CrossRef] [PubMed]
2. Clarke, S.M.; Mur, L.A.; Wood, J.E.; Scott, I.M. Salicylic acid dependent signaling promotes basal thermotolerance but is not essential for acquired thermotolerance in *Arabidopsis thaliana*. *Plant J.* **2004**, *38*, 432–447. [CrossRef] [PubMed]
3. Klessig, D.F.; Malamy, J. The salicylic acid signal in plants. *Plant Mol. Biol.* **1994**, *26*, 1439–1458. [CrossRef]
4. Metwally, A.; Finkemeier, I.; Georgi, M.; Dietz, K.J. Salicylic acid alleviates the cadmium toxicity in barley seedlings. *Plant Physiol.* **2003**, *132*, 272–281. [CrossRef] [PubMed]
5. Raskin, I. Salicylate, a New Plant Hormone. *Plant Physiol.* **1992**, *99*, 799–803. [CrossRef] [PubMed]
6. Delaney, E.J.; Sherrill, R.G.; Palaniswamy, V.; Sedergran, T.C.; Taylor, S.P. A Reinvestigation of the D-Homoannular Rearrangement and Subsequent Degradation Pathways of (11-Beta,16-Alpha)-9-Fluoro-11,16,17,21-Tetrahydroxypregna-1,4- Diene-3,20-Dione (Triamcinolone). *Steroids* **1994**, *59*, 196–204. [CrossRef]
7. Vlot, A.C.; Dempsey, D.A.; Klessig, D.F. Salicylic Acid, a Multifaceted Hormone to Combat Disease. *Annu. Rev. Phytopathol.* **2009**, *47*, 177–206. [CrossRef]
8. Yildirim, E.; Turan, M.; Guvenc, I. Effect of foliar salicylic acid applications on growth, chlorophyll, and mineral content of cucumber grown under salt stress. *J. Plant Nutr.* **2008**, *31*, 593–612. [CrossRef]
9. Sun, W.; Xu, X.; Zhu, H.; Liu, A.; Liu, L.; Li, J.; Hua, X. Comparative transcriptomic profiling of a salt-tolerant wild tomato species and a salt-sensitive tomato cultivar. *Plant Cell Physiol.* **2010**, *51*, 997. [CrossRef]
10. Noreen, S.; Ashraf, M. Alleviation of adverse effects of salt stress on sunflower (*Helianthus annuus* L.) by exogenous application of salicylic acid: Growth and photosynthesis. *Int. J. Bot.* **2008**, *40*, 1657–1663.
11. Brosche, M.; Vinocur, B.; Alatalo, E.R.; Lamminmaki, A.; Teichmann, T.; Ottow, E.A.; Djilianov, D.; Afif, D.; Bogeat-Triboulot, M.B.; Altman, A.; et al. Gene expression and metabolite profiling of *Populus euphratica* growing in the Negev desert. *Genome. Biol.* **2005**, *6*, R101. [CrossRef]
12. Jia, H.; Li, J.; Zhang, J.; Ren, Y.; Hu, J.; Lu, M. Genome-wide survey and expression analysis of the stress-associated protein gene family in desert poplar, *Populus euphratica*. *Tree Genet. Genomes* **2016**, *12*, 78. [CrossRef]
13. Wang, H.L.; Chen, J.H.; Tian, Q.Q.; Wang, S.; Xia, X.L.; Yin, W.L. Identification and validation of reference genes for *Populus euphratica* gene expression analysis during abiotic stresses by quantitative real-time PCR. *Physiol. Plant.* **2014**, *152*, 529–545. [CrossRef]
14. Chen, J.; Tian, Q.; Pang, T.; Jiang, L.; Wu, R.; Xia, X.; Yin, W. Deep-sequencing transcriptome analysis of low temperature perception in a desert tree, *Populus euphratica*. *BMC Genom.* **2014**, *15*, 326. [CrossRef]
15. Brinker, M.; Brosche, M.; Vinocur, B.; Abo-Ogiala, A.; Fayyaz, P.; Janz, D.; Ottow, E.A.; Cullmann, A.D.; Saborowski, J.; Kangasjarvi, J.; et al. Linking the salt transcriptome with physiological responses of a salt-resistant *Populus* species as a strategy to identify genes important for stress acclimation. *Plant Physiol.* **2010**, *154*, 1697–1709. [CrossRef] [PubMed]
16. Zhang, J.; Jia, H.X.; Li, J.B.; Li, Y.; Lu, M.Z.; Hu, J.J. Molecular evolution and expression divergence of the *Populus euphratica Hsf* genes provide insight into the stress acclimation of desert poplar. *Sci. Rep.* **2006**, *6*, 30050. [CrossRef] [PubMed]
17. Zhang, J.; Xie, P.H.; Lascoux, M.; Meagher, T.R.; Liu, J.Q. Rapidly Evolving Genes and Stress Adaptation of Two Desert Poplars, *Populus euphratica* and *P. pruinosa*. *PLoS ONE* **2013**, *8*, e66370. [CrossRef]
18. Ma, T.; Wang, J.; Zhou, G.; Yue, Z.; Hu, Q.; Chen, Y.; Liu, B.; Qiu, Q.; Wang, Z.; Zhang, J.; et al. Genomic insights into salt adaptation in a desert poplar. *Nat. Commun.* **2013**, *4*, 2797. [CrossRef]
19. Madhava Rao, K.V.; Sresty, T.V. Antioxidative parameters in the seedlings of pigeonpea (*Cajanus cajan* (L.) Millspaugh) in response to Zn and Ni stresses. *Plant Sci.* **2000**, *157*, 113–128. [CrossRef]
20. Wang, N.; Qi, H.; Qiao, W.; Shi, J.; Xu, Q.; Zhou, H.; Yan, G.; Huang, Q. Cotton (*Gossypium hirsutum* L.) genotypes with contrasting K^+/Na^+ ion homeostasis: implications for salinity tolerance. *Acta Physiol. Plant.* **2017**, *39*, 77. [CrossRef]

21. Meloni, D.A.; Oliva, M.A.; Martinez, C.A.; Cambraia, J. Photosynthesis and activity of superoxide dismutase, peroxidase and glutathione reductase in cotton under salt stress. *Environ. Exp. Bot.* **2003**, *49*, 69–76. [CrossRef]

22. Becana, M.; Aparicio-Tejo, P.; Irigoyen, J.J.; Sanchez-Diaz, M. Some enzymes of hydrogen peroxide metabolism in leaves and root nodules of Medicago sativa. *Plant Physiol.* **1986**, *82*, 1169–1171. [CrossRef]

23. Cock, P.J.; Fields, C.J.; Goto, N.; Heuer, M.L.; Rice, P.M. The Sanger FASTQ file format for sequences with quality scores, and the Solexa/Illumina FASTQ variants. *Nucleic Acids Res.* **2010**, *38*, 1767–1771. [CrossRef]

24. Kim, D.; Langmead, B.; Salzberg, S.L. HISAT: A fast spliced aligner with low memory requirements. *Nat. Methods* **2015**, *12*, 357–360. [CrossRef]

25. Langmead, B.; Salzberg, S.L. Fast gapped-read alignment with Bowtie 2. *Nat. Methods* **2012**, *9*, 357–359. [CrossRef]

26. Li, B.; Dewey, C.N. RSEM: Accurate transcript quantification from RNA-Seq data with or without a reference genome. *BMC Bioinform.* **2011**, *12*, 323. [CrossRef]

27. Wang, L.; Feng, Z.; Wang, X.; Wang, X.; Zhang, X. Degseq: An r package for identifying differentially expressed genes from rna-seq data. *Bioinformatics* **2010**, *26*, 136–138. [CrossRef]

28. Mistry, J.; Finn, R.D.; Eddy, S.R.; Bateman, A.; Punta, M. Challenges in homology search: HMMER3 and convergent evolution of coiled-coil regions. *Nucleic Acids Res* **2013**, *41*, e121. [CrossRef]

29. Livak, K.J.; Schmittgen, T.D. Analysis of Relative Gene Expression Data Using Real-Time Quantitative PCR and the 2^{-Schm} Method. *Methods* **2001**, *25*, 402–408. [CrossRef]

30. Shakirova, F.M.; Sakhabutdinova, A.R.; Bezrukova, M.V.; Fatkhutdinova, R.A.; Fatkhutdinova, D.R. Changes in the hormonal status of wheat seedlings induced by salicylic acid and salinity. *Plant Sci.* **2003**, *164*, 317–322. [CrossRef]

31. Sereshti, H.; Poursorkh, Z.; Aliakbarzadeh, G.; Zarre, S.; Ataolahi, S. An image analysis of TLC patterns for quality control of saffron based on soil salinity effect: A strategy for data (pre)-processing. *Food Chem.* **2018**, *239*, 831–839. [CrossRef] [PubMed]

32. Ding, M.; Hou, P.; Shen, X.; Wang, M.; Deng, S.; Sun, J.; Xiao, F.; Wang, R.; Zhou, X.; Lu, C.; et al. Salt-induced expression of genes related to Na(+)/K(+) and ROS homeostasis in leaves of salt-resistant and salt-sensitive poplar species. *Plant Mol. Biol.* **2010**, *73*, 251–269. [CrossRef]

33. Miller, G.; Suzuki, N.; Ciftci-Yilmaz, S.; Mittler, R. Reactive oxygen species homeostasis and signalling during drought and salinity stresses. *Plant Cell Environ.* **2010**, *33*, 453–467. [CrossRef] [PubMed]

34. Katarzyna, P.S.; Lidia, P.K.; Małgorzata, L.; Justyna, M.; Grażyna, D. SNF1-Related Protein Kinases SnRK2.4 and SnRK2.10 Modulate ROS Homeostasis in Plant Response to Salt Stress. *Int. J. Mol. Sci.* **2019**, *20*, 143–167.

35. Ashraf, M.A.; Akbar, A.; Parveen, A.; Rasheed, R.; Hussain, I.; Iqbal, M. Phenological application of selenium differentially improves growth, oxidative defense and ion homeostasis in maize under salinity stress. *Plant Physiol. Biochem.* **2018**, *123*, 268–280. [CrossRef]

36. Boscolo, P.R.; Menossi, M.; Jorge, R.A. Aluminum-induced oxidative stress in maize. *Phytochemistry* **2003**, *62*, 181–189. [CrossRef]

37. Bharti, A.; Garg, N. SA and AM symbiosis modulate antioxidant defense mechanisms and asada pathway in chickpea genotypes under salt stress. *Ecotoxicol. Environ. Saf.* **2019**, *178*, 66–78. [CrossRef] [PubMed]

38. Zsigmond, L.; Tomasskovics, B.; Deák, V.; Rigó, G.; Szabados, L.; Bánhegyi, G.; Szarka, A. Enhanced activity of galactono-1,4-lactone dehydrogenase and ascorbate–glutathione cycle in mitochondria from complex III deficient Arabidopsis. *Plant Physiol. Biochem.* **2011**, *49*, 809–815. [CrossRef]

39. Matteo, A.; Sacco, A.; Stefano, R.; Frusciante, L.; Barone, A. Comparative Transcriptomic Profiling of Two Tomato Lines with Different Ascorbate Content in the Fruit. *Biochem. Genet.* **2012**, *50*, 908–921. [CrossRef]

40. Noshi, M.; Hatanaka, R.; Tanabe, N.; Terai, Y.; Maruta, T.; Shigeoka, S. Redox regulation of ascorbate and glutathione by a chloroplastic dehydroascorbate reductase is required for high-light stress tolerance in *Arabidopsis*. *J. Agric. Chem. Soc. Jpn.* **2016**, *80*, 8.

41. Huang, S.; Van Aken, O.; Schwarzlander, M.; Belt, K.; Millar, A.H. The Roles of Mitochondrial Reactive Oxygen Species in Cellular Signaling and Stress Response in Plants. *Plant Physiol.* **2016**, *171*, 1551–1559. [CrossRef]

42. Rehman, R.U.; Zia, M.; Chaudhary, M.F. Salicylic acid and ascorbic acid retrieve activity of antioxidative enzymes and structure of Caralluma tuberculata calli on PEG stress. *Gener. Physiol. Biophys.* **2017**, *36*, 167–174. [CrossRef] [PubMed]

43. Perry, J.J.; Shin, D.S.; Getzoff, E.D.; Tainer, J.A. The structural biochemistry of the superoxide dismutases. *Biochim. Biophys. Acta* **2010**, *1804*, 245–262. [CrossRef]

44. Ullah, A.; Sun, H.; Hakim; Yang, X.; Zhang, X. A novel cotton *WRKY* gene, *GhWRKY6*-like, improves salt tolerance by activating the ABA signaling pathway and scavenging of reactive oxygen species. *Physiol. Plant* **2018**, *162*, 439–454. [CrossRef]

45. Gunes, A.; Inal, A.; Alpaslan, M.; Cicek, N.; Guneri, E.; Eraslan, F.; Guzelordu, T. Effects of exogenously applied salicylic acid on the induction of multiple stress tolerance and mineral nutrition in maize (*Zea mays* L.). *Arch. Agron. Soil Sci.* **2005**, *51*, 687–695. [CrossRef]

46. Ben Saad, R.; Fabre, D.; Mieulet, D.; Meynard, D.; Dingkuhn, M.; Al-Doss, A.; Guiderdoni, E.; Hassairi, A. Expression of the Aeluropus littoralis AlSAP gene in rice confers broad tolerance to abiotic stresses through maintenance of photosynthesis. *Plant Cell Environ.* **2012**, *35*, 626–643. [CrossRef]

47. Eraslan, F.; Inal, A.; Pilbeam, D.J.; Gunes, A. Interactive effects of salicylic acid and silicon on oxidative damage and antioxidant activity in spinach (*Spinacia oleracea* L. cv. Matador) grown under boron toxicity and salinity. *Plant Growth Regul.* **2008**, *55*, 207–219. [CrossRef]

48. Hamamoto, S.; Horie, T.; Hauser, F.; Deinlein, U.; Schroeder, J.I.; Uozumi, N. Hkt transporters mediate salt stress resistance in plants: From structure and function to the field. *Curr. Opin. Biotechnol.* **2015**, *32*, 113–120. [CrossRef]

49. Uozumi, N.; Kim, E.J.; Rubio, F.; Yamaguchi, T.; Muto, S.; Tsuboi, A.; Bakker, E.P.; Nakamura, T.; Schroeder, J.I. The Arabidopsis *HKT1* gene homolog mediates inward Na + currents in Xenopuslaevis oocytes and Na + uptake in Saccharomyces cerevisiae. *Plant Physiol.* **2000**, *122*, 1249–1259. [CrossRef]

50. Garriga, M.; Raddatz, N.; Véry, A.A.; Sentenac, H.; Rubiomeléndez, M.E.; González, W.; Dreyer, I. Cloning and functional characterization of *hkt1* and *akt1* genes of *fragaria* spp.-relationship to plant response to salt stress. *J. Plant Physiol.* **2017**, *210*, 9–17. [CrossRef]

51. Ye, C.Y.; Zhang, H.C.; Chen, J.H.; Xia, X.L.; Yin, W.L. Molecular characterization of putative vacuolar NHX-type Na(+)/H(+) exchanger genes from the salt-resistant tree *Populus euphratica*. *Physiol. Plant.* **2009**, *137*, 166–174. [CrossRef]

52. Bassil, E.; Tajima, H.; Liang, Y.C.; Ohto, M.A.; Ushijima, K.; Nakano, R.; Esumi, T.; Coku, A.; Belmonte, M.; Blumwald, E. The *Arabidopsis* Na⁺/H⁺ antiporters NHX1 and NHX2 control vacuolar pH and K⁺ homeostasis to regulate growth, flower development, and reproduction. *Plant Cell* **2011**, *23*, 3482–3497. [CrossRef]

53. Si, J.P.; Sun, Y.; Wang, L.; Qin, Y.; Wang, C.Y.; Wang, X.Y. Functional analyses of *Populus euphratica* brassinosteroid biosynthesis enzyme genes DWF4 (*PeDWF4*) and CPD (*PeCPD*) in the regulation of growth and development of *Arabidopsis thaliana*. *J. Biosci.* **2016**, *41*, 727–742. [CrossRef]

54. Balazadeh, S.; Riano-Pachon, D.M.; Mueller-Roeber, B. Transcription factors regulating leaf senescence in *Arabidopsis thaliana*. *Plant Biol. (Stuttg)* **2010**, *10*, 63–75. [CrossRef]

55. Iwasaki, T.; Yamaguchi-Shinozaki, K.; Shinozaki, K. Identification of a cis-regulatory region of a gene in *Arabidopsis thaliana* whose induction by dehydration is mediated by abscisic acid and requires protein synthesis. *Mol. Gen. Genet.* **1995**, *247*, 391–398. [CrossRef]

56. Yamaguchi-Shinozaki, K.; Urao, T.; Shinozaki, K. Regulation of Genes That are Induced by Drought Stress in *Arabidopsis thaliana*. *J. Plant Res.* **1995**, *108*, 127–136. [CrossRef]

57. Siahpirani, A.F.; Ay, F.; Roy, S. A multi-task graph-clustering approach for chromosome conformation capture data sets identifies conserved modules of chromosomal interactions. *Genome Biol.* **2016**, *17*, 114. [CrossRef]

58. Zheng, X.N.; Chen, B.; Lu, G.J.; Han, B. Overexpression of a NAC transcription factor enhances rice drought and salt tolerance. *Biochem. Biophys. Res. Commun.* **2009**, *379*, 985–989. [CrossRef]

59. Yang, K.Y.; Liu, Y.; Zhang, S. Activation of a Mitogen-Activated Protein Kinase Pathway is Involved in Disease Resistance in Tobacco. *PNAS* **2001**, *98*, 741–746. [CrossRef]

60. Tran, L.S.; Nakashima, K.; Sakuma, Y.; Simpson, S.D.; Fujita, Y.; Maruyama, K.; Fujita, M.; Seki, M.; Shinozaki, K.; Yamaguchi-Shinozaki, K. Isolation and functional analysis of *Arabidopsis* stress-inducible *NAC* transcription factors that bind to a drought-responsive cis-element in the early responsive to dehydration stress 1 promoter. *Plant Cell* **2004**, *16*, 2481–2498. [CrossRef]

61. Zhang, G.Y.; Chen, M.; Li, L.C.; Xu, Z.S.; Chen, X.P.; Guo, J.M.; Ma, Y.Z. Overexpression of the soybean *GmERF3* gene, an AP2/ERF type transcription factor for increased tolerances to salt, drought, and diseases in transgenic tobacco. *J. Exp. Bot.* **2009**, *60*, 3781–3796. [CrossRef] [PubMed]

62. Pasquali, G.; Biricolti, S.; Locatelli, F.; Baldoni, E.; Mattana, M. *Osmyb4* expression improves adaptive responses to drought and cold stress in transgenic apples. *Plant Cell Rep.* **2008**, *27*, 1677–1686. [CrossRef]

63. Jung, C.; Seo, J.S.; Han, S.W.; Koo, Y.J.; Kim, C.H.; Song, S.I.; Nahm, B.H.; Choi, Y.D.; Cheong, J.J. Overexpression of *atmyb44* enhances stomatal closure to confer abiotic stress tolerance in transgenic arabidopsis. *Plant Physiol.* **2008**, *146*, 623–635. [CrossRef]

64. Nakagami, H.; Pitzschke, A.; Hirt, H. Emerging MAP kinase pathways in plant stress signalling. *Trends Plant Sci.* **2005**, *10*, 339–346. [CrossRef] [PubMed]

65. Mizoguchi, T.; Irie, K. A gene encoding a mitogen-activated protein kinase kinase kinase is induced simultaneously with genes for a mitogen-activated protein kinase and an S6 ribosomal protein kinase by touch, cold, and water stress in *Arabidopsis thaliana*. *PNAS* **1996**, *93*, 765–769. [CrossRef]

66. Asai, T.; Tena, G.; Plotnikova, J.; Willmann, M.R.; Chiu, W.L.; Gomez-Gomez, L.; Boller, T.; Ausubel, F.M.; Sheen, J. MAP kinase signalling cascade in *Arabidopsis* innate immunity. *Nature* **2002**, *415*, 977–983. [CrossRef]

67. Ichimura, K.; Mizoguchi, T.; Yoshida, R.; Yuasa, T.; Shinozaki, K. Various abiotic stresses rapidly activate *Arabidopsis* MAP kinases ATMPK4 and ATMPK6. *Plant J.* **2000**, *24*, 655–665. [CrossRef]

68. Teige, M.; Scheikl, E.; Eulgem, T.; Doczi, F.; Ichimura, K.; Shinozaki, K.; Dangl, J.L.; Hirt, H. The MKK2 pathway mediates cold and salt stress signaling in *Arabidopsis*. *Mol. Cell* **2004**, *15*, 141–152. [CrossRef]

69. Jammes, F.; Yang, X.; Xiao, S.; Kwak, J.M. Two Arabidopsis guard cell-preferential *MAPK* genes, *MPK9* and *MPK12*, function in biotic stress response. *Plant Signal. Behav.* **2011**, *6*, 1875–1877. [CrossRef] [PubMed]

70. Shou, H.X.; Bordallo, P.; Wang, K. Expression of the Nicotiana protein kinase (NPK1) enhanced drought tolerance in transgenic maize. *J. Exp. Bot.* **2004**, *55*, 1013–1019. [CrossRef]

71. Marten, H.; Hyun, T.; Gomi, K.; Seo, S.; Hedrich, R.; Roelfsema, M.R.G. Silencing of *NtMPK4* impairs CO_2-induced stomatal closure, activation of anion channels and cytosolic $Ca^{(2+)}$ signals in Nicotiana tabacum guard cells. *Plant J.* **2008**, *55*, 698–708. [CrossRef]

72. Zhang, S.; Klessig, D.F. The tobacco wounding-activated mitogen-activated protein kinase is encoded by SIPK. *Proc. Natl. Acad. Sci. USA* **1998**, *95*, 7225–7230. [CrossRef] [PubMed]

73. Genot, B.; Lang, J.; Berriri, S.; Garmier, M.; Gilard, F.; Pateyron, S.; Haustraete, K.; Van Der Straeten, D.; Hirt, H.; Colcombet, J. Constitutively Active Arabidopsis MAP Kinase 3 Triggers Defense Responses Involving Salicylic Acid and SUMM2 Resistance Protein. *Plant Physiol.* **2017**, *174*, 1238–1249. [CrossRef] [PubMed]

74. Nicole, M.C.; Hamel, L.P.; Morency, M.J.; Beaudoin, N.; Ellis, B.E.; Seguin, A. MAP-ping genomic organization and organ-specific expression profiles of poplar MAP kinases and MAP kinase kinases. *BMC Genom.* **2006**, *7*, 223. [CrossRef] [PubMed]

75. Witon, D.; Gawronski, P.; Czarnocka, W.; Slesak, I.; Rusaczonek, A.; Sujkowska-Rybkowska, M.; Bernacki, M.J.; Dabrowska-Bronk, J.; Tomsia, N.; Szechynska-Hebda, M.; et al. Mitogen activated protein kinase 4 (MPK4) influences growth in *Populus tremula* L. x tremuloides. *Environ. Exp. Bot.* **2016**, *130*, 189–205. [CrossRef]

76. Chen, J.; Xue, B.; Xia, X.; Yin, W. A novel calcium-dependent protein kinase gene from *Populus euphratica*, confers both drought and cold stress tolerance. *Biochem. Biophys. Res. Commun.* **2013**, *441*, 630–636. [CrossRef] [PubMed]

77. Wang, L.L.; Yu, C.C.; Xu, S.L.; Zhu, Y.G.; Huang, W.C. OsDi19–4 acts downstream of *OsCDPK14* to positively regulate ABA response in rice. *Plant Cell Environ.* **2016**, *39*, 2740–2753. [CrossRef] [PubMed]

78. Sharma, A.; Komatsu, B. Involvement of a Ca^{2+}-Dependent Protein Kinase Component Downstream to the Gibberellin-Binding Phosphoprotein, RuBisCO Activase, in Rice. *Biochem. Biophys. Res. Commun.* **2001**, *290*, 690–695. [CrossRef]

79. Chang, W.J.; Su, H.S.; Li, W.J.; Zhang, Z.L. Expression profiling of a novel calcium-dependent protein kinase gene, *LeCPK2*, from tomato (*Solanum lycopersicum*) under heat and pathogen-related hormones. *Biosci. Biotechnol. Biochem.* **2009**, *73*, 2427–2431. [CrossRef]

 *forests*

Article

Drought-Affected *Populus simonii* Carr. Show Lower Growth and Long-Term Increases in Intrinsic Water-Use Efficiency Prior to Tree Mortality

Shoujia Sun [1,2], Lanfen Qiu [3], Chunxia He [1,2], Chunyou Li [4], Jinsong Zhang [1,2] and Ping Meng [1,2,*]

[1] Key Laboratory of Tree Breeding and Cultivation of State Forestry Administration, Research Institute of Forestry, Chinese Academy of Forestry, Beijing 100091, China; sunshj@caf.ac.cn (S.S.); hechunxia08@126.com (C.H.); zhangjs@caf.ac.cn (J.Z.)

[2] Collaborative Innovation Center of Sustainable Forestry in Southern China, Nanjing Forestry University, Nanjing 210037, China

[3] Beijing Key Laboratory of Ecologic Function Assessment and Regulation Technology of Green Space, Beijing Institute of Landscape Architecture, Beijing 100102, China; lanfenq@163.com

[4] College of Landscape and Travel, Agricultural University of Hebei Baoding, Baoding 071000, China; lchy0815@163.com

* Correspondence: mengping@caf.ac.cn; Tel.: +86-010-6288-9632

Received: 20 August 2018; Accepted: 12 September 2018; Published: 13 September 2018

Abstract: The Three-North Shelter Forest (TNSF) is a critical ecological barrier against sandstorms in northern China, but has shown extensive decline and death in *Populus simonii* Carr. in the last decade. We investigated the characteristics—tree-ring width, basal area increment (BAI), carbon isotope signature ($^{13}C_{cor}$), and intrinsic water-use efficiency (iWUE)—of now-dead, dieback, and non-dieback trees in TNSF shelterbelts of Zhangbei County. Results from the three groups were compared to understand the long-term process of preceding drought-induced death and to identify potential early-warning proxies of drought-triggered damage. The diameter at breast height (DBH) was found to decrease with the severity of dieback, showing an inverse relationship. In all three groups, both tree-ring width and BAI showed quadratic relationships with age, and peaks earlier in the now-dead and dieback groups than in the non-dieback group. The tree-ring width and BAI became significantly lower in the now-dead and dieback groups than in the non-dieback group from 17 to 26 years before death, thus, these parameters can serve as early-warning signals for future drought-induced death. The now-dead and dieback groups had significantly higher $\delta^{13}C_{cor}$ and iWUEs than the non-dieback group at 7–16 years prior to the mortality, indicating a more conservative water-use strategy under drought stress compared with non-dieback trees, possibly at the cost of canopy defoliation and long-term shoot dieback. The iWUE became significantly higher in the now-dead group than in the dieback group at 0–7 years before death, about 10 years later than the divergence of BAI. After the iWUE became significantly different among the groups, the now-dead trees showed lower growth and died over the next few years. This indicates that, for the TNSF shelterbelts studied, an abrupt iWUE increase can be used as a warning signal for acceleration of impending drought-induced tree death. In general, we found that long-term drought decreased growth and increased iWUE of poplar tree. Successive droughts could drive dieback and now-dead trees to their physiological limits of drought tolerance, potentially leading to decline and mortality episodes.

Keywords: *Populus simonii* Carr. (poplar); intrinsic water-use efficiency; tree rings; basal area increment; long-term drought

1. Introduction

Drought-induced plant mortality is increasing globally as the earth continues to warm. Large-scale forest decline is expected to fundamentally affect carbon and water cycles, biodiversity, and goods and environmental services to local residents [1,2]. Drought-induced tree death has been recognized as an important ecological issue. However, it is not fully understood why some trees survive drought effectively while other coexisting individuals do not [3]. This insufficient understanding has stimulated investigation into the mechanisms of plant death by hydraulic failure [4,5], carbon starvation [6], and biological attack [3,7,8]. Earlier studies have revealed that the probability of tree death is related to height or diameter [9], with smaller trees showing a higher death rate than larger ones [10,11]. However, few studies have addressed whether tree size is related to the probability of tree death after periods of drought. An examination of tree-ring records found that dead trees had typically experienced slower (but highly varied) growth and greater responsiveness to water deficit [12]. In response to water deficit, trees limit their vigor or growth, leading to their decline. To better characterize this phenomenon, the long period of growth before drought-induced tree death needs to be quantified by new approaches. Other studies have reported that comparison of past radial growth trends among dead, dieback (severely defoliated), and non-dieback (slightly or not defoliated) trees may help to identify early warning signals of drought-triggered mortality [13,14].

During growth, plants respond differently to drought events of different durations and intensities, which are consistent with why some plants survive while others succumb [3]. Under drought stress, a plant may reduce its stomatal conductance to avoid hydraulic failure; this concomitantly slows its photosynthesis and carbon assimilation [15], creating increased $\delta^{13}C$ and water-use efficiency (iWUE) [16]. Many studies have explored the relationship between iWUE and plant growth [17,18], water stress [19,20], and drought-induced plant mortality [14,21]. The iWUE was observed to increase substantially during dry spells, but increased iWUE could not improve tree growth sufficiently to compensate for water stress [18,22,23]. In a declined forest, some declining tree species had a lower iWUE compared with non-declining trees [24,25], but the reverse pattern was observed in other plants [14]. However, many research gaps still exist concerning variations in iWUE spanning the period from tree decline to death, or whether iWUE is an effective warning signal of accelerated drought-induced death.

The forests have served as an important ecological barrier in northern China since the initiation of the Three-North Shelter Forest (TNSF) program 1976. However, there has been large-scale decline of poplar trees in the TNSF in the past decade, and mortality since 2012. Recent surveys in Zhangbei County (Hebei Province, China) found that 80% of the TNSF stands planted in the county contained dieback poplar trees, with trees already dead or approaching death accounting for 1/3 of the area [26]. In Northern China, water availability is already limited and land-use changes are increasing the competition for water resources. These conditions will likely result in more serious water stress. This remarkable tree mortality has prompted studies on the characteristics of trees during the period prior to drought-induced death, with an aim to identify associated warning signals. Retrospective proxies (e.g., tree-ring data) are effective tools for analyzing past trends of tree growth [27]. Additionally, because tree-ring cellulose is a stable isotopic record of past environmental conditions during the assimilation of carbohydrates used for ring growth [22], the response of a tree to past water conditions is reflected in the $\delta^{13}C$ signature of its rings. Thus, tree-rings are a probable source for identifying warning signals of dieback and death.

Surveys of the TNSF forests in Zhangbei County found that, even in the same poplar stand, some trees were now-dead or had died back whereas others remained healthy (i.e., no sign of dieback). The different fates of these trees are, however, difficult to explain from their water utilization histories, because there are no records of the impact of past environmental conditions on the growth, decline, and death of these trees. However, isotopic signatures of tree rings may provide an effective tool to identify factors related to their different fates. We hypothesized that now-dead, dieback, and non-dieback trees had different growths and iWUE during the long period of growth before the mortality of poplar trees in TNSF shelterbelts. Our specific goals were as follows: (1) to analyze differences in δ^{13}C, iWUE, and basal area increment (BAI) among now-dead, die-back, and healthy (non-dieback) trees; (2) to reconstruct the past and recent growth trends of these three groups; and (3) to understand the long-term response of trees to drought, and identify early warning signals of drought-induced death. The overall aim of this study was to provide information for future studies on the growth sustainability, variability, and mortality of poplar trees in TNSF forests.

2. Materials and Methods

2.1. Experimental Sites and Sample Collection

Samples were collected from TNSF shelterbelts in Zhangbei County. Located at the southern edge of the Inner Mongolia Plateau, the county is characterized by a mid-temperate continental monsoon climate (a mean elevation of 1300 m, an annual mean temperature of 3.2 °C, annual precipitation of 300 mm, annual mean sunshine of 2897.8 h, an annual active accumulated temperature of 2448 °C, and a frost-free period of 90–110 day). Shelterbelts in the county comprise primarily *Populus simonii* Carr., a species tolerant to drought and cold. The trees typically sprout in May, grow rapidly in June–July, and enter defoliation–dormancy from September. They were planted as cutting seedlings in 1976, and the remaining trees are generally of the same age.

Twenty-six experimental sites of 100 m × 100 m were selected for analysis of poplar growth, dieback occurrence, and death. Following other studies [14,16,28], we used 'non-dieback' to describe healthy trees without dead branches, 'dieback' to describe poplar trees whose lower canopy was growing while the top was dead and dry, and 'now-dead' to describe those with no leaves and no living branches, which died in 2012. Samples were collected from three representative locations (Ertai Town Forest Farm, Xiaoertai Forest Farm, and Renjia Village). At each location, 18 trees (six non-dieback, six with 50% dieback shoots, six now-dead) were felled, and their diameters at breast height (DBH, 1.3 m) and heights were measured. Discs (5-cm thick) were collected at breast height for analyses of tree-ring width and carbon isotopes.

2.2. Tree-Ring Width and Carbon Isotope Analyses

The discs were dried, fixed, and surface-smoothed with the LignoTrim component of a LignoStation densitometry system (Rinntech, Heidelberg, Germany). The surfaces were scanned with the LignoScan component and the images were analyzed to determine tree width with the LignoVision component (precision, 20 μm). The ring series was cross-dated and checked using COFECHA software version 6.02P (Holmes, 1983) to eliminate potential errors. The BAI was calculated using Equation (1) [29]:

$$\mathrm{BAI} = \pi\left(RD_n^2 - RD_{n-1}^2\right) \tag{1}$$

where RD is the tree radius and n is the year of tree-ring formation.

The discs were dissected from bark to pith under a stereomicroscope to prepare tree-ring samples. To minimize carbon isotopic contamination, each sample was kept on a smooth glass slide during dissection. At least 0.5 g sample was obtained for each year. Samples were transferred into labeled tin capsules. Other studies have indicated that wholewood and cellulose have different isotopic values but similar trends in their variations [30], reflecting similar responses to climatic signals [31]. Therefore, wholewood δ^{13}C was measured in the present study. Briefly, the samples were dried (70 °C, 48 h), milled to a powder, and sieved through an 80-mesh screen. The sieved particles were oxidized in an elemental analyzer (Flash EA1112 HT; Thermo Scientific, Waltham, MA, USA) to CO_2 and analyzed with a mass spectrometer (DELTA V Advantage, Thermo Scientific; precision 0.1‰) to detect δ^{13}C. The δ^{13}C (‰) was calculated from Equation (2):

$$\delta^{13}\mathrm{C} = \left(\frac{R_{\mathrm{sample}}}{R_{\mathrm{standard}}} - 1\right) \times 1000 \tag{2}$$

where R is the ratio of $^{13}C/^{12}C$. The standard was Vienna Pee Dee Belemnite (VPDB).

Since the Industrial Revolution in 1850, the use of ^{13}C-deficient fossil fuel has caused an elevation in the atmospheric CO_2 concentration and a decrease in $\delta^{13}C$. To compensate for this effect, measured $\delta^{13}C$ values were corrected [31,32] by Equation (3):

$$\delta^{13}C_{cor} = \delta^{13}C_{tree} - (\delta^{13}C_{atm} + 6.4) \tag{3}$$

where $\delta^{13}C_{tree}$ represents the $\delta^{13}C$ measured from the tree-ring sample; $\delta^{13}C_{atm}$ is the value of the atmospheric background; and $\delta^{13}C_{cor}$ is the corrected value.

The $\delta^{13}C$ for the pre-Industrial Revolution atmospheric CO_2 was taken as $-6.4‰$. Data for atmospheric CO_2 and $\delta^{13}C$ between 1976 and 2003 were obtained from an earlier study [31], and data for 2004–2016 were obtained from the Earth System Research Laboratory (ESRL, National Oceanic & Atmospheric Administration of the United States; average of 22 monitoring sites) (http://www.esrl.noaa.gov/gmd/). After correction, the annual $\delta^{13}C$ series for tree rings was obtained.

2.3. iWUE Calculation

Carbon isotope discrimination, $\Delta^{13}C$, was calculated [33] using Equation (4):

$$\Delta^{13}C = \left(\frac{\delta^{13}C_{atm} - \delta^{13}C_{tree}}{1 + \delta^{13}C_{tree}/1000} \right) \tag{4}$$

where $\delta^{13}C_{atm}$ and $\delta^{13}C_{tree}$ represent $\delta^{13}C$ values for atmospheric CO_2 and tree rings, respectively.

For C_3 plants, $\Delta^{13}C$ follows a linear relationship described by Equation (5):

$$\Delta^{13}C = a + (b - a)\frac{C_i}{C_a} \tag{5}$$

where C_i is the intercellular CO_2 concentration, C_a is the atmospheric CO_2, a is the discrimination due to diffusion of $^{13}CO_2$ through stomata ($a = 4.4‰$), and b is fractionation discrimination by Rubisco against $^{13}CO_2$ ($b = 27‰$).

Subsequently, iWUE was determined from the relationship between $\Delta^{13}C$ and C_a [22,34] by Equation (6):

$$\text{iWUE} = \frac{A}{g_s} = \left(\frac{C_a - C_i}{1.6} \right) = \frac{C_a(b - \Delta^{13}C)}{1.6(b - a)} \tag{6}$$

where 1.6 represents the ratio of diffusivities between water vapor and CO_2 in the atmosphere.

2.4. Meteorological Data and Potential Evapotranspiration

Meteorological data (e.g., temperature, precipitation, relative humidity, wind rate, atmospheric pressure) recorded between 1976 and 2016 were retrieved from the Zhangbei County Meteorological Station ($40.15°$ N, $114.70°$ E; elevation: 1393 m) near the experimental sites. Data were checked by Kendall's test to ensure consistency, and were confirmed to be free of random variations, valid, and representative of the local climatic trend. Recording of the depth to groundwater at the town of Ertai only began in 1995, and we obtained these data from the City Water Resource Bureau.

Potential evapotranspiration (ET_0) was estimated using the Penman-Monteith Equation (7) [35]:

$$ET_0 = \frac{0.408\Delta(R_n - G) + \gamma\frac{900}{T+273}\mu_2(e_s - e_d)}{\Delta + \gamma(1 + 0.3\mu_2)} \tag{7}$$

where ET_0 represents potential evapotranspiration, R_n is net radiation from plant surfaces, G is soil heat flux, γ is the psychrometric constant, Δ is the slope of the saturated vapor pressure–temperature relationship, T is mean air temperature, μ_2 is wind speed at 2 m above the ground surface, e_s is saturation vapor pressure, and e_d is vapor pressure.

2.5. Data Analysis

Trees at the 26 experimental sites (Figure 1) were examined and classified into four levels of dieback (10–30%, 30–50%, 50–70%, and 70–90%) according to the proportion of dead crown relative to the whole crown. In the correlation analysis, median values (20%, 40%, 60%, and 80%) were used to represent the four levels. Dead trees had been observed at the beginning of our research in 2014, but their time of death had not been closely monitored. Plantation records showed that trees in the three locations had been planted in the same year. Therefore, the calendar year of death was determined to be 2013 by site investigation, and their tree-ring widths and ^{13}C signatures at the end of 2012 were analyzed. Ring width, BAI, $\delta^{13}C_{cor}$, $\Delta^{13}C$, and iWUE were analyzed by a repeated measures ANOVA, where year was the repeated factor, and subsequent least significant difference (LSD) test with SPSS software (IBM SPSS Statistics 20, Chicago, IL, USA). A p-value < 0.05 was considered statistically significant. The responses of the tree-ring width, BAI, and iWUE to climatic factors for non-dieback, dieback, and now-dead trees were calculated using Pearson's correlation coefficients based on monthly values in the growth season (May–September).

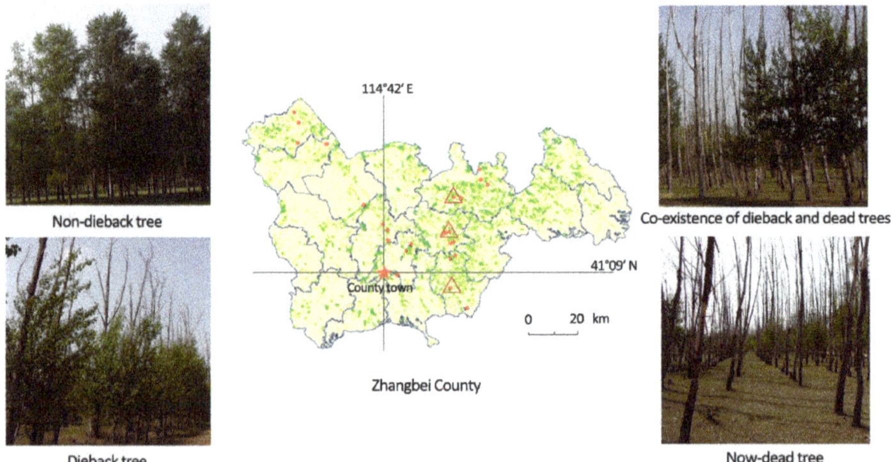

Figure 1. Location of Three-North Shelter Forest and experimental sites in Zhangbei County, Hebei Province, China. Red dots: sampling sites for analysis of relationship between poplar growth and dieback. Triangles: sampling sites for analysis of tree rings and carbon isotopes. Red star: County town.

3. Results

3.1. Meteorological Factors and Potential Evapotranspiration (ET$_0$)

Monthly mean precipitation (Figure 2A) was uneven during the study period (1976–2016), with approximately 65% falling as a rain during June–August. The mean monthly temperature showed wide variations, ranging from $-14.7\ °C$ in January to $19.5\ °C$ in July (Figure 2B). Relative humidity was lower in April–May and higher in July–August (Figure 2C), and exhibited a seasonal dry–wet cycle. Mean ET$_0$ was highest in May, coinciding with low monthly mean precipitation, thereby leading to a severe water imbalance (ET$_0$-P).

Between 1976 and 2016 (Figure 2F), total annual precipitation in Zhangbei County was consistently <540 mm (mean, 379.7 mm). Annual precipitation varied considerably during this period, creating alternate wet and drought years. An extreme drought (245.2 mm precipitation) occurred in 1997. The years 2006 and 2009 had only 292.2 and 276.8 mm precipitation, respectively. Overall, the annual precipitation tended to slightly decrease over time. The annual mean temperature (Figure 2G) was relatively low because of high local elevation, and fluctuated between 2.1 °C and

5.1 °C. However, it had been rising significantly (R^2 = 0.43, $p < 0.01$), increasing by 1.72 °C over the 41-year period. The relative humidity (Figure 2H) varied between 52% and 61%, with a weak decreasing trend over time. Rather than increasing with temperature, the ET_0 continued to decrease over time, which may have resulted from changes in vegetation cover associated with poplar die-off (Figure 2I). The ET_0 substantially exceeded the precipitation level, indicating this county is in an arid area (Figure 2J). Overall, the meteorological data indicated that the region had been undergoing severe aridification over the past four decades, driven by decreasing precipitation and increasing temperature.

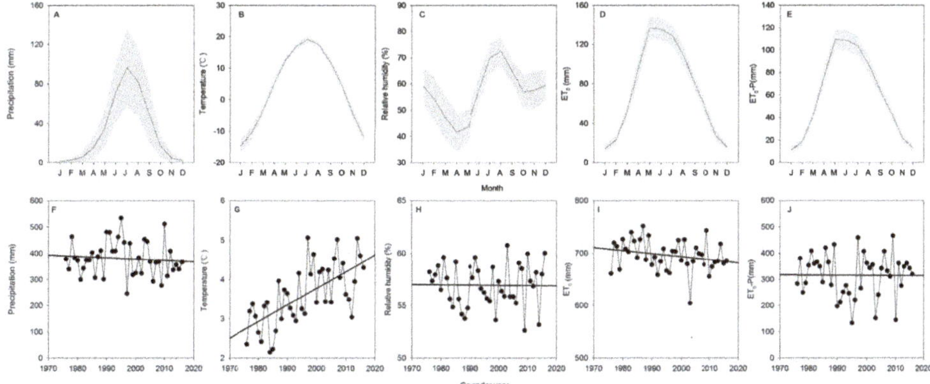

Figure 2. Variations in mean precipitation (**A**), temperature (**B**), relative humidity (**C**), potential evapotranspiration (ET_0) (**D**), and water balance (ET_0-P) (**E**) at a monthly timescale from 1976 to 2016. Also shown are variations in precipitation (**F**), temperature (**G**), relative humidity (**H**), ET_0 (**I**), and ET_0-P (**J**) during poplar tree growth.

The groundwater depth (Figure 3) was <-4 m before 1998. It began to increase in 1999 and accelerated between 2002 and 2016, increasing from -6 to -19 m. Groundwater depth increased the water imbalance in the experimental area and aggravated local droughts.

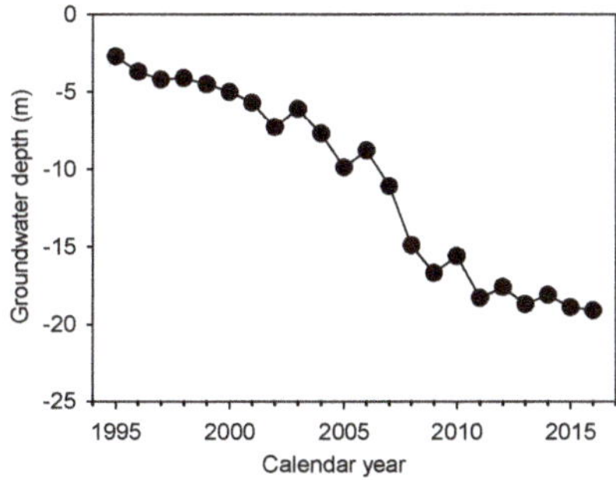

Figure 3. Changes in groundwater depth from 1995 to 2016.

3.2. Relationship between Poplar Growth and Dieback

Surveys found that the county had planted 1.02×10^5 hectares of shelterbelts. Of these, 8.11×10^4 hectares (79.5%) of stands experienced dieback, with 3.39×10^4 hectares (33.2%) now dead or nearly dead. The non-dieback group (Figure 4) had a mean DBH of 19.66 ± 2.36 cm, compared with 11.61 ± 2.31 cm in the die-off group. The trees with 20%, 40%, 60%, and 80% dieback levels had mean DBH of 13.05 ± 2.42, 14.52 ± 1.98, 15.43 ± 2.77, and 16.01 ± 2.70 cm, respectively. A linear correlation analysis revealed a highly significant ($p < 0.01$) inverse relationship between the DBH and the level of dieback, showing that the occurrence of dieback severely restricted tree growth.

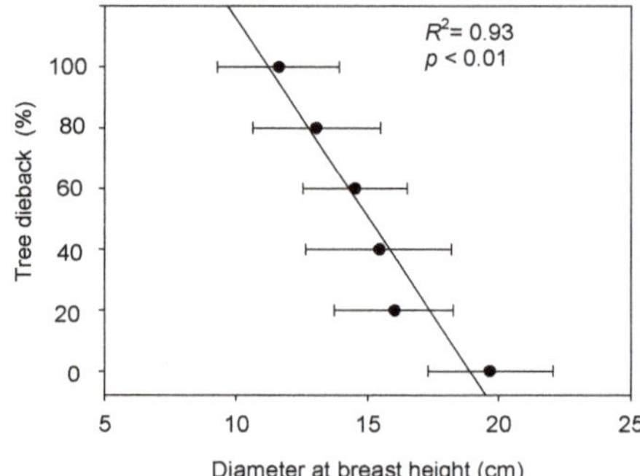

Figure 4. Relationship between diameter at breast height and dieback rate.

The tree-ring width in the now-dead group peaked in 1986 (Figure 5A) and decreased subsequently. The tree-ring width in the other two groups peaked in 1991 and then decreased. For all three groups, the tree-ring width showed a highly significant (all $p < 0.01$) quadratic relationship with age. The regression curves peaked at 1985 (now-dead, dieback) and 1990 (non-dieback), indicating that the radial growth began to slow down five years earlier in the now-dead and dieback groups than in the non-dieback group. The differences among the three groups were not statistically significant (all $p > 0.05$) during the first decade (1976–1985) after planting. The tree-ring width in the non-dieback group was significantly greater than that in the now-dead and dieback groups during the second decade (1986–1995), but did not differ between the now-dead and dieback trees. The tree-ring width differed significantly ($p < 0.05$) among the three groups from the third decade (1996–2005) onward (Figure 5B).

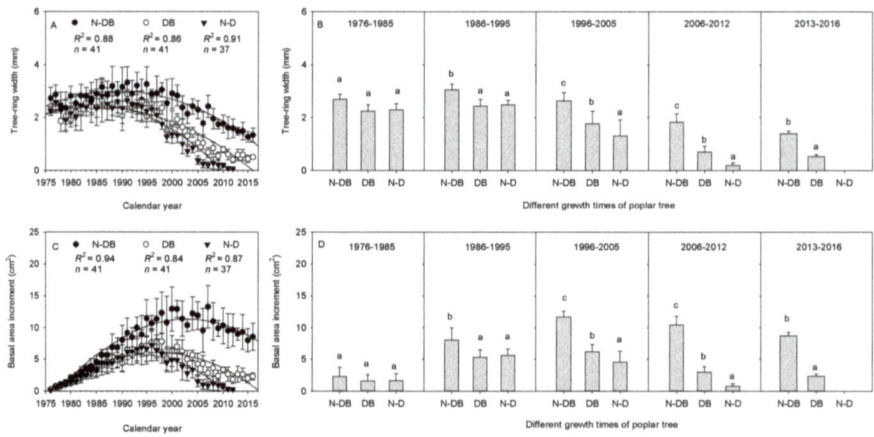

Figure 5. Differences in tree-ring width (**A,B**) and basal area increment (**C,D**) among now-dead (N-D), dieback (DB) and non-dieback (N-DB) poplar trees at different growth times. Lower case letters indicate significant differences ($p < 0.05$).

For all three groups, the BAI showed a highly significant (all $p < 0.01$) quadratic relationship with age. The regression curves peaked at 1994 (now-dead), 1997 (dieback), and 2003 (non-dieback), indicating that the BAI of the now-dead and dieback groups began to slow down nine and six years earlier, respectively, than did the BAI of the non-dieback group. The difference (Figure 5C) in BAI between the now-dead and non-dieback groups was not statistically significant during the first decade, but was significant ($p < 0.05$) from the second decade onward (Figure 5D). The difference in

BAI between the now-dead and dieback groups was statistically significant from the third decade (all $p < 0.05$).

3.3. Differences in Tree-Ring $\delta^{13}C$

After correction for the influence of the Industrial Revolution, the $\delta^{13}C_{cor}$ (Figure 6A) of the now-dead and dieback groups increased gradually over time, whereas that of the non-dieback group decreased over time. The variations in $\delta^{13}C_{cor}$ could be divided into four stages. In the first stage (1976–1995), the three groups had similar trends and no significant difference in $\delta^{13}C_{cor}$. In the second stage (1996–2005), the non-dieback group had significantly lower $\delta^{13}C_{cor}$ than the other two groups ($p < 0.05$), but the difference between the now-dead and dieback groups was not significant. In the third stage (2006–2012), the three groups had significant differences in $\delta^{13}C_{cor}$ ($p < 0.05$). The non-dieback group had significantly lower ($p < 0.05$) $\delta^{13}C_{cor}$ than the dieback after the now-dead trees died in 2013 (Figure 6B). The $\Delta^{13}C$ value of rings (Figure 6C) in the now-dead group varied between 15.06‰ and 17.76‰; that in the dieback group varied between 15.56‰ and 17.90‰; and that in the non-dieback group varied between 16.28‰ and 18.42‰. The $\Delta^{13}C$ of now-dead and dieback trees decreased over time, whereas that of non-dieback trees increased over time (Figure 6D). The changes in $\Delta^{13}C$ were also divided into four stages and the statistical significance was similar to that of $\delta^{13}C_{cor}$.

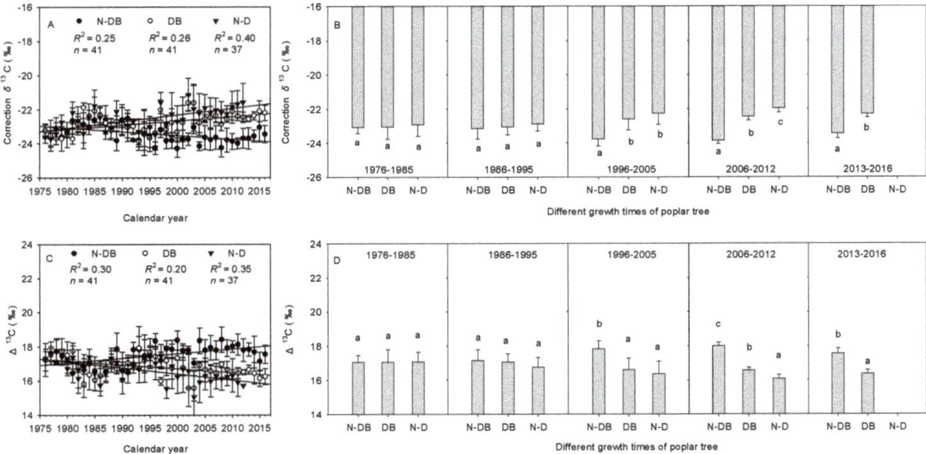

Figure 6. Difference in $\delta^{13}C_{cor}$ (**A,B**) and carbon isotope discrimination $\Delta^{13}C$ (**C,D**) among now-dead (N-D), dieback (DB), and non-dieback (N-DB) poplar trees at different growth times. Lower case letters indicate significant differences ($p < 0.05$).

3.4. Differences in iWUE

All three groups showed increasing iWUE with age, with highly significant (all $p < 0.01$) linear relationships (Figure 7A). The non-dieback group had a significantly lower regression slope (0.52) than those of the other two groups (now-dead: 1.17; dieback: 1.04). The variations in iWUE were divided into four stages. In the first stage (1976–1995), the three groups had similar iWUEs and no significant differences. In the second stage (1996–2005), the iWUEs were significantly higher ($p < 0.05$) in the now-dead and dieback groups than the non-dieback group, but the difference in iWUE between the now-dead and dieback groups was not significant. In the third stage (2007–2012), the three groups showed significant differences in iWUE. The iWUE was significantly lower ($p < 0.05$) in the non-dieback group than in the dieback group after now-dead trees died in 2013 (Figure 7B).

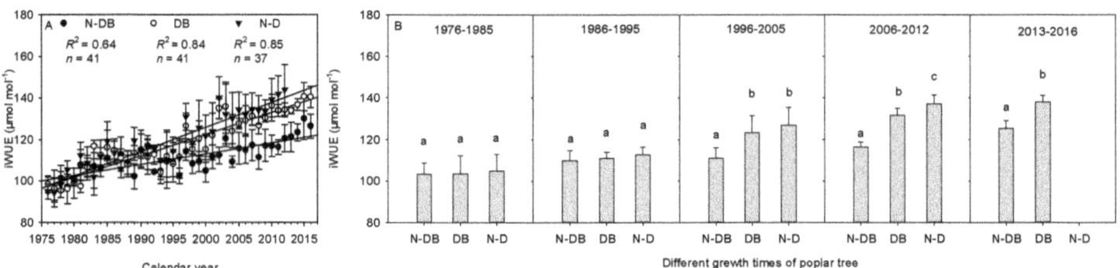

Figure 7. Difference in intrinsic water-use efficiency (**A,B**) among now-dead (N-D), dieback (DB), and non-dieback (N-DB) poplar trees at different growth times. Lower case letters indicate significant differences ($p < 0.05$).

3.5. Relationship between Tree-Ring Records and Environmental Factors

In all three groups, tree-ring width was negatively correlated with temperature (Figure 8). The correlation was highly significant ($p < 0.01$) from June to September for the now-dead group, significant ($p < 0.05$) in May, July, August, and September for the dieback group, and significant ($p < 0.05$) only in July and August for the non-dieback group. These findings indicated that radial growth was more sensitive to temperature in the now-dead and dieback groups than in the non-dieback group. The BAI of the non-dieback group was significantly positively correlated with temperature from June to September ($p < 0.01$). There was no significant correlation between BAI and temperature for now-dead and dieback groups, indicating that the now-dead and dieback groups did not respond positively to the temperature increase like the non-dieback group did. The correlation between iWUE and temperature for the now-dead group was highly significant ($p < 0.01$) from June to August, and that for the dieback group was significant ($p < 0.05$) from May to September, and highly significant ($p < 0.01$) in May, July and August; that for the non-dieback group was significant ($p < 0.05$) only in July and August. These results indicated that the iWUE of the now-dead group was the most sensitive to temperature, especially during the vigorous growth period of June, July, and August.

Relative humidity is one of the important factors affecting the growth of poplar trees. In the now-dead and dieback groups, the annual tree-ring width ($p < 0.05$) was positively associated with relative humidity in July and August, but the iWUE ($p < 0.01$) was significantly negatively associated with relative humidity in July and August. The annual tree-ring width and iWUE of the non-dieback group were only significantly affected by relative humidity in August ($p < 0.05$). The iWUE of all three groups was significantly correlated with precipitation, temperature, relative humidity, ET_0, and water balance (ET_0–P) in August ($p < 0.05$). These results indicated that, compared with tree-ring width and BAI, iWUE was more sensitive to environmental factors. Thus, during the entire growing season, the environmental factors in August played a key role in the water use of poplar trees.

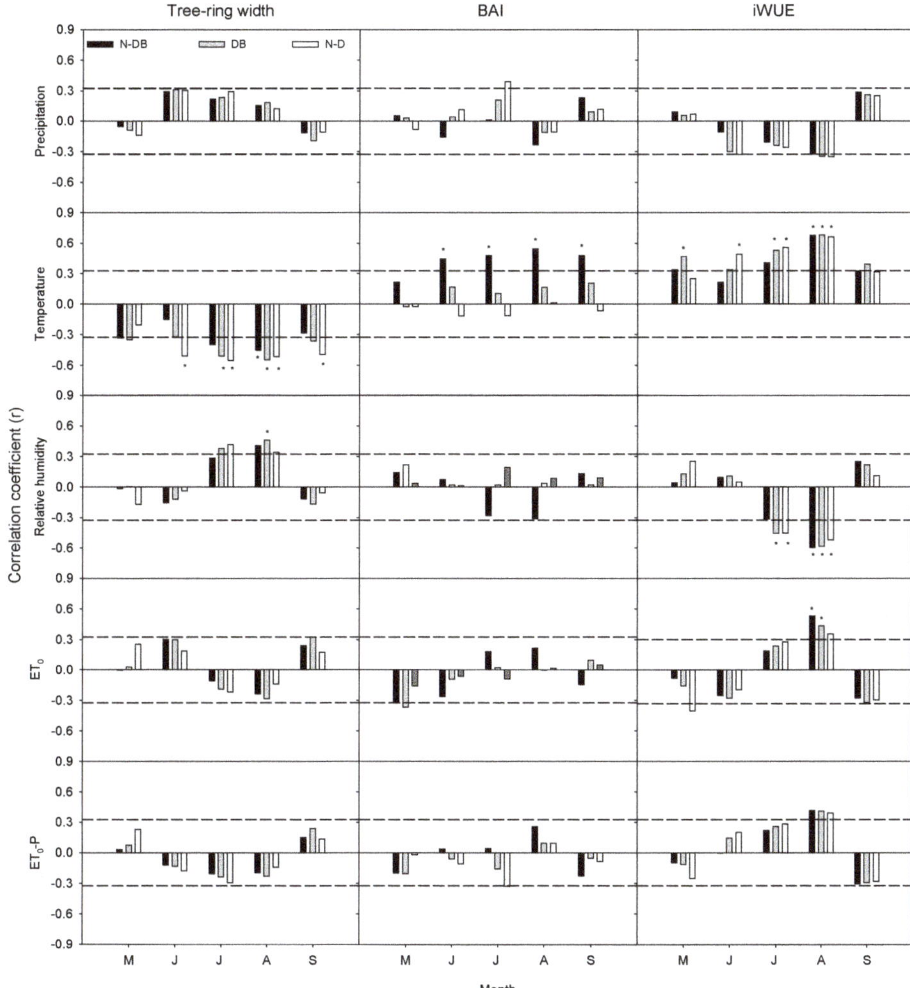

Figure 8. Relationships between tree-ring parameters (tree-ring width; BAI, iWUE) and monthly climatic factors for non-dieback (filled bar), dieback (grey bar), and now-dead (empty bar) tree groups in growth season (May–September). Months abbreviated in uppercase letters correspond to years during tree-ring formation. Dashed lines represent threshold values for statistical significance ($p < 0.05$). Asterisks (*) indicate significant correlation at $p < 0.01$.

3.6. Relationship between Radial Growth and iWUE

For all three groups, the tree-ring width (Figure 9A) showed a highly significant ($p < 0.01$) quadratic relationship with iWUE. This indicated that, with increasing iWUE, the radial growth slowed gradually after peaking. The iWUEs were much higher in the now-dead and dieback groups than in the non-dieback group, indicating slower growth of the former two groups. Nevertheless, all regression curves peaked at similar values (now-dead: 105.79, dieback: 106.35, non-dieback: 104.81 μmol mol^{-1}). The BAIs of the three groups (Figure 9B) also showed highly significant ($p < 0.01$) quadratic relationships with their iWUEs, with regression curves peaking at 117.79 (now-dead), 118.94 (dieback), and 118.06 (non-dieback) μmol mol^{-1}.

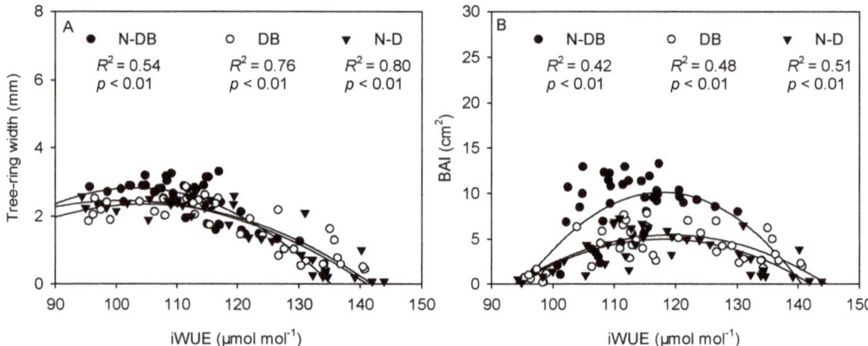

Figure 9. Differences in tree-ring width (**A**) and basal area increments (**B**) in relation to intrinsic water-use efficiency among now-dead (N-D), dieback (DB), and non-dieback (N-DB) poplar tree groups.

4. Discussion

4.1. Differences in Radial Growth before Drought-Induced Death

Earlier studies in many regions have found that increasing temperature and reducing precipitation can intensify drought stress and significantly increase the risk of death among drought-stressed plants [36,37]. The leaf area is a major determinant of the plant water requirement. Under drought stress, a plant can reduce its transpiration area by adjusting the leaf area [38]. This strategy is used by plants to reduce water transpiration from the canopy, and is used by poplar trees to survive drought. The leaf area index (LAI) decreases as drought becomes increasingly severe [39], and trees show progressive changes chronologically [40]: premature leaf senescence, partial dieback of the crown and shoots, and eventually tree mortality after successive droughts [37,41]. In the shelterbelts studied here, these three changes all appeared, and the severity of partial dieback and tree mortality increased steadily over time.

Under water stress, a tree may reduce its vigor or growth rate for long-term survival. A decrease in radial growth was observed before death in 84% of mortality events [42]. In the present study, the now-dead group had the lowest DBH, followed sequentially by the dieback and non-dieback groups. This trend indicates that drought had affected tree growth at the experimental sites. Our finding is consistent with those of an earlier study [27] reporting that Italian oaks killed by drought were smaller and had grown more slowly before death, compared with the surviving ones. In temperate and subalpine forests of northern China, the rate of tree death has frequently been reported to show a U-shaped relationship with tree size (e.g., DBH). The high death rate of smaller trees was explained by competition with taller ones for solar radiation, and that of taller ones was attributed to hydraulic failures associated with the longer distance of water transport [43]. Additionally, a temperature increase or precipitation decrease may create a water imbalance, affecting the local productivity of plants [44]. Consequently, biomass loss occurred before drought-induced death of poplar trees. In the present study, we detected a downward trend in BAI (Figure 5B) and significantly lower BAIs in the now-dead and dieback groups than in the non-dieback group. These are persuasive signs of retarded growth [29]. Plants routinely respond to changes in resource availability via gaining or losing biomass [40], but even a sharp decrease in biomass may not necessarily allow trees to avoid dieback or death. In Zhangbei County, 1997 was very dry (245 mm precipitation), and this extreme drought triggered a divergence in the subsequent fates of poplar trees. Extreme climatic events are known to affect plant growth and mortality [45]. Extreme summer drought has been shown to affect the water relationships and carbohydrate dynamics of woody angiosperms [46,47], causing dieback, hydraulic failure, and death. In the present study, the severity and duration of the drought in 1997 exceeded the tolerance of poplar trees, thus triggering long-term effects such as canopy defoliation and retarded growth. Gao reported that both the timing of onset and severity of drought increased "legacy effects" on tree stem radial growth, which reduced the drought resilience of trees [48]. Other studies have suggested that drought-induced stand changes may negatively affect the composition and ecological services of forests [49].

4.2. Differences in iWUE before Drought-Induced Death

Plants adapt to drought stress via multiple mechanisms such as adjusting growth [50] and increasing iWUE [15,22]. When environmental factors vary, trees change their discrimination among carbon isotopes. Under drought stress, trees close stomata to avoid hydraulic failure; this lowers the intracellular CO_2 concentration (C_i) and increases the $\delta^{13}C$, thereby affecting the ^{13}C signature of tree-ring cellulose. In our study, the now-dead group had significantly higher $\delta^{13}C_{cor}$ than the non-dieback group between 1997 and 2007 (Figure 6A); from 2008 onward, the now-dead and dieback groups showed significantly higher $\delta^{13}C_{cor}$ than the non-dieback group. These differences are in line with other reports that declining stands had higher $\delta^{13}C$ than healthy ones [20].

We indirectly estimated iWUE from $\delta^{13}C$ records of tree rings. The iWUE data can provide an annual archive of gas-exchange and growth. The trees may not only respond to environmental factors, but also their growth status, which would influence their stomatal conductance and photosynthetic rates, thereby altering the iWUE. Earlier studies have found that declining silver fir [14] and Scots pine [16] showed a higher iWUE compared with non-declining trees, but a reverse pattern was observed in *Quercus frainetto* Ten. [24] and *Pinus nigra* J.F.Arnold [25]. In our study, the non-dieback group showed a gradually increasing iWUE (Figure 7), whereas the other two groups had approximately two-fold greater regression slopes of increase (compared with that of the non-dieback group). Between 1996 and 2005, the iWUEs were significantly higher in the now-dead and dieback groups than in the non-dieback group. From 2006 to 2012, the three groups showed significant differences in iWUE. These findings suggested that the now-dead and dieback groups experienced more severe environmental stress and relied more on water-saving strategies than did the non-dieback group [51]. Some studies have suggested that drought stress in arid regions forces trees to close stomata, thus reducing CO_2 absorption. Consequently, tree growth decelerated despite the increased iWUE [17,52]. This theory is in agreement with the results of the present study.

The iWUE characterizing tree rings is affected by multiple environmental factors such as temperature, precipitation, relative humidity, and ambient CO_2 [53]. In all three groups, the iWUE showed a highly significant ($p < 0.01$) relationship with temperature, presumably because a higher temperature accelerated CO_2 assimilation and increased the vapor pressure deficit. Trees partially close stomata to reduce water loss [21], leading to increased $\delta^{13}C$ and iWUE. The now-dead and dieback groups had 2.2-fold greater regression slopes than that of the non-dieback group, indicating that the former two groups were more sensitive to temperature changes. Another study on Mediterranean plants found that iWUE was weakly related or unrelated to temperature [19].

All three groups exhibited highly significant quadratic relationships between iWUE and BAI. In comparison, other studies reported positive [17] or negative [54] correlations between the two parameters. When the iWUE was low, BAI increased with iWUE, primarily because of faster photosynthesis rather than reduced stomatal conductance [29]. With a further increase in iWUE, the BAIs of all three groups started to decrease, producing similar peak points (Figure 9B). The inverse relationship between iWUE and BAI reflects a deteriorating environment [52]. The declining trend of BAI was attributable to two factors: (1) the adverse impact of increased temperature on cell growth and proliferation; and (2) insufficient power of increased iWUE to counteract that impact [17].

4.3. Early-Warning Signals of Drought-Induced Death

External stimuli can drive an ecosystem to shift abruptly between alternative states when certain critical transitions or tipping points are exceeded [49]. The critical transitions can be anticipated from preceding warning signals. Global warming and its associated prolonged severe droughts can lead to retardation of tree growth and, beyond tipping points, trigger death. Drought-induced tree death can be considered as a nonlinear change in tree vigor and growth that may occur years after the causative lethal stress has exerted its effect [55]. Other researchers have compared the growth rates of now-dead and non-dieback trees under drought; they observed that the two groups showed statistically significant differences in growth rates five years before the mortality of now-dead trees, and differences were already detectable 10 years before mortality. Therefore, retarded growth for 5–50 years may be considered as a reliable indicator of impending tree death [56]. In our study, the tree-ring width of the now-dead and dieback groups peaked in 1985, whereas that of the non-dieback group peaked in 1990. The BAI quantifies the long-term trend of tree vigor, and therefore, it is a better indicator of biomass changes than is tree-ring width. Early warning signals may potentially be identified from a BAI series, even after tree death. The BAI of the now-dead, dieback, and non-dieback groups peaked in 1994, 1997, and 2003, respectively. The now-dead and dieback groups had remarkably lower tree-ring width and BAI than did the non-dieback group from 17–26 years (1986–1995) preceding death, while the difference between now-dead and dieback trees was significant from 7–16 years (1996–2005) preceding death. This indicated that tree-ring width and BAI can potentially be used as an early warning signal of tree death. Bigler and Veblen [57] reported that subalpine conifers that died during drought had increased early growth rates and large sizes, but shorter longevities. Another study suggested that a high annual growth rate and an abrupt decline were associated with a high death rate [56]. A study linking repeat forest inventories and satellite remote sensing results also indicated the potential use satellite Normalized Difference Vegetation Index data as early warning signals of tree mortality [58].

Saurer et al. [59] compared the experimentally observed changes in $\delta^{13}C$ and iWUE with predicted variations. They found that the stomatal conductance and photosynthesis of trees were affected by environmental factors, and that iWUE increased with increasing drought severity [19,60]. In a deteriorating environment, the iWUE of plants increased as a result of reduced stomatal conductance instead of accelerated photosynthesis [52]. In our study, the now-dead and dieback groups exhibited significantly higher iWUEs than those of the non-dieback group from 7 to 16 years (1996–2005) prior to death, and the difference between now-dead and dieback groups was significant from 0 to 6 years (2006–2012) prior to death. Furthermore, divergence of iWUE occurred later about 10 years later than the divergences of tree-ring width and BAI. Poplar tree death occurred about 16 years after the divergence and escalation of iWUE. These findings indicate that an abrupt escalation in iWUE can serve as a warning signal for acceleration of tree decline and drought-induced death in the TNSF shelterbelts of Zhangbei County. Similar research results showed the growth rate of Scots pine had already reduced to some extent already several decades earlier, while iWUE derived from $\delta^{13}C$ values was higher [16], indicating a more conservative water-use strategy of now-dead trees compared with that of living trees.

5. Conclusions

Long-term drought may enhance tree decline and mortality. The large-scale decline of poplar trees in the TNSF Program has provided an opportunity to examine the effects of long-term drought on growth and iWUE prior to tree mortality. Our results showed that the DBH was inversely related to the severity of dieback. Among now-dead, dieback, and non-dieback trees, both tree-ring width and BAI showed quadratic relationships with age, with the now-dead and dieback trees reaching peaks earlier than the non-dieback group. The now-dead and dieback groups had significantly higher iWUEs than did the non-dieback trees, indicating a more conservative water-use strategy under drought stress than that of non-dieback trees, possibly at the cost of canopy defoliation and shoot dieback. The drought-affected poplar trees showed lower growth and long-term increases in iWUE prior to tree mortality. Therefore, tree-width, BAI, and iWUE can serve as early warning signals for drought-induced death.

Author Contributions: Conceptualization, S.S., J.Z. and P.M.; Data curation, L.Q., C.H. and C.L.; Formal analysis, S.S.; Funding acquisition, S.S. and P.M.; Investigation, S.S., L.Q. and C.H.; Methodology, S.S., L.Q., C.H. and C.L.; Project administration, S.S., J.Z. and P.M.; Writing—original draft, S.S., L.Q. and P.M.; Writing—review & editing, S.S. and P.M.

Funding: This research was funded by the National Natural Science Foundation of China (grant number: 31470705), the Special Fund for Forest Scientific Research in the Public Welfare (grant number: 201404206) and the Project of Co-Innovation Center for Sustainable Forestry in Southern China of Nanjing Forestry University.

Acknowledgments: The China Meteorological Data Service Center provided meteorological data. We thank Elaine Monaghan, Econ, and Jennifer Smith, from Liwen Bianji, Edanz Editing China (www.liwenbianji.cn/ac), for editing the English text of a draft of this manuscript. Finally, we acknowledge Hongyan Sun for reviewing and improving the manuscript.

Conflicts of Interest: The authors declare no conflict of interest.

References

1. O'Brien, M.J.; Engelbrecht, B.M.J.; Joswig, J.; Pereyra, G.; Schuldt, B.; Jansen, S.; Kattge, J.; Landhäusser, S.M.; Levick, S.R.; Preisler, Y.; et al. A synthesis of tree functional traits related to drought-induced mortality in forests across climatic zones. *J. Appl. Ecol.* **2017**, *54*, 1669–1686. [CrossRef]

2. Trumbore, S.; Brando, P.; Hartmann, H. Forest health and global change. *Science* **2015**, *349*, 814–818. [CrossRef] [PubMed]

3. McDowell, N.; Pockman, W.T.; Allen, C.D.; Breshears, D.D.; Cobb, N.; Kolb, T.; Plaut, J.; Sperry, J.; West, A.; Williams, D.G. Mechanisms of plant survival and mortality during drought: Why do some plants survive while others succumb to drought? *New Phytol.* **2008**, *178*, 719–739. [CrossRef] [PubMed]

4. Pangle, R.E.; Limousin, J.M.; Plaut, J.A.; Yepez, E.A.; Hudson, P.J.; Boutz, A.L.; Gehres, N.; Pockman, W.T.; McDowell, N.G. Prolonged experimental drought reduces plant hydraulic conductance and transpiration and increases mortality in a piñon-juniper woodland. *Ecol. Evol.* **2015**, *5*, 1618–1638. [CrossRef] [PubMed]

5. Plaut, J.A.; Yepez, E.A.; Hill, J.; Pangle, R.; Sperry, J.S.; Pockman, W.T.; McDowell, N.G. Hydraulic limits preceding mortality in a piñon-juniper woodland under experimental drought. *Plant Cell Environ.* **2012**, *35*, 1601–1617. [CrossRef] [PubMed]

6. Trifilò, P.; Casolo, V.; Raimondo, F.; Petrussa, E.; Boscutti, F.; Lo Gullo, M.A.; Nardini, A. Effects of prolonged drought on stem non-structural carbohydrates content and post-drought hydraulic recovery in *Laurus nobilis* L.: The possible link between carbon starvation and hydraulic failure. *Plant Physiol. Biochem.* **2017**, *120*, 232–241. [CrossRef] [PubMed]

7. Colangelo, M.; Camarero, J.; Ripullone, F.; Gazol, A.; Sánchez-Salguero, R.; Oliva, J.; Redondo, M. Drought decreases growth and increases mortality of coexisting native and introduced tree species in a temperate floodplain Forest. *Forests* **2018**, *9*, 205. [CrossRef]

8. McAvoy, T.; Régnière, J.; St-Amant, R.; Schneeberger, N.; Salom, S. Mortality and recovery of hemlock woolly adelgid (*Adelges tsugae*) in response to winter temperatures and predictions for the future. *Forests* **2017**, *8*, 497. [CrossRef]

9. Grote, R.; Gessler, A.; Hommel, R.; Poschenrieder, W.; Priesack, E. Importance of tree height and social position for drought-related stress on tree growth and mortality. *Trees* **2016**, *30*, 1467–1482. [CrossRef]

10. Bennett, A.C.; Mcdowell, N.G.; Allen, C.D.; Anderson-Teixeira, K.J. Larger trees suffer most during drought in forests worldwide. *Nat. Plants* **2015**, *1*, 15139. [CrossRef] [PubMed]

11. Holzwarth, F.; Kahl, A.; Bauhus, J.; Wirth, C. Many ways to die—Partitioning tree mortality dynamics in a near-natural mixed deciduous forest. *J. Ecol.* **2013**, *101*, 220–230. [CrossRef]

12. Cailleret, M.; Bigler, C.; Bugmann, H.; Camarero, J.J.; Cufar, K.; Davi, H.; Meszaros, I.; Minunno, F.; Peltoniemi, M.; Robert, E.M.; et al. Towards a common methodology for developing logistic tree mortality models based on ring-width data. *Ecol. Appl.* **2016**, *26*, 1827–1841. [CrossRef] [PubMed]

13. Gentilesca, T.; Camarero, J.; Colangelo, M.; Nolè, A.; Ripullone, F. Drought-induced oak decline in the western Mediterranean region: An overview on current evidences, mechanisms and management options to improve forest resilience. *iFor. Biogeosci. For.* **2017**, *10*, 796–806. [CrossRef]

14. Pellizzari, E.; Camarero, J.J.; Gazol, A.; Sangüesa-Barreda, G.; Carrer, M. Wood anatomy and carbon-isotope discrimination support long-term hydraulic deterioration as a major cause of drought-induced dieback. *Glob. Chang. Biol.* **2016**, *22*, 2125–2137. [CrossRef] [PubMed]

15. Lévesque, M.; Siegwolf, R.; Saurer, M.; Eilmann, B.; Rigling, A. Increased water-use efficiency does not lead to enhanced tree growth under xeric and mesic conditions. *New Phytol.* **2014**, *203*, 94–109. [CrossRef] [PubMed]

16. Timofeeva, G.; Treydte, K.; Bugmann, H.; Rigling, A.; Schaub, M.; Siegwolf, R.; Saurer, M. Long-term effects of drought on tree-ring growth and carbon isotope variability in Scots pine in a dry environment. *Tree Physiol.* **2017**, *37*, 1028–1041. [CrossRef] [PubMed]

17. Urrutia-Jalabert, R.; Malhi, Y.; Barichivich, J.; Lara, A.; Delgado-Huertas, A.; Rodríguez, C.G.; Cuq, E. Increased water use efficiency but contrasting tree growth patterns in *Fitzroya cupressoides* forests of southern Chile during recent decades. *J. Geophys. Res. Biogeosci.* **2015**, *120*, 2505–2524. [CrossRef]

18. Wang, W.; Liu, X.; An, W.; Xu, G.; Zeng, X. Increased intrinsic water-use efficiency during a period with persistent decreased tree radial growth in northwestern China: Causes and implications. *For. Ecol. Manag.* **2012**, *275*, 14–22. [CrossRef]

19. Battipaglia, G.; de Micco, V.; Brand, W.A.; Saurer, M.; Aronne, G.; Linke, P.; Cherubini, P. Drought impact on water use efficiency and intra-annual density fluctuations in *Erica arborea* on Elba (Italy). *Plant Cell Environ.* **2014**, *37*, 382–391. [CrossRef] [PubMed]

20. Linares, J.C.; Camarero, J.J. From pattern to process: Linking intrinsic water-use efficiency to drought-induced forest decline. *Glob. Chang. Biol.* **2012**, *18*, 1000–1015. [CrossRef]

21. Hartmann, H.; Moura, C.F.; Anderegg, W.R.L.; Ruehr, N.K.; Salmon, Y.; Allen, C.D.; Arndt, S.K.; Breshears, D.D.; Davi, H.; Galbraith, D.; et al. Research frontiers for improving our understanding of drought-induced tree and forest mortality. *New Phytol.* **2018**, *218*, 15–28. [CrossRef] [PubMed]

22. Nock, C.A.; Baker, P.J.; Wanek, W.; Leis, A.; Grabner, M.; Bunyavejchewin, S.; Hietz, P. Long-term increases in intrinsic water-use efficiency do not lead to increased stem growth in a tropical monsoon forest in western Thailand. *Glob. Chang. Biol.* **2011**, *17*, 1049–1063. [CrossRef]

23. Zhang, X.; Liu, X.; Zhang, Q.; Zeng, X.; Xu, G.; Wu, G.; Wang, W. Species-specific tree growth and intrinsic water-use efficiency of Dahurian larch (*Larix gmelinii*) and Mongolian pine (*Pinus sylvestris* var. mongolica) growing in a boreal permafrost region of the Greater Hinggan Mountains, Northeastern China. *Agric. For. Meteorol.* **2018**, *248*, 145–155. [CrossRef]

24. Colangelo, M.; Camarero, J.J.; Battipaglia, G.; Borghetti, M.; de Micco, V.; Gentilesca, T.; Ripullone, F. A multi-proxy assessment of dieback causes in a Mediterranean oak species. *Tree Physiol.* **2017**, *37*, 617–631. [CrossRef] [PubMed]

25. Petrucco, L.; Nardini, A.; von Arx, G.; Saurer, M.; Cherubini, P. Isotope signals and anatomical features in tree rings suggest a role for hydraulic strategies in diffuse drought-induced die-back of *Pinus nigra*. *Tree Physiol.* **2017**, *37*, 523–535. [PubMed]

26. Sun, S.; He, C.; Qiu, L.; Li, C.; Zhang, J.; Meng, P. Stable isotope analysis reveals prolonged drought stress in poplar plantation mortality of the Three-North Shelter Forest in Northern China. *Agric. For. Meteorol.* **2018**, *252*, 39–48. [CrossRef]

27. Colangelo, M.; Camarero, J.J.; Borghetti, M.; Gazol, A.; Gentilesca, T.; Ripullone, F. Size Matters a Lot: Drought-affected Italian oaks are smaller and show lower growth prior to tree death. *Front. Plant Sci.* **2017**, *8*, 135. [CrossRef] [PubMed]

28. Hoffmann, W.A.; Marchin, R.M.; Abit, P.; Lau, O.L. Hydraulic failure and tree dieback are associated with high wood density in a temperate forest under extreme drought. *Glob. Chang. Biol.* **2011**, *17*, 2731–2742. [CrossRef]

29. Silva, L.C.; Anand, M.; Leithead, M.D. Recent widespread tree growth decline despite increasing atmospheric CO_2. *PLoS ONE* **2010**, *5*, e11543. [CrossRef] [PubMed]

30. Schleser, G.H.; Anhuf, D.; Helle, G.; Vos, H. A remarkable relationship of the stable carbon isotopic compositions of wood and cellulose in tree-rings of the tropical species *Cariniana micrantha* (Ducke) from Brazil. *Chem. Geol.* **2015**, *401*, 59–66. [CrossRef]

31. McCarroll, D.; Loader, N.J. Stable isotopes in tree rings. *Quat. Sci. Rev.* **2004**, *23*, 771–801. [CrossRef]

32. Gagen, M.; Finsinger, W.; Wagner-Cremer, F.; McCarroll, D.; Loader, N.J.; Robertson, I.; Jalkanen, R.; Young, G.; Kirchhefer, A. Evidence of changing intrinsic water-use efficiency under rising atmospheric CO_2 concentrations in Boreal Fennoscandia from subfossil leaves and tree ring $\delta^{13}C$ ratios. *Glob. Chang. Biol.* **2011**, *17*, 1064–1072. [CrossRef]

33. Farquhar, G.D.; O'Leary, M.H.; Berry, J.A. On the relationship between carbon isotope discrimination and the intercellular carbon dioxide concentration in leaves. *Funct. Plant Biol.* **1982**, *9*, 121–137. [CrossRef]

34. Farquhar, G.; Richards, R. Isotopic composition of plant carbon correlates with water-use efficiency of wheat genotypes. *Funct. Plant Biol.* **1984**, *11*, 539–552. [CrossRef]

35. Allen, R.G.; Pereira, L.S.; Raes, D.; Smith, M. Crop evapotranspiration-guidelines for computing crop water requirements-FAO Irrigation and drainage paper 56. *FAO Rome* **1998**, *300*, D05109.

36. Greenwood, S.; Ruiz-Benito, P.; Martínez-Vilalta, J.; Lloret, F.; Kitzberger, T.; Allen, C.D.; Fensham, R.; Laughlin, D.C.; Kattge, J.; Bönisch, G.; et al. Tree mortality across biomes is promoted by drought intensity, lower wood density and higher specific leaf area. *Ecol. Lett.* **2017**, *20*, 539–553. [CrossRef] [PubMed]

37. Navarro-Cerrillo, R.; Rodriguez-Vallejo, C.; Silveiro, E.; Hortal, A.; Palacios-Rodríguez, G.; Duque-Lazo, J.; Camarero, J. Cumulative drought stress leads to a loss of growth resilience and explains higher mortality in planted than in naturally regenerated *Pinus pinaster* stands. *Forests* **2018**, *9*, 358. [CrossRef]

38. Delucia, E.H.; Maherali, H.; Carey, E.V. Climate-driven changes in biomass allocation in pines. *Glob. Chang. Biol.* **2000**, *6*, 587–593. [CrossRef]

39. Luo, Y.; Su, B.; Currie, W.S.; Dukes, J.S.; Finzi, A.; Hartwig, U.; Hungate, B.; McMurtrie, R.E.; Oren, R.; Parton, W.J.; et al. Progressive nitrogen limitation of ecosystem responses to rising atmospheric carbon dioxide. *BioScience* **2004**, *54*, 731–739. [CrossRef]

40. Jump, A.S.; Ruiz-Benito, P.; Greenwood, S.; Allen, C.D.; Kitzberger, T.; Fensham, R.; Martínez-Vilalta, J.; Lloret, F. Structural overshoot of tree growth with climate variability and the global spectrum of drought-induced forest dieback. *Glob. Chang. Biol.* **2017**, *23*, 3742–3757. [CrossRef] [PubMed]

41. Galiano, L.; Martínez-Vilalta, J.; Lloret, F. Carbon reserves and canopy defoliation determine the recovery of Scots pine 4 yr after a drought episode. *New Phytol.* **2011**, *190*, 750–759. [CrossRef] [PubMed]

42. Cailleret, M.; Jansen, S.; Robert, E.M.R.; Desoto, L.; Aakala, T.; Antos, J.A.; Beikircher, B.; Bigler, C.; Bugmann, H.; Caccianiga, M.; et al. A synthesis of radial growth patterns preceding tree mortality. *Glob. Chang. Biol.* **2017**, *23*, 1675–1690. [CrossRef] [PubMed]

43. Chen, H.; Fu, S.; Monserud, R.; Gillies, I. Relative size and stand age determine *Pinus banksiana* mortality. *For. Ecol. Manag.* **2008**, *255*, 3980–3984. [CrossRef]

44. Juday, G.P.; Alix, C.; Grant, T.A. Spatial coherence and change of opposite white spruce temperature sensitivities on floodplains in Alaska confirms early-stage boreal biome shift. *For. Ecol. Manag.* **2015**, *350*, 46–61. [CrossRef]

45. Niu, S.; Luo, Y.; Li, D.; Cao, S.; Xia, J.; Li, J.; Smith, M.D. Plant growth and mortality under climatic extremes: An overview. *Environ. Exp. Bot.* **2014**, *98*, 13–19. [CrossRef]

46. Churakova, O.; Lehmann, M.; Saurer, M.; Fonti, M.; Siegwolf, R.; Bigler, C. Compound-specific carbon isotopes and concentrations of carbohydrates and organic acids as indicators of tree decline in mountain pine. *Forests* **2018**, *9*, 363. [CrossRef]

47. Nardini, A.; Casolo, V.; Dal Borgo, A.; Savi, T.; Stenni, B.; Bertoncin, P.; Zini, L.; McDowell, N.G. Rooting depth, water relations and non-structural carbohydrate dynamics in three woody angiosperms differentially affected by an extreme summer drought. *Plant Cell Environ.* **2016**, *39*, 618–627. [CrossRef] [PubMed]

48. Gao, S.; Liu, R.; Zhou, T.; Fang, W.; Yi, C.; Lu, R.; Zhao, X.; Luo, H. Dynamic responses of tree-ring growth to multiple dimensions of drought. *Glob. Chang. Biol.* **2018**, 1–12. [CrossRef] [PubMed]

49. Camarero, J.J.; Gazol, A.; Sangüesa-Barreda, G.; Oliva, J.; Vicente-Serrano, S.M. To die or not to die: Early warnings of tree dieback in response to a severe drought. *J. Ecol.* **2015**, *103*, 44–57. [CrossRef]

50. McDowell, N.; Allen, C.D.; Marshall, L. Growth, carbon-isotope discrimination, and drought-associated mortality across a *Pinus ponderosa* elevational transect. *Glob. Chang. Biol.* **2010**, *16*, 399–415. [CrossRef]

51. Martin-Benito, D.; Anchukaitis, K.; Evans, M.; del Río, M.; Beeckman, H.; Cañellas, I. Effects of drought on xylem anatomy and water-use efficiency of two co-occurring pine species. *Forests* **2017**, *8*, 332. [CrossRef]

52. Lévesque, M.; Rigling, A.; Bugmann, H.; Weber, P.; Brang, P. Growth response of five co-occurring conifers to drought across a wide climatic gradient in Central Europe. *Agric. For. Meteorol.* **2014**, *197*, 1–12. [CrossRef]

53. Rezaie, N.; D'Andrea, E.; Bräuning, A.; Matteucci, G.; Bombi, P.; Lauteri, M. Do atmospheric CO_2 concentration increase, climate and forest management affect iWUE of common beech? Evidences from carbon isotope analyses in tree rings. *Tree Physiol.* **2018**. [CrossRef] [PubMed]

54. Silva, L.C.R.; Anand, M. Probing for the influence of atmospheric CO_2 and climate change on forest ecosystems across biomes. *Glob. Ecol. Biogeogr.* **2013**, *22*, 83–92. [CrossRef]

55. Huang, M.; Wang, X.; Keenan, T.F.; Piao, S. Drought timing influences the legacy of tree growth recovery. *Glob. Chang. Biol.* **2018**, *24*, 3546–3559. [CrossRef] [PubMed]

56. Das, A.J.; Battles, J.J.; Stephenson, N.L.; van Mantgem, P.J. The relationship between tree growth patterns and likelihood of mortality: A study of two tree species in the Sierra Nevada. *Can. J. For. Res.* **2007**, *37*, 580–597. [CrossRef]

57. Bigler, C.; Veblen, T.T. Increased early growth rates decrease longevities of conifers in subalpine forests. *Oikos* **2009**, *118*, 1130–1138. [CrossRef]

58. Rogers, B.M.; Solvik, K.; Hogg, E.H.; Ju, J.; Masek, J.G.; Michaelian, M.; Berner, L.T.; Goetz, S.J. Detecting early warning signals of tree mortality in boreal North America using multiscale satellite data. *Glob. Chang. Biol.* **2018**, *24*, 2284–2304. [CrossRef] [PubMed]

59. Saurer, M.; Siegwolf, R.T.W.; Schweingruber, F.H. Carbon isotope discrimination indicates improving water-use efficiency of trees in northern Eurasia over the last 100 years. *Glob. Chang. Biol.* **2004**, *10*, 2109–2120. [CrossRef]

60. Sangüesa-Barreda, G.; Linares, J.C.; Julio Camarero, J. Drought and mistletoe reduce growth and water-use efficiency of Scots pine. *For. Ecol. Manag.* **2013**, *296*, 64–73. [CrossRef]

 *forests*

Article

Low Water Availability Increases Necrosis in *Picea abies* after Artificial Inoculation with Fungal Root Rot Pathogens *Heterobasidion parviporum* and *Heterobasidion annosum*

Eeva Terhonen [1,*], Gitta Jutta Langer [2], Johanna Bußkamp [2], David Robert Răscuţoi [1] and Kathrin Blumenstein [1]

[1]   Forest Pathology Research Group, Büsgen-Institute, Department of Forest Botany and Tree Physiology, Faculty of Forest Sciences and Forest Ecology, University of Göttingen, Büsgenweg 2, 37077 Göttingen, Germany; d.rascutoi@stud.uni-goettingen.de (D.R.R.); kathrin.blumenstein@uni-goettingen.de (K.B.)

[2]   Northwest German Forest Research Institute, Grätzelstraße 2, 37079 Göttingen, Germany; gitta.langer@nw-fva.de (G.J.L.); johanna.busskamp@nw-fva.de (J.B.)

*   Correspondence: terhonen@uni-goettingen.de; Tel.: +49-0551-39-10380

Received: 19 December 2018; Accepted: 10 January 2019; Published: 12 January 2019

Abstract: Research Highlights: Dedicated experiments to investigate how disturbances will affect *Heterobasidion* sp.—Norway spruce pathosystems are important, in order to develop different strategies to limit the spread of *Heterobasidion annosum* s.l. under the predicted climate change. Here, we report on a greenhouse experiment to evaluate the effects of water availability on the infection severity of *Heterobasidion parviporum* or *Heterobasidion annosum*, respectively, on *Picea abies* saplings. Background and Objectives: Changes in climatic conditions and intense logging will continue to promote *H. annosum* s.l. in conifer forests, increasing annual economic losses. Thus, our aim was to test if disease severity in Norway spruce was greater after infection with *H. parviporum* or *H. annosum* in low water availability conditions, compared to seedlings with high water availability. Materials and Methods: We performed inoculation studies of three-year-old saplings in a greenhouse. Saplings were treated as high (+) or low (−) water groups: High water group received double the water amount than the low water group. The necrosis observed after pathogen inoculation was measured and analyzed. Results: The seedling growth was negatively influenced in the lower water group. In addition, the water availability enhanced the necrosis length of *H. parviporum* in phloem and sapwood (vertical length) in the low water group. *H. annosum* benefited only in horizontal length in the phloem. Conclusions: Disturbances related to water availability, especially low water conditions, can have negative effects on the tree host and benefit the infection ability of the pathogens in the host.

Keywords: *Heterobasidion parviporum*; *Heterobasidion annosum*; Norway spruce; disturbance; water availability; pathogen; infection

---

1. Introduction

Projected climate change will increase the disturbances related to water availability in forests e.g., drought, wind and snow [1]. In addition to their direct negative effects on trees, such as timber quality losses, the weakening of the tree's immune system or early decay or death, the impacts of pests and pathogens in changing environments represent some of the most important threats to global forest health [1]. To mitigate these climate change induced disturbances, such as drought, it is important to understand the factors that contribute to the development of forest tree disease epidemics and host susceptibility [2].

Norway spruce (*Picea abies* (L.) H. Karst.), together with Scots pine (*Pinus sylvestris* L.), forms the basis for raw materials of the forest sector in Europe, contributing to several billion EUR net income annually. Currently, Norway spruce covers 25% of the forest area in Germany [3]. Thus, interest in the biology and ecology of plant pathogenic fungi of conifers in forest ecosystems, with special emphasis on sustainable management strategies for forestry, is high. The main fungal pathogens causing devastating losses of wood quality in conifers are the root rot pathogens of the species complex *Heterobasidion annosum sensu lato* (s.l.) [4,5]. Consequently, *H. annosum* s.l. is considered the most dangerous and economically most significant root rot pathogen in coniferous forests [4,5]. *Heterobasidion* species induce root and stem rot and bark necrosis. Resin flow on the stem is one of the visible symptoms with which *H. annosum* s.l. can be detected in living trees (Figure 1). In the above-average warm years (2012 and 2018), when precipitation was low (average precipitation in Oerrel, Germany 2012: 201−250 mm; 2017: 301−350 mm) [6], these symptoms could be observed more regularly in Germany (Figure 1) [7]. The water balance as well was lower in this area in the year 2012 (0−49 mm) compared to the year 2017 (51−100 mm) [6]. The anamorph stage *Spiniger meineckellus* (A.J. Olson) Stalpers could be observed, proving infestation of roots, stems and other host tissues with *H. annosum* s.l. (Figure 1) [8,9].

Figure 1. Resin flow on the stem is one of the symptoms that indicates the possible infection of *H. annosum* s.l. Living Norway spruce, confirmed to be infected by *Heterobasidion annosum* s.l.; (**a**) resinating stem necrosis; (**b**) stem necrosis with removed bark. These symptoms were observed more in the year 2012 in a forest stand close to Oerrel, Germany.

Extensive logging of cultivated conifer forests has created a habitat favourable for this pathogen, which has dramatically increased the occurrence of infections. Primary infections by *Heterobasidion* species are established when airborne basidiospores land and germinate on freshly cut stump surfaces and wounds [10–12]. Spore deposition is followed by rapid colonisation of the wood material.

In Europe, three species of native *Heterobasidion* exist: *H. abietinum* Niemelä and Korhonen, *H. annosum* (Fr.) Bref. and *H. parviporum* Niemelä and Korhonen. All *Heterobasidion* species have different, but overlapping, host preferences, mainly associated with fir (*H. abietinum*), pine (*H. annosum*), and spruce (*H. parviporum*) [5,13,14]. In North-West Germany, only *H. annosum* and *H. parviporum* are present [9,15,16]. Differentiation between these two species can be performed with species-specific primers developed for boreal species of *H. parviporum* and *H. annosum* ([17], this study). Silvicultural control of *Heterobasidion* species is difficult because they cause secondary infections by spreading through root contacts to neighbouring trees [18,19]. This is the main pathway for infections to spread inside stands [9,16,20]. Furthermore, *Heterobasidion* species can remain viable and infective in stumps for decades [21–23], resulting in an inoculum source for new tree generations [23,24].

Changes in climatic conditions and intense logging will continue to promote *H. annosum* s.l. in conifer forests [1,25,26], increasing the economic losses on a yearly basis. Linnakoski et al. [2] could show already, that the fungal strain of *Endoconidiophora polonica* (Siemaszko) Z.W. de Beer, T.A. Duong and M.J. Wingf. caused greater disease severity in *P. abies* seedlings with low water availability compared to those with high water availability. Similarly, Gori et al. [27] found indications that *H. parviporum* infection makes *P. abies* trees more susceptible to drought stress in the field. Thus, dedicated experiments to investigate how disturbances will affect the *Heterobasidion* sp.—Norway spruce pathosystems are important, in order to develop different strategies to limit the spread of *H. annosum* s.l. under the predicted climate change.

Here, we report an in vivo experiment, conducted in a greenhouse in Germany, to evaluate the effects of water availability on the infection severity of *H. parviporum* or *H. annosum* in *P. abies* saplings during the growing season of 2018. Dimitri and Schumann [28] found that the infection incidence did not change with the age of inoculated ramets among clones of *P. abies*. This indicates that results from inoculation experiments (for at least some tree species) with seedlings may be applicable to trees. The aim of this research was to test if the low tolerance of Norway spruce to water scarcity will lead to greater disease severity caused by *H. parviporum* and *H. annosum*, when compared to seedlings with optimal water availability. Drought disturbances in European forests are predicted to increase [1]. Based on our observations in the field in the year 2012 (Figure 1), we hypothesize that *H. annosum* s.l. can cause greater necrosis in *P. abies* saplings under low water availability.

2. Materials and Methods

2.1. Plant and Fungal Material

Plant material consisted of 355 three-year-old, apparently healthy and vital, Norway spruce saplings purchased from the nursery Schlegel and Co Gartenprodukte (provenience of the saplings: No. 840 11, Thüringer Wald and Frankenwald, montane zone 600 m). Seedlings were potted onto 3-liter plastic pots filled with fertilized peat (Flora gard, TKS®2 Instant Plus, Hermann Meyer KG, Rellingen, Germany). Pots were placed into tables covered with plastic sheets where excess water could accumulate and be absorbed later. The potted saplings were acclimatized to the greenhouse conditions for 16-days prior to the water experiment, during which time they received tap water, as required, to maintain moist soil. No additional fertilization was given during the experiment. Before and after the experiment, the sapling height was measured to the nearest 0.1 cm.

2.2. Pre-Experiment Detection of Heterobasidion Species

Three of the saplings were transferred to North West German Forest Research Institute (Göttingen) in order to examine the pre-colonization in the internal tissues for pathogens and endophytes. Two different methods were used to detect *Heterobasidion* species: (1) Incubation method after Langer and Bressem [16]; (2) isolation of fungi on artificial agar media. Firstly, stem-disks (10 mm) cut from saplings (Figure 2) and two randomly chosen root segments were pre-incubated in a refrigerator for four weeks and subsequently wrapped in moist newspaper. After incubation at room temperature for 14 days, the samples were analyzed under a binocular microscope for the presence of *S. meineckellus* [6,16]. Secondly, surface sterilized pieces of the trunk were placed on petri-dishes with Malt-Yeast-Peptone (MYP-agar, after Langer [29]). After 7 and 14 days they were checked for the presence of *H. annosum* s.l. conidiophores and conidiospores.

Figure 2. Pre-experiment detection of *Heterobasidion* species from Norway spruce saplings. The disks from spruce trunk were cut 3.5 cm above and 3.5 cm underneath the root neck, which is shown in the figure as a red line. The topmost 1 cm and the lowermost 1 cm of the trunk were stored in the refrigerator for pre-incubation.

The remaining trunk parts and root segments were surface sterilized (1 min in 70% ethanol/5 min 4% sodium hypochlorite/1 min 70% ethanol) and cut into units (0.5 cm). Three units each were plated on a petri dish filled with 1.5% MYP-agar. The presence for *H. annosum* s.l. and other fungi were checked after 7 and 14 days.

2.3. Fungal Material

Heterobasidion sp. strains were received from the North-West German Forest Research Institute, both collected by G. Langer and colleagues. *H. annosum* (strain NW-FVA 3492) was isolated from *P. abies* in 2016 (Germany, Lower Saxony, forest department Oerrel, Karrenbusch). *H. parviporum* (strain NW-FVA 0459) was isolated from *P. abies* in 2010 (Germany, Lower Saxony, forest department Oerrel, Bobenwald). Before inoculations, the fungal strains used in this study were tested for their specificity with species-specific primers developed for *H. parviporum* and *H. annosum* [17]. Similarly, to make sure the pathogen cultures were not contaminated, the DNA of both cultures were extracted and PCR was performed with ITS1-F and ITS4 primers and sequenced.

2.4. DNA Extractions, PCR and Sequencing

DNA of *H. parviporum* and *H. annosum* was extracted from 150 mg of the homogenized mycelium sample using the "innuPREP Plant DNA Kit" (Analytik Jena AG, Jena, Germany), according to the manufacturer's instructions. Species-specific primers (*H. annosum*, MJ-F and MJ-R [17]; *H. parviporum* KJ-F and KJ-R [17]) were used to confirm *Heterobasidion* species [17]. In brief, DNA template (100 ng), buffer (KCl extra buffer, 1×), 1.5 mM MgCl2, primers (each concentration of 0.5 µM), a dNTP-mix (each deoxynucleotide in a concentration of 200 µM) and 20 U/mL of DNA-polymerase (VWR) was adjusted to 25 µL reaction with autoclaved MQ H$_2$O. The cycling conditions used were: Initial denaturation 10 min at 95 °C, followed by 3 step cycling: Denaturation 30 s 95 °C; annealing 35 s 67 °C; extension 1 min 72 °C, number of cycles 40; final extension 7 min in 72 °C. Both species were run with both primer pairs. Taq DNA polymerase (Qiagen) was used for PCR amplification of ITS regions with the primer pair, ITS1-F and ITS4 [30,31]. Briefly, the PCR protocol was as follows: 1X CoralLoad PCR Buffer, 200 µM dNTP, 0.5 µM primer 1, 0.5 µM primer 2, 100ng template DNA, 0.2 U/µL DNA polymerase;

the reaction was adjusted to 25 µL with autoclaved MQ H_2O. The PCR conditions used were 94 °C for 3 min; 30 cycles of 94 °C for 30 s, 55 °C for 1 min, 72 °C for 1 min, and 72 °C for 10 min. Possible contaminations were determined with a negative control using sterile water as a template in both PCR protocols. StainIN™ RED Nucleic Acid Stain was used to confirm DNA amplicons on a 2% agarose gel and the visual detection was made by ultraviolet transillumination. ITS region PCR products were purified and sequenced using the ITS4 primer at Microsynth SEQLAB (Göttingen, Germany). The FASTA files thus obtained were checked with BioEdit [32] to confirm that the pathogen was not contaminated with other fungi.

2.5. Experimental Design

The experiment was conducted at the Forest Botany and Tree Physiology greenhouses, Göttingen, Germany (51°33′28.4″ N 9°57′30.5″ E) from early April until early August 2018. The 352 seedlings were randomly block-assigned to waterproof tables with either high water (high water group = +group) or low water (low water group = −group) availability treatments [2], and experimentally inoculated with *H. parviporum* (66 per treatment group), *H. annosum* (66 per treatment group), mock-inoculated controls (25 per treatment group) or left entirely untreated (19 per treatment group). The water treatment experiment was running for 35 days before the inoculations were performed. Fungal isolates were plated on 2% Malt Extract Agar (MEA) and grown at 21 °C for two weeks prior to the experimental inoculations. Inoculations were made approximately mid-way up the stem of a first-year shoot (distance of the inoculation from the stem base was ∼5 cm). A sterile 5mm cork borer was used to punch through the bark to reach the sapwood surface. Equal sized plugs from pure culture of *Heterobasidion* sp. or control (2% MEA) were placed onto the exposed surface and sealed with Parafilm® [2]. The inoculation experiment was run for 70 days.

Since seedling water intake varied with the ambient temperature, it was necessary to continuously monitor and adjust the watering amounts, in order to maintain the essential level in high water availability treatment (moist soil). The water amounts needed to be continuously adjusted to the increasing temperatures during the growing season 2018. Overall, high water availability seedlings received double the water of low water availability seedlings. This was considered the suitable level of drought in low water treatment [2]. At the beginning of the experiment, each high water availability seedling was given 60 mL of water three times per week (Monday, Wednesday and Friday), and each low water availability seedling was given 30 mL of water three times per week. After two-weeks, the watering regimes were modified to 120 mL × 3 (+group), and 60 mL × 3 (−group). After five weeks, the water level was modified again to 180 mL × 3 (+group) and 90 mL × 3 (−group). Water quantities were increased in July (384 mL × 3 (+group), and 192 mL × 3 (−group)), and maintained at that level until August. Seedlings were rotated on tables monthly, in order to minimize potential variation in water requirements, due to positioning in the greenhouse. Throughout the experiment, temperatures were measured inside the greenhouse every watering day using digital thermometers. The average daily temperatures during the experiment period are presented in Table 1.

Table 1. The average, minimum and maximum temperatures in each month measured inside a greenhouse.

Month	Average, °C	Min, °C	Max, °C
April	17.9	13.3	20.7
May	21.8	15.6	25.7
June	24.8	20.4	31.1
July	24.2	17.4	30.9
August	32.5	32.5	32.8

2.6. Data Collection and Post-Experiment Detection of Heterobasidion Species

Seedling height was measured at the beginning and the end of the experiment, in order to determine seedling growth (height$_{\text{end}}$ − height$_{\text{start}}$) during the experiment. At the end of the experiment, the lesion lengths (nearest 0.001 mm) were measured with a stereomicroscope (Stemi 508, Zeiss, Omnilab-laborzentrum GmbH & Co.KG, Gehrden, Germany) with an attached camera (Axiocam ERc5s, Zeiss, Omnilab-laborzentrum GmbH & Co.KG, Gehrden, Germany) by using the freely available software Labscope (Carl Zeiss Microscopy GmbH, Jena, Germany). First, the bark was gently peeled to expose the necrosis in phloem (Figure 3a), measured, and then further sapwood (Figure 3b) was exposed and measured. The lesion length was measured in horizontal and vertical directions. After measurements, DNA was extracted directly from the inoculated sapwood (Figure 3b,c) (100 mg) using the "innuPREP Plant DNA Kit" (Analytik Jena AG, Jena, Germany), according to the manufacturer's instructions. From each treatment, 10 randomly selected inoculated saplings (in total 60 saplings) was selected. To confirm infection, species-specific primers were used as described above to detect *Heterobasidion* in inoculated stems [17]. Additionally, five inoculated stems (Figure 3b,c) from each treatment (30 samples), were placed inside a plastic bag and incubated in darkness for five days. The presence of *H. annosum* s.l. was identified based on conidiophores and conidiospores under stereomicroscopy.

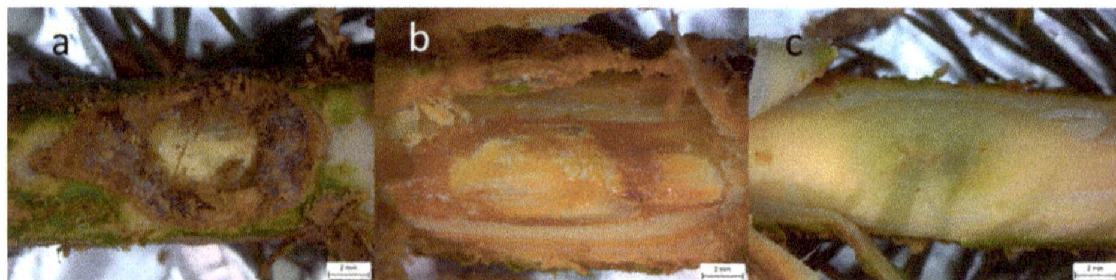

Figure 3. Measuring the necrosis in phloem and sapwood of Norway spruce saplings. (**a**) Necrosis in phloem after inoculation with *H. parviporum*; (**b**) necrosis in sapwood after inoculation with *H. parviporum*; (**c**) no sign of necrosis in the sapwood of mock-inoculated control.

2.7. Data Analysis

Data were analyzed using the SPSS version 25.0 (IBM Corporation, New York, NY, USA). A generalized linear model was constructed to evaluate the fixed effects of necrosis length (mock-inoculated-control, *H. annosum*, *H. parviporum*) and water availability treatments on sapling growth (height$_{end}$ − height$_{start}$). Initial fixed explanatory variables in the seedling growth model included water treatment group, inoculation group, lesion length (sapwood/phloem; vertical/horizontal), and all their interactions. Table number was set as a random factor in the model. In addition, non-treated saplings were included in this model. A generalized linear model was also constructed to evaluate the fixed effects of inoculation method (mock-inoculated-control, *H. annosum*, *H. parviporum*) under different water treatment (high, low) on necrosis length in phloem and sapwood (vertical/horizontal). Initial fixed explanatory variables in the necrosis length model included inoculation method in the water treatment group, sapling growth and sapling start height. Non-treated saplings were not included in this model. Table number was set as a random factor in the model. The length of necrosis caused by *H. parviporum* and *H. annosum* (horizontal/vertical) in phloem and sapwood per water treatment was further assessed by ONE-WAY-ANOVA and pairwise differences were found using Dunnett's C tests. Differences were considered statistically significant if the p-value was equal or below the threshold of 0.01. Pearson's correlation analysis was used to determine correlations of seedling growth (heigth$_{start}$ and growth) with the sizes of necrotic lesions (phloem/sapwood and vertical/horizontal) caused by *Heterobasidion* species. Similarly, necrosis length (vertical/horizontal) in sapwood and phloem were analyzed with Pearson's correlation. Correlations were considered as statistically significant if the *p*-value was below the threshold of 0.05.

3. Results

3.1. Necrosis Length

The extention growth of lesions and necroses occuring in pathogen inoculated stems was significantly higher than in mock-inoculated controls in sapwood (vertical and horizontal) (Figure 4a,c) and phloem (vertical and horizontal) (Figure 4b,d), except for the horizontal extension of *H. annosum* in phloem, which did not differ from control (Figure 4d). *H. parviporum* outcompeted *H. annosum* in vertical necrosis length in sapwood (Figure 4a) and phloem (Figure 4b), and in horizontal necrosis in phloem (Figure 4d). The damage caused by *H. parviporum* was statistically greater in the low water group (−group) in the vertical direction in sapwood (Figure 4a) and phloem (Figure 4b). Necroses, due to *H. annosum* were statistically higher only for the low water group in the horizontal direction in phloem (Figure 4d). Even though both pathogens were affected by the low water availibility to the seedlings, *H. parviporum* was able to cause greater vertical necrosis. The vertical necrosis in sapwood (Figure 4a) is also presented as necrosis length (mm) as a porportion of seedling growth (height$_{end}$ − height$_{start}$, mm) during the experiment (Figure 5). In the untreated saplings there were no lesions and necroses observed.

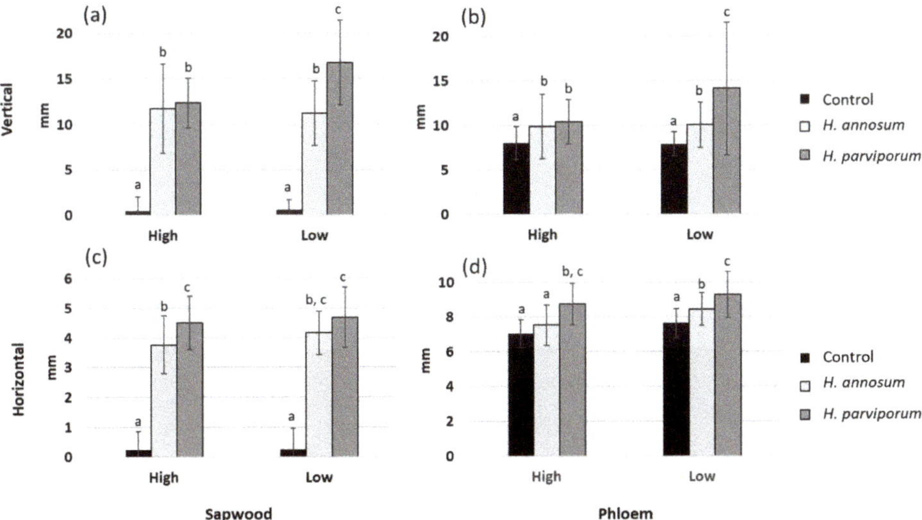

Figure 4. Length of necrosis (mm) observed after inoculation experiment with mock-inoculated control, *H. annosum* and *H. parviporum* in Norway spruce saplings in different water availability groups. Averages of vertical (**a,b**) and horizontal (**c,d**) length (mm) of necrosis in sapwood (**a,c**) and phloem (**b,c**) measured to the nearest 0.001 mm.

The necrosis length model for pathogens (GLM) showed that seedling starting height ($p < 0.01$) affected horizontal necrosis length in phloem and sapwood (Table 2). Based on Pearson correlation coefficients (r), there were significant positive correlations between the seedling height and horizontal pathogen necrosis in phloem ($p < 0.01$, $r = 0.3$) and sapwood ($p < 0.05$, $r = 0.1$) (Figure 6). Higher seedling height indicated higher horizontal necrosis length (Figure 6). However, the strength of these associations was low ($0.1 < r < 0.3$). Statistically significant ($p < 0.01$) and strong positive correlations were found between sapwood vertical and horizontal necrosis length ($r = 0.8$), as well as between vertical phloem and sapwood necrosis length ($r = 0.6$) (Figure 7a). When vertical necrosis in sapwood increased, the necrosis in horizontal direction increased (Figure 7a). Similarly, when necrosis in phloem increased, it also increased in sapwood (Figure 7a). Significant positive correlations ($p < 0.01$) were found also between horizontal phloem and sapwood necrosis length ($r = 0.5$) (Figure 7b) and between horizontal and vertical necrosis in phloem ($r = 0.4$) (Figure 7b). The strength of these associations was moderate ($0.4 < r < 0.5$).

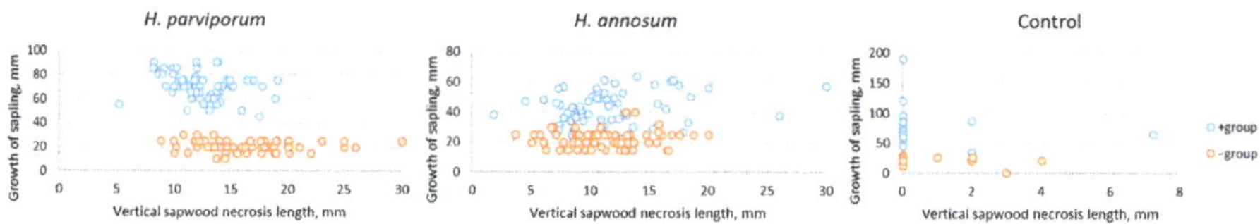

Figure 5. Vertical necrosis (mm) observed in sapwood after inoculation with *H. parviporum*, *H. annosum* and mock-inoculated control in high (+group) and low (−group) water groups as proportions of the sapling growth (mm). The Norway spruce saplings in low water group (−group) in each treatment grew less during the experiment than in high water group (+group). *H. parviporum* caused higher necrosis in the low water group (−group).

Table 2. Estimates from the generalized linear models (GML), statistically significant effect ($p < 0.01$) are bolded.

Variable	Fixed Effects	Std. Error	F	Sig.
	Water treatment	0.281	341.586	**<0.001**
	Inoculation treatment	0.529	0.751	0.603
Seedling growth	Phloem, vertical	0.027	1.032	0.311
	Phloem, horizontal	0.084	5.004	0.026
	Sapwood, vertical	0.032	3.866	0.05
	Sapwood, horizontal	0.114	2.156	0.143
	Inoculation × water treatment	1.117	106.57	**<0.001**
Sapwood, vertical	Height$_{start}$	0.028	1.802	0.181
	Seedling growth	0.151	2.236	0.136
	Inoculation × water treatment	0.36	186.452	**<0.001**
Sapwood, horizontal	Height$_{start}$	0.007	13.592	**<0.001**
	Seedling growth	0.035	0.009	0.926
	Inoculation × water treatment	1.56	11.542	**<0.001**
Phloem, vertical	Height$_{start}$	0.032	0.569	0.451
	Seedling growth	0.172	0.03	0.863
	Inoculation × water treatment	0.426	15.252	**<0.001**
Phloem, horizontal	Height$_{start}$	0.008	11.164	**<0.001**
	Seedling growth	0.044	2.012	0.157

3.2. Saplings Growth

Low water availability negatively affected the growth of Norway spruce saplings during the experiment. Sapling growth (height$_{end}$ − heigth$_{start}$, +group = 7.2 ± 2.0 cm; −group = 2.2 ± 0.6 cm) was statistically ($p < 0.01$) affected by water availability, (Figure 5, Table 2), but was not affected by the fungal infections or necrosis length (horizontal/vertical) of the phloem or sapwood, or by seedling starting height (Table 2). In total, 11 seedlings died during the experiment; one seedling in *H. parviporum* (+group), one in *H. annosum* (−group), six in *H. annosum* (+group) groups and three in non-treated (−group). The death of seedlings was assumed to be random, and they were removed from data analysis.

3.3. Plant and Fungal Material

H. annosum s.l. could not be detected in the Norway spruce saplings sampled and investigated by the Northwest German Forest Research Institute before the experimet started. The endophyte-community was isolated from the roots, which included species of *Diaporthe* sp., *Epicoccum nigrum* Link, *Sydowia polyspora* (Bref. and Tavel) E. Müll. and *Alternaria* sp. (data not shown). ITS region sequencing indicated no contamination in the cultures of *Heterobasidion* species. The species-specific primers [17] also showed species speficity and worked only for the target species. The PCR run from inoculated sapwood samples (altogether 60 samples) with species-specific primers detected the *H. parviporum* and *H. annosum* from 70% (14/20: Ten from +group; four from −group) and 70% (14/20: Ten from +group; 4 from −group), respectively. In mock-inoculated controls (20 samples) no *Heterobasion* was detected (data not shown). From inoculated sapwood (30 stems of saplings) incubated in darkness, *H. parviporum* could be detected from nine samples, (four from +group; five from −group) and *H. annosum* from seven samples (three from +group; four from −group). From mock-inoculated controls (10) no *Heterobasidion* was detected.

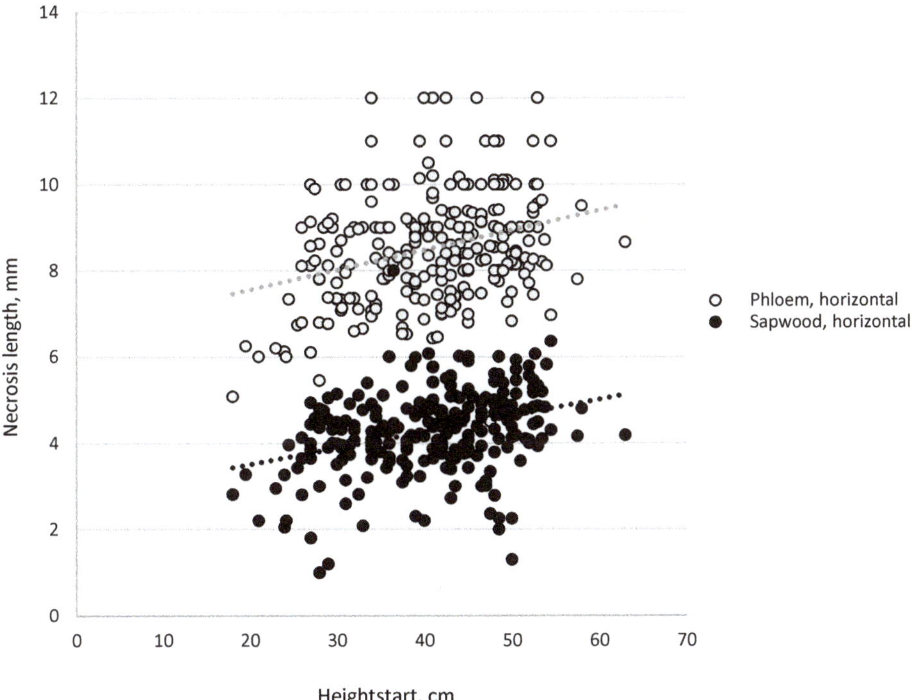

Figure 6. Combined horizontal necrosis length of *H. parviporum* and *H. annosum* correlated positively with seedling height at the start of the experiment. Higher seedling height indicates higher horizontal necrosis length.

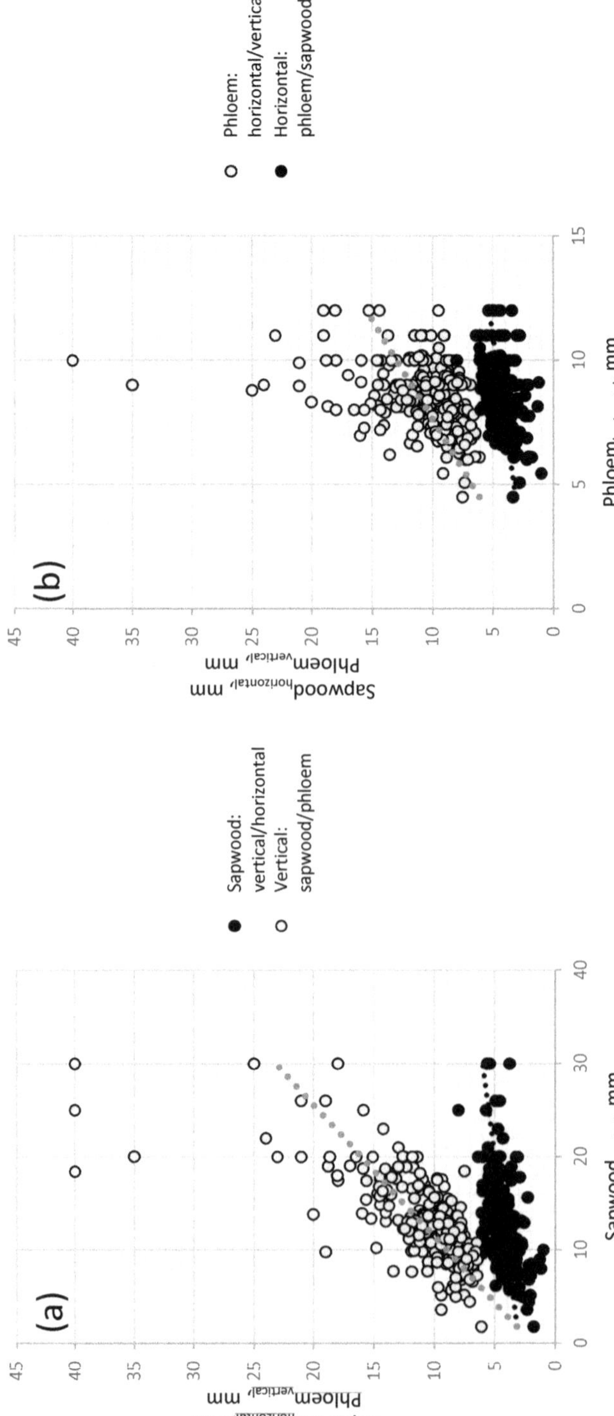

Figure 7. (**a**) Statistically significant ($p < 0.01$) positive correlations between vertical and horizontal necrosis length in sapwood (black), and vertical necrosis length between sapwood and phloem (grey); (**b**) Statistically significant ($p < 0.01$) positive correlations between vertical and horizontal necrosis length in phloem (grey), and horizontal necrosis length between sapwood and phloem (black). Positive correlations indicate that when one variable increases the other variable simultaneously increases.

4. Discussion

Pathogen growth, here expressed by necrosis length (vertical/horizontal) in phloem and sapwood, was shown to increase under abiotic disturbance (drought) in *P. abies* saplings. This result is consistent with Linnakoski et al. [2], who found that blue stain fungi *E. polonica* caused greater necrosis and mortality in *P. abies* seedlings with low water availability compared to those with high water availability. Madmony et al. [33] inoculated *H. parviporum* into two-year-old branches of two different Norway spruce clonal ramets (4-year old) and subjected them to well-watered and drought environments. They observed increased pathogen growth in well-watered seedlings [33], which is in contrast to our results. However, the duration of water availability treatment (22 days) in Madmony et al. [33] was much shorter than in this study (105 days). These results highlight the need for well-established inoculation studies for this pathosystem, when studying effects of abiotic disturbances. In our study, the necrosis length did not affect the growth of the saplings, but horizontal necrosis had a significant correlation with the sapling height (height$_\text{start}$) measured at the start of the experiment (Figure 6). A similar observation was made in an inoculation study of three-year-old Norway spruce ramets with *H. parviporum* [34]. However, this is in contrast to other studies, where Norway spruce saplings' height were negatively correlated with the necrosis length caused by *H. parviporum* [35]. Based on these results, saplings with lower values of growth parameters (in this case height) are not automatically considerably more sensitive to fungal inoculation, rather the necrosis development most likely involves multiple, interacting factors, e.g., different experimental settings and origins of pathogens and trees. However, the possibility that the duration of pathogen exposure to the host and the damage caused by increasing necrosis has an influence on the vitality of the host cannot be ruled out. The latter could have affected height growth increment (Table 2).

In nature, disturbances caused by water limitations leads to reduced establishment of aboveground tissues [2,36]. A similar phenomenon was observed in this study, as Norway spruce growth was significantly lower under the low water availability treatment than high. Necrotrophic fungal infections activate defensive responses in trees that aim to stop infection [37]. As previously mentioned, the necrosis length did not affect the growth of the seedlings during the experiment, indicating that the defense activated in saplings did not lead to a trade-off between resources allocated to growth. Neilson et al. [38] opined that calculations of the costs of the trade-off between growth and chemical defense related secondary metabolites can be overestimated and should be investigated in parallel with the identification of new supplemental functions. In this study, the sapling growth was not influenced by lesion length indicating that the infection with fungal pathogen did not require more resources compared to a mock-inoculated wound [2]. Pairwise comparison between non-treated saplings and inoculated saplings indicated that water treatment was the only parameter affecting growth (data not shown). These results suggest that low water availability (drought) affected the growth of Norway spruce saplings, making them more susceptible to root rot pathogens (Figures 4 and 5).

The adaptation of a tree species to changes in local climate may be too slow to successfully respond to the present rapid rate of climate change [1,39]. Further uncertainty in the climate change predictions is the possibility of change in tree resistance to fungal pathogens at abiotic disturbances, e.g., drought and higher temperatures [1]. *Heterobasidion* species growth is optimum at temperatures above 20 °C and below 30 °C [40–43]. Wood moisture content is not critical for *H. annosum* stump infection, whereas the further colonization of sapwood is decreased significantly with increasing stump moisture [44]. Müller et al. [43] stated that after having invaded living woody tissue, *Heterobasidion* species could expect to have sufficient moisture to continue growing as long as the tree is alive [43]. In this sense, the growth of *Heterobasidion* may not be affected by decreased water availability [43,44], in contrast to the host [36,45]. The probability of longer summer drought periods [46] may decrease the defense ability of trees towards increasing activity of root rot fungi [27,36,45]. Our results support this hypothesis, as the growth of Norway spruce saplings was decreased in the lower water group, leading to higher necrosis length (Figures 4 and 5).

Linares et al. [45] found that infected *Abies pinsapo* Boiss. trees have decreased ability to withstand drought stress, and root rot infections (*H. abietinum*) act as predisposing factors of forest decline and mortality. Gori et al. [27] found that *H. parviporum* infection made *P. abies* more susceptible to drought stress at a low elevation site. The artificial fungal inoculation of seedlings and saplings of trees has been widely practiced under controlled greenhouse conditions, in order to investigate hosts susceptibility to *Heterobasidion* infection [34,35,47,48]. For the current experiment, three-year-old saplings were used instead of mature trees. This enabled us to employ larger sample numbers with manipulation treatments, all within a relatively short time. While previous inoculation studies indicate that young saplings and seedlings are an effective model for larger tree health in *Heterobasidion*—Norway spruce pathosystems [28], we highlight the importance of research on mature trees, which will be the most accurate reflection of *Heterobasidion* effects on spruce trees in nature. Fine roots response negatively by decreasing their biomass during drought [49]. This effects the vitality of Norway spruce and causes more environmental stress. As *Heterobasidion* species are not dependent on the fine roots we suggest that these pathogens can benefit from the stress of their host trees caused by drought. Consequently, combined effects to the host root system by drought and *Heterobasidion* in more detail remains to be investigated.

5. Conclusions

This study found significant increase in the necrosis length after inoculation of *H. parviporum* (vertical necrosis in phloem and sapwood) and *H. annosum* (horizontal necrosis in phloem) in *P. abies* saplings that were stressed, due to lower water availability. This highlights the fact that root rot pathogens in the genus *Heterobasidion* can benefit from disturbances caused by projected climate change. The number of long drought periods in summer is expected to increase in Europe in the next decades [1,39,46,50], highlighting the need for the development of new adaptation strategies in forestry [50,51]. Here we provide experimental evidence that reduced water availability can enhance necrosis length in Norway spruce saplings after inoculation with *H. parviporum* or *H. annosum*. This study contributes to experimental research on interactions between biotic and abiotic disturbances in forest trees. Further empirical and theoretical research on mature trees under these disturbance conditions (drought and root rot) are required to better understand the genetic measures of host resistance and pathogen virulence, which can ultimately lead to different control strategies via resistance breeding.

In the course of global climate change, the next generation of Norway spruce forests in Germany are exposed to several risks besides *H. annosum* s.l. Norway spruce is the most vulnerable to wind damage, which is predicted to increase due to climate change [1], of all tree species in Germany [52]. This highlights the need for more applied studies of how different abiotic (wind and drought) disturbances can benefit the biotic (*H. annosum* s.l.) disturbance in Norway spruce dominated forests. In future, in German lowlands with limited water supply, *P. abies* will be a high-risk tree species with respect to forest protection and abiotic or biotic risk factors [53–55].

Author Contributions: E.T. designed the experimental protocol, conceived the experiment, measured the necrosis, performed PCRs, analyzed the results and wrote the first draft. K.B. conceived the experiment and advised on the experimental protocol, analysis and contributed to the writing of the manuscript. D.R.R. designed the watering protocol in the greenhouse, maintained the greenhouse trial and performed the DNA extractions. G.J.L. and J.B. performed the seedling health experiments, advised on the experimental protocol, analysis and contributed to the writing of the manuscript.

Funding: This research was funded by Faculty of Forest Sciences and Forest Ecology, Georg-August-Universität Göttingen, Germany.

Acknowledgments: Muhammad Rafaqat, Linda Rigerte and Wilhelmine Bach are highly acknowledged by their contribution in the greenhouse experimental set up. We are obliged to Robert Larkin for kindly improving the manuscript as a native speaker.

Conflicts of Interest: The authors declare no conflict of interest.

References

1. Seidl, R.; Thom, D.; Kautz, M.; Martin-Benito, D.; Peltoniemi, M.; Vacchiano, G.; Wild, J.; Ascoli, D.; Petr, M.; Honkaniemi, J.; et al. Forest disturbances under climate change. *Nat. Clim. Chang.* **2017**, *7*, 395–402. [CrossRef] [PubMed]
2. Linnakoski, R.; Sugano, J.; Junttila, S.; Pulkkinen, P.; Asiegbu, F.O.; Forbes, K.M. Effects of water availability on a forestry pathosysstem: Fungal strain-specific variation in disease severity. *Sci. Rep.* **2017**, *7*, 13501. [CrossRef] [PubMed]
3. Thünen-Institut, Dritte Bundeswaldinventur—Ergebnisdatenbank. Available online: https://bwi.info (accessed on 3 December 2018).
4. Asiegbu, F.O.; Adomas, A.; Stenlid, J. Conifer root and butt rot caused by *Heterobasidion annosum* (Fr.) Bref. s.l. *Mol. Plant Pathol.* **2005**, *6*, 395–409. [CrossRef] [PubMed]
5. Garbelotto, M.; Gonthier, P. Biology, epidemiology, and control of *Heterobasidion* species worldwide. *Annu. Rev. Phytopathol.* **2013**, *51*, 39–59. [CrossRef] [PubMed]
6. Deutscher Wetterdienst. Available online: https://www.dwd.de/DE/leistungen/klimakartendeutschland/klimakartendeutschland.html?nn=480164 (accessed on 9 January 2019).
7. Langer, G.J.; Northwest German Forest Research Institute, Göttingen, Germany. Personal communication, 2018.
8. Rishbeth, J. Observations on the Biology of *Fomes annosus*, with particular reference to East Anglian pine plantations: II. Spore production, stump infection, and saprophytic activity in stumps. *Ann. Bot.* **1951**, *15*, 1–22. [CrossRef]
9. Langer, G.J.; Bressem, U.; Haberman, M. Vermehrt Pilzkrankheiten an Bergahorn in Nordwestdeutschland. *AFZ/Der Wald* **2013**, *6*, 22–26.
10. Rishbeth, J. Dispersal of *Fomes annosus* Fr. and *Peniophora gigantea* (Fr.) Massee. *Trans. Br. Mycol. Soc.* **1959**, *42*, 243–260. [CrossRef]
11. Isomäki, A.; Kallio, T. Consequences of injury caused by timber harvesting machines on the growth and decay of spruce (*Picea abies* (L.) Karst.). *Acta For. Fennica* **1974**, *136*, 1–25. [CrossRef]
12. Redfern, D.B.; Stenlid, J. Spore dispersal and infection. In *Heterobasidion annosum: Biology, Ecology, Impact and Control*; Woodward, S., Stenlid, J., Karjalainen, R., Hüttermann, A., Eds.; CAB International: Wallingford, UK; New York, NY, USA, 1998; pp. 105–124.
13. Korhonen, K. Intersterility groups of *Heterobasidion annosum*. *Commun. Inst. For. Fenn.* **1978**, *94*, 1–25.
14. Capretti, P.; Korhonen, K.; Mugnai, L.; Romagnioli, C. An intersterility group of *Heterobasidion annosum*, specialized to *Abies alba*. *Eur. J. For. Pathol.* **1990**, *20*, 257–262. [CrossRef]
15. Metzler, B.; Langer, G.J.; Heydeck, P.; Peters, F.; Scham, J.; Langer, E. Survey on *Heterobasidion* species and perspectives of butt rot control in Germany. In *XIII Conference Root and Butt Rot of Forest Trees IUFRO*; Firenze University Press: Firenze, Italy, 2011; pp. 206–208.
16. Langer, G.J.; Bressem, U. *Phlebiopsis gigantea* als Antagonist des Wurzelschwamms. *AFZ/Der Wald* **2017**, *72*, 39–43.
17. Hantula, J.; Vainio, E. Specific primers for the differentiation of *Heterobasidion annosum* (s.str.) and *H. parviporum* infected stumps in northern Europe. *Silva Fenn.* **2003**, *37*, 181–187. [CrossRef]
18. Rishbeth, J. Stump protection against *Fomes annosus*. Treatment with substances other than creosote. *Ann. Appl. Biol.* **1959**, *47*, 529–541. [CrossRef]
19. Oliva, J.; Bendz-Hellgren, M.; Stenlid, J. Spread of *Heterobasidion annosum* s.s. and *Heterobasidion parviporum* in *Picea abies* 15 years after stump inoculation. *FEMS Microbiol. Ecol.* **2011**, *75*, 414–429. [CrossRef] [PubMed]
20. NW-FVA 2018 Gemeiner Wurzelschwamm (*Heterobasidion annosum* s.l.) Praxis Information Nr. 5—Oktober 2018. Available online: https://www.nw-fva.de/index.php?id=173 (accessed on 1 December 2018).
21. Piri, T. The spreading of the S type of *Heterobasidion annosum* from Norway spruce stumps to the subsequent tree stand. *Eur. J. For. Pathol.* **1996**, *26*, 193–204. [CrossRef]
22. Stenlid, J.; Redfern, D. Spread within the tree and stand. In *Heterobasidion annosum: Biology, Ecology, Impact and Control*; Woodward, S., Stenlid, J., Karjalainen, R., Hüttermann, A., Eds.; CAB International: Wallingford, UK; New York, NY, USA, 1998; pp. 125–142.
23. Piri, T.; Korhonen, K. Spatial distribution and persistence of *Heterobasidion parviporum* genets on a Norway spruce site. *For. Pathol.* **2007**, *37*, 1–8. [CrossRef]

24. Piri, T. Early development of root rot in young Norway spruce planted on sites infected by *Heterobasidion* in southern Finland. *Can. J. For. Res.* **2003**, *33*, 604–611. [CrossRef]

25. La Porta, N.; Capretti, P.; Thomsen, I.M.; Kasanen, R.; Hietala, A.M.; Von Weissenberg, K. Forest pathogens with higher damage potential due to climate change in Europe. *Can. J. Plant Pathol.* **2008**, *30*, 177–195. [CrossRef]

26. Trishkin, M.; Lopatin, E.; Gavrilova, O. The potential impact of climate change and forest management practices on *Heterobasidion* spp. infection distribution in northwestern Russia—A case study in the Republic of Karelia. *J. For. Sci.* **2016**, *62*, 529–536. [CrossRef]

27. Gori, Y.; Cherubini, P.; Camin, F.; La Porta, N. Fungal root pathogen (*Heterobasidion parviporum*) increases drought stress in Norway spruce stand at low elevation in the Alps. *Eur. J. For. Res.* **2013**, *132*, 607. [CrossRef]

28. Dimitri, L.; Schumann, G. Further experiments on the host-parasite relationship between Norway spruce and *Heterobasidion annosum*. In Proceedings of the 7th International Conference on Root and Butt Rots, Vernon and Victoria, BC, Canada, 9–16 August 1988; pp. 171–179.

29. Langer, G. Die Gattung Botryobasidium Donk (Corticiaceae, Basidiomycetes). *Bibl. Mycol.* **1994**, *158*, 459.

30. Gardes, M.; Bruns, T.D. ITS primers with enhanced specificity for higher fungi and basidiomycetes: Application to identification of mycorrhizae and rusts. *Mol. Ecol.* **1993**, *2*, 113–118. [CrossRef]

31. White, T.J.; Bruns, T.D.; Lee, S.B.; Taylor, J.W. Amplification and direct sequencing of fungal ribosomal RNA genes for phylogenetics. In *PCR Protocols—A Guide to Methods and Applications*; Innis, M.A., Gelfand, D.H., Sninsky, J.J., White, T.J., Eds.; Academic Press: San Diego, CA, USA, 1990; pp. 315–322.

32. Hall, T.A. BioEdit: A user-friendly biological sequence alignment editor and analysis program for Windows 95/98/NT. *Nucleic Acids Symp. Ser.* **1999**, *41*, 95–98.

33. Madmony, A.; Tognetti, R.; Zamponi, L.; Capretti, P.; Michelozzi, M. Monoterpene responses to interacting effects of drought stress and infection by the fungus *Heterobasidion parviporum* in two clones of Norway spruce (*Picea abies*). *Environ. Exp. Bot.* **2018**, *152*, 137–148. [CrossRef]

34. Karlsson, B.; Tsopelas, P.; Zamponi, L.; Capretti, P.; Soulioti, N.; Swedjemark, G. Susceptibility to *Heterobasidion parviporum* in *Picea abies* clones grown in different environments. *For. Pathol.* **2008**, *38*, 83–89. [CrossRef]

35. Mukrimin, M.; Kovalchuk, A.; Neves, L.G.; Jaber, E.H.A.; Haapanen, M.; Kirst, M.; Asiegbu, F.O. Genome-wide exon-capture approach identifies genetic variants of Norway spruce genes associated with susceptibility to *Heterobasidion parviporum* infection. *Front. Plant. Sci.* **2018**, *9*, 793. [CrossRef] [PubMed]

36. Lévesque, M.; Saurer, M.; Siegwolf, R.; Eilmann, B.; Brang, P.; Bugmann, H.; Rigling, A. Drought response of five conifer species under contrasting water availability suggests high vulnerability of Norway spruce and European larch. *Glob. Chang. Biol.* **2013**, *19*, 3184–3199. [CrossRef]

37. Kovalchuk, A.; Keriö, S.; Oghenekaro, A.O.; Jaber, E.; Raffaello, T.; Asiegbu, F.O. Antimicrobial defenses and resistance in forest trees: Challenges and perspectives in a genomic era. *Annu. Rev. Phytopathol.* **2013**, *51*, 221–244. [CrossRef]

38. Neilson, E.H.; Goodger, J.Q.; Woodrow, I.E.; Møller, B.L. Plant chemical defense: At what cost? *Trends Plant Sci.* **2013**, *5*, 250–258. [CrossRef]

39. Allen, C.D.; Macalady, A.K.; Chenchouni, H.; Bachelet, D.; McDowell, N.; Vennetier, M.; Kitzberger, T.; Rigling, A.; Breshears, D.D.; Hogg, E.H.; et al. A global overview of drought and heat induced tree mortality reveals emerging climate change risks for forests. *For. Ecol. Manag.* **2010**, *259*, 660–684. [CrossRef]

40. Gooding, G.V.; Hodges, C.S., Jr.; Ross, E.W., Jr. Effect of temperature on growth and survival of *Fomes annosus*. *For. Sci.* **1966**, *12*, 325–333. [CrossRef]

41. Brown, A.V.; Webber, J.F. Biocontrol of decay in seasoning utility poles. I. Growth rate and colonizing ability of bluestain and decay fungi in vivo and in vitro. *For. Pathol.* **2009**, *39*, 145–156. [CrossRef]

42. Scirè, M.; Motta, E.; D'Amico, L. Behaviour of *Heterobasidion annosum* and *Heterobasidion irregulare* isolates from central Italy in inoculated *Pinus pinea* seedlings. *Mycol. Prog.* **2011**, *10*, 85–91. [CrossRef]

43. Müller, M.M.; Sievänen, R.; Beuker, E.; Meesenburg, H.; Kuuskeri, J.; Hamberg, L.; Korhonen, K. Predicting the activity of *Heterobasidion parviporum* on Norway spruce in warming climate from its respiration rate at different temperatures. *For. Pathol.* **2014**, *44*, 325–336. [CrossRef]

44. Bendz-Hellgren, M.; Stenlid, J. Effects of clear-cutting, thinning, and wood moisture content on the susceptibility of Norway spruce stumps to *Heterobasidion annosum*. *Can. J. For. Res.* **1998**, *28*, 759–765. [CrossRef]

45. Linares, J.C.; Camarero, J.J.; Bowker, M.A.; Ochoa, V.; Carreira, J.A. Stand-structural effects on *Heterobasidion abietinum*-related mortality following drought events in *Abies pinsapo*. *Oecologia* **2010**, *164*, 1107–1119. [CrossRef] [PubMed]
46. Ruosteenoja, K.; Tuomenvirta, H.; Jylhä, K. GCM-based regional temperature and precipitation change estimates for Europe under four SRES scenarios applying a super-ensemble pattern-scaling method. *Clim. Chang.* **2007**, *81*, 193–208. [CrossRef]
47. Swedjemark, G.; Stenlid, J. Variation in spread of *Heterobasidion annosum* in clones of *Picea abies* grown at different vegetation phases under greenhouse conditions. *Scand. J. For. Res.* **1996**, *11*, 137–144. [CrossRef]
48. Swedjemark, G.; Stenlid, J.; Karlsson, B. Variation in growth of *Heterobasidion annosum* among clones of *Picea abies* incubated for different periods of time. *For. Pathol.* **2001**, *31*, 163–175. [CrossRef]
49. Gaul, D.; Hertel, D.; Borken, W.; Matzner, E.; Leuschner, C. Effects of experimental drought on the fine root system of mature Norway spruce. *For. Ecol. Manag.* **2008**, *256*, 1151–1159. [CrossRef]
50. Lindner, M.; Maroschek, M.; Netherer, S.; Kremer, A.; Barbati, A.; Garcia-Gonzalo, J.; Seidl, R.; Delzon, S.; Corona, P.; Kolström, M.; et al. Climate change impacts, adaptive capacity, and vulnerability of European forest ecosystems. *For. Ecol. Manag.* **2010**, *259*, 698–709. [CrossRef]
51. Subramanian, N.; Bergh, J.; Johansson, U.; Nilsson, U.; Sallnäs, O. Adaptation of forest management regimes in southern Sweden to increased risks associated with climate change. *Forests* **2016**, *7*, 8. [CrossRef]
52. Schmidt, M.; Hanewinkel, M.; Kändler, G.; Kublin, E.; Kohnle, U. An inventory-based approach for modelling single tree storm damage—Experiences with the winter storm 1999 in south-western Germany. *Can. J. For. Res.* **2010**, *40*, 1636–1652. [CrossRef]
53. Overbeck, M.; Schmidt, M. Modelling infestation risk of Norway spruce by *Ips typographus* (L.) in the Lower Saxon Harz Mountains (Germany). *For. Ecol. Manag.* **2012**, *266*, 115–125. [CrossRef]
54. Spellmann, H.; Sutmöller, J.; Meesenburg, H. Risikovorsorge im Zeichen des Klimawandels. Vorläufige Empfehlungen der NW-FVA am Beispiel des Fichtenanbaus. *AFZ/Der Wald* **2007**, *62*, 1246–1249.
55. Spellmann, H.; Albert, M.; Schmidt, M.; Sutmöller, J.; Overbeck, M. Waldbauliche Anpassungsstrategien für veränderte Klimaverhältnisse. *AFZ/Der Wald* **2011**, *11*, 19–23.

 *forests*

Article

Identification and Analysis of a CPYC-Type Glutaredoxin Associated with Stress Response in Rubber Trees

Kun Yuan [1], Xiuli Guo [2], Chengtian Feng [1], Yiyu Hu [1], Jinping Liu [2,*] and Zhenhui Wang [1]

[1] Key Laboratory of Biology and Genetic Resources of Rubber Tree, Ministry of Agriculture, Rubber Research Institute, Chinese Academy of Tropical Agricultural Sciences, Haikou 571101, China; yuankun628@126.com (K.Y.); fengchengtian@126.com (C.F.); huyy2009@gmail.com (Y.H.); wzh-36@163.com (Z.W.)

[2] Hainan Key Laboratory for Sustainable Utilization of Tropical Bioresources, Tropical Agriculture and Forestry Institute, Hainan University, Haikou 570228, China; aura.guo@vivachek.com

* Correspondence: 990785@hainu.edu.cn; Tel.: +86-0898-6696-1263

Received: 17 January 2019; Accepted: 7 February 2019; Published: 12 February 2019

Abstract: Glutaredoxins (GRXs) are a class of small oxidoreductases which modulate various biological processes in plants. Here, we isolated a GRX gene from the rubber tree (*Hevea brasiliensis* Müll. Arg.), named as *HbSRGRX1*, which encoded 107 amino acid residues with a CPYC active site. Phylogenetic analysis displayed that *HbSRGRX1* was more correlated with GRXs from *Manihot esculenta* Crantz. and *Ricinus communis* L. *HbSRGRX1* was localized in the nuclei of tobacco cells, and its transcripts were preferentially expressed in male flowers and in the high-yield variety Reyan 7-33-97 with strong resistance against cold. The expression levels of *HbSRGRX1* significantly decreased in tapping panel dryness (TPD) trees. Furthermore, *HbSRGRX1* was regulated by wounding, hydrogen peroxide (H_2O_2), and multiple hormones. Altogether, these results suggest important roles of *HbSRGRX1* in plant development and defense response to TPD and multiple stresses.

Keywords: glutaredoxin; subcellular localization; expression; tapping panel dryness; defense response; rubber tree

1. Introduction

The rubber tree (*Hevea brasiliensis* Müll. Arg.) from the Euphorbiaceae family, as the major source of natural rubber, is thought to be one of the important industrial trees. The production of natural rubber is facing a serious threat caused by tapping panel dryness (TPD), which is identified by a part or complete cessation of latex flow. It is found that over exploitation, namely over tapping or excessive stimulation by ethephon (ET), might lead to the onset of TPD [1]. Although many researches have tried to reveal the mechanism of TPD, it is still unclear. Numerous genes related to TPD have been identified [2–4]. The onset of TPD is thought to be closely associated with reactive oxygen species (ROS) signaling [4]. The excessive generation of ROS can elicit oxidative damage to lipids, DNA, and proteins, leading to the plant cell death [5]. Plants have evolved various ROS scavenging systems, of which, gutaredoxins (GRXs) are important members that can maintain and modulate cellular redox status with the reducing power of glutathione (GSH) [6]. Increasing evidence indicates that GRXs have antioxidant functions in plant responses to oxidative stress [7–10]. According to the sequences of active sites, plant GRXs contain three main groups, CPY(F)C-, CGFS- and CC-type [6,11]. They take part in the regulation of growth and development, stress responses, and iron-sulfur cluster assembly [9,12–15]. There are 31 GRX members in *Arabidopsis thaliana* (L.) Heynh., 48 in *Oryza sativa*

L., and 36 in *Populus trichocarpa* Torr. & A.Gray ex. Hook. Many researches into CGFS- and CC-type GRX genes have been carried out in plants [8,12,16,17]. However, there are few data about CPYC-type GRXs. Several CPYC-type GRX genes in *Populus trichocarpa* are reported to show differential expression patterns in various organs [11]. Recently, it was found that various treatments could induce the expression of rice *OsGRX20*, a CPYC-type GRX. Further analysis indicates that overexpression of *OsGRX20* in the rice sensitive genotype strikingly increases resistance to bacterial blight [10]. Although many plant GRX genes have been characterized, the biological functions of GRXs in rubber treesremain unknown.

The transcript of the *GRXC9* gene belonging to CC-type class shows decreased expression in TPD trees as compared to healthy ones [4,18], but their biological functions have not been researched in rubber trees. Previously, our proteomic analysis indicated that a GRX protein (ABZ88803.1/ EU295478.1) decreased in the latex of TPD plants compared with healthy ones [19]. Therefore, we postulate that this GRX gene might have critical roles during TPD onset, and it is essential to be further study its function. Here, we sequenced a stress-responsive GRX gene in rubber trees, named as *HbSRGRX1*, and the phylogenetic tree was constructed in the present research. Meanwhile, the expression profiles of *HbSRGRX1* were systematically analyzed in various tissues and varieties, different degrees of TPD trees, wounding, hydrogen peroxide (H_2O_2), and various hormone treatments. The results demonstrated that *HbSRGRX1* participated in response to TPD and other various stimuli, suggesting its crucial function in rubber trees.

2. Materials and Methods

2.1. Plant Materials

The rubber tree clone Reyan 7-33-97 from an experimental field of the Chinese Academy of Tropical Agricultural Sciences in China was used. Seven different tissues from 10-year-old trees were collected to study the tissue-specific expression of *HbSRGRX1*. This type of tree was similarly selected to analyze the influence of H_2O_2 and various hormones on *HbSRGRX1* expression. To examine the variety-specific expression of *HbSRGRX1*, 5 different varieties, including Reyan 7-33-97, 7-20-59, 8-79, Reken 523, and PR107 were selected from 10-year-old trees. To study the effect of wounding on *HbSRGRX1* expression, virgin trees from 8-years-old were used. To detect the influence of TPD on *HbSRGRX1*, 24-year-old trees with different TPD degrees (Grade 1, degree <25% tapping panel dry; Grade 2, degree 25% < tapping panel dry < 50%; and Grade 3, degree >50% tapping panel dry) were used. The healthy ones (Grade 0) were used as control (the images of rubber trees from different TPD degrees shown in Figure S1).

2.2. Wounding, H_2O_2 and Hormones Treatments

For the wounding treatment, eight stainless drawing pins were used to stick into the bark of each tree and left in place as described [20]. Four batches of 10 virgin trees were selected, 3 of which were wounded at 6, 12, 24, and 48 h before the first tapping, and the fourth batch was unwounded as the control. The H_2O_2 and hormone treatments were arranged according to the methods of Deng et al. [21] and Long et al. [22], respectively. Three batches (five trees each) were treated with 2% H_2O_2, 1.5% ET, 200 μmol/L abscisic acid (ABA), 0.005% methyl jasmonate (MeJA), 200 μmol/L salicylic acid (SA), 66 μmol/L 2,4-dichlorophenoxyacetic acid (2,4-D), 100 μmol/L gibberellic acid (GA₃), 200 μmol/L 6-Benzylaminopurine (6-BA) and 100 μmol/L indole-3-acetic acid (IAA). Another one was used as the control. The H_2O_2 and hormones were treated at 6, 12, 24, and 48 h.

2.3. RNA Isolation and cDNA Synthesis

All latex total RNA was isolated as described [23]. RNA of other tissues was extracted using the RNAprep pure Plant Kit (TIANGEN, Beijing, China). The integrity and concentration of RNA was examined by agarose gel electrophoresis, and a spectrophotometer (Thermo, New York, NY, USA).

The synthesis of First-strand cDNA was conducted with the RevertAid™ First Strand cDNA Synthesis Kit (Fermentas, Waltham, MA, USA)).

2.4. ORF Cloning of HbSRGRX1

According to the sequence of EU295478.1, several pairs of primers were designed (Table 1). The open reading frame (ORF) of *HbSRGRX1* was flanked by the pair of primers HbSRGRX1-F (5′-ATGGCGATGACCAAGGCCAAG-3′) and HbSRGRX1-R (5′-TTTAAGCAGAAGCCTTAGCAAGAGCTCC-3′). The product was cloned into the pMD18-T vector, and then sequenced.

Table 1. Primer sequences.

Primer Name	Primer Sequence (5′→3′)	Use
HbSRGRX1-F	ATGGCGATGACCAAGGCCAAG	ORF cloning
HbSRGRX1-R	TTTAAGCAGAAGCCTTAGCAAGAGCTCC	ORF cloning
1302-HbSRGRX1-F	CTCCCATGGATGGCGATGACCAAGGCCAAG	Subcellular localization analysis
1302-HbSRGRX1-R	CGCACTAGTTTAAGCAGAAGCCTTAGCAAGAGCTCC	Subcellular localization analysis
HbSRGRX1-QF	CGTTTCTTCCAATTCTGTTGTCGTT	Real-time PCR analysis
HbSRGRX1-QR	CAATGTGCTTGCCACTGATG	Real-time PCR analysis
Hb18SrRNA-QF	GCTCGAAGACGATCAGATACC	Real-time PCR analysis
Hb18SrRNA-QR	TTCAGCCTTGCGACCATAC	Real-time PCR analysis

Underline represents the restriction enzyme sites.

2.5. Sequence Analyses

The molecular weight (Mw) and isoelectric point (pI) of *HbSRGRX1* were predicted by the ExPASy compute pI/Mw tool. The protein conserved domain was identified with the NCBI (National Center for Biotechnology Information) CDD (Conserved Domain Database) and SMART (Simple Modular Architecture Research Tool) [24]. The protein sequence was aligned by DNAMAN 6. The phylogenetic tree was constructed by MEGA (Molecular Evolutionary Genetics Analysis) 6.06 with neighbor-joining method [25]. A bootstrap test was performed using 1000 replicates.

2.6. Subcellular Localization

HbSRGRX1 was localized as described [21]. A pair of primers 1302-HbSRGRX1-F (5′-CTCCCATGGATGGCGATGACCAAGGCCAAG-3′) and 1302-HbSRGRX1-R (5′-CGCACTAGTTTAAGCAGAAGCCTTAGCAAGAGCTCC-3′) was used to amplify the coding sequence of *HbSRGRX1* (Table 1). The product was cloned into the pMD18-T vector, the correct plasmids and pCAMBIA1302-GFP vector were digested with Nco I and Spe I, respectively. The products were ligated with T_4-DNA ligase to obtain the recombinant vector pCAMBIA1302-HbSRGRX1-GFP, which was then transformed into *A. tumefaciens* strain EHA105. The *A. tumefaciens* harboring the recombinant vector was infiltrated into the *N. benthamiana* leaves. Fluorescence signals were examined using a confocal microscope (Fluo View™ FV1000).

2.7. Real-time Quantitative PCR (qPCR)

The expression patterns of *HbSRGRX1* were examined by qPCR using the primers HbSRGRX1-QF and HbSRGRX1-QR (Table 1). The 18S rRNA gene (primers: Hb18SrRNA-QF and Hb18SrRNA-QR; Table 1) was used as the internal control. The qPCR was carried out on a LightCycler 2.0 system (Roche Diagnostics, Switzerland) using SYBR Premix Ex Taq™ II (Takara, Dalian, China). The PCR procedures and the calculation of the relative abundance of transcripts were performed as described [20]. All qPCR experiments were reproduced in triplicate, and the values were presented as mean ± SD (Standard Deviation). Figures were drawn by OriginPro 9.0 software (OriginLab Corporation, Northampton, MA, USA).

3. Results

3.1. ORF Cloning, Sequence Alignment and Phylogenetic Analysis of HbSRGRX1

Primers HbSRGRX1-F and HbSRGRX1-R were designed to clone the ORF of *HbSRGRX1* using latex cDNA of the clone Reyan 7-33-97 as the template. *HbSRGRX1* contained a 324-bp ORF encoding 107 amino acid residues. The putative molecular mass was 11.3 kDa, and the pI was 6.71. HbSRGRX1 contained a conserved motif of the thioredoxin_like superfamily, a CPYC active site at 23–26 amino acids at the N-terminus and GSH binding sites, belonging to CPYC-type class (Figure 1A).

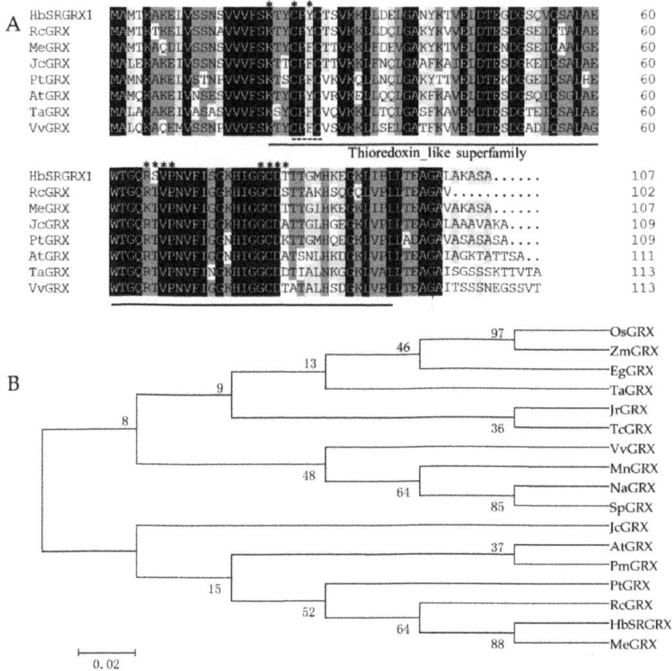

Figure 1. Sequence alignment and phylogenetic analyses of *HbSRGRX1* and other plant GRXs. (**A**) Dark and gray indicate identical and similar amino acids, respectively. The straight line, asterisk, and dotted line represent the conserved motif of the thioredoxin_like superfamily, GSH binding site, and active site, respectively. (**B**) The phylogenetic tree was generated by MEGA (Molecular Evolutionary Genetics Analysis) 6.06 using the neighbor-joining method with 1000 bootstrap tests. The scale bar represents the estimated number of amino acid substitutions per site. The accession numbers of GRX proteins are as follows: *Manihot esculenta* (MeGRX, XP 021594601.1), *Ricinus communis* (RcGRX, XP 002524673.1), *Populus trichocarpa* (PtGRX, XP 002298529.2), *Prunus mume* (PmGRX, XP 008230572.1), *Arabidopsis thaliana* (AtGRX, NP 198853.1), *Jatropha curcas* (JcGRX, NP 001295635.1), *Theobroma cacao* (TcGRX, XP 007031532.1), *Juglans regia* (JrGRX, XP 018842397.1), *Solanum pennellii* (SpGRX, XP 015077482.1), *Nicotiana attenuate* (NaGRX, XP 019224739.1), *Morus notabilis* (MnGRX, XP 024029530.1), *Vitis vinifera* (VvGRX, XP 002276266.1), *Triticum aestivum* (TaGRX, AAP80853.1), *Elaeis guineensis* (EgGRX, XP 010940165.1), *Zea mays* (ZmGRX, NP 001158948.1), *Oryza sativa* (OsGRX, XP 015626005.1).

Multiple sequence alignment of *HbSRGRX1* with its related CPY (F) C-type ones from several other plants revealed high identities with the GRXs from *Manihot esculenta* Crantz. *Ricinus communis* L. and *Jatropha curcas* L. (84.11%, 77.57%, and 75.23%, respectively). HbSRGRX1 had 73.39%, 71.17%, 68.42%, and 64.60% identity with GRXs from *Populus trichocarpa*, *Arabidopsis thaliana*, *Vitis vinifera* L., and *Triticum aestivum* L., respectively (Figure 1A). To establish the phylogenic relationships among plant GRXs, a phylogenetic tree was constructed between HbSRGRX1 and another 16 plant GRXs in CPY (F) C-type class, using the neighbor-joining method. As shown in Figure 1B, HbSRGRX1 was more closely related to MeGRX (*Manihot esculenta*, XP 021594601.1) and RcGRX (*Ricinus communis*, XP 002524673.1), which belonged to the same family as *Hevea brasiliensis* of Euphorbiaceae. This result demonstrated that the GRXs were highly conserved during evolution.

3.2. Subcellular Localization of HbSRGRX1

To further investigate the subcellular localization of *HbSRGRX1* in cells, the *N. benthamiana* leaves were infiltrated with the *A. tumefaciens* harboring pCAMBIA1302-HbSRGRX1-GFP vector and observed with a confocal microscope. The results demonstrated that the fluorescent signal was examined in the nuclei (Figure 2), indicating that HbSRGRX1 was a nucleus located protein.

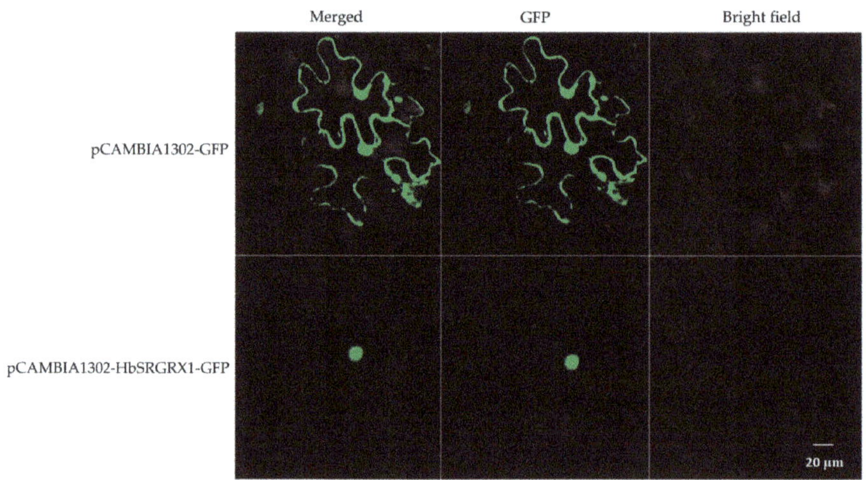

Figure 2. Subcellular localization of HbSRGRX1. GFP alone (top row) was localized throughout the whole cell and pCAMBIA1302-HbSRGRX1-GFP (bottom row) in the nuclei of tobacco epidermal cells. Bar = 20 µm.

3.3. Expression of HbSRGRX1 in Different Tissues and Varieties

The expression of *HbSRGRX1* was examined in seven tissues, including latex, leaf, bark, male flower, female flower, xylem, and petiole. As shown in Figure 3A, *HbSRGRX1* displayed a tissue-specific expression profile. The expression of *HbSRGRX1* was highest in male flowers, which was strikingly higher than in other tissues ($p < 0.05$), 38.9- and 617.5-fold over its expression in latex and leaf, respectively. In addition, *HbSRGRX1* was differentially expressed in five different varieties of rubber trees, including Reyan 7-33-97, 7-20-59, 8-79, Reken 523, and PR107. The expression level of *HbSRGRX1* was significantly higher in Reyan 7-33-97 than other varieties, the lowest in Reyan 7-20-59 and 8-79 (Figure 3B). Together, the above results indicated that *HbSRGRX1* might have particular functions in different tissues and varieties.

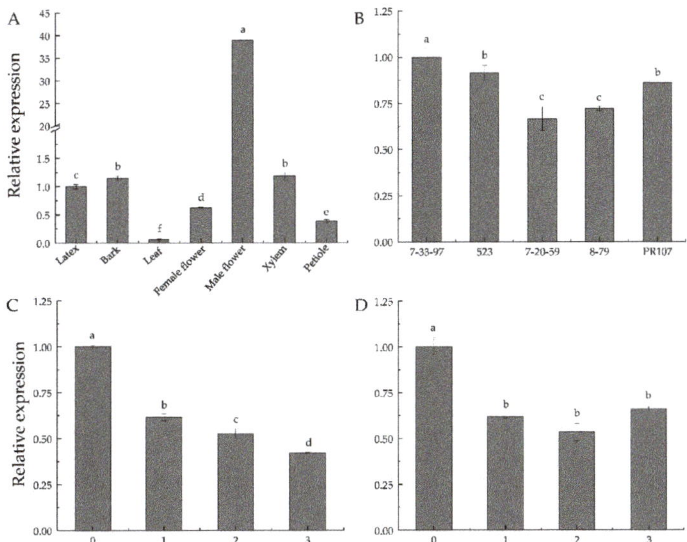

Figure 3. Expression profiles of *HbSRGRX1* gene in different tissues and varieties, and in bark and latex from different tapping panel dryness (TPD) levels. Expression profiles of *HbSRGRX1* in different tissues (**A**), barks from different varieties (**B**), bark (**C**), and latex (**D**) from different TPD levels. 0, 1, 2, and 3 indicated healthy trees (Grade 0), Grade 1 (degree <25% tapping panel dry), Grade 2 (degree 25% < tapping panel dry < 50%) and Grade 3 (degree >50% tapping panel dry), respectively. Relative expression was normalized using 18S rRNA gene. Values represent mean ± SD of three independent replicates. Different letters indicate significant difference ($p < 0.05$).

3.4. Expression of HbSRGRX1 in Different Degrees of TPD Trees

The TPD of rubber tree is a complicated physiological disorder caused by over tapping or excessive stimulation by ET, seriously affecting natural rubber production [26]. To further analyze the expression of *HbSRGRX1* at the transcriptional level, the bark and latex from different degrees of TPD trees (Grade 0, healthy trees; Grade 1, degree <25% tapping panel dry; Grade 2, degree 25% < tapping panel dry < 50%; and Grade 3, degree >50% tapping panel dry) were collected, respectively. In general, the expression of *HbSRGRX1* was markedly higher in healthy plants than TPD ones (Figure 3C,D). Remarkably, the expression of *HbSRGRX1* in bark showed a trend of constantly significant decline with the degree of TPD rising (Figure 3C), while did not change in latex of different degrees of TPD (1–3) except the healthy control (0) (Figure 3D). These results indicated that the expression levels of *HbSRGRX1* were closely correlated with TPD severities.

3.5. Expression of HbSRGRX1 in Response to Different Treatments

Tapping is a kind of mechanical wounding, over tapping or excessive stimulation by ET can elicit oxidative stresses of laticifer cells and the balance between ROS production and scavenging is broken, resulting in ROS burst and the onset of TPD in *Hevea brasiliensis* [1]. Multiple hormones are reported to regulate the expression of GRX genes in rice [10,27]. Thus, the expression profiles of *HbSRGRX1* in latex were systematically investigated in response to wounding, H_2O_2, ET, MeJA, ABA, SA, GA_3, 2, 4-D, 6-BA, and IAA using qPCR method. As shown in Figure 4, all the treatments regulated the expression of *HbSRGRX1*, but the expression profiles showed obvious variations. After wounding and H_2O_2 treatments, *HbSRGRX1* expression indicated an irregular fluctuation and was generally suppressed, with the lowest expression at 6 h and 24 h, respectively (Figure 4A,B). With ET treatment, the expression of *HbSRGRX1* significantly decreased to the lowest level at 24 h (Figure 4C).

Similarly, its expression was rapidly inhibited at 6 h after MeJA treatment and kept a similar expression level from 6 h to 48 h (Figure 4D). *HbSRGRX1* showed similar expression profiles and was quickly induced, all rising to the highest point at 6 h, and then markedly declining at 12 h after ABA, SA, and GA$_3$ treatments (Figure 4E–G). As for 2, 4-D, 6-BA, and IAA treatments, *HbSRGRX1* exhibited a similar expression trend of first rising and then declining, reaching the peaks at 6 h, 12 h and 24 h, respectively (Figure 4H–J). Taken together, these results demonstrated that *HbSRGRX1* might have distinct functions in response to wounding, H$_2$O$_2$, and hormones.

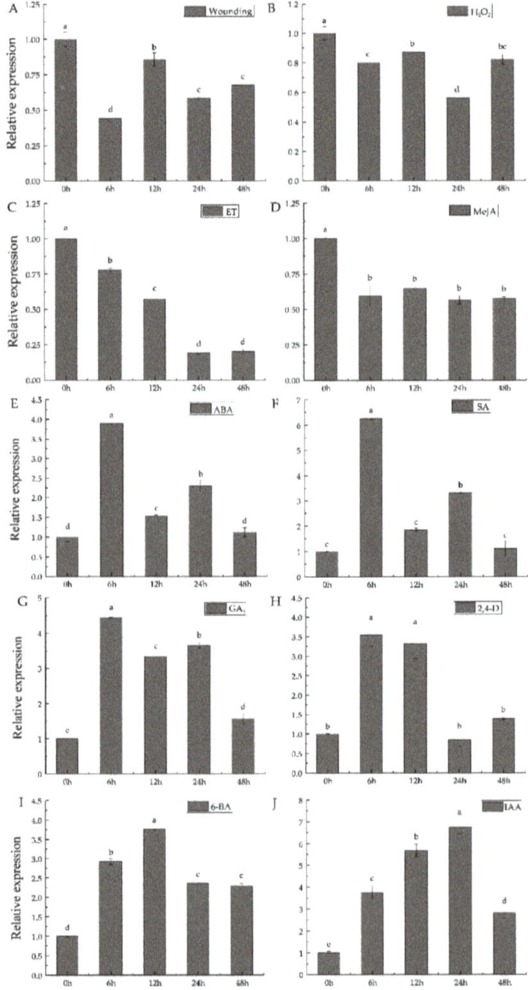

Figure 4. Expression profiles of the *HbSRGRX1* gene under different treatments. Expression profiles of *HbSRGRX1* in latex at 6, 12, 24, and 48h after wounding (**A**), hydrogen peroxide (H$_2$O$_2$) (**B**), ethephon (ET) (**C**), methyl jasmonate (MeJA) (**D**), abscisic acid (ABA) (**E**), salicylic acid (SA) (**F**), gibberellic acid (GA$_3$) (**G**), 2,4-dichlorophenoxyacetic acid (2,4-D) (**H**), 6-Benzylaminopurine (6-BA) (**I**), and indole-3-acetic acid (IAA) (**J**) treatments. Untreated trees were used as the control (0 h). Relative expression was normalized using 18S rRNA gene. Values represent mean ± SD of three independent replicates. Different letters indicate significant difference ($p < 0.05$).

4. Discussion

Glutaredoxins (GRXs) are a class of small oxidoreductases whose size is commonly 10 to 15 kDa. Many studies have suggested that plant GRXs perform various roles, such as modulating organ development, response to oxidative stress, and hormone signaling [7,12,27]. However, there are few data about the rubber tree GRXs, and their functions are still unclear. In this study, a CPYC-type GRX gene *HbSRGRX1*, encoding 107 amino acids, was isolated from rubber trees. Sequence alignment and phylogenetic analysis indicated that *HbSRGRX1* had high homology with MeGRX and RcGRX of Euphorbiaceae family, suggesting its conservation in the evolution. This research demonstrated that *HbSRGRX1* was localized in the nuclei of tobacco cells. Nuclear localization of GRXs were previously found in other plants [10,27–29]. It is indicated that nuclear localization of a GRX protein belonging to CC-type class from *Arabidopsis thaliana*, ROXY1, is pivotal in controlling petal development [29]. The rice *OsGRX8*, localized in the nuclei and cytosol, is found to take part in response to osmotic, salinity and oxidative stresses [27].

In the present study, the qPCR method was used to systematically investigate the expression profiles of the rubber tree *HbSRGRX1* gene in various tissues and varieties, different degrees of TPD trees, and under multiple treatments. It is reported that the GRXs genes are differentially expressed in various tissues in rice and poplar [10,11,30]. The rubber tree *HbSRGRX1* gene had a tissue-specific expression profile, with the strongest expression in male flowers and weakest in leaves (Figure 3A), suggesting its vital function in the development of male flowers. The CYPC-type *PtrcGrxC3* from *Populus trichocarpa* shows the highest expression in flowers and the lowest in leaves, which is similar to our results [11]. Additionally, the *HbSRGRX1* gene had differential expression in five various varieties, with the highest level in Reyan 7-33-97, followed by Reken 523, PR107, Reyan 8-79, and 7-20-59 (Figure 3B). In the five varieties, Reyan 7-33-97 is a high-yield variety with strong resistance against cold, PR107 and 7-20-59 against wind, while the resistance of Reken 523 and Reyan 8-79 against cold and wind is the weakest. The differential expression of *HbSRGRX1* in various varieties suggested its specific function in defense response of rubber trees.

ROS signaling is reported to be involved in the process of TPD onset, and some ROS-scavenging genes have been identified [4,31]. GRX, as a scavenger of ROS, is found to show down-regulated expression in TPD trees by using transcriptome analysis [4]. Consistently, the expression of the *HbSRGRX1* gene significantly declined in TPD trees in this study (Figure 3C,D). Furthermore, the *HbSRGRX1* expression was repressed by H_2O_2 treatment (Figure 4B). According to these results, we speculated that *HbSRGRX1* might play an antioxidant role in TPD response.

Here, we found that the wounding and diverse hormones treatments could also regulate the expression of the *HbSRGRX1* gene (Figure 4). Hormones are known to modulate plant development and environmental stress response. Many studies have demonstrated that the expression of GRX genes is affected by a number of hormones in plants. The *Arabidopsis* GRX480 (also known as GRXC9) may play a key role in SA/JA cross-talk [13], and its expression can be activated by UVB exposure through an SA-dependent and NPR1-independent pathway [32]. Overexpression of rice CC-type *OsGRX8* reduces the sensitivity to plant hormones, ABA, and IAA [27]. The transcripts of rice GRX genes are reported to participate in response to a variety of hormones, including IAA, SA, JA, ABA, cytokinin, and ethylene derivatives [30]. In addition, the expression of the rice *OsGRX20* gene significantly rises after 2, 4-D, JA, SA, and ABA treatments [10]. ET (an ethylene releaser) has been generally used in stimulating latex regeneration in rubber trees. However, the mechanism of ET regulating latex production remains poorly understood. In our study, *HbSRGRX1* showed a significantly down-regulated expression after ET treatment (Figure 4C), suggesting its important function in the ethylene signaling pathway. In plants, JA is known to be a vital hormone involved in controlling a variety of physiological responses [33–36]. It has been demonstrated that laticifer differentiation can be induced by exogenous JA in *H. brasiliensis*, and the number of laticifers is closely correlated with latex production [37]. JA is also indicated to be a key signal molecule in the modulation of rubber biosynthesis [38]. The present results displayed that JA markedly down-regulated the expression of the *HbSRGRX1* gene (Figure 4D). It was possible that *HbSRGRX1* played an important role in JA-regulating rubber biosynthesis. Besides ET and JA, *HbSRGRX1* expression was induced by hormones ABA, SA, GA_3, 2, 4-D, 6-BA, and IAA, reaching the highest level at different times after treatments (Figure 4E-J). Altogether, our results suggest a critical function of *HbSRGRX1* in plant development and response to TPD, wounding, H_2O_2, and various hormones, and would make a foundation for further characterizing the function of *HbSRGRX1* in *Hevea brasiliensis*.

5. Conclusions

HbSRGRX1 encoded a protein for CPYC-type GRX with CPYC active site. *HbSRGRX1* was significantly down-regulated in TPD trees and might play important roles during the onset of TPD. *HbSRGRX1* was preferentially expressed in male flower and might regulate the development of the male flower. *HbSRGRX1* was involved in defense response to wounding and other various stresses. To better understand the relation between *HbSRGRX1* and TPD, it is necessary to study the function of *HbSRGRX1* in-depth in the future.

Supplementary Materials: The following are available online at http://www.mdpi.com/1999-4907/10/2/158/s1, Figure S1: Rubber trees from different TPD degrees.

Author Contributions: The experiments were designed by Z.W. and J.L. The manuscript was written by K.Y. The experiments were carried out by X.G. The experimental materials were collected by C.F. and Y.H.

Funding: This research was funded by the China Agriculture Research System-Natural Rubber (CARS-34-GW5) and the Ministry of Agriculture of China (1630022018009).

Conflicts of Interest: The authors declare no conflict of interest.

References

1. Faridah, Y.; Siti Arija, M.; Ghandimathi, H. Changes in some physiological latex parameters in relation to over exploitation and onset of induced tapping panel dryness. *J. Nat. Rubber Res.* **1996**, *10*, 182–186.

2. Chen, S.; Peng, S.; Huang, G.; Wu, K.; Fu, X.; Chen, Z. Association of decreased expression of a Myb transcription factor with the TPD (tapping panel dryness) syndrome in *Hevea brasiliensis*. *Plant Mol. Biol.* **2003**, *51*, 51–58. [CrossRef] [PubMed]

3. Venkatachalam, P.; Thulaseedharan, A.; Raghothama, K. Identification of expression profiles of tapping panel dryness (TPD) associated genes from the latex of rubber tree (*Hevea brasiliensis* Muell. Arg.). *Planta* **2007**, *226*, 499–515. [CrossRef] [PubMed]

4. Li, D.; Deng, Z.; Chen, C.; Xia, Z.; Wu, M.; He, P.; Chen, S. Identification and characterization of genes associated with tapping panel dryness from *Hevea brasiliensis* latex using suppression subtractive hybridization. *BMC Plant Biol.* **2010**, *10*, 140. [CrossRef] [PubMed]

5. Apel, K.; Hirt, H. Reactive oxygen species: Metabolism, oxidative stress, and signal transduction. *Annu. Rev. Plant Biol.* **2004**, *55*, 373–399. [CrossRef] [PubMed]

6. Rouhier, N.; Gelhaye, E.; Jacquot, J.P. Plant glutaredoxins: Still mysterious reducing systems. *Cell Mol. Life Sci.* **2004**, *61*, 1266–1277. [CrossRef] [PubMed]

7. Cheng, N.H. AtGRX4, an Arabidopsis chloroplastic monothiol glutaredoxin, is able to suppress yeast grx5 mutant phenotypes and respond to oxidative stress. *FEBS Lett.* **2008**, *582*, 848–854. [CrossRef] [PubMed]

8. Laporte, D.; Olate, E.; Salinas, P.; Salazar, M.; Jordana, X.; Holuigue, L. Glutaredoxin GRXS13 plays a key role in protection against photooxidative stress in *Arabidopsis*. *J. Exp. Bot.* **2012**, *63*, 503–515. [CrossRef] [PubMed]

9. Wu, Q.Y.; Lin, J.; Liu, J.Z.; Wang, X.F.; Lim, W.; Oh, M.; Park, J.; Rajashekar, C.B.; Whitham, S.A.; Cheng, N.H.; Hirschi, K.D.; Park, S. Ectopic expression of Arabidopsis glutaredoxin *AtGRXS17* enhances thermotolerance in tomato. *Plant Biotech. J.* **2012**, *10*, 945–955. [CrossRef]

10. Ning, X.; Sun, Y.; Wang, C.; Zhang, W.; Sun, M.; Hu, H.; Liu, J.; Yang, L. A rice CPYC-type glutaredoxin OsGRX20 in protection against bacterial blight, methyl viologen and salt stresses. *Front. Plant Sci.* **2018**, *9*, 111. [CrossRef] [PubMed]

11. Rouhier, N.; Couturier, J.; Jacquot, J.P. Genome-wide analysis of plant glutaredoxin systems. *J. Exp. Bot.* **2006**, *57*, 1685–1696. [CrossRef] [PubMed]

12. Xing, S.; Rosso, M.G.; Zachgo, S. ROXY1, a member of the plant glutaredoxin family, is required for petal development in *Arabidopsis thaliana*. *Development* **2005**, *132*, 1555–1565. [CrossRef] [PubMed]

13. Ndamukong, I.; Al Abdallat, A.; Thurow, C.; Fode, B.; Zander, M.; Weigel, R.; Gatz, C. SA-inducible Arabidopsis glutaredoxin interacts with TGA factors and suppresses JA-responsive PDF1.2 transcription. *Plant J.* **2007**, *50*, 128–139. [CrossRef] [PubMed]

14. Bandyopadhyay, S.; Gama, F.; Molina-Navarro, M.M.; Gualberto, J.M.; Claxton, R.; Naik, S.G.; Huynh, B.H.; Herrero, E.; Jacquot, J.P.; Johnson, M.K.; et al. Chloroplast monothiol glutaredoxins as scaffold proteins for the assembly and delivery of [2Fe-2S] clusters. *EMBO J.* **2008**, *27*, 1122–1133. [CrossRef] [PubMed]

15. Yang, F.; Bui, H.T.; Pautler, M.; Llaca, V.; Johnston, R.; Lee, B.H.; Kolbe, A.; Sakai, H.; Jackson, D. A maize glutaredoxin gene, *Abphyl2*, regulates shoot meristem size and phyllotaxy. *Plant Cell* **2015**, *27*, 121–131. [CrossRef] [PubMed]

16. Cheng, N.H.; Liu, J.Z.; Liu, X.; Wu, Q.; Thompson, S.M.; Lin, J.; Chang, J.; Whitham, S.A.; Park, S.; Cohen, J.D.; et al. Arabidopsis monothiol glutaredoxin, AtGRXS17, is critical for temperature-dependent postembryonic growth and development via modulating auxin response. *J. Biol. Chem.* **2011**, *286*, 20398–20406. [CrossRef] [PubMed]

17. Xing, S.; Zachgo, S. ROXY1 and ROXY2, two Arabidopsis glutaredoxin genes, are required for another development. *Plant J.* **2008**, *53*, 790–801. [CrossRef] [PubMed]

18. Li, D.; Wang, X.; Deng, Z.; Liu, H.; Yang, H.; He, G. Transcriptome analyses reveal molecular mechanism underlying tapping panel dryness of rubber tree (*Hevea brasiliensis*). *Sci. Rep.* **2016**, *6*, 23540. [CrossRef]

19. Yuan, K.; Wang, Z.; Zhou, X.; Zou, Z.; Yang, L. The identification of differentially expressed latex proteins in healthy and tapping panel dryness (TPD) *Hevea brasiliensis* trees by iTRAQ and 2D LC-MS/MS. *Acta Agric. Univ. Jiangxiensis* **2014**, *36*, 650–655.

20. Tang, C.; Huang, D.; Yang, J.; Liu, S.; Sakr, S.; Li, H.; Zhou, Y.; Qin, Y. The sucrose transporter HbSUT3 plays an active role in sucrose loading to laticifer and rubber productivity in exploited trees of *Hevea brasiliensis* (para rubber tree). *Plant Cell Environ.* **2010**, *33*, 1708–1720. [CrossRef]

21. Deng, Z.; Zhao, M.; Liu, H.; Wang, Y.; Li, D. Molecular cloning, expression profiles and characterization of a glutathione reductase in *Hevea brasiliensis*. *Plant Physiol. Bioch.* **2015**, *96*, 53–63. [CrossRef] [PubMed]

22. Long, X.; He, B.; Wang, C.; Fang, Y.; Qi, J.; Tang, C. Molecular identification and characterization of the pyruvate decarboxylase gene family associated with latex regeneration and stress response in rubber tree. *Plant Physiol. Bioch.* **2015**, *87*, 35–44. [CrossRef] [PubMed]

23. Tang, C.; Qi, J.; Li, H.; Zhang, C.; Wang, Y. A convenient and efficient protocol for isolating high-quality RNA from latex of *Hevea brasiliensis* (para rubber tree). *J. Biochem. Bioph. Meth.* **2007**, *70*, 749–754. [CrossRef] [PubMed]

24. Schultz, J.; Milpetz, F.; Bork, P.; Ponting, C.P. SMART, a simple modular architecture research tool: Identification of signaling domains. *Proc. Natl. Acad. Sci. USA* **1998**, *95*, 5857–5864. [CrossRef] [PubMed]

25. Tamura, K.; Stecher, G.; Peterson, D.; Filipski, A.; Kumar, S. MEGA6: Molecular evolutionary genetics analysis version 6.0. *Mol. Biol. Evol.* **2013**, *30*, 2725–2729. [CrossRef]

26. Jacob, J.L.; Prevot, J.C.; Laccrotte, R. Tapping panel dryness in *Hevea brasiliensis*. *Plant Rech. Dev.* **1994**, *2*, 15–21.

27. Sharma, R.; Priya, P.; Jain, M. Modified expression of an auxin responsive rice CC-type glutaredoxin gene affects multiple abiotic stress responses. *Planta* **2013**, *238*, 871–884. [CrossRef]

28. Hong, L.; Tang, D.; Zhu, K.; Wang, K.; Li, M.; Cheng, Z. Somatic and reproductive cell development in rice anther is regulated by a putative glutaredoxin. *Plant Cell* **2012**, *24*, 577–588. [CrossRef]

29. Li, S.; Lauri, A.; Ziemann, M.; Busch, A.; Bhave, M.; Zachgo, S. Nuclear activity of ROXY1, a glutaredoxin interacting with TGA factors, is required for petal development in *Arabidopsis thaliana*. *Plant Cell* **2009**, *21*, 429–441. [CrossRef]

30. Garg, R.; Jhanwar, S.; Tyagi, A.K.; Jain, M. Genome-wide survey and expression analysis suggest diverse roles of glutaredoxin gene family members during development and response to various stimuli in rice. *DNA Res.* **2010**, *17*, 353–367. [CrossRef]

31. Putranto, R.A.; Herlinawati, E.; Rio, M.; Leclercq, J.; Piyatrakul, P.; Gohet, E.; Sanier, C.; Oktavia, F.; Pirrello, J.; Kuswanhadi; et al. Involvement of ethylene in the latex metabolism and tapping panel dryness of *Hevea brasiliensis*. *Int. J. Mol. Sci.* **2015**, *16*, 17885–17908. [CrossRef] [PubMed]

32. Herrera-Vásquez, A.; Carvallo, L.; Blanco, F.; Tobar, M.; Villarroel-Candia, E.; Vicente-Carbajosa, J.; Salinas, P.; Holuigue, L. Transcriptional control of glutaredoxin GRXC9 expression by a salicylic acid-dependent and NPR1-Independent pathway in *Arabidopsis*. *Plant Mol. Biol. Report.* **2015**, *33*, 624–637. [CrossRef] [PubMed]

33. McConn, M.; Creelman, R.A.; Bell, E.; Mullet, J.E.; Browse, J. Jasmonate is essential for insect defense in *Arabidopsis*. *Proc. Natl. Acad. Sci. USA* **1997**, *94*, 5473–5477. [CrossRef] [PubMed]

34. Yuan, Z.; Zhang, D. Roles of jasmonate signalling in plant inflorescence and flower development. *Curr. Opin. Plant Biol.* **2015**, *27*, 44–51. [CrossRef] [PubMed]

35. Farooq, M.A.; Gill, R.A.; Islam, F.; Ali, B.; Liu, H.; Xu, J.; He, S.; Zhou, W. Methyl jasmonate regulates antioxidant defense and suppresses arsenic uptake in *Brassica napus* L. *Front Plant Sci.* **2016**, *7*, 468. [CrossRef] [PubMed]

36. Hu, Y.; Jiang, Y.; Han, X.; Wang, H.; Pan, J.; Yu, D. Jasmonate regulates leaf senescence and tolerance to cold stress: Crosstalk with other phytohormones. *J. Exp. Bot.* **2017**, *68*, 1361–1369. [CrossRef] [PubMed]

37. Hao, B.Z.; Wu, J.L. Laticifer differentiation in *Hevea brasiliensis*: Induction by exogenous jasmonic acid and linolenic acid. *Ann. Bot.* **2000**, *85*, 37–43. [CrossRef]

38. Zeng, R.Z.; Duan, C.F.; Li, X.Y.; Tian, W.M.; Nie, Z.Y. Vacuolar-type inorganic pyrophosphatase located on the rubber particle in the latex is an essential enzyme in regulation of the rubber biosynthesis in *Hevea brasiliensis*. *Plant Sci.* **2009**, *176*, 602–607.

 *forests*

Article

The Positive Effect of Different 24-epiBL Pretreatments on Salinity Tolerance in *Robinia pseudoacacia* L. Seedlings

Jianmin Yue [1,2], **Zhiyuan Fu** [1], **Liang Zhang** [1], **Zihan Zhang** [1,3] **and Jinchi Zhang** [1,*]

[1] Co-Innovation Center for Sustainable Forestry in Southern China of Jiangsu Province, Key Laboratory of Soil and Water Conservation and Ecological Restoration, Nanjing Forestry University, 159 Longpan Road, Nanjing 21037, Jiangsu, China; jianminyue@njfu.edu.cn (J.Y.); lixiangqian@jit.edu.cn (Z.F.); zhangliang001@jit.edu.cn (L.Z.); zhangzh@njfu.edu.cn (Z.Z.)

[2] Department of Forest and Conservation Sciences, Faculty of Forestry, University of British Columbia, 2424 Main Mall, Vancouver, BC V6T 1Z4, Canada

[3] State Key Laboratory of Tree Genetics and Breeding & Key Laboratory of Tree Breeding and Cultivation, State Forestry Administration, Research Institute of Forestry, Chinese Academy of Forestry, Bejing 100091, China

* Correspondence: zhang8811@njfu.edu.cn; Tel.: +86-854-272-02

Received: 28 November 2018; Accepted: 18 December 2018; Published: 20 December 2018

Abstract: As a brassinosteroid (BR), 24-epibrassinolide (24-epiBL) has been widely used to enhance the resistance of plants to multiple stresses, including salinity. Black locust (*Robinia pseudoacacia* L.) is a common species in degraded soils. In the current study, plants were pretreated with three levels of 24-epiBL (0.21, 0.62, or 1.04 µM) by either soaking seeds during the germination phase (Sew), foliar spraying (Spw), or root dipping (Diw) at the age of 6 months. The plants were exposed to salt stress (100 and 200 mM NaCl) via automatic drip-feeding (water content ~40%) for 45 days after each treatment. Increased salinity resulted in a decrease in net photosynthesis rate (P_n), stomatal conductance (G_s), intercellular:ambient CO_2 concentration ratio (C_i/C_a), water-use efficiency (WUE_i), and maximum quantum yield of photosystem II (PSII) (F_v/F_m). Non-photochemical quenching (NPQ) and thermal dissipation (H_d) were elevated under stress, which accompanied the reduction in the membrane steady index (MSI), water content (RWC), and pigment concentration (Chl a, Chl b, and Chl). Indicators of oxidative stress (i.e., malondialdehyde (MDA) and antioxidant enzymes (peroxidase (POD) and superoxide dismutase (SOD)) in leaves and Na^+ content in chloroplasts increased accompanied by a reduction in chloroplastid K^+ and Ca^{2+}. At 200 mM NaCl, the chloroplast and thylakoid ultrastructures were severely disrupted. Exogenous 24-epiBL improved MSI, RWC, K^+, and Ca^{2+} content, reduced Na^+ levels, maintained chloroplast and thylakoid membrane structures, and enhanced the antioxidant ability in leaves. 24-epiBL also substantially alleviated stress-induced limitations of photosynthetic ability, reflected by elevated chlorophyll fluorescence, pigment levels, and P_n. The positive effects of alleviating salt stress in *R. pseudoacacia* seedlings in terms of treatment application was Diw > Sew > Spw, and the most positive impacts were seen with 1.04 µM 24-epiBL. These results provide diverse choice for 24-epiBL usage to defend against NaCl stress of a plant.

Keywords: 24-epiBL application; salt stress; ion contents; chloroplast ultrastructure; photosynthesis; *Robinia pseudoacacia* L.

---

1. Introduction

A current report states that more than 1 billion hectares of land have been damaged by salinity as a result of anthropogenic activities [1]. With current practices, salinization could affect half of the currently cultivated land by 2050 [2]. China has the third largest area of saline soils (366,500 km^2) in the world [3], resulting in significant agricultural losses and limiting economic development and quality of life. As the most densely populated province in China, located at the lower reaches of Yangtze River and along the coast of the Yellow Sea, Jiangsu province is at risk of soil salinity resulting from evaporation from the soil surface, especially in the ~15 km^2 of degraded land that has been reclaimed along the coastline annually [4].

Black locust (*Robinia pseudoacacia* L.), an arboreal member of the *Fabaceae* indigenous to the eastern USA, is the third most-planted tree in reforestation schemes because of its ability to adapt to a range of environmental conditions in China [5]. As a salt-tolerant species, *R. pseudoacacia* is able to withstand salinity levels up to ~50 mM Na$^+$ in laboratory conditions, or ~100 mM Na$^+$ in field conditions [6]. It is also widely used as a major tree species to improve degraded soils [7], including saline soils, and has been widely cultivated in China since the early 20th century [8,9]. In Jiangsu, the soils most affected by salinity are along the coast, and these areas are also negatively impacted by hurricanes. It has been reported that stands of *R. pseudoacacia* could eliminate or reduce to <5% the impact of hurricanes on such stands and could also significantly increase soil nitrate concentrations along the coast [10]. Thus, there is a need to improve the adaptation of *R. pseudoacacia* to salinity-affected soils so that they can be planted along the coast in Jiangsu province.

Salt stress always results in water deficit, ion toxicities and imbalances, and oxidative stress, damaging plant cells and organs, limiting plant growth, and even causing death. For example, 85.73 mM NaCl increased the water saturation deficit in weak roots and limited the biomass accumulation of individual *R. pseudoacacia* [11]. The growth of the medicinal plant, *L. japonica*, was reduced by 200 mM NaCl because of negative effects on the ionic uptake and their distribution to plant organs [12]. Oxidative factors are prevalent in both lower and higher salinity conditions. For instance, 50 mM NaCl caused a significant reduction in plant growth and photosynthetic parameters accompanied by lipid peroxidation and hydrogen peroxide in *Capsicum annuum* L. [13]. In addition, 250 mM NaCl and 500 mM NaCl limited the growth of *R. pseudoacacia* by reducing the antioxidant enzyme activity, combined with a reduction in the photosynthetic pigment contents, damaging the chloroplast ultrastructure [14].

As a brassinosteroid (BR), 24-epibrassinolide (24-epiBL) has been used in many plant species to regulate their tolerance to stress; for example, against chilling stress in young grapevine seedlings (*Vitis vinifera* L.) [15], against Ca (NO$_3$)$_2$ stress in cucumber, *Cucumis sativus* (L., cv. 'Jinyou No. 4') [16], and against low temperatures and poor light intensities in tomato (*Lycopersicon esculentum* Mill.) [17]. In addition, 24-epiBl has also been used to modify salinity stress, such as, 24-epiBL ameliorated saline stress, and improved the productivity of wheat (*Triticum aestivum* L.) sprayed with 0.105 μM and 0.21 μM 24-epiBL by modifying the photosynthesis, chlorophyll, and nitrate levels in wheat leaves [18]. 24-epiBL ameliorated salinity-induced injuries by reducing the K$^+$ efflux in barley seedlings (*Hordeum vulgare* cv. 'Franklin') when the seeds were soaked in 0.21, 0.53, or 1.05 μM 24-epiBL solutions [19].

Therefore, we hypothesized that soaking seeds, spraying leaves, and dipping roots of *R. pseudoacacia* in 0.21–1.04 μM 24-epiBL could improve the ability of seedlings to adapt to saline soils. This adaptation might result from response mechanisms, including maintaining the regular structure and function of chloroplasts and modifying photosynthesis. Our approach could also reveal the most suitable method for 24-epiBL application. Thus, we applied NaCl stress and 24-epiBL treatments to greenhouse seedlings and then measured: (a) Photosynthesis parameters; (b) chlorophyll fluorescence; (c) water content and membrane stability of leaves; (d) photosynthetic pigments in leaves; (e) malondialdehyde (MDA), and antioxidant enzyme activities (peroxidase (POD) and superoxide dismutase (SOD)) in leaves; (f) ions in chloroplasts; and (g) chloroplast ultrastructure. The results obtained could help shed light on the most appropriate treatment to use to improve the salinity tolerance of *R. pseudoacacia* seedlings for use in reforestation programs in areas of high salinity, both along the coastline in Jiangsu and elsewhere.

2. Materials and Methods

2.1. Plant Material and 24-epibrassinolide Treatments

The experiments ran from September 2014 to July 2015. 24-epiBL (Sigma-Aldrich, St. Louis, MO, USA) was applied at the concentrations of 0.21, 0.62, and 1.04 μM for each treatment type: Seed soaking, foliar spraying, and root dipping. Seeds were sown into trays containing quartz sand moistened with either distilled water or the relevant concentration 24-epiBL. When the cotyledons were fully expanded, the seedlings were transplanted into seedling bags. Seedlings were grown with no stress for 6 months until Spring 2015, when they were ~45 cm tall. All seedlings were then washed to remove the rooting medium. Seedlings grown from seeds pretreated with 24-epiBL (0.21, 0.62, or 1.04 μM) or distilled water were replanted directly into plastic pots (40 cm × 25 cm × 15 cm; 1 plant per pot) containing coarse sand and vermiculite 2:1 (*v/v*). After 1 week, seedlings were partly picked up of the seedlings from seeds with distilled water for foliar spraying with 0.21, 0.62, or 1.04 μM 24-epiBL every 7 days, each lasting 28 days. Other seedlings from untreated seeds were split into two groups and root dipped for 48 h in either distilled water or 0.21, 0.62, or 1.04 μM 24-epiBL (refreshed every 12 h) before being replanted using the methodology described earlier. Seedlings were watered to the drip-point every 7 days with a modified Hoagland's nutrient solution [20]. The details of materials used to cultivate the seedlings were as described by our previous paper [21].

2.2. Groups and Salt Stress Treatments

After a further 2 weeks of growth, seedlings of uniform height (~50 cm) were selected for the salt stress treatments. There were four groups: (i) Seedlings never exposed to 24-epiBL (CK); (ii) seedlings from seeds treated with 24-epiBL (Sew); (iii) seedlings foliar sprayed with 24-epiBL (Spw); and (iv) seedlings root-dipped in 24-epiBL (Diw). Fifteen seedlings of each group were subjected to each of two levels of salt stress (100 and 200 mM NaCl, both in Hoagland's solution) and non-stress (Hoagland's solution). Pots were first flushed with their respective nutrient solutions and the water content was then maintained near to saturation (40%) by automatic drip-feeding with fresh solution delivered by an Intelligent Automatic Watering System (patent CN 201398356 Y). There were two salt stress treatments (100 and 200 mM NaCl, designated SS-S1 and SS-S2, respectively) plus a zero-salt control (CK), resulting in 21 treatment groups overall (Table 1).

Data for gas exchange, chlorophyll fluorescence, antioxidant systems, pigment content, ion content, and chloroplast ultrastructure were collected as described below, and measurements began 45 days after the treatments had started. Over the measurement period, the day/night temperature regime was 29 °C/18 °C, relative humidity varied from 45%–80%, and natural lighting provided a photoperiod of ~14 h and a mid-day photosynthetic photon flux density (PPFD) of ~1000 μmol m^{-2} s^{-1}.

2.3. Photosynthesis

The net photosynthetic rate (P_n), stomatal conductance (G_s), and intercellular CO$_2$ concentration (C_i) were measured on 0.7 cm × 3.0 cm (2.1 cm^2) of the sunlight-exposed leaves using an ambient CO$_2$ concentration (C_a) of 380 μmol mol^{-1} by gas exchange analyzer (LI-6400, LI-COR Inc., Lincoln, NE, USA). The ratio of intercellular to ambient CO$_2$ concentration was calculated as C_i/C_a and the intrinsic water-use efficiency (WUE_i) was calculated as P_n/G_s. The measurement details were as described in our previous paper [21].

Chlorophyll fluorescence was measured using a portable fluorometer (149 PAM-2500; Walz, Effeltrich, Germany). We report the maximum quantum yield of photosystem II (PSII) as F_v/F_m, thermal dissipation (H_d) as $1 - (F_v'/F_m')$, and non-photochemical quenching (NPQ) as $(F_m - F_m')/F_m'$ in this research. The measurement details were as described in our previous paper [21].

Table 1. The different treatments' details.

Name	Seeds Pretreatment	Seedling Pretreatment		Salt Stress
	24-epiBL(μM) by Seeds Soaking	24-epiBL(μM) by Foliar Spraying	24-epiBL(μM) by Roots Dipping	NaCl (mM)
CK	—	—		—
SS-S1	—	—		100
SS-S2	—	—		200
Sew-0.21S1	0.21	—		100
Sew-0.62S1	0.62	—		100
Sew-1.04S1	1.04	—		100
Spw-0.21S1		0.21		100
Spw-0.62S1		0.62		100
Spw-1.04S1		1.04		100
Diw-0.21S1			0.21	100
Diw-0.62S1			0.62	100
Diw-1.04S1			1.04	100
Sew-0.21S2	0.21	—		200
Sew-0.62S2	0.62	—		200
Sew-1.04S2	1.04	—		200
Spw-0.21S2		0.21		200
Spw-0.62S2		0.62		200
Spw-1.04S2		1.04		200
Diw-0.21S2			0.21	200
Diw-0.62S2			0.62	200
Diw-1.04S2			1.04	200

Abbreviations: Control (CK); 100 mM NaCl treatment (SS-S1); 200 mM NaCl treatment (SS-S2); seeds soaking with 24-epiBL (0.21, 0.62, 1.04 μM) under 100 mM NaCl stress (Sew-0.21S1/0.62S1/1.04S1); foliar spraying with 24-epiBL (0.21, 0.62, 1.04 μM) under 100 mM NaCl stress (Spw-0.21S1/0.62S1/1.04S1); roots dipping with 24-epiBL (0.21, 0.62, 1.04 μM) under 100 mM NaCl stress (Diw-0.21S1/0.62S1/1.04S1); seeds soaking with 24-epiBL (0.21, 0.62, 1.04 μM) under 200 mM NaCl stress (Sew-0.21S2/0.62S2/1.04S2); foliar spraying with 24-epiBL (0.21, 0.62, 1.04 μM) under 200 mM NaCl stress (Spw-0.21S2/0.62S2/1.04S2); roots dipping with 24-epiBL (0.21, 0.62, 1.04 μM) under 200 mM NaCl stress (Diw-0.21S2/0.62S2/1.04S2).

2.4. Estimation of Membrane Stability Index and Relative Water Content

The membrane stability index (MSI) and relative water content (RWC) were estimated following the method from Yue et al. [21]. MSI and RWC was calculated using the formula given by Sairam [22] and Hayat et al. [23] as follows:

MSI = $[1 - C_1/C_2 \times 100]$ and RWC = $(FW - DW)/(TW - DW) \times 100\%$. Electrical conductivity bridge measured at 40 °C for 30 min (C_1), electrical conductivity bridge measured at 100 °C for 100 min (C_2); fresh weight of leaves (FW), dry weight of leaves (DW) and turgor weight of leaves (TW).

2.5. Observation of Chloroplast Ultrastructure

The chloroplasts were observed using a Hitachi transmission electron microscope (Carl Zeiss, Göttingen, Germany) and the details are described by Yue et al. [21].

2.6. Assays for Malondialdehyde and Antioxidant Enzyme Activities

200 mg of fresh leaf samples ($n = 3$) taken from the plant were pulverized with a mortar and pestle in liquid nitrogen before adding 0.05M phosphate buffer (pH = 7.0) for further grinding. Then, the suspension was collected in a test tube and diluted with the same buffer to 10 mL for testing enzyme activity [24]. Superoxide dismutase (SOD) was assayed with the nitroblue tetrazolium (NBT) method of Fridovich (1975) [25] and expressed in terms of units $min^{-1} g^{-1}$ fresh weight (FW). One unit (U) of SOD was defined as the amount of enzyme required to cause 50% inhibition of the reduction of NBT as monitored at 560 nm. Peroxidase (POD) was measured according to Hammerschmidt et al. (1982) [26] by monitoring the rate of guaiacol oxidation at 470 nm. The standard curve was constructed using 4-methoxyphenol and activity was expressed as μg oxidized $min^{-1} g^{-1}$ FW. Lipid peroxidation

was determined by measuring the amount of malondialdehyde (MDA) produced per g FW by the thiobarbituric acid reaction [27].

2.7. Assays for Chlorophyll Pigment Concentration

Chlorophyll pigment was obtained by 0.2 g of laminal tissue from three leaves ground in ice-cold 1:1 (v/v) ethanol: acetone using a mortar and pestle. The extracts were centrifuged at 6000g for 10 min, washed with extractant, then centrifuged again. The absorbance of the combined supernatants was at 645 nm and 663 nm. Total chlorophyll (Chl), chlorophyll a (Chl a), and chlorophyll b (Chl b) were calculated according to our previous paper [21].

2.8. Isolation of Chloroplasts for Testing

Chloroplasts were isolated from fully expanded leaves by differential and density gradient centrifugation as described by Cerović and Plesnicar (1984) [28] and Song and others (2006) [29] with some modifications. The details were the same as Yue et al. [21]. For chloroplast Na^+, Ca^{2+}, and K^+ content, pelleted chloroplasts from a final 1 mL aliquot were extracted with 5 mL of 0.5 M HCl by shaking on a water-bath at 50 °C for 45 min. The volume was brought up to 10 mL with 0.5 M HCl before filtering and diluting for inductively coupled plasma optical spectroscopy (ICP-OES) as elemental analysis on a ICP-OES spectrometer (SPECTRO CIRO CCD, GmbH & Co KG, Kleve, Germany) at wavelengths of 226-502 nm [30].

2.9. Statistical Analysis

All analyses had at least three biological replicates. The data of control, salt stress, the seeds soaking, and roots dipping with 1.04 μM 24-epiBL in P_n, G_s, F_v/F_m, NPQ, RWC, and MSI was shown in our previous paper [21], and other data were initial. Data were statistically analyzed with the Statistical Package for the Social Sciences (SPSS) 19.0 (SPSS Inc., Chicago, IL, USA). One-way analyses of variance (ANOVA) were employed to test the effects of treatment with 24-epiBL under salt stress. Pairwise comparisons were by a Duncan test and considered significant at $p < 0.05$. Three-way analyses of variance (ANOVA) were employed to test the effects of salt stress, 24-epiBL application methods, 24-epiBL usage concentration, and their interaction, and means were separated using Duncan's multiple range tests, with the significance considered as $p < 0.05$.

3. Results

3.1. Photosynthesis

Compared with the CK group, salt stress in the absence of any pretreatments with 24-epiBL obviously reduced photosynthesis (P_n) and stomatal conductance (G_s), 4.4 and 1.67 times under 100 NaCl mM stress (SS-S1) and 6.3 and 5.2 times under 200 mM NaCl stress (SS-S2), respectively. The change in P_n was not significant, but the difference in G_s was obvious between the two levels of NaCl (Figure 1A,B). Consequently, C_i/C_a was decreased at both salt stress levels, as was WUE_i, although there were no significant difference in C_i/C_a between SS-S1 and CK and in WUE_i between SS-S2 and CK (Figure 1C,D). There were no significant differences in P_n, C_i/C_a, or WUE_i in the SS-S1 versus SS-S2. This inhibition of photosynthesis was partially prevented by pretreatment with 24-epiBL and the effect of 24-epiBL in SS-S1 was better than in SS-S2; P_n in Diw-1.04S1 > Diw-0.62S1 > Sew-1.04S1 > Spw-1.04S1 > Diw-0.21S1 > Sew-0.62S1 were significantly higher than in SS-S1 (Figure 1A), and P_n in SS-S2 was obviously increased only by Diw-1.04S2 and Sew-1.04S2 (Figure 1A). Under 100 mM NaCl stress, pretreatment with 24-epiBL did not significantly affect G_s relative to SS-S1 combined with the decrease in C_i/C_a (except for Sew-0.21S1, Spw-0.21S1, and Spw-0.62S1) and increase in WUE_i (Figure 1B–D). The significant improvement in G_s following pretreatment with 24-epiBL was in contrast to the lack of an effect on C_i/C_a or WUE_i in the 200 mM NaCl groups, and the significant difference of G_s existed in Diw-1.04S2 versus Spw-0.21/0.62S2; the increase of C_i/C_a was only in the Diw group

(Figure 1B–D). All these changes were modified by increasing the concentration of 24-epiBL; the order of the concentration effect was 1.04 > 0.62 > 0.21 µM. The effects of salt concentration (Sc), the 24-epiBL application method (Am), and 24-epiBL application concentration (BRc) were obvious ($p < 0.01$), but the interactions in the above three factors were not significant ($p > 0.05$) in P_n, G_s, and WUE_i. The obvious role of C_i/C_a was only shown in BRc and the interaction between the salt concentration and 24-epiBL application method ($Sc*Am$) ($p < 0.01$).

Indexes	Sc	Am	BRc	$Sc*Am$	$Sc*BRc$	$Am*BRc$	$Sc*Am*BRc$
P_n	<0.01	<0.01	<0.01	0.45	0.67	1.00	0.65
G_s	<0.01	<0.01	<0.01	0.19	0.33	0.99	0.68
C_i/C_a	0.30	0.14	<0.01	<0.01	0.07	0.08	0.15
WUE_i	<0.01	<0.01	<0.01	0.17	0.12	0.99	0.97

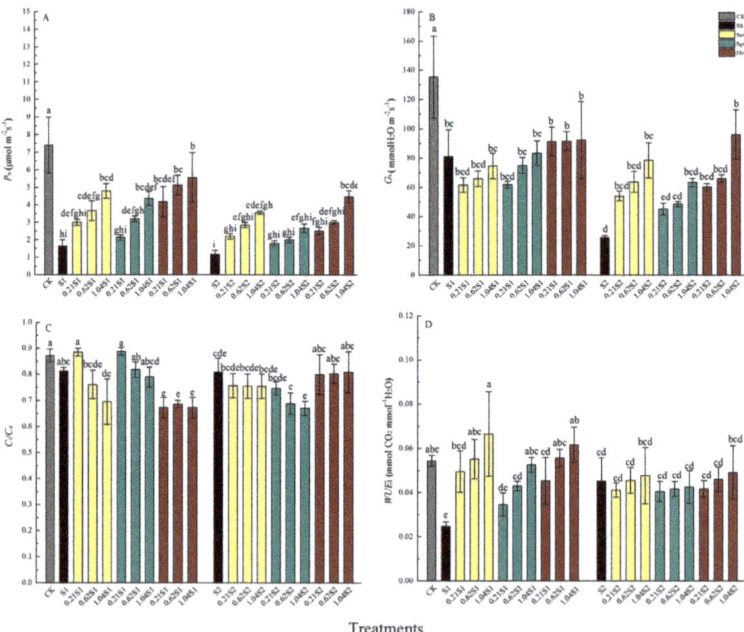

Figure 1. Effects of exogenous 24-epibrassinolide (24-epiBL) on (**A**) net photosynthetic rate (P_n), (**B**) stomatal conductance (G_s), (**C**) ratio of intercellular to ambient CO_2 concentration (C_i/C_a), and (**D**) intrinsic water use efficiency (WUE_i) of leaves of black locust seedlings at 100 (SS-S1) or 200 (SS-S2) mM NaCl. Each panel shows data for the zero salt control (CK, grey bar) and salt-stressed control (S1 & S2, black bar) and salt-stressed plants pretreated with either 0.21, 0.62, or 1.04 mM 24-epiBL by seed soaking (Sew, yellow bars), foliar spraying (Spw, green bars), or roots dipping (Diw, red bars). Each bar represents the mean (± standard deviation (SE)) of three replicates. Different letters above the bars show significant differences between means within each panel ($p < 0.05$). The p values are for the salt concentration effect (Sc), the 24-epiBL application method effect (Am), the 24-epiBL application concentration effect (BRc), the interaction effect between the salt concentration and 24-epiBL application method ($Sc*Am$), the interaction effect between the salt concentration and 24-epiBL application concentration ($Sc*BRc$), the interaction effect between the 24-epiBL application method and 24-epiBL application concentration ($Am*BRc$), and the interaction effect among the salt concentration, 24-epiBL application method, and 24-epiBL application concentration ($Sc*Am*BRc$), respectively. The significant effect was $p < 0.05$. Control (CK); 100 mM NaCl treatment (S1); 200 mM NaCl treatment (S2); treatments with 24-epiBL (0.21, 0.62, 1.04 µM) under 100 mM NaCl stress (0.21S1, 0.62S1, 1.04S1); treatments with 24-epiBL (0.21, 0.62, 1.04 µM) under 200 mM NaCl stress (0.21S2, 0.62S2, 1.04S2).

3.2. Chlorophyll Fluorescence

Maximum quantum yield of photosystem II (PSII) (F_v/F_m) decreased in response to saline treatments, whereas nonphotochemical quenching (NPQ) and thermal dissipation (H_d) significantly increased at both levels of salt stress compared with unstressed controls (Figure 2). There were no significant differences in F_v/F_m, NPQ, or H_d between the SS-S1 and SS-S2. Pretreatment with 24-epiBL by either Sew, Spw, or Diw had no effect on F_v/F_m, where F_v/F_m was equal to the CK level at both salinity stress levels (98.8% and 97.6% of CK) (Figure 2A). H_d improvement was reduced by all pretreatments with 24-epiBL, with the most significant effect in the Diw group at a 24-epiBL concentration of 1.04 µM (Figure 2B). NPQ decreased with increasing concentrations of 24-epiBL in the Sew, Spw, and Diw groups, although the values were higher than in SS-S1 and SS-S2. The treatment groups, Diw-0.21S1 (93.7% of SS-S1), Diw-0.62S1 (82.0% of SS-S1), Diw-1.04S1 (78.1% of SS-S1), Sew-1.04S2 (99.3% of SS-S2), and Diw-1.04S2 (90.9% of SS-0S2), showed a smaller decrease in NPQ in response to the two levels of NaCl (Figure 2C). The effects of *Sc*, *Am*, and *BRc* were obvious ($p < 0.05$), but the interactions in the above three factors were not significant ($p > 0.05$) in F_v/F_m, H_d, and NPQ.

Indexes	Sc	Am	BRc	Sc*Am	Sc*BRc	Am*BRc	Sc*Am*BRc
F_v/F_m	<0.01	<0.01	<0.01	0.36	0.19	0.38	0.99
H_d	<0.01	<0.01	<0.01	0.10	0.36	0.74	0.96
NPQ	<0.01	<0.01	0.02	0.22	0.23	0.86	0.98

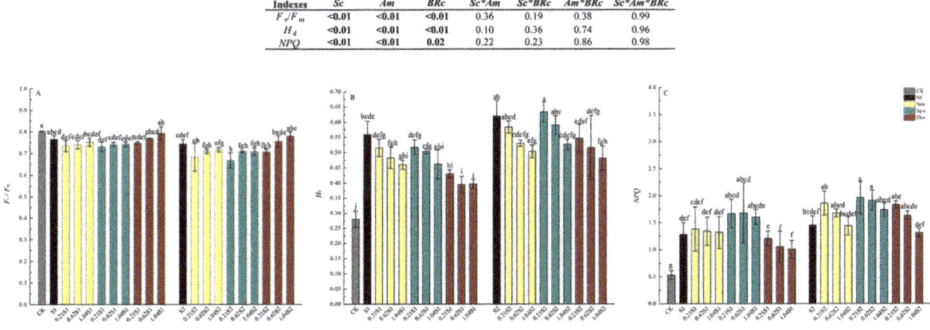

Treatments

Figure 2. Effects of exogenous 24-epiBL on (**A**) maximum yield of photosystem II (PSII) (F_v/F_m), (**B**) thermal dissipation (H_d), and (**C**) non-photochemical quenching (NPQ) of leaves of black locust seedlings at 100 (SS-S1) or 200 (SS-S2) mM NaCl. Other details as in Figure 1.

3.3. RWC and MSI

Salinity stress resulted in reductions in RWC and MSI, in particular in MSI at 100 and 200 mM NaCl (Figure 3). Additionally, the significant reduction that compared 200 mM NaCl to 100 mM NaCl only appeared in MSI. In general, 24-epiBL pretreatments reversed these effects. At 100 mM NaCl, RWC was significantly increased in the Sew-1.04S1 and all Diw treatment groups compared with salt-stressed plants (SS-S1), whereas the mean RWC was similar to the CK group. Responses to 200 mM NaCl were similar and RWC was significantly greater than SS-S2 in the Sew-1.04S2 and Diw-1.04S2 groups (Figure 3A); the lower content of 24-epiBL in the Diw groups was obviously less effective in SS-S2 than in SS-S1. For MSI at 100 mM NaCl, only the Sew-0.21S1, Sew-0.62S1, and Spw-0.21S1 pretreatments showed no differences with the salt-stressed control, whereas, at 200 mM NaCl, all nine pretreatments were effective, and Sew-1.04S1, Diw-0.62S2, and Diw-1.04S2 showed values similar to the CK group (Figure 3B). The effects of *Sc*, *Am*, and *BRc* were obvious ($p < 0.01$), but the interactions in the above three factors were not significant ($p > 0.05$) in RWC and MSI.

Indexes	Sc	Am	BRc	Sc*Am	Sc*BRc	Am*BRc	Sc*Am*BRc
RWC	<0.01	<0.01	<0.01	0.86	0.36	0.94	0.90
MSI	<0.01	<0.01	<0.01	0.94	0.32	0.34	0.40

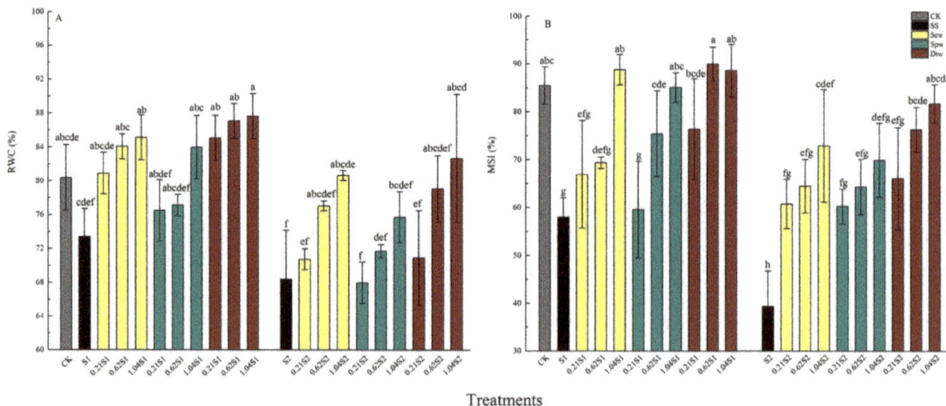

Figure 3. Effects of exogenous 24-epiBL on (**A**) relative water content (RWC) and (**B**) membrane stability index (MSI) of leaves of black locust seedlings at 100 (SS-S1) or 200 (SS-S2) mM NaCl. Other details as in Figure 1.

3.4. Photosynthetic Pigments

Salinity stress had negative effects on pigment concentrations (Figure 4), although there were no statistically significant differences at different salinity stress levels. Relative to the unstressed control (CK), Chl a, Chl b, and total chlorophyll (Chl) were all reduced by 24%–28% under mild stress (SS-S1), and by 36%–40% at the higher stress level (SS-S2). All 24-epiBL pretreatments resulted in complete retention of these pigments at levels that were either statistically greater than the stressed control or even greater than the unstressed control (i.e., Diw-0.62 and Diw-1.04 at 100 mM NaCl) except for Chl b in the Sew-0.21S1 group, but the improvements were not obvious except Diw-1.04S1 in Chl a, Sew-1.04S1 and Diw-0.64/1.04S1 in Chl b, and Spw-1.04S1 and Diw-0.62/1.04S1 in Chl. Overall, the effects of regulation were Diw > Sew > Spw and the optimum concentration of 24-epiBL was 1.04 μM. The effect of *Sc* was obvious in Chl a, Chl b, and Chl ($p < 0.01$). The effects in *Am* ($p = 0.03$), *BRc* ($p = 0.01$), and *Sc*Am* ($p = 0.01$) were significant only in Chl b.

Indexes	Sc	Am	BRc	Sc*Am	Sc*BRc	Am*BRc	Sc*Am*BRc
Chl a	<0.01	0.20	0.23	0.26	0.97	0.95	0.96
Chl b	<0.01	0.03	0.01	0.01	0.16	0.65	0.70
Chl	<0.01	0.06	0.11	0.11	0.91	0.94	0.98

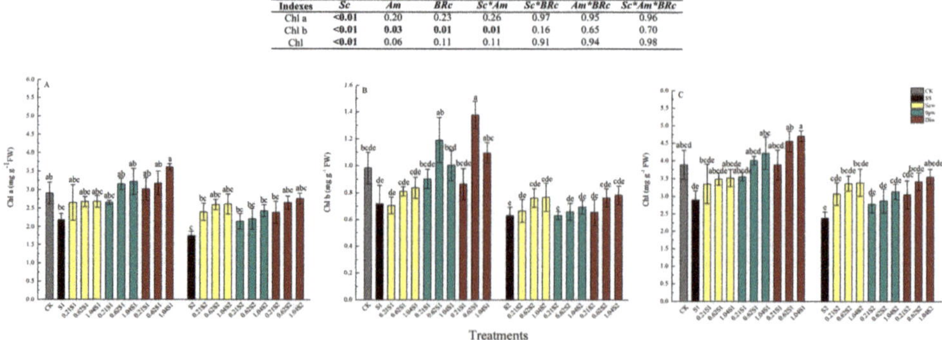

Figure 4. Effects of exogenous 24-epiBL on (**A**) chlorophyll a (Chl a), (**B**) chlorophyll b (Chl b), and (**C**) total chlorophyll (Chl) of leaves of black locust seedlings at 100 (SS-S1) or 200 (SS-S2) mM NaCl. Other details as in Figure 1.

3.5. MDA, POD, and SOD in Leaves

Compared with the unstressed control group (CK), salt stress obviously increased the concentration of MDA by ~1.64-fold at 100 mM NaCl and by 2.04-fold at 200 mM NaCl (Figure 5A; SS-S1 and SS-S2 were significantly different from each other). Pretreatment with 24-epiBL was ineffective at reducing this increase to within just 1.18–1.92 times the unstressed value in all groups, with clear distinction between the concentration or developmental stage of application except the Sew-0.21S1,Spw-0.21S1, and Spw-0.21/0.62S2, which were no different to the salinity treatment separately; the optimal application method and 24-epiBL concentration were Diw and 1.04 μM, respectively, which resulted in a percentage modification of MDA of 72.4% of that of SS-S1 and 60.1% of that of SS-S2. The effect of *Sc* ($p < 0.01$), *Am* ($p < 0.01$), *BRc* ($p < 0.01$), and *Sc*Am* ($p = 0.02$) were significant in MDA (Figure 5A).

Salinity stress resulted in improved activities of POD, but these were only obvious in SS-S1 and the improvement of POD activities in SS-S1 was significantly better than SS-0S2. In general, 24-epiBL pretreatments increased POD activity and the regulated effect was much better in SS-S1 than SS-S2. The effective role of 24-epiBL pretreatments under 100 mM NaCl were Diw-1.04S1 > Spw-1.04S1 > Sew-1.04S1 > Diw-0.62S1, and only Diw-1.04S1 was significantly different than the other effective treatments. Under 200 mM NaCl, only Sew-0.21S2 and Spw-0.21S2 had no difference to SS-S2; other treatments in POD activities had no significant difference between the application method (Sew, Spw, Diw) and 24-epiBL contents (0.21, 0.62, 1.04 μM). The 24-epiBL pretreatment methods and concentrations in order of obvious impact on POD activity in SS-S2 were as follows: Sew-1.04S2 > Diw-1.04S2 > Sew-0.62S2 > Spw-1.04S2 > Spw-0.62S2 > Diw-0.62S2 > Diw-0.21S2. The effect of *Sc* ($p < 0.01$), *Am* ($p < 0.01$), *BRc* ($p < 0.01$), *Sc*Am* ($p = 0.05$), and the interaction in salt concentration and 24-epiBL application concentration (*Sc*BRc*) ($p < 0.01$) were significant in POD (Figure 5B).

NaCl treatment significantly increased SOD activity, although this was obviously lower in the SS-S2 than in the SS-S1, and the activities in SOD by 24-epiBL treatments in SS-S1 was higher than in SS-S2. Under 100 mM NaCl stress, the three 24-epiBL pretreatment approaches significantly increased SOD activity in contrast to CK, but only 1.04 μM 24-epiBL treatments resulted in higher SOD activity compared with SS-S1 (Diw-1.04S1 > Sew-1.04S1 > Spw-1.04S1). No significant effects were seen in the Sew-0.21S2, Spw-0.21S2, and Diw-0.21S2 groups compared with SS-S2. The optimum effects occurred in the Sew-1.04S2 (1.58-fold of SS-S2) and Diw-1.04S2 (1.74-fold of SS-S2) and they had no difference between each other. The effect of *Sc* ($p < 0.01$), *Am* ($p < 0.01$), *BRc* ($p < 0.01$), *Sc*Am* ($p < 0.01$), and the interaction in the 24-epiBL application method and 24-epiBL application concentration (*Am*BRc*) ($p = 0.01$) were significant in SOD (Figure 5C).

Indexes	*Sc*	*Am*	*BRc*	*Sc*Am*	*Sc*BRc*	*Am*BRc*	*Sc*Am*BRc*
MDA	<0.01	<0.01	<0.01	0.02	0.78	0.84	0.88
POD	<0.01	<0.01	<0.01	0.05	<0.01	0.49	0.27
SOD	<0.01	<0.01	<0.01	<0.01	0.47	0.01	0.39

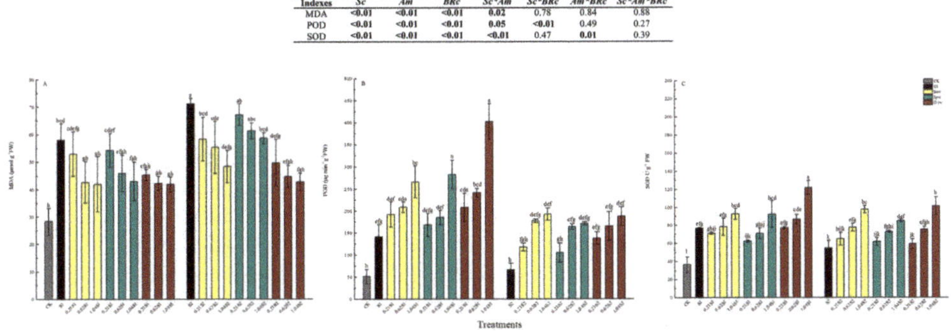

Figure 5. Effects of exogenous 24-epiBL on (**A**) malondialdehyde (MDA), (**B**) peroxidase (POD), and (**C**) superoxide dismutase (SOD) of leaves of black locust seedlings at 100 (SS-S1) or 200 (SS-S2) mM NaCl. Other details as in Figure 1.

3.6. Variation in Na⁺, K⁺, and Ca²⁺ Content in Chloroplasts

3.6. Variation in Na^+, K^+, and Ca^{2+} Content in Chloroplasts

There was a significant increase in Na^+ content and decrease in K^+ and Ca^{2+} accumulation in chloroplasts with increasing NaCl concentration (Figure 6). 24-epiBL reduced Na^+ accumulation, although not significantly in each treatment, at both salinity levels, with Diw pretreatments resulting in more of a decrease in Na^+ compared with Sew and Spw, especially at 200 mM NaCl. The significant effect only appeared in *Sc* in the Na^+ increase ($p < 0.01$) (Figure 6A).

The higher the concentration of 24-epiBL, the greater the increase in K^+, particularly in 200 mM NaCl, with the order of effect being Diw-1.04S2 > Sew-1.04S2 > Spw-1.04S2 > Diw-0.62S2. In 100 mM NaCl, although higher concentrations of 24-epiBL reduced the decrease in K^+ under all 24-epiBL treatments, there was no significant difference compared with SS-S1. The effects of *Sc* (p < 0.01), *Am* (*p* = 0.03), *BRc* (*p* < 0.01), and *Sc*BRc* (*p* < 0.01) were significant in K^+ reduction (Figure 6B).

The effect of 24-epiBL on regulating the decrease of Ca^{2+} was better in SS-S1 than in SS-S2. Under 100 mM NaCl stress, the three 24-epiBL application methods inhibited the reduction in Ca^{2+}. The significantly effective inhibition was in the order compared to SS-S1: Diw-1.04S1 > Sew-1.04S1 > Diw-0.62S1. In the 200 mM NaCl treatment, most 24-epiBL treatment groups showed a reduction in the decrease of Ca^{2+} except Spw-0.21S2 and Spw-0.62S2, with the obvious effects compared to SS-S2 following the order: Diw-1.04S2 > Diw-0.62S2 > Sew-1.04S2 > Sew-0.62S2 > Diw-0.21S2 > Spw-1.04S2 (Figure 6C). The effects of *Sc*, *Am*, *BRc*, *Sc*Am*, and *Sc*BRc* were significant ($p < 0.01$ & $p = 0.01$) in regulating Ca^{2+} reduction.

Indexes	Sc	Am	BRc	Sc*Am	Sc*BRc	Am*BRc	Sc*Am*BRc
Increased Na⁺	<0.01	0.11	0.28	0.61	0.95	0.99	0.99
Decreased K⁺	<0.01	0.03	<0.01	0.36	<0.01	0.87	0.87
Decreased Ca²⁺	<0.01	<0.01	<0.01	<0.01	0.01	0.56	0.74

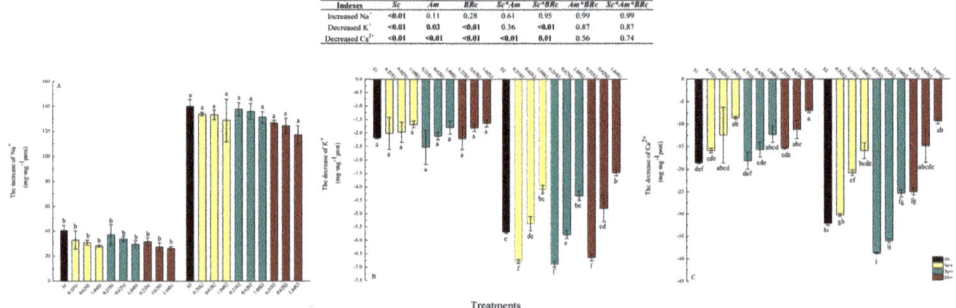

Figure 6. Effects of exogenous 24-epiBL on (**A**) the increase of Na^+, (**B**) the decrease of K^+, and (**C**) decrease of Ca^{2+} in chloroplasts of leaves of black locust seedlings at 100 (SS-S1) or 200 (SS-S2) mM NaCl. Other details as in Figure 1.

3.7. Ultrastructure of Chloroplasts and Thylakoids

3.7. Ultrastructure of Chloroplasts and Thylakoids

Based on the physiological results detailed earlier, the optimal concentration of 24-epiBL for application was 1.04 μM. Therefore, ultrastructure analyses were only performed on plants grown using Sew, Spw, and Diw with 1.04 μM 24-epiBL. Alterations in the structure of chloroplasts and thylakoids induced by salt stress are shown in Figures 7 and 8 relative to controls. The chloroplasts of salt-stressed plants were swollen and partly separated from the plasma membrane, with a more irregular lamellar structure to the thylakoids, particularly at 200 mM NaCl. Chloroplasts changed from being elliptical in shape to almost round, particularly at 200 mM NaCl. Plastoglobuli were swollen, but fewer in number compared with CK plants. Under higher salt stress (200mM NaCl), the cell wall was bent, and starch granules were smaller and less well defined (Figures 7 and 8A–C).

These severe impacts on chloroplast and thylakoid structure were partly alleviated by 24-epiBL pretreatments. Application of 24-epiBL increased the number of plastoglobuli relative to stressed controls (Figure 7D–I). At 100 mM NaCl, all 24-epiBL pretreatments decreased the swelling of chloroplasts and resulted in more plastoglobuli, particularly in the Diw groups (compare panels D–F with B in Figure 7). The swelling of chloroplasts at 200 mM NaCl was significantly modified by Diw pretreatments

(Figure 7G–I). Under 100 mM NaCl stress, the lamellae were more similar to CK plants in the Sew and Diw groups than in the Spw group. However, at higher salinity (200 mM NaCl stress), the loose lamellae were not modified significantly by 24-epiBL and Diw pretreatment resulted in less lamellar modification compared with Sew and Spw (compare panels D–F and G–I with B,C in Figure 8).

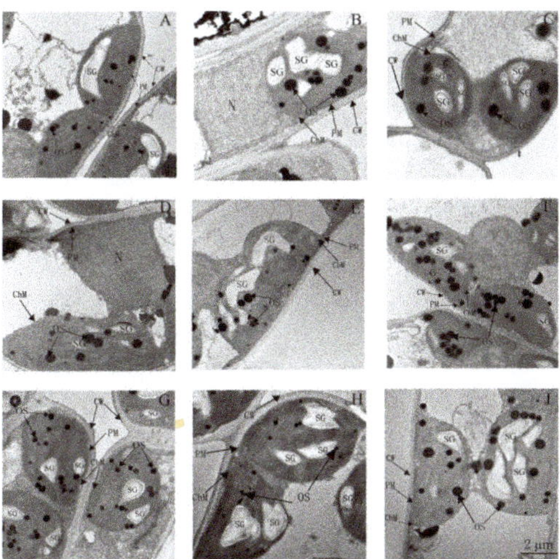

Figure 7. Ultrastructure of chloroplasts of leaves of black locust seedlings at three levels of salt stress (0, 100, or 200 mM NaCl), or with salt stress imposed after pretreatment with 1.04 µM 24-epiBL by either seed soaking, foliar spraying, or roots dipping (see Table 1 for treatment designation). (**A**) CK; (**B**) SS-S1; (**C**) SS-S2; (**D**) Sew-1.04S1; (**E**) Spw-1.04S1; (**F**) Diw-1.04S1; (**G**) Sew-1.04S2; (**H**) Spw-1.04S2; (**I**) Diw-1.04S2. CW: cell wall, PM: plasma membrane, ChM: chloroplast membrane, SG: starch granule, OS: osmiophilic plastoglobuli. Scale bars for chloroplasts are 2µm. Other details as in Figure 1.

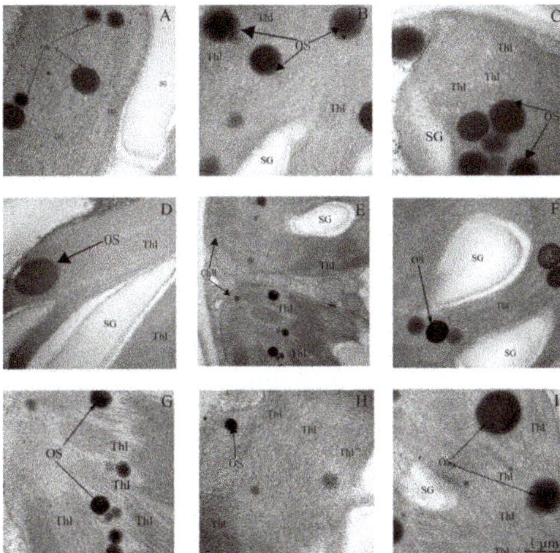

Figure 8. Ultrastructure of thylakoid membranes of leaves of black locust seedlings at three levels of salt stress (0, 100, or 200 mM NaCl), or with salt stress imposed after pretreatment with 1.04 µM 24-epiBL by either seed soaking, foliar spraying, or roots dipping (see Table 1 for treatment designation). (**A**) CK; (**B**) SS-S1; (**C**) SS-S2; (**D**) Sew-1.04S1; (**E**) Spw-1.04S1; (**F**) Diw-1.04S1; (**G**) Sew-1.04S2; (**H**) Spw-1.04S2; (**I**) Diw-1.04S2. Thl: thylakoid lamella, SG: starch granule, OS: osmiophilic plastoglobuli. Scale bars for thylakoids are 1µm. Other details are the same as Figure 1.

4. Discussion

Many studies have reported the benefits of 24-epiBL applications on plants, although studies have focused on herbaceous crops rather than woody plants and have used different doses and application methods. For example, foliar spraying of melon cultivars (*Cucumis melo* L.) with 2.1 μM 24-epiBL enabled the plants to adapt to higher temperatures [31], soaking radish seeds (*Raphanus sativus* L. cv. 'Japanese White') with 0.5–2 μM 24-epiBL enabled seedlings to adapt to zinc stress [32], and dipping the roots of wheat seedlings in 0.052–0.156 μM 24-epiBL improved their salt tolerance [33]. Thus, in the current study, we investigated the effects of 0.21–1.04 μM 24-epiBL pretreatments (seed soaking, foliar spraying, and root dipping) on the adaptation to different levels of salinity by *R. pseudoacacia* seedlings. Given that the number of potential treatment combinations was unwieldy, we focused on a key subset (Table 1) to examine the effects of 24-epiBL and to select the optimal pretreatment combination to inform the use of such treatments in the field.

4.1. Photosynthesis

NaCl inhibited photosynthesis (evidenced by a decrease in P_n, G_s, C_i/C_a, and WUE_i compared with CK) combined with a reduction in RWC and pigment content (Chl a, Chl b, Chl). There was no significant difference between the impact of the two NaCl levels on the RWC and pigment content. G_s was less severely impacted than P_n by 100 mM NaCl, leading to a large decrease in WUE_i; the reduction on G_s of 200 mM NaCl was much lower than that of 100 mM NaCl, P_n was similar at both salinity levels, whereas WUE_i was obvious higher in 200 mM NaCl compared with SS-0S1.This suggests that mesophyll conductance (g_m) at both levels of NaCl stress and stomatal conductance under 200 mM NaCl negatively impacted photosynthesis. This result is similar to other studies that showed that G_m was more seriously affected than G_s in poplars (*Populus cathayana* Rehder) under alkaline (75mM Na_2CO_3) stress compared with control plants [34]. The decrease in F_v/F_m observed in salt-stressed plants indicates a progressive reduction in the maximum quantum yield of photosynthesis [35]. The increases in NPQ and H_d were consistent with reduced rates of carbon fixation and greater energy dissipation as heat [36]. The above results were similar to the significant inhibition of P_n, G_s, and WUE_i caused by 200 mM NaCl in *Brassica napus* leaves accompanied with the reduction in Chl a, Chl b, and total Chl after three days of salt stress [37]. 24-epiBL pretreatments improved the photosynthetic rate by regulating the limitation on G_m and G_s given the significant reduction in C_i/C_a and increase in WUE_i in response to 100 mM NaCl and the change in G_s in response to 200 mM NaCl. The most positive effects on photosynthesis were seen in the Diw-1.04S1 and Diw-1.04S2 groups, also indicated by the reduction in H_d and changes in F_v/F_m and NPQ. 24-epiBL application also modified the changes in pigment contents (21.1%–51.1% increase in Cha, 12.5%–91.6% in Chb, and 15.5%–62.1% in Chl compared with SS-S1; and 23%–58.6% in Cha, 4.8%–25.8% in Chb, and 16.9%–49.4% in Chl compared with SS-S2). In grape seedlings (*V. vinifera* L.) with 10% PEG stress, the application of 0.105–0.42 μM 24-epiBL significantly increased the chlorophyll content (Chl a, Chl b, and Chl), maximal fluorescence, F_v/F_m, and NPQ, to improve drought tolerance [38]. Thussagunpanit and co-workers sprayed the foliage of rice (*Oryza sativa* L.) with 0.1 μM 24-epiBL and reported that 24-epiBL improved P_n, G_s, F_v/F_m, and Chl a, Chl b, and Chl contents with a reduction in C_i to protect the plants from heat stress (47 °C) [39].

4.2. Antioxidative Effect in Leaves

NaCl increased the MDA content (1.64~2.04-fold compared with CK) and the activity of POD and SOD (1.05~1.92-fold and 1.23~1.36-fold compared with CK, respectively), although POD activity was not significantly different under 200 mM NaCl stress, whereas that of SOD and MDA was significant at both levels of NaCl stress compared with controls. These results suggest that NaCl was harmful to the leaves and induced oxidative stress at both levels of NaCl. A previous study reported that 150 mM NaCl increased the MDA content and SOD activity in European larch needles (*Larix decidua* Mill.), also suggesting that salinity induces oxidative stress that negatively impacts on plant growth (i.e., reduced the increases in stem length, fresh weight, and leaf water content compared with controls) [40]. Application of 24-epiBL modified the MDA content (reduced by 6.3%–27.6% in SS-S1 and 5.6%–39.8% in SS-S2), antioxidant enzyme activity (increased the activities of SOD and POD by 0.8%–12% and 39.4%–90.4%, respectively, compared with SS-S1 and by 1.4%–41% and 43.1%–116.8% compared with SS-S2, respectively). The weaker changes recorded at 100 mM NaCl and stronger changes at 200 mM NaCl suggested the moderate tolerance of *R. pseudoacacia* seedlings to 100 mM NaCl stress and that 24-epiBL could increase the tolerance of the plant to both levels of salt stress, although this was more effective under the 200 mM NaCl stress. The optimal effects were seen with Diw pretreatments and 1.04 μmol L^{-1} 24-epiBL. These results indicate that 24-epiBL is able to modify the oxidative stress induced by abiotic stressors. The antioxidant effects of BRs were also reported by Xia et al., who showed that overexpression of the BR biosynthetic gene, *Dwarf*, enhanced the ratio of reduced/oxidized 2-cysteine peroxiredoxin (2-Cys Prx) and activated antioxidant enzymes, including APX, MDAR, DHAR, and GR, in correlation with a reduction in H_2O_2 in tomatos (*Solanum lycopersicum* L.) under chilling stress [41]. Spraying 1μM 24-epiBL onto a single-cross maize hybrid PMH 3 (LM 17 $\times$ LM 14) improved its adaptation to heat stress by arresting protein degradation and improving the cell membrane stability, accompanied by modulation of the biochemical activities of antioxidant enzymes (CAT, SOD, and POD), highlighting BRs as anti-stress agents [42]. The presence of endogenous 28-norbrassinolide in a resistant (~4.8 pg mg^{-1} fresh mass) genotype (CE704) of maize (*Zea mays* L.) resulted in higher chlorophyll and proline contents and lower MDA content, and reduced membrane injury compared with a sensitive maize genotype (2023; ~1.5 pg mg^{-1} fresh mass) under drought stress [43], also confirming the ability of BRs to regulate antioxidative effects under different stressors, including salinity.

4.3. Ion Toxicity and Chloroplast Ultrastructure

NaCl induces an ion imbalance in leaves, with significant Na$^+$ accumulation in plant organs in most cases accompanying decreases in K$^+$ and Ca^{2+} and damage to chloroplasts. The improved salt tolerance of barley (*H. vulgare*) seedlings compared with wheat (*T. turgidum* L. ssp. *durum* Desf.) correlated with a higher photosynthetic capacity and higher K$^+$ and lower Na$^+$ in the cytoplasm of mesophyll cells [44]. In the current study, salinity caused 31.7%–54.1% and 9.2%–15.4% reductions in MSI and RWC, respectively, and a 40–140 mg· mg^{-1}prot increase in Na$^+$ accompanied by a significant decrease in K$^+$ (2~5.5 mg·mg^{-1}prot) and Ca^{2+} (18-33 mg·mg^{-1}prot) in chloroplasts regardless of NaCl concentration (Figure 6). These results were similar to the increased Na$^+$ content reported in chloroplasts of poplar leaves (*P. euphratica* and *P. popularis*) accompanied by reduced P_n under salt stress [45]. These results support the occurrence of ion toxicity in leaves and chloroplasts in response to saline stress. Na$^+$ accumulation also induced changes to the chloroplasts, such as swollen chloroplasts partly separated from the plasma membrane and the more irregular lamellar structure of thylakoids (Figs 7 and 8). This was similar to the swollen chloroplasts and the looser thylakoid lamellar structures reported in diploid (2$\times$) *R. pseudoacacia* L., as well as increases in Na$^+$ and decreases in K$^+$/Na$^+$ in response to 500 mM NaCl stress for 15 days [14].

Three 24-epiBL pretreatment types reversed the changes to the chloroplasts (i.e., decreased swelling, maintained normal size of starch granules, increased the number of plastoglobuli, and kept a normal lamellar structure) and resulted in higher P_n, effects that were similar to putrescine (Put) in regulating the salt tolerance of *C. sativus* by maintaining regular thylakoid membrane structures with normal P_n and the photochemical efficiency of PSII, and reducing the ion toxicity (Na$^+$ contents in chloroplasts) under 75 mM NaCl [46]. These results were also similar to the ability of silicon (Si) to maintain a normal chloroplast structure and normal thylakoid lamellae, accompanied by higher Chl a and Chl b contents in leaves and lower electrolyte leakage in response to drought in tomatos (*S. lycopersicum*) [47]. A study also reported the ability of exogenous K_2SiO_3 to maintain chloroplast lamellae and reduce foliar Na$^+$ levels to increase P_n and G_s of *L. japonica* under 100 and 200 mM NaCl stress [12]. In the current study, application of 24-epiBL inhibited the increase in Na^+ and the reduction of K^+ and Ca^{2+} particularly in response to 200 mM NaCl. Ca^{2+} levels showed obvious changes in response to the different 24-epiBL treatments, whereas K^+ only showed significant differences in response to 200 mM NaCl, and the Diw pretreatment group showed more significant responses overall than either the Sew or Spw groups. The most effective pretreatments were Diw-1.04 S1 (64.2% Na^+ increase, 75.2% K^+ decrease, and 37.8% Ca^{2+} decrease compared with SS-S1) and Diw-1.04 S2 (87.2% Na^+ increase, 60.9% K^+ decrease, and 39.7% Ca^{2+} decrease compared with SS-S2). Although the concentration of 24-epiBL in our pretreatment groups differed, the effects were similar to a study in which foliage was sprayed with 0.01–0.1 μM 24-epiBL to improve the salt tolerance (to 250 mM NaCl) of perennial ryegrass (*Lolium perenne* L.), resulting in higher concentrations of leaf proline, K^+, Mg^{2+}, Ca^{2+}, and chlorophyll, increased P_n, and a weak reduction in Na^+ [48]. In addition, soaking seeds in 0.0001~1μM 24-epiBL improved the tolerance of *R. sativus* L. seedlings to mercury (Hg) by increasing Na^+ and K^+ and activating the antioxidative ability of the plant (i.e., reduced MDA and increased activities of POD, SOD, glutathione reductase (GR), and dehydroascorbate reductase (DHAR)) [49]. Thus, we can conclude that 24-epiBL has the ability to improve the tolerance of plants to abiotic stress by maintaining the ion balance and overcoming any ion toxicity.

5. Conclusions

Our results showed that salt stress greatly induced photoinhibition in *R. pseudoacacia* seedlings as a result of oxidative stress and ion toxicity. Active oxygen can be generated from the interaction of O_2 and an obstructed electron transport chain [50], damaging various photosynthetic components [51]. Ion toxicity also harms leaves and chloroplasts when combined with oxidative effects caused by stressors [45,52]. Exogenous 24-epiBL pretreatments alleviated this photoinhibition and stabilized the chloroplast structure, by reducing ion toxicity and increasing the activities of antioxidant enzymes. Application of 24-epiBL by either Sew, Spw, or Diw was effective, although the Diw approach resulted in the optimal overall performance, with the effects of the 24-epiBL concentration being in the order 1.04 > 0.62 > 0.21μM.

Author Contributions: J.Y. led to the submission, acquired data, wrote the draft and revised the manuscript; Z.F. did the part of experiment and got some data used in this paper, L.Z. did the part of experiment and got some data used in this paper, Z.Z. did the modification of manuscript and J.Z. conceived and designed the work and applied all the funding.

Funding: Financial support for this study was provided by the Jiangsu Agriculture Science and Technology Innovation Fund (Grant No. CX(17)1004), National Special Fund for Forestry Scientific Research in the Public Interest (Grant No. 201504406), Major Fund for Natural Science of Jiangsu Higher Education Institutions (Grant No. 15KJA220004), National Foundation of Forestry Science and Technology Popularization (Grant No. [2015]17), Priority Academic Program Development of Jiangsu Higher Education Institutions (PAPD), and the Doctorate Fellowship Foundation of Nanjing Forestry University.

Acknowledgments: Financial support for this study was provided by the Jiangsu Agriculture Science and Technology Innovation Fund (Grant No. CX(17)1004), National Special Fund for Forestry Scientific Research in the Public Interest (Grant No. 201504406), Major Fund for Natural Science of Jiangsu Higher Education Institutions (Grant No. 15KJA220004), National Foundation of Forestry Science and Technology Popularization (Grant No. [2015]17), Priority Academic Program Development of Jiangsu Higher Education Institutions (PAPD), and the Doctorate Fellowship Foundation of Nanjing Forestry University. We thank International Science Editing (http://www.internationalscienceediting.com) for editing this manuscript.

References

1. Shahid, S.A. Developments in spoil salinity assessment, modeling, mapping, and monitoring from regional to submicroscopic scales. In *Developments in Soil Salinity Assessment and Reclamation*; Shahid, S.A., Abelfattah, M.A., Taha, F.K., Eds.; Springer: Dordrecht, The Netherland, 2013; pp. 3–44. [CrossRef]
2. Wang, W.; Vinocur, B.; Altman, A. Plant responses to drought, salinity and extreme temperatures: Towards genetic engineering for stress tolerance. *Planta* **2003**, *218*, 1–14. [CrossRef]

3. Li, J.; Pu, L.; Han, M.; Zhu, M.; Zhang, R.; Xiang, Y. Soil salinization research in China: Advances and prospects. *J. Geogra. Sci.* **2014**, *24*, 943–960. [CrossRef]

4. Chen, X.; Shen, Q.; Xu, Y. Hydraulic properties of typical salt-affected soils in Jiangsu Province, China. *Front. Environ. Sci. Eng. China* **2007**, *1*, 443–447. [CrossRef]

5. DeGomez, T.; Wagner, M.R. Culture and use of black locust. *HortTechnology* **2001**, *11*, 279–288. [CrossRef]

6. Liu, W.N.; Jiang, H.Y.; Yang, M.S. Selection of salt—tolerance on *Robinia pseudoacacia* seedings and impact of salt stress on seedlings' physiological characteristics. *J. Agric. Univ. Hebei* **2010**, *3*, 62–66.

7. Sitzia, T.; Campagnaro, T.; Dainese, M.; Cierjacks, A. Forest Ecology and Management Plant species diversity in alien black locust stands: A paired comparison with native stands across a north-Mediterranean range expansion. *For. Ecol. Manag.* **2012**, *285*, 85–91. [CrossRef]

8. Tateno, R.; Tokuchi, N.; Yamanaka, N.; Du, S.; Otsuki, K.; Shimamura, T.; Xue, Z.; Wang, S.; Hou, Q. Comparison of litterfall production and leaf litter decomposition between an exotic black locust plantation and an indigenous oak forest near Yan'an on the Loess Plateau, China. *For. Ecol. Manag.* **2007**, *241*, 84–90. [CrossRef]

9. Akatov, V.V.; Akatova, T.V.; Shadzhe, A.E. Robinia pseudoacacia L. in the Western Caucasus. *Rus. J. Biol. Invasion* **2016**, *7*, 105–118. [CrossRef]

10. Holle, B.V.; Neill, C.; Largay, E.F.; Budreski, K.A.; Ozimec, B.; Clark, S.A.; Lee, K. Ecosystem legacy of the introduced N2-fixing tree *Robinia pseudoacacia* in a coastal forest. *Oecologia* **2013**, *172*, 915–924. [CrossRef]

11. Mao, P.; Zhang, Y.; Cao, B.; Guo, L.; Shao, H.; Cao, Z.; Jiang, Q.; Wang, X. Effects of salt stress on eco-physiological characteristics in *Robinia pseudoacacia* based on salt-soil rhizosphere. *Sci. Total Environ.* **2016**, *568*, 118–123. [CrossRef]

12. Zhao, G.; Li, S.; Sun, X.; Wang, Y.; Chang, Z. The role of silicon in physiology of the medicinal plant (*Lonicera japonica* L.) under salt stress. *Sci. Rep.* **2015**, *5*, 12696. [CrossRef]

13. Manivannan, A.; Soundararajan, P.; Muneer, S.; Ko, C.H.; Jeong, B.R. Silicon mitigates salinity stress by regulating the physiology, antioxidant enzyme activities, and protein expression in Capsicum annuum "Bugwang". *BioMed Res. Int.* **2016**, *2016*. [CrossRef]

14. Wang, Z.; Wang, M.; Liu, L.; Meng, F. Physiological and proteomic responses of diploid and tetraploid black locust (*Robinia pseudoacacia* L.) subjected to salt stress. *Int. J. Mol. Sci.* **2013**, *14*, 20299–20325. [CrossRef]

15. Xi, Z.; Wang, Z.; Fang, Y.; Hu, Z.; Hu, Y.; Deng, M.; Zhang, Z. Effects of 24-epibrassinolide on antioxidation defense and osmoregulation systems of young grapevines (*V. vinifera* L.) under chilling stress. *Plant Growth Regul.* **2013**, *71*, 57–65. [CrossRef]

16. An, Y.; Zhou, H.; Zhong, M.; Sun, J.; Shu, S. Root proteomics reveals cucumber 24-epibrassinolide responses under Ca (NO$_3$)$_2$ stress. *Plant Cell Rep.* **2016**, *35*, 1081–1101. [CrossRef]

17. Cui, L.; Zou, Z.; Zhang, J.; Zhao, Y.; Yan, F. 24-Epibrassinoslide enhances plant tolerance to stress from low temperatures and poor light intensities in tomato (*Lycopersicon esculentum* Mill.). *Funct. Integr. Genomic.* **2016**, *16*, 29–35. [CrossRef]

18. Talaat, N.B.; Shawky, B.T. 24-Epibrassinolide ameliorates the saline stress and improves the productivity of wheat (*Triticum aestivum* L.). *Environ. Exp. Bot.* **2012**, *82*, 80–88. [CrossRef]

19. Azhar, N.; Su, N.; Shabala, L.; Shabala, S. Exogenously applied 24-epibrassinolide (EBL) ameliorates detrimental effects of salinity by reducing K$^+$ efflux via depolarization-activated K$^+$ channels. *Plant Cell Physiol.* **2017**, *58*, 802–810. [CrossRef]

20. Hoagland, D.R.; Arnon, D.I. The water-culture method for growing plants without soil. *Circ. Calif. Agr. Ex. Stn.* **1950**, *347*, 1–32.

21. Yue, J.M.; You, Y.H.; Zhang, L.; Fu, Z.Y.; Wang, J.P.; Zhang, J.C.; Guy, R.D. Exogenous 24-epibrassinolide alleviates effects of salt stress on chloroplasts and photosynthesis in *Robinia pseudoacacia* L. seedlings. *J. Plant Growth Reg.* **2018**, 1–14. [CrossRef]

22. Sairam, R.K. Effects of homobrassinolide application on plant metabolism and grain yield under irrigated and moisture-stress conditions of two wheat varieties. *Plant Growth Regul.* **1994**, *306*, 173–181. [CrossRef]

23. Hayat, S.; Ali, B.; Aiman Hasan, S.; Ahmad, A. Brassinosteroid enhanced the level of antioxidants under cadmium stress in Brassica juncea. *Environ. Exp. Bot.* **2007**, *60*, 33–41. [CrossRef]

24. He, M.; Shi, D.; Wei, X.; Hu, Y.; Wang, T.; Xie, Y. Gender-related differences in adaptability to drought stress in the dioecious tree Ginkgo biloba. *Acta Physiol. Plant.* **2016**, *38*. [CrossRef]

25. Fridovich, I. Superoxide Dismutases. *Annu. Rev. Biochem.* **1975**, *44*, 147–159. [CrossRef]

26. Hammerschmidt, R.; Nuckles, E.M.; Kuć, J. Association of enhanced peroxidase activity with induced systemic resistance of cucumber to Colletotrichum lagenarium. *Physiol. Plant Pathol.* **1982**, *20*, 73–82. [CrossRef]

27. Heath, R.L.; Packer, L. Photoperoxidation in isolated chloroplasts. *Arch. Biochem. Biophys.* **1968**, *125*, 189–198. [CrossRef]

28. Cerović, Z.G.; Plesnicar, M. An improved procedure for the isolation of intact chloroplasts of high photosynthetic capacity. *Biochem. J.* **1984**, *223*, 543–545. [CrossRef]

29. Song, X.S.; Tiao, C.L.; Shi, K.; Mao, W.H.; Ogweno, J.O.; Zhou, Y.H.; Yu, J.Q. The response of antioxidant enzymes in cellular organelles in cucumber (*Cucumis sativus* L.) leaves to methyl viologen-induced photo-oxidative stress. *Plant Growth Regul.* **2006**, *49*, 85–93. [CrossRef]

30. Abo-Ogiala, A.; Carsjens, C.; Diekmann, H.; Fayyaz, P.; Herrfurth, C.; Feussner, I.; Polle, A. Temperature-induced lipocalin (TIL) is translocated under salt stress and protects chloroplasts from ion toxicity. *J. Plant Physiol.* **2014**, *171*, 250–259. [CrossRef]

31. Zhang, Y.P.; Zhu, X.H.; Ding, H.D.; Yang, S.J.; Chen, Y.Y. Foliar application of 24-epibrassinolide alleviates high-temperature-induced inhibition of photosynthesis in seedlings of two melon cultivars. *Photosynthetica* **2013**, *51*, 341–349. [CrossRef]

32. Ramakrishna, B.; Rao, S.S.R. 24-Epibrassinolide alleviated zinc-induced oxidative stress in radish (*Raphanus sativus* L.) seedlings by enhancing antioxidative system. *Plant Growth Regul.* **2012**, *68*, 249–259. [CrossRef]

33. Ali, Q.; Athar, H.U.R.; Ashraf, M. Modulation of growth, photosynthetic capacity and water relations in salt stressed wheat plants by exogenously applied 24-epibrassinolide. *Plant Growth Regul.* **2008**, *56*, 107–116. [CrossRef]

34. Xu, G.; Huang, T.F.; Zhang, X.L.; Duan, B.L. Significance of mesophyll conductance for photosynthetic capacity and water-use efficiency in response to alkaline stress in Populus cathayana seedlings. *Photosynthetica* **2013**, *51*, 438–444. [CrossRef]

35. Kalaji, H.M.; Govindjee; Bosa, K.; Koscielniak, J.; Zuk-Golaszewska, K. Effects of salt stress on photosystem II efficiency and CO_2 assimilation of two Syrian barley landraces. *Environ. Exp. Bot.* **2011**, *73*, 64–72. [CrossRef]

36. Oxborough, K. Imaging of chlorophyll a fluorescence: Theoretical and practical aspects of an emerging technique for the monitoring of photosynthetic performance. *J. Exp. Bot.* **2004**, *55*, 1195–1205. [CrossRef] [PubMed]

37. Jia, H.; Shao, M.; He, Y.; Guan, R.; Chu, P.; Jiang, H. Proteome dynamics and physiological responses to short-term salt stress in brassica napus leaves. *PLoS ONE* **2015**, *10*. [CrossRef]

38. Wang, Z.; Zheng, P.; Meng, J.; Xi, Z. Effect of exogenous 24-epibrassinolide on chlorophyll fluorescence, leaf surface morphology and cellular ultrastructure of grape seedlings (*Vitis vinifera* L.) under water stress. *Acta Physiol. Plant.* **2015**, *37*, 1–12. [CrossRef]

39. Thussagunpanit, J.; Jutamanee, K.; Kaveeta, L.; Chai-arree, W.; Pankean, P.; Homvisasevongsa, S.; Suksamrarn, A. Comparative effects of brassinosteroid and brassinosteroid mimic on improving photosynthesis, lipid peroxidation, and rice seed set under heat stress. *J. Plant Growth Regul.* **2015**, *34*, 320–331. [CrossRef]

40. Plesa, I.; González-Orenga, S.; Al Hassan, M.; Sestras, A.; Vicente, O.; Prohens, J.; Sestras, R.; Boscaiu, M. Effects of Drought and Salinity on European Larch (*Larix decidua* Mill.) Seedlings. *Forests* **2018**, *9*, 320. [CrossRef]

41. Xia, X.J.; Fang, P.P.; Guo, X.; Qian, X.J.; Zhou, J.; Shi, K.; Zhou, Y.H.; Yu, J.Q. Brassinosteroid-mediated apoplastic H_2O_2-glutaredoxin 12/14 cascade regulates antioxidant capacity in response to chilling in tomato. *Plant Cell Environ.* **2018**, *41*, 1052–1064. [CrossRef]

42. Yadava, P.; Kaushal, J.; Gautam, A.; Parmar, H.; Singh, I. Physiological and biochemical effects of 24-epibrassinolide on heat-stress adaptation in Maize (*Zea mays* L.). *Nat. Sci.* **2016**, *8*, 171–179. [CrossRef]

43. Tůmová, L.; Tarkowská, D.; Řřová, K.; Marková, H.; Kočová, M.; Rothová, O.; čečetka, P.; Holá, D. Drought-tolerant and drought-sensitive genotypes of maize (*Zea mays* L.) differ in contents of endogenous brassinosteroids and their drought-induced changes. *PLoS ONE* **2018**, *13*, 1–22. [CrossRef] [PubMed]

44. James, R.A.; Munns, R.; Von Caemmerer, S.; Trejo, C.; Miller, C.; Condon, T. Photosynthetic capacity is related to the cellular and subcellular partitioning of Na^+, K^+ and Cl^- in salt-affected barley and durum wheat. *Plant Cell Environ.* **2006**, *29*, 2185–2197. [CrossRef] [PubMed]

45. Wang, R.; Chen, S.; Deng, L.; Fritz, E.; Hüttermann, A.; Polle, A. Leaf photosynthesis, fluorescence response to salinity and the relevance to chloroplast salt compartmentation and anti-oxidative stress in two poplars. *Trees Struct. Funct.* **2007**, *21*, 581–591. [CrossRef]

46. Shu, S.; Yuan, Y.; Chen, J.; Sun, J.; Zhang, W.; Tang, Y.; Zhong, M.; Guo, S. The role of putrescine in the regulation of proteins and fatty acids of thylakoid membranes under salt stress. *Sci. Rep.* **2015**, *5*, 1–16. [CrossRef]

47. Cao, B.L.; Ma, Q.; Zhao, Q.; Wang, L.; Xu, K. Effects of silicon on absorbed light allocation, antioxidant enzymes and ultrastructure of chloroplasts in tomato leaves under simulated drought stress. *Sci. Hortic.* **2015**, *194*, 53–62. [CrossRef]

48. Wu, W.; Zhang, Q.; Ervin, E.H.; Yang, Z.; Zhang, X. Physiological mechanism of enhancing salt stress tolerance of Perennial ryegrass by 24-epibrassinolide. *Fron. Plant Sci.* **2017**, *8*, 1–11. [CrossRef]

49. Kapoor, D.; Rattan, A.; Gautam, V.; Bhardwaj, R. Mercury-induced changes in growth, metal & ions uptake, photosynthetic pigments, osmoprotectants and antioxidant defence system in *Raphanus sativus* L. seedlings and role of steroid hormone in stress amelioration. *J. Pharmacogn. Phytochem.* **2016**, *5*, 259–265.

50. Fisher, A.B. Redox signaling across cell membranes. *Antioxid. Redox Signal.* **2009**, *11*, 1349–1356. [CrossRef]

51. Gill, S.S.; Tuteja, N. Reactive oxygen species and antioxidant machinery in abiotic stress tolerance in crop plants. *Plant Physiol. Biochem.* **2010**, *48*, 909–930. [CrossRef]

52. Wali, M.; Gunsè, B.; Llugany, M.; Corrales, I.; Abdelly, C.; Poschenrieder, C.; Ghnaya, T. High salinity helps the halophyte Sesuvium portulacastrum in defense against Cd toxicity by maintaining redox balance and photosynthesis. *Planta* **2016**, *244*, 333–346. [CrossRef] [PubMed]

Article

Ecophysiological Responses of *Carpinus turczaninowii* L. to Various Salinity Treatments

Qi Zhou [1,2], Man Shi [1,2], Zunling Zhu [1,2,3,]* and Longxia Cheng [1,2]

1   Co-Innovation Center for Sustainable Forestry in Southern China, Nanjing Forestry University, Nanjing 210037, China; zhouqi514@njfu.edu.cn (Q.Z.); shiman1031@126.com (M.S.); zwahzchenglx@163.com (L.C.)
2   College of Landscape Architecture, Nanjing Forestry University, Nanjing 210037, China
3   College of Arts & Design, Nanjing Forestry University, Nanjing 210037, China
*   Correspondence: zhuzunling@njfu.edu.cn; Tel.: +86-025-6963-8089

Received: 10 December 2018; Accepted: 22 January 2019; Published: 25 January 2019

Abstract: *Carpinus turczaninowii* L., commonly known as hornbeam, has significant economic and ornamental importance and is largely distributed in the northern hemisphere, including parts of China and Korea, with high adaptation to harsh conditions in very unfertile soils. In this study, the ecophysiological responses of *C. turczaninowii* seedlings to various salinity stress treatments (NaCl: 0, 17, 34, 51, 68, and 85 mM) were studied for 42 days by determining stress-induced changes in growth parameters and biochemical markers. Salinity stress affected the values of all the examined parameters, both morphological and physiological, and caused the inhibition of plant growth, the degradation of photosynthetic capacity and stomatal behavior, a decrease in the photosynthetic pigments contents and relative water content, an increase in the Malondialdehyde (MDA) content and relative electrolytic conductivity, and the accumulation of Na^+ and Cl^- content. The presence of relatively high concentrations of organic osmolytes, the activation of antioxidant enzymes, and the ionic transport capacity from the root to shoots may represent a constitutive mechanism of defence against stress in *C. turczaninowii* seedlings. Our results suggest that *C. turczaninowii* can tolerate salinity at low and moderate concentrations (17–51 mM) under nursery conditions and can be widely used in roadsides, gardens, parks, and other urban areas.

Keywords: *Carpinus turczaninowii*; salinity treatments; ecophysiology; photosynthetic responses; organic osmolytes; ion homeostasis; antioxidant enzymes

---

1. Introduction

During growth, plants are often exposed to a variety of abiotic stresses that will affect their growth and production [1]. Among these stresses, high soil salinity is a widespread problem, and soil salinization has been one of the major reasons for the decrease of crop yield in arid and semi-arid areas [2]. Salinity affects approximately one-third of the irrigated land, restricting the vegetative and reproductive growth of plants by causing severe physiological disorders and direct or indirect damaging effects [3]. In many temperate regions, de-icing salts are commonly used to keep the roads dry and safe in winter, involving large amounts of NaCl. Every year, the releases of de-icing salt are 400–1400 t in France, 2000 t in Germany, 2.2 million t in England, 4–5 million t in Canada, and 13.6–18 million t in the USA, and the use of de-icing salt is increasing in China; this country has recently become the largest producer in the world [4]. Salt can increase soil salinity via runoff, splashing, and aerosols and can modify roadside soil properties and plant communities [5]. Moreover, the airborne deposition and distribution of de-icing salts extend the scope of impact areas [6]. Plants near the road are therefore suppressed due to salinity stress [7]. Their symptoms depend on whether the damage is caused by the salinity of the soil or by deposition directly to the aboveground parts of the plants [8].

Many studies have been conducted on herbaceous halophytes [9,10], sensitive or tolerant crop species [11], and tolerant woody species [12]. However, there have been fewer studies on the physiological responses of woody plants sensitive to salinity stress compared to those on other plants. Salinity levels can be expressed by the electrical conductivity (EC) of the irrigation water or an aqueous extract of the soil. According to Richards (1954) [13], the levels of soils were: 0–2 dS m^{-1}, not saline; 2–4 dS m^{-1}, slightly saline; 4–8 dS m^{-1}, saline; 8–16 dS m^{-1}, strongly saline; >16 dS m^{-1}, extremely saline. Different plants may differ in their sensitivity to salt stress, and the ability of plants to adapt to stress depends on the type, intensity and duration of stress as well as the species of plants and stage of stress [14]. Salinity stress will have multiple adverse effects on plants, including negative effects on morphology, photosynthesis, and physiological and biochemical processes [15]. High concentrations of Na$^+$ and Cl$^-$ ions will induce water stress and ion deficiencies, impair membrane function, decrease chlorophyll content and some enzyme activities, and disturb metabolic processes, such as photosynthesis, stomatal behavior, protein synthesis, and carbohydrate and lipid metabolism [16]. It has been confirmed that many salt-stressed plants can participate in the inhibition of stress throughout a plant's life cycle; endurance can also occur at the cellular level, for example, compatible organic solutes can be synthesized, such as prolines, sugars, and proteins, which will contribute to the protection of cell structure against dehydration [17]. The activation of antioxidant systems can help to prevent or reduce oxidative damage to proteins and membranes, which is also a general reaction to abiotic stresses that cause oxidative stress through the generation of reactive oxygen species (ROS) [18]. Antioxidant enzymes include superoxide dismutase (SOD), peroxidase (POD), ascorbate peroxidase (APX), and glutathione reductase (GR) [19]. Most forest tree species growing alongside roads and highways in temperate regions are described as salt sensitive, and the impact of salt on these species needs to be assessed.

Carpinus turczaninowii L. is a member of the Betulaceae family, a well-known deciduous tree mainly distributed in the Northern Hemisphere as a dominant forest species [20]. This plant can reach a height of 15 m and grows in forests on hills and in valleys between 500 and 2000 m above sea level. *C. turczaninowii* has strong adaptability, which can be resistant to drought and barren soil, and can tolerate very low temperatures, even below −20 °C [21]. The fallen leaves of this species can display a beautiful orange-red color in the autumn. They are commonly used for bonsai and landscaping in gardens or along roadsides in China, Korea, and Japan. The wood of *C. turczaninowii* is very hard, dense, and fine textured, and has often been used for making agricultural tools and furniture [22]. Previous chemical investigations have indicated that this plant has anti-inflammatory and antioxidative properties in extracts from its leaves and branches, making it a potential candidate for use in pharmaceuticals and cosmetics [23,24]. Even though *C. turczaninowii* does not grow naturally in saline environments, as a type of ornamental plant, it is now used in many places. It may be affected by relatively high salinity concentrations in locations close to mountain roads because of the use of de-icing salts in winter, a widespread practice worldwide. In addition, *C. turczaninowii* is often planted in parks or gardens for landscaping, where these trees may be irrigated with low-quality water containing different concentrations of salts. However, little research has been conducted on the responses of *C. turczaninowii* to salinity stress, experimental data are very scarce, and its salt tolerance is not fully understood. The aim of this study was to evaluate the resistance to salinity stress in *C. turczaninowii* seedlings.

We hypothesize that *C. turczaninowii* seedlings will be affected by salinity stress and that the plants may develop relevant response mechanisms to stress, such as the biosynthesis of osmolytes and the activation of antioxidant mechanisms. To test this hypothesis, we evaluated the growth and ecophysiological responses of *C. turczaninowii* seedlings to different levels of salinity stress under controlled greenhouse conditions. These results may help to better design, implement conservation and cultivation practices for *C. turczaninowii*. We also can provide useful reference data for the selection of ornamental plant species for landscaping in urban areas.

2. Materials and Methods

2.1. Plant Material

The *C. turczaninowii* seeds were obtained from the Ta-pieh Mountains, Henan, China. The seeds were sown in containers to initiate growth in March 2012 at the landscape experimental teaching centre of Nanjing Forestry University in Nanjing, Jiangsu, China. After one year of growth, in March 2013, we selected well-grown seedlings that were approximately 0.8 cm in diameter with 30.0 cm-tall stems and transplanted them into pots with dimensions of 10 cm in diameter and 15 cm in height and filled with well-mixed loamy soil, peat, vermiculite, and pearlite (1:1:1:1, v/v/v/v) with a pH of 6.5. The content of soil nutrients was determined by a TFW-VI soil nutrient and moisture tester (TFW-VI, Wuhan, China). The soil contained 50.18 mg kg^{-1} available N, 12.15 mg kg^{-1} available P, 145 mg kg^{-1} available K. Each pot contained one seedling and 0.5 kg of soil. The seedlings received 0.5 L of full-strength Hoagland's nutrient solution biweekly before salt stress treatment according to the method of Seth D. Hothem (2003) [25]. The *C. turczaninowii* seedlings were grown in an artificially controlled greenhouse under natural sunlight conditions at a temperature of 25–27 °C and relative humidity of 65–75%.

2.2. Salt Stress Treatments

The trials were conducted when the *C. turczaninowii* seedlings had grown in the greenhouse for one month. In April 2013, we chose uniform seedlings that were approximately 30.0 cm tall and had similar leaf numbers. The seedlings were subjected to a 0 (control), 17, 34, 51, 68, and 85 mM NaCl solution. The electrical conductivity of the substrates was 1.6, 3.3, 4.4, 6.2, 8.4, and 10.5 dS m^{-1}, respectively. These concentrations were selected according to our preliminary experiments on *C. turczaninowii* seedlings. Each treatment had three replicates with 25 random seedlings per replicate, and one seedling per pot. They were grown in the greenhouse with same plant maintenance and management. The control plants were watered with 200 mL of water per plant, whereas for the salt treatment plants, the same volume of 17, 34, 51, 68, or 85 mM NaCl solution was used. To avoid osmotic shock, the NaCl solution was added in three equal parts on alternating days until the target concentration was reached. We also placed a plastic tray at the bottom of the pot to retain any excess solution that had been applied, and this was placed back into the pot to ensure the accuracy of the experimental design. The stress treatments were conducted for 42 days, and after the midpoint (21 days) and the end time (42 days) of growth, measurements were carried out, and samples were collected for different physiological analyses. The whole plants were harvested after 42 days of treatment. Each measured parameter had three replicates.

2.3. Determination of Plant Growth Parameters

The following growth parameters were recorded, with three replicates: stem length increase and diameter increase (measured on the first day and the last day of the salt stress treatments, the location on the seedlings of the stem diameter measurements was 2.0 cm above the soil surface); seedling survival at the end of the trial; biomass of different tissues (dry weight, DW; the plants were separated into roots, stems and leaves, which were dried at 80 °C in an oven for 48 h); taproot length (roots were washed with distilled water, blotted dry on filter paper, measured the taproot length with a ruler); leaf area, measured with a portable leaf area meter (LI-3000C, LI-COR, Lincoln, NE, USA); and root/shoot ratio, which was calculated as the dry mass of the aboveground parts/belowground parts.

2.4. Measurement of Photosynthetic Parameters and Chlorophyll Fluorescence

Photosynthetic parameters were evaluated using a portable photosynthetic system (Ciras-2, Shanghai, China). The net photosynthetic rate (*Pn*), transpiration rate (*Tr*), stomatal conductance (*Gs*), and intercellular CO_2 concentration (Ci) were measured in the first fully expanded leaves of each plant on the final day of the salinity treatments from 8:00 a.m. to 11:00 a.m. on a sunny, cloudless day. The water use efficiency (WUE_i) was calculated as Pn/Tr. To obtain a stable photosynthetic rate, the air temperature was 25 °C, the intensity of the light was 1000 μmol m^{-2}s^{-1}, the relative humidity was 65%, and the concentration of reference CO_2 was 380 μmol mol^{-1}. Chlorophyll fluorescence parameters were determined on the same day that the gas exchange measurements were taken with the fluorescence leaf chamber of the Ciras-2 photosynthetic system, including the photochemical quantum efficiency (Φ_{PSII}), maximum quantum yield efficiency of photosystem II (Fv/Fm), photochemical quenching parameter (qP), electron transfer rate (ETR), and non-photochemical quenching parameter (NPQ) on the dark-adapted leaves. All the parameter measurements were performed on three leaves per plant.

2.5. Scanning Electron Microscopy

The anatomical features (including the leaf surface and cross-section) of the leaves of *C. turczaninowii* seedlings in the control and salinity treatments were analyzed using scanning electron microscopy (SEM) (Quanta 200, FEI, Hillsborough, OR, USA). On the final day of the salinity treatments, the leaves were collected, properly cleaned with tap water and cut into pieces (5 × 5 mm) with a sharp blade. Formalin acetic acid (FAA) was used for one week for fixation. Then, the samples were dehydrated with a graded ethanol series, placed in isoamyl acetate aldehyde, and subjected to critical-point drying (K850, Emitech, London, UK), mounting, and gold coating using an ion sputtering apparatus (E1010, Hitachi, Tokyo, Japan). Finally, SEM was performed to observe the blade surface and cross-section of the leaves. The stomatal characteristics, including the stomatal density (stomata mm^{-2}) and size (length and width), were evaluated following the method reported by Camposeo et al. [26]. The leaf thickness and palisade tissue thickness of the leaves were measured by the photomicroscope system with a computer attachment.

2.6. Photosynthetic Pigments

The chlorophyll a (Chl a), chlorophyll b (Chl b), and total carotenoids (Caro) contents of the leaves were measured with the spectrophotometric method. Fresh leaves were cleaned with distilled water, and chlorophyll was extracted with a mixture of acetone and 95% ethanol (v:v = 1:1). After mixing overnight in an orbital shaker and following 10 min of centrifugation at 9,000 rpm, the absorbance of the supernatant was measured at wavelengths of 645, 663, and 470 nm by ultraviolet-visible spectrophotometry (Lambda25, PerkinElmer, Waltham, MA, USA). The chl and carotenoid contents were expressed as mg g^{-1} fresh weight according to the method of H.K. Lichtenthaler [27].

2.7. Relative Water Content (RWC)

The relative water content was calculated on the basis of the fresh and dry weight of the leaves according to the method of Vivekanandan [28] using the equation RWC = $[(FW - DW)/(TW - DW)]$ × 100%, where FW is the fresh leaf weight, DW is the dry weight, and TW is the turgid weight.

2.8. Organic Osmolytes

Three main types of organic osmolytes were analyzed: soluble sugars, soluble proteins, and proline. The soluble sugar content was quantified according to the method of Magné et al. [29]. Samples (0.1 g) were mixed with 10 mL of deionized water and incubated in water for 30 min at 100 °C, and then the extracts were cooled and filtered into a 25-mL volumetric flask filled with distilled water. One-half millilitre of supernatant was mixed with 1.5 mL of distilled water, 0.5 mL of anthrone-ethyl acetate, and 5 mL of 98% sulfuric acid. The tube was thoroughly shaken with an oscillator and immediately placed into boiling water for 1 min. After cooling, the absorbance of the supernatant was measured at 630 nm. The value for the soluble sugars was input into a glucose standard curve to identify the appropriate content.

To determine the soluble protein, samples were homogenized in phosphate buffer solution (pH 7.8) at 4 °C after centrifugation at 9000 rpm for 20 min, and the supernatant was used to determine the soluble protein. Supernatant and a volume of 0.1 mL was mixed with 5 mL of Coomassie brilliant blue G-250 solution. The absorbance of the supernatant was measured at 595 nm after 5 min. The leaf soluble protein was determined according to the method of Bradford [30]. Bovine serum albumin (BSA) was used to generate the standard curve.

Free proline was determined according to the method of Steinert et al. [31]: Extracts were prepared in 3% sulfosalicylic acid solution and then centrifuged at 4000 rpm at 4 °C for 20 min. The supernatant was mixed with acetic acid and acid ninhydrin and incubated in water for 1 h at 100 °C. The absorbance of the organic phase was measured at 520 nm. Proline was used to generate a standard curve. The proline content was calculated in mg g^{-1}.

2.9. Measurements of the Na^+, Cl^-, K^+, Ca^{2+}, and Mg^{2+} Contents of Different Organs

All plants were harvested at the end of the experiment. The contents of sodium (Na^+), chloride (Cl^-), potassium (K^+), calcium (Ca^{2+}), and magnesium (Mg^{2+}) were determined in the roots, stems, and leaves. The different tissues of the samples were dried at 85 °C for 48 h and weighed. Then, the dried samples were digested with a mixture of HNO_3 acid and $HClO_4$ acid (5:1 v:v). The Na^+, Cl^-, K^+, Ca^{2+}, and Mg^{2+} ion concentrations in the digested samples were measured using a plasma emission spectrometer (OPTIMA PE-4300DV, Waltham, MA, USA).

2.10. Malondialdehyde Concentration (MDA) and Cell Membrane Stability

An enzyme solution was prepared: 0.3 g leaves were ground at 4 °C in a mortar in 6 mL of a pH 7.8 phosphate buffer solution. The supernatants were collected from the homogenate after centrifugation at 9000 rpm at 4 °C for 20 min. The MDA content was quantified according to the method described by Hodges et al. [32]. Two millilitres of the enzyme solution was mixed with 3 mL of 0.5% thiobarbituric acid (TBA) and 8% trichloroacetic acid (TCA), which was then incubated at 100 °C for 20 min. After cooling the extracts on ice and centrifugation at 9000 rpm for 30 min at 4 °C, the absorbance of the supernatant was measured at 532 and 600 nm.

Cell membrane stability was expressed as the relative electrolytic conductivity (REC). Fresh leaves were cleaned with deionized water and cut into segments and incubated in hermetic tubes containing 20 mL of deionized water for 6 h at room temperature. The first electrical conductivity (EC_1) of the leaf solution was determined with a conductivity meter (DDS-307, Shanghai, China). Then, the maximum conductivity (EC_2) was obtained by placing the tubes containing leaf samples in water at 100 °C for 20 min to release all electrolytes. Leaf electrolyte leakage was calculated as follows: REC = $(EC_1/EC_2) \times 100\%$, according to Dionisio-Sese and Tobita [33].

2.11. Antioxidant Enzyme Activities

The activity of superoxide dismutase (SOD), peroxidase (POD), ascorbate peroxidase (APX), and glutathione reductase (GR) were determined in *C. turczaninowii* leaf extracts. Frozen leaves (0.3 g) were ground in liquid N_2 using a mortar with 6 mL of phosphate buffer solution (pH 7.8). After centrifugation at 9000 rpm at 4 °C for 20 min, the supernatant was collected as the enzyme extract for measurements. Enzyme activities were assayed using a spectrophotometer.

The SOD activity was assayed according to Beyer et al. [34]. The reaction mixture contained 0.1 mL of enzyme solution, 1.5 mL of sodium phosphate buffer, 0.3 mL of 130 mM methionine, 0.3 mL of 20 μM riboflavin, 0.3 mL of 100 μM EDTA, 0.3 mL of 750 μM nitroblue tetrazolium (NBT), and 0.5 mL of distilled water. The reaction was started by exposing the mixture to white fluorescent light for 15 min. The reaction was stopped by turning the light off. Reduced NBT was measured at 560 nm, and one unit of SOD activity was defined as the amount of enzyme causing 50% inhibition of NBT reduction per min under the assay conditions.

The POD activity in the enzyme extracts was assayed by the method of Civello et al. [35]. The reaction mixtures contained 1 mL of enzyme solution, 3.8 mL of 0.3% guaiacol reaction solution, and 0.1 mL of 3% H_2O_2. The increase in absorbance was recorded at 470 nm three times per min.

APX activity was monitoring at the decrease in absorbance at 290 nm according to Nakano et al. [36]. The reaction system contained 1.8 mL of 50 mM phosphate buffer (pH 7.0), 0.1 mL of 5 mM ascorbate (ASA), 1 mL of 30 mM H_2O_2, and 0.1 mL enzyme extracts. H_2O_2 was added to start the reaction.

The method of Fengwang Ma [37] was used for GR assays by monitoring the decrease in absorbance at 340 nm. The reaction system contained 0.1 mL of 1 mM NADPH, 2.7 mL of 0.1 mM phosphate buffer (pH 7.8), 0.1 mL of 5 mM GSSG, and 0.1 mL enzyme extracts. The reaction was initiated by adding NADPH.

2.12. Statistical Analysis

All the results are presented as the means $\pm$ standard deviation (S.D., $n = 3$). One-way analysis of variance (ANOVA) followed by Tukey's multiple comparison test at the 5% probability level were performed with the SPSS statistical package version 22.0 (IBM Corp., Armonk, NY, USA).

3. Results

3.1. Plant Growth

Salinity stress inhibited the growth of *C. turczaninowii* seedlings, as shown by the reduction of growth parameters in the stressed plants compared to the control plants (Table 1). ANOVA results showed that salinity treatments had significant effects on growth parameters ($p < 0.05$). The effect of the salinity treatments on the plant survival rate was significant at high salinities (68–85 mM); more than 55% of the plants died after being subjected to 85 mM salinity stress after 42 days. The reductions in stem, root, and leaf growth compared to the control plants became more pronounced with an increase in the salinity levels. The root/shoot ratio of the treatment group was significantly higher than that of the control group under 51–68 mM NaCl.

Table 1. Effects of various salinities on the survival rate, stem length increase, diameter increase, stem biomass, taproot length, root biomass, root/shoot ratio, leaf area, leaf biomass and total biomass of *Carpinus turczaninowii* L. after 42 days.

Salinity Level (mM)	Survival Rate (%)	Stem Length Increase (cm)	Diameter Increase (cm)	Stem Biomass (DW, g)	Taproot Length (cm)
CK	100 ± 0.00 a	9.03 ± 0.76 a	0.103 ± 0.10 a	2.21 ± 0.22 a	22.33 ± 1.86 a
17	100 ± 0.00 a	8.90 ± 0.79 a	0.095 ± 0.04 ab	2.02 ± 0.09 ab	21.24 ± 1.54 a
34	100 ± 0.00 a	8.57 ± 0.60 a	0.089 ± 0.07 ab	1.72 ± 0.04 bc	20.01 ± 1.06 ab
51	92 ± 2.00 ab	7.87 ± 0.78 a	0.079 ± 0.09 b	1.51 ± 0.15 cd	18.65 ± 0.25 b
68	75 ± 1.67 b	5.07 ± 0.74 b	0.060 ± 0.06 c	1.38 ± 0.22 cd	15.58 ± 0.88 bc
85	45 ± 2.33 c	4.67 ± 0.76 b	0.048 ± 0.04 c	1.18 ± 0.20 d	12.91 ± 0.75 c

Salinity Level (mM)	Root Biomass (DW, g)	Leaf Area (cm^2)	Leaf Biomass (DW, g)	Root/Shoot Ratio (R/S)	Total Biomass (DW, g)
CK	1.63 ± 0.10 a	9.82 ± 0.48 a	2.13 ± 0.24 a	0.38 ± 0.02 b	5.96 ± 0.55 a
17	1.41 ± 0.09 ab	9.32 ± 1.22 a	1.88 ± 0.07 ab	0.36 ± 0.03 c	5.31 ± 0.15 ab
34	1.27 ± 0.02 b	8.89 ± 1.56 ab	1.60 ± 0.04 bc	0.38 ± 0.01 b	4.60 ± 0.09 bc
51	1.19 ± 0.12 b	8.52 ± 1.01 ab	1.33 ± 0.09 cd	0.42 ± 0.01 a	4.03 ± 0.35 cd
68	1.12 ± 0.18 bc	7.14 ± 1.33 b	1.23 ± 0.09 cd	0.43 ± 0.02 a	3.72 ± 0.49 cd
85	0.83 ± 0.15 c	6.53 ± 0.68 c	1.05 ± 0.21 d	0.37 ± 0.01 bc	3.05 ± 0.55 d

Data in the table are the means $\pm$ S.D. ($n = 3$); different lowercase letters in each column indicate significant differences among treatments ($p < 0.05$).

3.2. Photosynthesis, Chlorophyll Fluorescence, and Stomatal Behavior

ANOVA results indicated that photosynthesis, chlorophyll fluorescence parameters of *C. turczaninowii* differed among salinity treatments ($p < 0.05$). The *Pn*, *Gs*, and WUE_i decreased with increasing salt concentration after 42 days of growth, while the *Ci* increased with increasing salinity (Table 2). There was a significant difference in the *Pn* between the 51–85 mM treatments and the control, especially under 85 mM NaCl, with the *Pn* reduced by 84.00% compared to that of the control. A significant decrease in the *Tr* occurred in the plants treated with high (68–85 mM) salinity. We found that 51–85 mM NaCl had the most pronounced effect on the *Gs*, *Ci*, and WUE_i. The *Ci* increased by 47.17% compared to that in the control treatment under 85 mM salinity. Salinity stress did not negatively influence the chlorophyll fluorescence parameters, including the Φ_{PSII}, *Fv/Fm*, qP, and ETR, at NaCl concentrations less than 85 mM. These chlorophyll fluorescence parameters were lower than those in the control under high salinity stress, except for the NPQ, which increased with the salt concentration and reached a maximum at 85 mM, 1.75 times that of the control.

Table 2. Effects of various salinities on photosynthesis and chlorophyll fluorescence parameters in *C. turczaninowii* after 42 days. *Pn*, net photosynthetic rate; *Tr*, transpiration rate; *Gs*, stomatal conductance; *Ci*, intercellular CO_2 concentration; WUE_i, water use efficiency; Φ_{PSII}, photochemical quantum efficiency; *Fv/Fm*, maximum quantum yield of photosystem II; qP, photochemical quenching parameter; ETR, electron transfer rate; NPQ, non-photochemical quenching parameter.

Salinity Level (mM)	*Pn* (μmol m^{-2}s^{-1})	*Tr* (mmol m^{-2}s^{-1})	*Gs* (μmol m^{-2}s^{-1})	*Ci* (μmol m^{-2}s^{-1})	WUE_i
CK	7.50 ± 0.70 a	2.21 ± 0.09 a	86.33 ± 6.03 a	212.00 ± 19.67 d	3.40 ± 0.39 a
17	7.20 ± 0.66 a	2.47 ± 0.08 a	82.33 ± 8.08 ab	227.67 ± 11.93 cd	2.92 ± 0.33 a
34	6.10 ± 0.46 ab	2.31 ± 0.07 a	70.00 ± 11.14 abc	257.00 ± 14.93 bc	2.64 ± 0.17 ab
51	4.90 ± 0.50 b	2.15 ± 0.11 a	61.67 ± 6.11 bc	264.33 ± 20.13 bc	2.28 ± 0.19 b
68	3.40 ± 0.40 c	1.67 ± 0.11 b	58.00 ± 8.54 c	277.33 ± 10.07 ab	2.05 ± 0.41 b
85	1.20 ± 0.20 d	1.47 ± 0.12 c	51.33 ± 7.10 c	312.00 ± 6.56 a	0.82 ± 0.16 c

Salinity Level (mM)	Φ_{PSII}	*Fv/Fm*	qP	ETR (μmol m^{-2}s^{-1})	NPQ
CK	0.50 ± 0.08 a	0.64 ± 0.04 a	0.91 ± 0.06 a	274.99 ± 19.11 a	0.81 ± 0.06 c
17	0.47 ± 0.03 a	0.66 ± 0.05 ab	0.94 ± 0.08 a	287.89 ± 25.37 a	0.95 ± 0.08 b
34	0.52 ± 0.04 a	0.58 ± 0.03 abc	0.96 ± 0.06 a	266.40 ± 21.18 ab	1.12 ± 0.13 ab
51	0.45 ± 0.04 a	0.53 ± 0.04 abc	0.87 ± 0.07 ab	256.81 ± 15.50 ab	1.25 ± 0.07 ab
68	0.41 ± 0.03 ab	0.49 ± 0.10 bc	0.81 ± 0.09 ab	225.57 ± 22.38 b	1.34 ± 0.10 a
85	0.38 ± 0.06 b	0.47 ± 0.06 c	0.72 ± 0.06 b	217.86 ± 25.25 b	1.42 ± 0.12 a

Data in the table are the means ± S.D. (n = 3); different lowercase letters in each column indicate significant differences among treatments ($p < 0.05$).

The stomatal density and size as a function of the soil salinity levels are shown in Table 3 and Figure 1a. At 17–51 mM salinity, the stomatal density of the *C. turczaninowii* leaves was not significantly different from that in the control but was pronounced under high (68–85 mM) salinity stress. Salinity levels of 51–85 mM caused a significant reduction in stomatal length and stomatal width, leading to smaller stomatal sizes. Finally, the stomata was nearly closed under high salinity (68–85 mM) treatments, with the guard cells becoming gradually deformed, and the epidermal tissue shrank.

The leaf thickness was higher than that in the control at low salinities (17–34 mM) but decreased significantly at 84 mM NaCl (Table 3). The palisade tissue thickness and palisade tissue thickness/leaf thickness of the *C. turczaninowii* leaves were noticeably higher than those in the control treatment at 51–85 mM salinity. Figure 1b shows a typical scanning electron micrograph of a *C. turczaninowii* leaf, including the upper and lower epidermis, epidermal hairs, and palisade and spongy tissues, which were densely arranged. However, the leaf cross-section characteristics were significantly different from those in the control under the high salinity treatments (68–85 mM), and the palisade and spongy tissues were loosely arranged. The spaces between the tissues became larger, and some of the tissues degraded.

Table 3. Effects of various salinities on the leaf stomatal density, stomatal length, stomatal width, leaf thickness, palisade tissue thickness, and palisade tissue thickness/leaf thickness of *Carpinus turczaninowii* L. after 42 days.

Treatment (mM)	Stomatal Density (number mm^{-2})	Stomatal Length (μm)	Stomatal Width (μm)	Leaf Thickness (μm)	Palisade Tissue Thickness (μm)	Palisade Tissue Thickness/Leaf Thickness (%)
CK	242.40 ± 22.12 ab	15.54 ± 1.21 a	12.69 ± 1.98 a	69.23 ± 5.02 ab	18.46 ± 3.21 c	26.67 ± 3.68 c
17	264.58 ± 25.38 a	13.23 ± 1.43 a	11.02 ± 2.06 a	84.62 ± 6.87 a	22.08 ± 2.60 bc	26.09 ± 2.02 c
34	230.06 ± 20.08 ab	11.15 ± 1.51 ab	9.68 ± 1.32 ab	73.85 ± 6.93 ab	24.34 ± 2.88 bc	32.96 ± 4.41 bc
51	190.65 ± 17.33 b	9.87 ± 1.06 bc	7.12 ± 1.58 b	64.62 ± 7.11 bc	25.69 ± 3.21 ab	39.76 ± 4.12 ab
68	162.58 ± 15.11 c	8.23 ± 1.22 bc	5.33 ± 1.82 c	61.54 ± 5.04 bc	27.54 ± 4.08 a	44.75 ± 5.05 ab
85	135.62 ± 12.49 d	7.85 ± 1.18 c	4.58 ± 0.95 c	59.61 ± 5.36 c	28.77 ± 4.65 a	48.26 ± 3.82 a

Data in the table are the means ± S.D. (n = 3); different lowercase letters in each column indicate significant differences among treatments ($p < 0.05$).

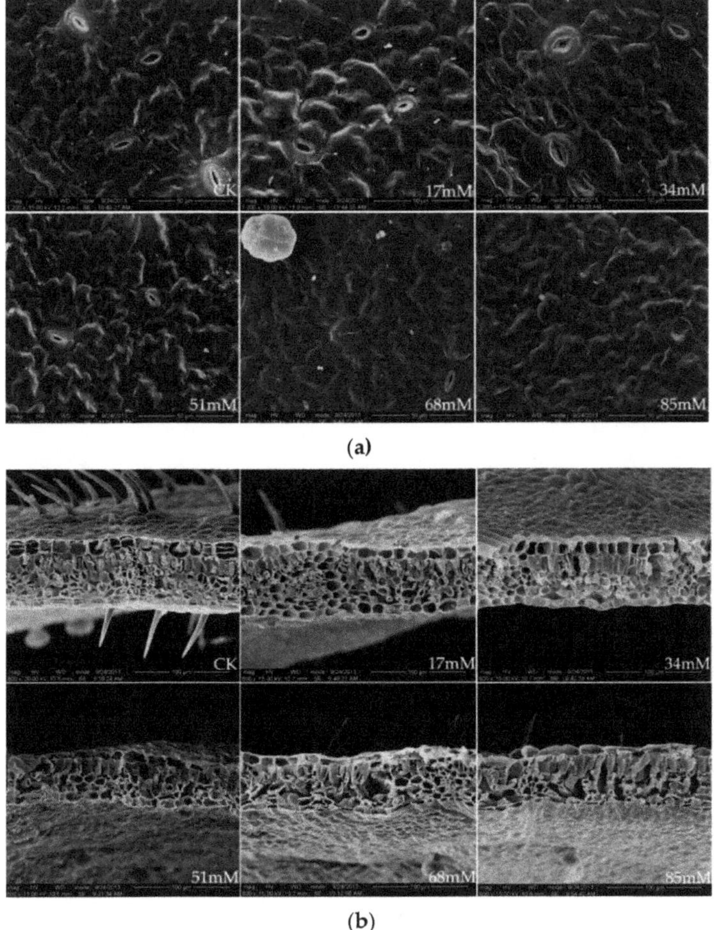

Figure 1. Stomatal structure on the leaf surface (**a**), ×1200, and leaf cross-section (**b**), ×600, of *C. turczaninowii* under salinity treatments.

3.3. Photosynthetic Pigments

The results showed that the contents of photosynthetic pigments (chl a and b, total chlorophyll, and carotenoids) in *C. turczaninowii* plants decreased when the plants were subjected to salinity ($p < 0.05$), and the differences between the control and 51–85 mM salinity treatments were significant after 42 days of growth (Figure 2a–d). At a concentration of 85 mM NaCl, the total chlorophyll content was reduced by 28.92% after 42 days of growth.

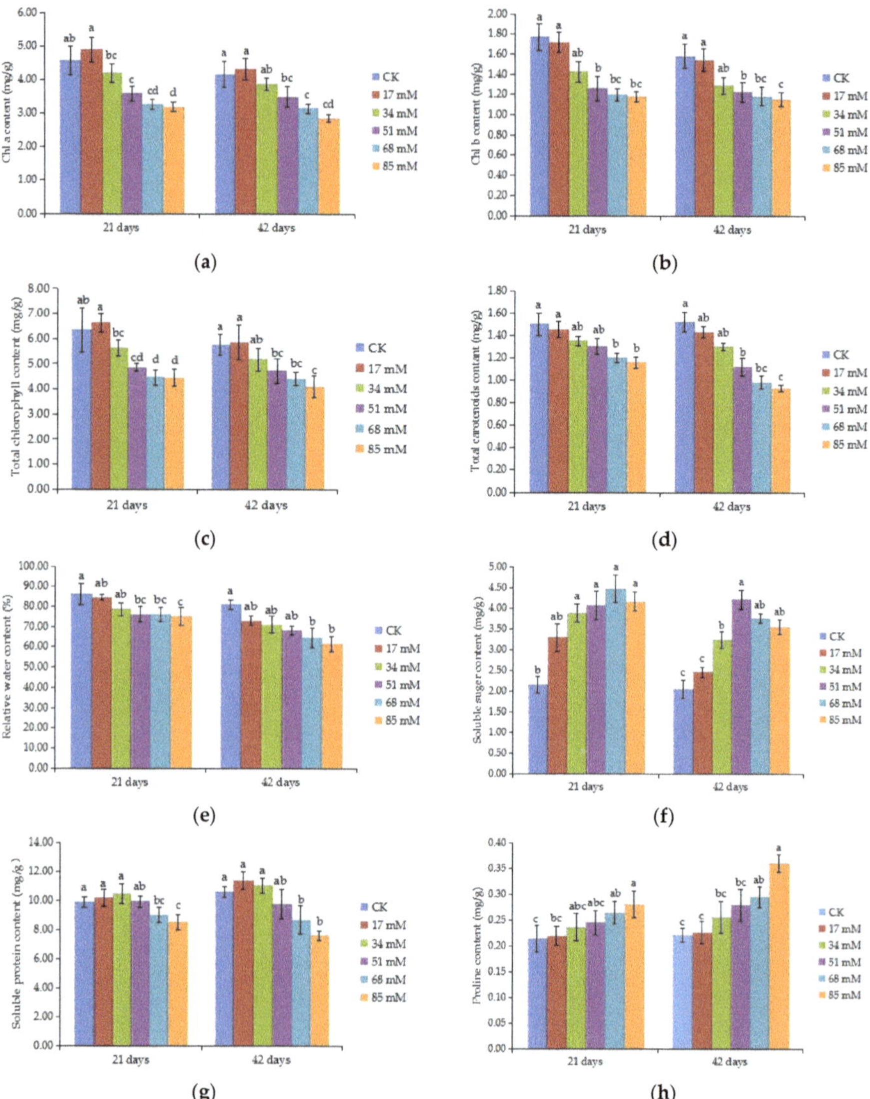

Figure 2. Effect of increasing levels of soil salinity on the content of chlorophyll a (**a**), chlorophyll b (**b**), total chlorophyll (**c**), and total carotenoids (**d**), relative water content (**e**), soluble sugars (**f**), soluble proteins (**g**), and proline content (**h**) in leaves of *C. turczaninowii* plants grown for 21 and 42 days. Different lowercase letters indicate significant differences among treatments ($p < 0.05$), $n = 3$.

3.4. Relative Water Content (RWC)

The relative water content of the *C. turczaninowii* leaves decreased with increasing salinity concentrations (Figure 2e). There was no difference of the RWC between the low salt stress treatment (17 mM) and the control plants, while a significant decrease was observed in the 68–85 mM salt treatments after 42 days ($p < 0.05$), with the RWC decreasing by 20.36% and 23.88%, respectively.

3.5. Organic Osmolytes

ANOVA results indicated that salinity treatments had significant effects on the organic solutes of *C. turczaninowii* seedlings ($p < 0.05$). With the increase in salinity stress, the soluble sugar content showed a tendency of first increasing and then falling, with a significant change in leaves in the 34~85 mM salinity treatments after 21 and 42 days of growth. After 42 days of growth, the soluble sugar content reached a maximum when the salt concentration was 51 mM, with a value of more than double of that in the control plants (Figure 2f). The soluble protein content did not change significantly under the slight (17–34 mM) and moderate (51 mM) salinity treatments, but a great reduction in soluble protein content was found in the high (68–85 mM) salinity treatments (Figure 2g). Proline is a type of amino acid that is easily accumulated under osmotic stress. The proline content of *C. turczaninowii* seedlings gradually increased with increasing salt concentration after 21 and 42 days of growth and was significantly higher than that of the control plants in the high (68–85 mM) salinity treatments (Figure 2h).

3.6. Inorganic Ion Content of Various Organs

The changes in ionic contents in *C. turczaninowii* seedlings are shown in Figure 3a–e. The Na^+ and Cl^- content of the root, stem, leaf, and the whole plant prominently increased with the salinity levels ($p < 0.05$), and mainly accumulated in the root and then in the stem (Figure 3a,b). In 85 mM NaCl treatment, the Na^+ content in the stem and leaf increased remarkably, 19.06 and 20.63 times, respectively, compared to that in the control group, and Cl^- content also reached a maximum value in this salt concentration. The K^+ content increased in the stem and leaf but decreased in the root and mostly accumulated in the leaf, followed by the stem (Figure 3c). In the whole plant, the K^+ ion content in the treatment groups was higher than that in the control ($p < 0.05$). The Ca^{2+} content remained high in the stem and leaf (Figure 3d). The differences in the Ca^{2+} content in the root among the various salinity treatments were not obvious, but the Ca^{2+} contents were all higher than that in the control. Under high salinity stress (68–85 mM), the Ca^{2+} content increased by 24.68% and 18.29%, respectively, in the whole plant. Differences in the Mg^{2+} content in the different tissues of *C. turczaninowii* seedlings were not obvious (Figure 3e). In the stem, the Mg^{2+} content in the 34–85 mM treatments was noticeably higher than that in the control group, and in the whole plant, little change was found in the 34 mM salinity stress treatment.

The K/Na, Mg/Na, and Ca/Na ratios in the *C. turczaninowii* seedlings under salinity treatments are shown in Table 4. It was found that all ion ratios significantly decreased under salinity stress in different tissues. In addition, the ion ratios in the aboveground parts (stem and leaf) were higher than the underground part (root), and the Ca/Na ratio was higher than the K/Na and Mg/Na ratios. These results indicate that the ability of *C. turczaninowii* to absorb Na^+ ions increased under salinity stress, and the absorption of K^+, Ca^{2+}, and Mg^{2+} by the leaves was higher than other tissues.

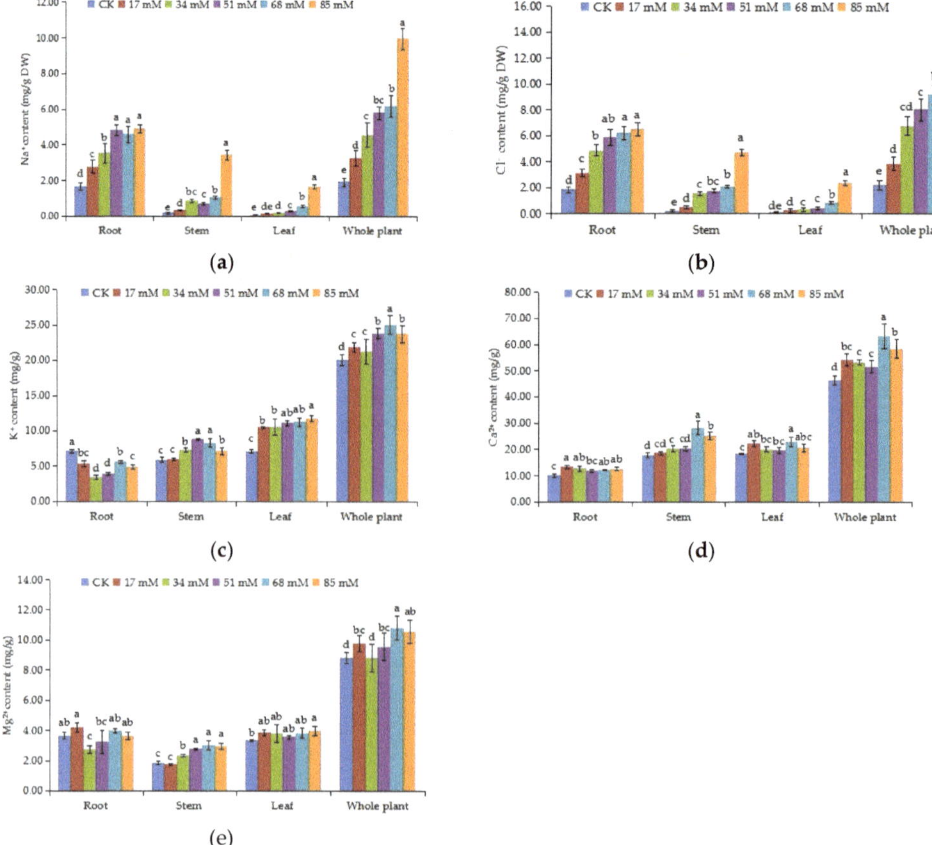

Figure 3. Effects of increasing levels of soil salinity on the Na$^+$ (**a**), Cl$^-$ (**b**), K$^+$ (**c**), Ca^{2+} (**d**) and Mg^{2+} (**e**) content in root, stem, leaf, and whole plant of *C. turczaninowii* seedlings after 42 days. Different lowercase letters indicate significant differences among treatments ($p < 0.05$), $n = 3$.

Table 4. Effects of various salinities on the K/Na, Ca/Na, and Mg/Na ratios in the root, stem, leaf, and whole plant of *C. turczaninowii* seedlings after 42 days.

	Salinity Level (mM)	K/Na	Ca/Na	Mg/Na
	CK	4.34 ± 0.56 a	6.20 ± 0.99 a	2.23 ± 0.35 a
	17	1.97 ± 0.36 b	4.89 ± 0.74 a	1.53 ± 0.27 b
Root	34	0.98 ± 0.13 c	3.62 ± 0.35 b	0.77 ± 0.10 c
	51	0.82 ± 0.01 c	2.45 ± 0.04 b	0.67 ± 0.12 c
	68	1.23 ± 0.15 bc	2.70 ± 0.31 b	0.87 ± 0.11 c
	85	1.00 ± 0.10 c	2.58 ± 0.13 b	0.75 ± 0.04 c
	CK	33.77 ± 4.87 a	101.68 ± 14.97 a	10.52 ± 1.44 a
	17	18.52 ± 0.88 b	57.57 ± 2.65 b	5.41 ± 0.20 b
Stem	34	8.45 ± 0.50 c	23.59 ± 1.54 bc	2.70 ± 0.16 c
	51	12.54 ± 0.72 c	28.79 ± 0.94 c	3.94 ± 0.23 bc
	68	7.92 ± 0.24 c	26.93 ± 0.27 c	2.89 ± 0.01 c
	85	2.09 ± 0.03 d	7.34 ± 0.16 d	0.86 ± 0.04 d

Table 4. *Cont.*

	Salinity Level (mM)	K/Na	Ca/Na	Mg/Na
Leaf	CK	90.77 ± 15.49 a	236.54 ± 39.71 a	42.86 ± 6.34 a
	17	71.10 ± 12.75 ab	150.01 ± 24.05 b	26.05 ± 4.08 b
	34	59.19 ± 3.42 bc	112.56 ± 3.62 bc	21.23 ± 0.87 b
	51	37.55 ± 2.41 cd	66.32 ± 3.21 cd	12.02 ± 0.74 c
	68	19.81 ± 1.04 de	40.44 ± 2.13E de	6.75 ± 0.22 dc
	85	7.10 ± 0.19 e	12.52 ± 0.10 e	2.40 ± 0.03 d
Whole plant	CK	128.88 ± 20.92 a	344.43 ± 55.67 a	55.61 ± 8.13 a
	17	91.58 ± 13.99 b	212.47 ± 27.45 b	33.00 ± 4.55 b
	34	68.63 ± 4.05 c	139.77 ± 5.51 c	24.70 ± 1.12 bc
	51	50.91 ± 3.13 c	97.56 ± 4.19 cd	16.64 ± 1.09 cd
	68	28.96 ± 1.43 d	70.07 ± 2.70 d	10.52 ± 0.35 d
	85	10.19 ± 0.32 d	22.44 ± 0.39 e	4.01 ± 0.11 e

Data in the table are the means ± S.D. (n = 3); different lowercase letters in each column indicate significant differences among treatments ($p < 0.05$).

3.7. Malondialdehyde Content (MDA) and Cell Membrane Stability

The MDA content can be used to measure the damage to the leaf cell membrane. The MDA content of *C. turczaninowii* continued to increase with the salinity concentration (Figure 4a). After 21 days, MDA content was not affected under the low salinity (17–34 mM) treatments, but 51–85 mM NaCl had a prominent effect on the MDA content ($p < 0.05$). After 42 days, the MDA content increased by 35.60% and 45.74% in the 68 mM and 85 mM salinity treatments, respectively, compared to the control. The relative electrolytic conductivity is an index of cell membrane stability. As shown in Figure 4b, the relative electrolytic conductivity increased with salinity levels and was noticeably higher among the treatment and control groups after 42 days of growth ($p < 0.05$). These results indicated that salt stress caused damage to the membranes of *C. turczaninowii* seedlings and that the damage was more serious at high concentrations.

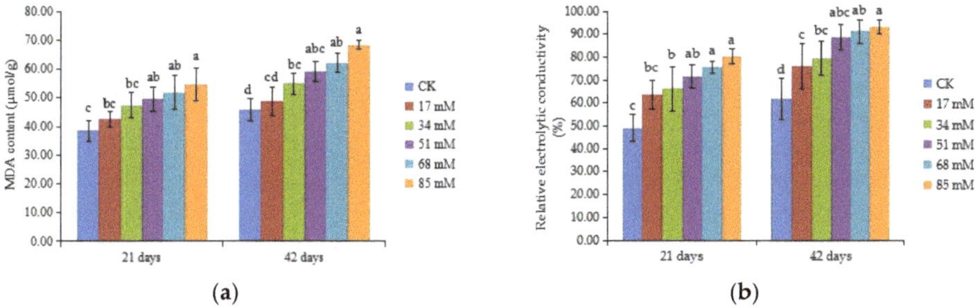

(a)   (b)

Figure 4. *Cont.*

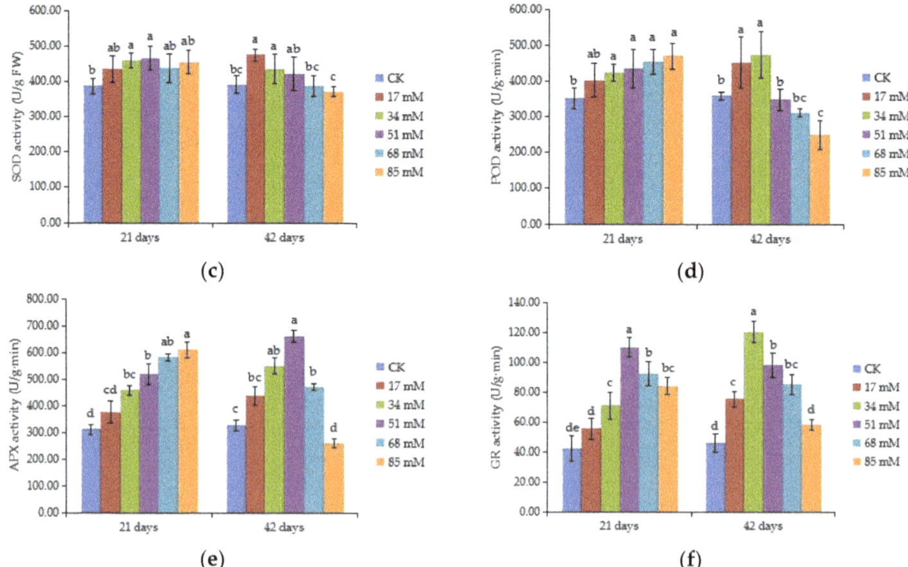

Figure 4. Effects of increasing levels of soil salinity on the MDA content (**a**), cell membrane stability (**b**), SOD activity (**c**), POD activity (**d**), APX (**e**), and GR (**f**) in leaves of *C. turczaninowii* grown for 21 and 42 days. Different lowercase letters indicate significant differences among treatments ($p < 0.05$), $n = 3$.

3.8. Antioxidant Enzyme Activities

ANOVA results indicated that antioxidant enzyme activities differed among salinity treatments ($p < 0.05$). The SOD and GR activities of *C. turczaninowii* leaves first increased then decreased with the level of salt stress (Figure 4c,f). The SOD activity was significantly higher than that in the control at 34–51 mM NaCl after 21 days. After 42 days, the SOD activity reached a maximum value of 475.17 U/g at 17 mM, increasing by 21.81% compared to the control, and little change was found at 51–85 mM NaCl. Salt treatments caused significant increase in the activity of GR after 21 days, and a maximum value was found at 34 mM after 42 days. The POD and APX activities increased as the results of the salinity treatments after 21 days, and the differences were remarkable at 34~85 mM salinity (Figure 4d,e). However, after 42 days, the POD activity was obviously higher than that in the control at low salinity (17–34 mM) and decreased at high salinity (85 mM), being only 69.15% that of the control, and APX actvity was significant higher than the control at 34–68 mM salinity, also decreased at 85 mM NaCl.

4. Discussion

The present study provides valuable information about *C. turczaninowii* traits that may affect seedling survivorship and growth under salt stress in nursery conditions. Salinity stress induced depressive effects on all the examined parameters, including the changes in survival rate, the production parameters, the morphological and physiological parameters. This research was the first attempt to look for the salt tolerance of *C. turczaninowii* seedlings, and confirm that salinity is one of the major abiotic stresses limiting the growth and productivity of this species.

4.1. Seedling Growth

Salinity affects plants in many different ways, the most obvious being a decrease in growth, which may be related to a reduction in water uptake, photosynthetic activity, and other affected physiological pathways [38]. Our results show that salinities over 68 mM inhibited the growth of

C. turczaninowii seedlings, including the survival rate, stem length and diameter increase, taproot length, biomass, and leaf area. Many studies have also described similar responses of plants to high salinity, a decrease in leaf area or root and stem length, and a change in the root/shoot ratio was also observed in oak (*Quercus robur* L.) seedlings [39]. Mulberry (*Morus alba* L.) seedlings suffered great reductions in root, shoot, and leaf growth under various levels of NaCl stress [40]. These results confirm that salinity is one of the major environmental factors that restricts the growth and productivity of plants. As expected, salinity caused a significant reduction in leaf area at 68~85 mM NaCl, which may have been caused by the inhibition of cell expansion after salt stress [41]. The same phenomenon was observed for stomatal behavior; the stomatal density and stomatal size of *C. turczaninowii* leaves were pronounced decreased under high salinity stress, with nearly closed stomata. Stomata in the epidermis of the leaf provide an important channel that allows plants to exchange materials such as CO_2 and H_2O with the external environment. Salinity stress can cause the close of stomata, the decrease of stomatal size and leaf thickness, and the atrophy of leaf tissues [42]. In our previous study on *Carpinus betulus* L. [43], we found the stomata, palisade, and spongy parenchyma tissues were seriously influenced over 51 mM NaCl, worse than *C. turczaninowii*. These changes in leaf morphology can adversely affect photosynthesis in plants. Under salinity stress, plants often face water deficiencies, which will lose water, wither, and even die. The RWC of the leaves decreased with increasing salinity levels and reached a minimum value at 85 mM NaCl after 42 days. This occurs by a reduction in the external water potential due to high salt concentrations in the solution of the soil and its accumulation in the extracellular region of the plants. Under 0–240 mM NaCl, the RWC in *Phoenix dactylifera* L. also significantly declined [44]. Alterations to the water content of plants may induce reductions in the photosynthetic rate and the growth of seedlings. According to Tuteja [45], salt sensitive plants cannot grow at 100 mM NaCl or above, and are called glycophytes, we believed that *C. turczaninowii* was a kind of salt sensitive plants.

4.2. Physiological Parameters

High salinity is associated hyperosmotic stress, inducing secondary effects or pathologies and ion toxicity [46]. The mechanism of photosynthesis response to salt stress is complex, including the influence of photochemistry, CO_2 diffusion, and the activities of enzymes involved in photosynthesis. In our study, the leaf gas exchange and chlorophyll fluorescence parameters were restricted with increasing levels of salinity in *C. turczaninowii* leaves. The *Pn*, *Tr*, *Gs*, and WUE_i under 51~85 mM salinity were lower than those in the control, while the *Ci* increased with the salt concentration. Similar behaviors were reported in *Pistacia lentiscus* and *Salsola inermis*, with reductions in stomatal conductance and the net photosynthetic rate [47]. Stomatal behavior was adversely affected under salinity stress, which inhibited the transpiration and photosynthetic rates of the plants. The chlorophyll fluorescence parameters, including the Φ_{PSII}, *Fv/Fm*, qP, and ETR, were negatively influenced at 85 mM NaCl concentration, and were lower than those in the control. The NPQ increased with salt stress, which may be a mechanism to protect the photosystems from the damage caused by photoinhibition under high salt stress [48]. A reduction of photosynthetic pigments contents in the leaves of plants is also a general response to salinity stress. In this study, the photosynthetic pigments contents of *C. turczaninowii* were not notably reduced at a salinity of 17~34 mM; however, salinities over 51 mM significantly reduced the content of Chl, and carotenoids after 42 days of treatment. This is consistent with previous reports on European Larch [49], *Phragmites australis* and *Spartina alterniflora* [50]. The photosynthetic apparatus of the plants was likely impaired and the chloroplast pigments were likely degraded under high salinity stress, thus inhibiting photosynthesis in the seedlings.

Osmotic adjustment is one of the cellular responses for the maintenance of turgor in saline environments to maintain stomatal opening and growth expansion [51]. A high concentration of organic solutes in the leaves of *C. turczaninowii* was also detected as the salinity levels increased, including the soluble sugar and proline contents, likely because the seedlings produce some organic molecules in response to salinity stress. The high production of these organic solutes decrease the leaf osmotic potential, increase the leaf turgor pressure, and provide additional evidence of osmotic adjustment [52]. According to other studies, salt stress can increase the content of soluble sugars (sucrose, glucose, and fructose) in plant leaves to act as a factor in maintaining turgor and reducing osmotic pressure, as occurs in apples [53]. A large number of studies have verified a positive correlation between the proline content and environmental stress [54], and our results confirm this viewpoint. *C. turczaninowii* can maintain its leaf osmotic potential and turgor pressure at a normal level when treated with salinities of less than 51 mM, but a significant difference was found in the high (68–85 mM) salinity treatments, indicating insufficient resistance to salt stress. According to Zidan [55], salt stress can promote protein synthesis through the transformation of free amino acids into protein in plants. However, the content of soluble protein in *C. turczaninowii* was significantly lower than that in the control at high salinity treatments. The same results were found in pear (*Pyrus communis*), the protein content also decreased with the salinity levels (40–160 mM), which may be due to a reduction in N synthesis, the degradation of protein structure, damage to biosynthesis pathways, and a reduction in the synthesis of protein [56].

4.3. Mineral Content and Uptake

Na^+ and Cl^- ions are the most destructive elements when plants are exposed to salinity. With respect to Na^+ and Cl^- content in the *C. turczaninowii* seedlings, a trend of higher values was observed with increasing salinity levels, confirming the findings obtained by Giuseppe Cristiano in *Pistacia lentiscus* [57]. A higher content of Na^+ was found in the roots under salt treatments in *C. turczaninowii* seedlings, as well as Cl^- content, while K^+ mostly accumulated in leaves, suggesting a defence mechanism for sodium toxicity through the compartmentalization of this ion in the roots. This mechanism has already been observed in other plants, such as *Salvadora oleoides* [58]. Controlling the uptake and transport of Na^+ and Cl^- ions to the shoot is a defence mechanism used by plants against excessive salt. A similar result was found in maize, the Na^+ and Cl^- contents were also higher in the root than in the leaf when growing under 34–102 mM NaCl salinity stress [59]. The ability of *C. turczaninowii* seedlings to maintain adequate selectivity in the uptake of K^+ in saline environment is associated with salt tolerance. The uptake of Ca^{2+} and Mg^{2+} did not change obviously based on the different levels of salinity, in agreement with findings in olive trees (*Olea europaea*) [60]. The ion ratios (K/Na, Ca/Na, and Mg/Na) decreased significantly with salinity levels but were higher in the stem and leaf than in the root, which mainly occurred because of the accumulation of Na^+ in accordance with the salinity levels in different tissues, especially in the roots. We can see that the capability of maintaining ion homeostasis in *C. turczaninowii* was limited under high salt stress (68–85 mM), causing a decline in biomass and even plant death. High salt stress can induce both metabolic changes and ion toxicity in plants resulting from ion imbalance with the accumulation of Na^+ and a reduction in K^+ content [61]. Because K^+ is an important activator of many enzymes, a reduction in K^+ may inhibit the growth of *C. turczaninowii* seedlings.

4.4. Oxidative Stress and Antioxidant Mechanisms

MDA is used as a marker for evaluating lipid peroxidation, which increases under saline stress conditions [62]. The relative electrolytic conductivity is usually used as an index for cell membrane stability, which can reflect the extent of lipid peroxidation caused by ROS [63]. In our results, the MDA content and the relative electrolytic conductivity of *C. turczaninowii* continued to increase with salinity concentration and were noticeably higher than those in the control group under moderate and high salt stress after 42 days, indicating that more lipid peroxidation products were formed due to cell damage under severe salinity stress. The increase in lipid peroxidation may be due to the incapability of antioxidants to scavenge ROS under salt stress. Antioxidant enzymes activities can be induced as a response to salinity stress, which can help alleviate oxidative stress. In our study, the activities of four important antioxidant enzymes first increased then decreased with the degree of stress in *C. turczaninowii* seedlings at last. At a NaCl concentration over 51 mM, there was a lower efficiency to eliminate ROS from the seedlings due to the imbalance between ROS and antioxidant formation, thus leading to the aggravation of oxidative stress. In *Lavandula angustifolia*, the same changes were found in SOD and POD activities at 25–100 mM salt concentrations [62]. While in European Larch seedlings, the SOD, APX, and GR activities were remarkable higher than the control at high salt stress (150 mM) [49], the results differed with ours. The increased antioxidant activity of enzymes under low salinity stress conditions suggests that the activation of an efficient free radical scavenging system can minimize the negative effects of peroxidation, thus contributing to the maintenance of membrane stability. However, under moderate and high salt stress, the activity of protective enzymes was limited, causing a negative effect on the survival and growth of the seedlings. Plant tolerance to salinity is correlated with the stimulation of antioxidant enzymes and their enhanced ability to scavenge ROS [64], our research supports these views.

5. Conclusions

This study presents new information on *C. turczaninowii* seedlings regarding their ecophysiological responses to salinity stress. *C. turczaninowii* was affected by salinity stress through the inhibition of growth, photosynthetic capacity, and stomatal behavior, the increase in MDA content and relative electrolytic conductivity, and the accumulation of Na^+ and Cl^- content, suggesting a limited degree of salt tolerance. Our results show that *C. turczaninowii* seedlings can activate relevant mechanisms to adapt to salinity stress, such as accumulating relatively high concentrations of organic solutes to maintain cellular osmotic balance and improving the activity of antioxidant enzymes and ionic transport capacity from the root to shoots in defence against salinity stress. A reduction in the growth of *C. turczaninowii* seedlings was observed under high (68~85 mM) NaCl stress. Our study is relevant for a better understanding of the responses of *C. turczaninowii* to salinity stress and may help in the administration of conservation and reforestation programs for this species. The cultivation of this species as an ornamental tree has the potential to be expanded in areas with low and moderate salinization conditions.

Author Contributions: Q.Z. and Z.Z. conceived and designed the experiments; Q.Z., M.S., and L.X. performed the experiments and analyzed the data; and Q.Z. wrote the paper. All authors read and approved the final paper.

Funding: This research was financially supported by the National Natural Science Foundation of China (no. 31770752), the Jiangsu Science and Technology Support Program (BY2015006-01), the Jiangsu Province Six Big Talent Peak Project (NY-029), the Fifth Stage Funded Research Projects of 333 in Jiangsu Province and the Excellent Doctoral Dissertation of Nanjing Forestry University.

Acknowledgments: We would like to thank Jing Yang, a laboratory specialist at Advanced Analysis Testing Center (AATC), Nanjing Forestry University, China, for assistance with stomatal measurements. We express our gratitude to Feibing Wang, a research assistant at the College of Landscape Architecture, Nanjing Forestry University, China, for her help in the cultivation of the seedlings.

Conflicts of Interest: The authors declare no conflict of interest.

References

1. Dobbertin, M. Tree growth as indicator of tree vitality and of tree reaction to environmental stress: A review. *Eur. J. For. Res.* **2006**, *125*, 89. [CrossRef]
2. Amira, M.S.; Abdul, Q. Effect of salt stress on plant growth and metabolism of bean plant *Vicia faba* (L.). *J. Saudi Soc. Agric. Sci.* **2011**, *10*, 7–15. [CrossRef]
3. Dashti, A.; Khan, A.A.; Collins, J.C. Effects of salinity on growth, ionic relations and solute content of *Sorghum Bicolor* (L.) Monench. *J. Plant Nutr.* **2009**, *32*, 1219–1236. [CrossRef]
4. Houska, C. *Deicing Salt-Recognizing the Corrosion Threat*; TMR Consulting: International Molybdenum Association: Pittsburgh, PA, USA, 2007; pp. 1–11.
5. Cunningham, M.A.; Snyder, E.; Yonkin, D.; Ross, M.; Elsen, T. Accumulation of deicing salts in soils in an urban environment. *Urban Ecosyst.* **2008**, *11*, 17–31. [CrossRef]
6. Lundmark, A.; Olofsson, B. Chloride Deposition and distribution in soils along a deiced highway- assessment using different methods of measurement. *Water Air Soil Pollut.* **2007**, *182*, 173–185. [CrossRef]

7. Kayama, M.; Quoreshi, A.M.; Kitaoka, S.; Kitahashi, Y.; Sakamoto, Y.; Maruyama, Y. Effects of deicing salt on the vitality and health of two spruce species, *Picea abies* Karst., and *Picea glehnii* Masters planted along roadsides in northern Japan. *Environ. Pollut.* **2003**, *124*, 127–137. [CrossRef]

8. Randrup, T.B. Differences in salt sensitivity of four deciduous tree species to soil or airborne salt. *Physiol. Plant.* **2010**, *114*, 223–230. [CrossRef]

9. Rozentsvet, O.A.; Nesterov, V.N.; Bogdanova, E.S. Membrane-forming lipids of wild halophytes growing under the conditions of Prieltonie of South Russia. *Phytochemistry* **2014**, *105*, 37–42. [CrossRef]

10. Jinbiao, X.I.; Zhang, F.; Chen, Y.; Mao, D.; Yin, C.; Tian, C. A preliminary study on salt contents of soil in root-canopy area of halophytes. *Chin. J. Appl. Ecol.* **2004**, *15*, 53–58. (In Chinese) [CrossRef]

11. Ashraf, M.; Orooj, A. Salt stress effects on growth, ion accumulation and seed oil concentration in an arid zone traditional medicinal plant ajwain (*Trachyspermum ammi* [L.] Sprague). *J. Arid Environ.* **2006**, *64*, 209–220. [CrossRef]

12. Schiop, S.T.; Hassan, M.A.; Sestras, A.F.; Boscaiu, N.; Mónica, T.; Vicente, M.; Óscar. Identification of salt stress biomarkers in Romanian Carpathian populations of *Picea abies* (L.) Karst. *PLoS ONE* **2015**, *10*, 980–981. [CrossRef] [PubMed]

13. Richards, L.A. Diagnosis and improvement of saline and alkaline soils. *Soil Sci.* **1954**, *78*, 154. [CrossRef]

14. Volkmar, K.M.; Hu, Y.; Steppuhn, H. Physiological responses of plants to salinity: A review. *Can. J. Plant Sci.* **1998**, *78*, 19–27. [CrossRef]

15. Ashraf, M.; Ozturk, M.; Ahmad, M.S.A. *Structural and Functional Adaptations in Plants for Salinity Tolerance*; Springer: Dordrecht, The Netherlands; Berlin, Germany, 2010; pp. 151–170. [CrossRef]

16. Negrão, S.; Schmöckel, S.M.; Tester, M. Evaluating physiological responses of plants to salinity stress. *Ann. Bot.* **2017**, *119*, 1–11. [CrossRef] [PubMed]

17. Singh, M.; Kumar, J.; Singh, S.; Singh, V.P.; Prasad, S.M. Roles of osmoprotectants in improving salinity and drought tolerance in plants: A review. *Rev. Environ. Sci. Bio/Technol.* **2015**, *14*, 407–426. [CrossRef]

18. Fardus, J.; Matin, M.A.; Hasanuzzaman, M.; Hossain, M.S.; Nath, S.D.; Hossain, M.A. Exogenous salicylic acid-mediated physiological responses and improvement in yield by modulating antioxidant defense system of wheat under salinity. *Not. Sci. Biol.* **2017**, *9*, 219–232. [CrossRef]

19. Esfandiari, E.; Gohari, G. Response of ROS-scavenging systems to salinity stress in two different wheat (*Triticum aestivum* L.) cultivars. *Not. Bot. Horti Agrobot.* **2017**, *45*, 287–291. [CrossRef]

20. Chen, Z.D.; Xing, S.P.; Liang, H.X.; Lu, A.M. Morphogenesis of female reproductive organs in *Carpinus turczaninowii* and *Ostryopsis davidiana* (Betulaceae). *Acta Bot. Sin.* **2001**, *43*, 1110–1114. (In Chinese) [CrossRef]

21. Cheng, L.X.; Jin, C.Z.; Zhu, Z.L. Growth characteristics of *Carpinus turczaninowii* and chlorophyll fluorescence annual variation. *J. Shanghai Jiaotong Univ.* **2014**, *32*, 21–26. (In Chinese) [CrossRef]

22. Ko, H.N.; Kim, J.M.; Bu, H.J.; Lee, N.H. Chemical constituents from the branches of *Carpinus turczaninowii* with antioxidative activities. *J. Korean Chem. Soc.* **2013**, *57*, 520–524. [CrossRef]

23. Kang, J.M.; Kim, J.E.; Lee, N.H. Anti-melanogenesis active constituents from the extracts of *Carpinus turczaninowii* leaves. *J. Soc. Cosmet. Sci. Korea* **2017**, *43*, 35–41. [CrossRef]

24. Yeo, J.H.; Son, Y.K.; Bang, W.Y.; Kim, O.Y. *Carpinus truczaninowii* extract showed anti-inflammatory response on human aortic vascular smooth muscle cells. *Planta Med.* **2016**, *82*, S1–S381. [CrossRef]

25. Hothem, S.D.; Marley, K.A.; Larson, R.A. Photochemistry in Hoagland's Nutrient Solution. *J. Plant Nutr.* **2003**, *26*, 845–854. [CrossRef]

26. Camposeo, S.; Palasciano, M.; Vivaldi, G.A.; Godini, A. Effect of increasing climatic water deficit on some leaf and stomatal parameters of wild and cultivated almonds under Mediterranean conditions. *Sci. Hortic.* **2011**, *127*, 234–241. [CrossRef]

27. Lichenthaler, H.K.; Wellburn, A.R. Determinations of total carotenoids and chlorophylls a and b of leaf extracts in different solvents. *Biochem. Soc. Trans.* **1983**, *11*, 591–592. [CrossRef]

28. Vivekanandan, M. Drought-induced responses of photosynthesis and antioxidant metabolism in higher plants. *J. Plant Physiol.* **2004**, *161*, 1189–1202. [CrossRef]

29. Magné, C.; Saladin, G.; Clément, C. Transient effect of the herbicide flazasulfuron on carbohydrate physiology in *Vitis vinifera* L. *Chemosphere* **2006**, *62*, 650–657. [CrossRef]

30. Bradford, M.M. A rapid and sensitive method for the quantitation of microgram quantities of protein utilizing the principle of protein-dye binding. *Anal. Biochem.* **1976**, *72*, 248–254. [CrossRef]

31. Steinert, A.; Bielka, S. Determination of free proline in stressed plants. *Arch. Züchtungsforsch.* **1990**, *20*, 199–204.

32. Hodges, D.M.; Delong, J.M.; Forney, C.F.; Prange, P.K. Improving the thiobarbituric acid-reactive-substances assay for estimating lipid peroxidation in plant tissues containing anthocyanin and other interfering compounds. *Planta* **1999**, *207*, 604–611. [CrossRef]

33. Dionisio-Sese, M.L.; Tobita, S. Antioxidant responses of rice seedlings to salinity stress. *Plant Sci.* **1998**, *135*, 1–9. [CrossRef]

34. Beyer, W.F.; Fridovich, I. Assaying for superoxide dismutase activity: Some large consequences of minor changes in conditions. *Anal. Biochem.* **1987**, *161*, 559–566. [CrossRef]

35. Civello, P.M.; Martinez, G.A.; Chaves, A.R.; Anon, M.C. Peroxidase from strawberry fruit (*Fragaria ananassa* Duch.): Partial purification and determination of some properties. *J. Agric. Food Chem.* **1995**, *43*, 2596–2601. [CrossRef]

36. Nakano, Y.; Asada, K. Hydrogen peroxide is scavenged by ascorbate specific peroxidase in spinach chloroplasts. *Plant Cell Physiol.* **1981**, *22*, 867–880. [CrossRef]

37. Ma, F.; Cheng, L. The sun-exposed peel of apple fruit has higher xanthophyll cycle-dependent thermal dissipation and antioxidants of the ascorbate–glutathione pathway than the shaded peel. *Plant Sci.* **2003**, *165*, 819–827. [CrossRef]

38. Niu, G.H.; Cabrera, R.I. Growth and physiological responses of landscape plants to saline water irrigation: A review. *Hortscience* **2010**, *45*, 1605–1609. [CrossRef]

39. Laffray, X.; Alaoui-Sehmer, L.; Bourioug, M.; Bourgeade, P.; Alaoui-Sossé, B.; Aleya, L. Effects of sodium chloride salinity on ecophysiological and biochemical parameters of oak seedlings (*Quercus robur* L.) from use of de-icing salts for winter road maintenance. *Environ. Monit. Assess.* **2018**, *190*, 266. [CrossRef]

40. Lu, N.; Luo, Z.; Ke, Y.; Dai, L.; Duan, H.D.; Hou, R.X.; Cui, B.B.; Dou, S.H.; Zhang, Y.D.; Sun, Y.H. Growth, physiological, biochemical, and ionic responses of *Morus alba* L. seedlings to various salinity levels. *Forests* **2017**, *8*, 488. [CrossRef]

41. Christian, Z.R.; Mühling, K.H.; Ulrich, K.; Christoph-Martin, G. Salinity stiffens the epidermal cell walls of salt-stressed maize leaves: Is the epidermis growth-restricting? *PLoS ONE* **2015**, *10*, e118406. [CrossRef]

42. Wang, B.X.; Zeng, Y.H.; Wang, D.Y.; Zhao, R.; Xu, X. Responses of leaf stomata to environmental stresses in distribution and physiological characteristics. *Agric. Res. Arid Areas* **2010**, *28*, 112–122. (In Chinese)

43. Zhou, Q.; Zhu, Z.L.; Shi, M.; Cheng, L.X. Growth and physicochemical changes of *Carpinus betulus* L. influenced by salinity treatments. *Forests* **2018**, *9*, 354. [CrossRef]

44. Kharusi, L.A.; Assaha, D.; Al-Yahyai, R.; Yaish, M.W. Screening of Date Palm (*Phoenix dactylifera* L.) cultivars for salinity tolerance. *Forests* **2017**, *8*, 136. [CrossRef]

45. Tuteja, N. Mechanisms of high salinity tolerance in plants. *Methods Enzymol.* **2007**, *428*, 419–438. [CrossRef] [PubMed]

46. Hasegawa, P.M.; Bressan, R.A.; Zhu, J.K.; Bohnert, H.J. Plant cellular and molecular responses to high salinity. *Annu. Rev. Plant Physiol. Plant Mol. Biol.* **2000**, *51*, 463–499. [CrossRef] [PubMed]

47. Barazani, O.; Golan-Goldhirsh, A. Salt-driven interactions between *Pistacia lentiscus* and *Salsola inermis*. *Environ. Sci. Pollut. Res. Int.* **2009**, *16*, 855. [CrossRef] [PubMed]

48. Müller, P.; Li, X.P.; Niyogi, K.K. Non-photochemical quenching. A response to excess light energy. *Plant Physiol.* **2001**, *125*, 1558–1566. [CrossRef] [PubMed]

49. Ioana, M.P.; Sara, G.O.; Mohamad, A.H.; Adriana, F.S.; Oscar, V.; Jaime, P.; Radu, E.S.; Monica, B. Effects of drought and salinity on European Larch (*Larix decidua* Mill.) Seedlings. *Forests* **2018**, *9*, 320. [CrossRef]

50. Li, S.H.; Ge, Z.M.; Xie, L.N.; Chen, W.; Yuan, L.; Wang, D.Q. Ecophysiological response of native and exotic salt marsh vegetation to waterlogging and salinity: Implications for the effects of sea-level rise. *Sci. Rep.* **2018**, *8*, 2441. [CrossRef]

51. Silveira, J.A.G.; Araujo, S.A.M. Roots and leaves display contrasting osmotic adjustment mechanisms in response to NaCl-salinity in *Atriplex nummularia*. *Environ. Exp. Bot.* **2009**, *66*, 1–8. [CrossRef]

52. Abdullakasim, S.; Kongpaisan, P.; Thongjang, P.; Saradhuldhat, P. Physiological responses of potted *Dendrobium orchid* to salinity stress. *Hortic. Environ. Biotechnol.* **2018**, *59*, 491–498. [CrossRef]

53. Sotiropoulos, T.E. Effect of NaCl and $CaCl_2$ on growth and contents of minerals, chlorophyll, proline and sugars in the apple rootstock M 4 cultured in vitro. *Biol. Plant.* **2007**, *51*, 177–180. [CrossRef]

54. Hayat, S.; Hayat, Q.; Alyemeni, M.N.; Wani, A.S.; Pichtel, J.; Ahmad, A. Role of proline under changing environments: A review. *Plant Signal Behav.* **2012**, *7*, 1456–1466. [CrossRef]

55. Zidan, M.A.; Elewa, M.A. Effect of salinity on germination, seedling growth and some metabolic changes in four plant species (Umbelliferae). *Indian J. Plant Physiol.* **1995**, *38*, 57–61. [CrossRef]

56. Fatemeh, Z.; Mohammad, E.A.; Alireza, N.; Parviz, A. Physiological and morphological responses of the 'Dargazi' pear (*Pyrus communis*) to in vitro salinity. *Agric. Conspec. Sci.* **2018**, *2*, 169–174.

57. Cristiano, G.; Camposeo, S.; Fracchiolla, M.; Vivaldi, G.A.; Lucia, B.D.; Cazzato, E. Salinity differentially affects growth and ecophysiology of two Mastic tree (*Pistacia lentiscus* L.) accessions. *Forests* **2016**, *7*, 156. [CrossRef]

58. Ramoliya, P.J.; Pandey, A.N. Effect of increasing salt concentration on emergence, growth and survival of seedlings of *Salvadora oleoides* (Salvadoraceae). *J. Arid Environ.* **2002**, *51*, 121–132. [CrossRef]

59. Hajlaoui, H.; Ayeb, N.E.; Garrec, J.P.; Denden, M. Differential effects of salt stress on osmotic adjustment and solutes allocation on the basis of root and leaf tissue senescence of two silage maize (*Zea mays* L.) varieties. *Ind. Crop. Prod.* **2010**, *31*, 122–130. [CrossRef]

60. Aragüés, R.; Puy, J.; Royo, A.; Espada, J.L. Three-year field response of young olive trees (*Olea europaea* L. cv. Arbequina) to soil salinity: Trunk growth and leaf ion accumulation. *Plant Soil* **2005**, *271*, 265–273. [CrossRef]

61. Munns, R.; Tester, M. Mechanisms salinity tolerance. *Annu. Rev. Plant Biol.* **2008**, *59*, 651–681. [CrossRef]

62. Chrysargyris, A.; Michailidi, E.; Tzortzakis, N. Physiological and biochemical responses of *Lavandula angustifolia* to salinity under mineral foliar application. *Front. Plant Sci.* **2018**, *9*, 1–23. [CrossRef]

63. Ashraf, M.; Ali, Q. Relative membrane permeability and activities of some antioxidant enzymes as the key determinants of salt tolerance in canola (*Brassica napus* L.). *Environ. Exp. Bot.* **2008**, *63*, 266–273. [CrossRef]

64. Tarchoune, I.; Sgherri, C.; Izzo, R.; Lachaal, M.; Navari-Izzo, F.; Ouerghi, Z. Changes in the antioxidative systems of *Ocimum basilicum* L. (cv. Fine) under different sodium salts. *Acta Physiol. Plant.* **2012**, *34*, 1873–1881. [CrossRef]

Article

Growth and Physicochemical Changes of *Carpinus betulus* L. Influenced by Salinity Treatments

Qi Zhou [1,2], Zunling Zhu [1,2,3,*], Man Shi [1,2] and Longxia Cheng [1,2]

1   Co-Innovation Center for Sustainable Forestry in Southern China, Nanjing Forestry University, Nanjing 210037, China; zhouqi514@njfu.edu.cn (Q.Z.); shiman1031@126.com (M.S.); zwahzchenglx@163.com (L.C.)
2   College of Landscape Architecture, Nanjing Forestry University, Nanjing 210037, China
3   College of Arts & Design, Nanjing Forestry University, Nanjing 210037, China
*   Correspondence: zhuzunling@njfu.edu.cn; Tel.: +86-025-6963-8089

Received: 7 May 2018; Accepted: 12 June 2018; Published: 14 June 2018

Abstract: *Carpinus betulus* L. is a deciduous tree widely distributed in Europe with strong adaptation, and it plays a key role in landscaping and timbering because of its variety of colors and shapes. Recently introduced to China for similar purposes, this species needs further study as to its physiological adaptability under various soil salinity conditions. In this study, the growth and physicochemical changes of *C. betulus* seedlings cultivated in soil under six different levels of salinity stress (NaCl: 0, 17, 34, 51, 68, and 85 mM) were studied for 14, 28 and 42 days. The plant growth and gas exchange parameters were not changed much by 17 and 34 mM NaCl, but they were significantly affected after treatments with 51 ~ 85 mM NaCl. The chlorophyll content was not significantly affected at 17 and 34 mM salinity, and the relative water content, malondialdehyde content and cell membrane stability of *C. betulus* did not change obviously under the 17 and 34 mM treatments, indicating that *C. betulus* is able to adapt to low-salinity conditions. The amount of osmotic adjustment substances and the antioxidant enzyme activity of *C. betulus* increased after 14 and 28 days and then decreased with increasing salinity gradients, but the proline content was increased during the entire time for different salinities. The Na content of different organs increased in response to salinity, and the K/Na, Ca/Na, and Mg/Na ratios were significantly affected by salinity. These results suggest that the ability of *C. betulus* to synthesize osmotic substances and enzymatic antioxidants may be impaired under severe saline conditions (68 ~ 85 mM NaCl) but that it can tolerate and accumulate salt at low salinity concentrations (17 ~ 34 mM NaCl). Such information is useful for land managers considering introducing this species to sites with various soil salinity conditions.

Keywords: salinity; *Carpinus betulus*; morphological indices; gas exchange; osmotic adjustment substances; antioxidant enzyme activity; ion relationships

---

1. Introduction

Salinity stress is one of the main abiotic stresses affecting the growth of plants. Due to the influence of human population growth, increased industrial pollution, and improper irrigation, the existing salinized land in the world covers approximately 9.5×10^8 hm^2, approximately 22% of agricultural land worldwide [1,2]. The area of saline-alkaline land in China is approximately 9.9×10^7 hm^2, which accounts for 25% of the arable land in China [3]. Soil salinization has caused problems over vast areas in the world. In addition to soil salinity, de-icing salts have become a serious constraint for plant growth, particularly in cities [4]. Although de-icing salts help to keep pavements dry and safe during ice and snow, their extensive use can cause damage to plants along sidewalks, walkways, and driveways. Trees and shrubs can be injured by the dissolved salt that spreads into the soil. High levels of salinity will negatively affect the morphology, photosynthesis, metabolism, and physiological and biochemical processes of plants [5,6].

To resist salt stress, plants have evolved complex mechanisms to adapt to the living environment [7]. The mechanisms include osmotic adjustments by the accumulation of compatible solutes (proline, betaine, polyols, and soluble sugars), scavenging the reactive oxygen species (ROS) by increasing the activity of antioxidant enzymes, such as superoxide dismutase (SOD), peroxidase (POD), and catalase (CAT), and maintaining a balance of intracellular ions by lowering the toxic concentration of ions in the cytoplasm. In addition to these responses, some plants can also change their morphological structure to adjust to saline conditions [8]. Plant species do vary in their sensitivity to salt damage [9,10]. Therefore, research of the changes in the morphology, physiological and biochemical processes of the plant in relation to the various salinity levels of the environment, the mechanisms involved in the plant response to salt stress, and the identification of trees more tolerant to salt for planting purposes, are of great significance [11].

Carpinus betulus L. (European hornbeam) is a well-known deciduous tree that originated in central Europe and Asia Minor as a dominant species in the forest canopy and has recently been introduced to China [12]. *C. betulus* is long-lived and has strong wood, which can tolerant a wide range of soil conditions, from coarse sand to clay, as well as acidic or alkaline soil pH levels [13,14]. These trees are very important landscaping trees in private and public green areas due to their rustic nature, beautiful shapes and strong adaptability, and they can readily be found in urban parks, gardens and along roadsides [15,16]. The wood of *C. betulus* is suitable for making pianos, violins, joinery, flooring, batons, pulleys, wooden gears and so on [14]. Recent studies have found remarkable antioxidant- and anticancer-related properties of *C. betulus* leaf extracts, making it a possible raw material for medicine [17]. The high ornamental and economic value of *C. betulus* makes it extremely popular all over the world. A considerable number of studies have been conducted related to *C. betulus*, including studies on breeding [18], seed biology [19], hybridization [20], heat resistance [21] and drought tolerance [22,23]. However, information on the response of *C. betulus* to salinity is scarce. Whether *C. betulus* can grow well in saline areas in China still remains unknown [24]. Therefore, the main objective of the study is to make a comprehensive assessment of the influence of salinity stress on *C. betulus* and to establish the mechanisms of adaptation it employs to tolerate salinity stress. We tested the hypothesis that the growth, biomass accumulation and leaf gas exchanges of the seedlings will decrease under the effect of salinity. We examine the growth and physicochemical changes of the seedlings in response to salinity stress to provide scientific data and findings related to the cultivation of *C. betulus* and to make better use of Carpinus species in landscaping.

2. Materials and Methods

2.1. Plant Material and Growth Conditions

The trials were conducted in the landscape experimental teaching center of Nanjing Forestry University in April 2013, in Nanjing (33°04′ N, 118°47′ E), Jiangsu Province, China, in a warm and humid subtropical monsoon climate, with annual rainfall of 1047 mm. The average annual, maximum and minimum temperatures are 15.7 °C, 40.7 °C, and −14 °C, respectively.

The *C. betulus* seeds were obtained from Hungary (imported by the China National Tree Seed Corporation). The seeds were treated with variable temperature stratification (23 °C, 30 days, then 5 °C, 4 months in moist sand) in November 2011 and sown in containers during March of 2012 to initiate growth. During March 2013, we selected well-grown seedlings, approximately 0.5 cm in diameter at ground level with 20-cm tall stems, and transplanted them into pots (10 cm in diameter and 15 cm tall) for uniform management. The experimental loamy clay soil was a mixture of equal amounts of soil, peat, vermiculite, and pearlite and had an acidic pH of 6.5. TFW-VI soil nutrient and moisture tester (TFW-VI, Wuhan, China) was used to determine the content of soil nutrients. The soil nutrient status was 32.05 mg g^{-1} total N, 12.15 ppm available P, 145 ppm available K, 687 ppm Ca, and 260 ppm Mg. Each pot had 500 g soil and one seedling. The potted plants received 0.5 L of Hoagland's nutrient solution at 2-week intervals. The potted plants were kept in a greenhouse under natural sunlight conditions at a temperature of 25 °C and relative humidity of 75%.

2.2. Salt Treatments and Experimental Design

Salt treatments were carried out when the seedlings were approximately 20 cm long with similar leaf number and leaf area, in April 2013. Each treatment had three replicates, and each replicate consisted of 25 random basins, with one seedling per pot. The pots were subjected to S0 (0 mM, control) and S1 ~ S5 (17, 34, 51, 68, and 85 mM) NaCl concentrations. Different salinity levels (S1 ~ S5) were developed by dissolving sodium chloride (NaCl) in distilled water as g L^{-1} and then recalculated into mM. The electrical conductivity of the substrates was 1.6 (S0), 3.3 (S1), 4.4 (S2), 6.2 (S3), 8.4 (S4) and 10.5 (S5) dS m^{-1}, respectively. In each pot, 200 mL of treatment water was applied based on the requirements for each treatment. To avoid osmotic shock, salt solutions were added in three equal parts on alternating days until the expected concentration was reached. Standard agronomic practices were adopted, and tree protection measures were carried out as necessary. To ensure the accuracy of the experimental design, we put a plastic tray at the bottom of the pot to retain any overflow solution, and this was placed back into the pot. The plants were monitored under the treatments for 42 days. After 14, 28 and 42 days of growth, measurements were carried out and samples collected for various physiological analyses, with three replicates for the measured parameters. Whole plants were harvested after 42 days of treatment.

2.3. Growth Parameters

Before the salt stress treatment, three seedlings were selected in each group to calculate their seedling height (H_0), diameter (D_0) and leaf number (L_0). At the final day of treatment, the height (H_1), diameter (D_1) and leaf number (L_1) were determined again. Height growth = $H_1 - H_0$; diameter growth = $D_1 - D_0$; leaf number increment = $L_1 - L_0$. After the treatment of 42 days of salt stress, the plants were separated into roots, stems and leaves, the leaf number was calculated, and the root length was measured; then, the samples were dried at 80 °C in an oven for 48 h, and their dry weights were recorded. The leaf area of each plant was determined using a portable leaf area meter (LI-3000C, LI-COR, Lincoln, NE, USA). Portions of these samples were frozen in liquid nitrogen stored at −80 °C.

2.4. Leaf Stomatal and Section Characteristics

Leaves of *C. Betulus* seedlings were collected at the final day of salinity treatment. The properly cleaned samples were cut into small pieces (approximately 5 × 5 mm) with a sharp blade. The excised leaves were placed in formalin acetic acid (FAA), dehydrated in a graded ethanol series of 30%, 50%, 70%, 90%, 95%, and 100% for 30 min each, penetrated with isoamyl acetate aldehyde, and dried in a critical point drying apparatus (K850, Emitech, London, UK). The tissues were mounted on stubs and coated with gold using an ion sputtering apparatus (E1010, Hitachi, Tokyo, Japan); then, the blade surface and cross-section were viewed under a scanning electron microscope (Quanta 200, FEI, Hillsborough, OR, USA), and the images were taken using the same instrument.

The stomatal density was calculated by the number of stomata per mm^2, and the stomatal size was reflected by the length between the junctions of the guard cells of each stoma. The number of open stomata, leaf thickness and palisade tissue thickness for each sample was measured under a photomicroscope system with a computer attachment.

2.5. Leaf Gas Exchanges and Chlorophyll Fluorescence Parameters

Photosynthetic parameters were determined in the third fully expanded leaves of each plant at the final day of salt treatments using a portable photosynthetic system (Ciras-2, Shanghai, China). The temperature was 25 °C, the intensity of the light was 1000 $\mu mol\ m^{-2}s^{-1}$, and the concentration of reference CO_2 was 380 $\mu mol\ mol^{-1}$. The parameters recorded included the net photosynthetic rate (Pn), transpiration rate (Tr), stomatal conductance (Gs), and intercellular CO_2 concentration (Ci). The water use efficiency (WUE_i) was calculated as follows: $WUE_i = Pn/Tr$. Chlorophyll fluorescence parameters were determined by the fluorescence leaf chamber of the Ciras-2 photosynthetic system.

The photochemical quantum efficiency (Φ_{PSII}), maximum photochemical efficiency of photosystem II (*Fv/Fm*), photochemical quenching parameter (qP), non-photochemical quenching parameter (NPQ), and electron transfer rate (ETR) were measured on dark-adapted leaves. Three seedlings were randomly determined for each treatment, and each seedling was measured 3 times.

2.6. Chlorophyll Content and Relative Water Content (RWC)

For the chlorophyll content of the leaves, chlorophyll was extracted from 4 ~ 5 pieces of fresh leaves with a mixture of acetone and 95% ethanol (*v:v* = 1:1), and the absorbance of the resulting extracts was measured by ultraviolet-visible spectrophotometry (Lambda25, PerkinElmer, Waltham, MA, USA) at wavelengths of 645 nm and 663 nm. The chlorophyll concentration (mg g^{-1}) was then calculated following the method of Porra et al. (1989) [25].

The relative water content (RWC) was measured according to Vivekanandan (2004) [26]. The fresh leaf tissues were weighed to obtain the fresh weight (W_1); then, they were placed in water and incubated at room temperature for 24 h to absorb water to reach a saturated state. The leaves were later taken out, blotted dry, and weighed to obtain the turgid weight (W_2). Finally, the tissues were dried at 80 °C for 24 h and weighed to obtain the dry weight (W_3). The relative water content (RWC) was then calculated according to the following formula:

$$RWC = (W_1 - W_3/W_2 - W_3) \times 100\% \tag{1}$$

2.7. Malondialdehyde Concentration (MDA) and Cell Membrane Stability

MDA content was measured as reported by Hodges et al. (1999) [27]. First, 0.3 g of leaves were ground at 4 °C in a mortar in 6 mL of a pH 7.8 phosphate buffer solution. The homogenate was centrifuged at 9000 rpm at 4 °C for 20 min. Then, 2 mL of an enzyme solution were homogenized in 3 mL of 0.5% thiobarbituric acid and 8% trichloroacetic acid, and the extracts were incubated at 100 °C for 20 min. After a brief passage in ice, the samples were centrifuged at 9000 rpm for 30 min at 4 °C, and the absorbance of the supernatant was measured at 532 and 600 nm.

The cell membrane stability was measured by the relative electrolytic conductivity (REC): 0.2 g sample of fresh leaves was rinsed three times with deionized water and incubated in hermetic tubes containing 20 mL deionized water for 5 h at 25 °C. The electrical conductivity of the leaf solution (S_1) was determined with a conductivity meter (DDS-307, Shanghai, China). Then, the tubes containing leaf samples were put in boiling water at 100 °C for 20 min to determine the electrical conductivity after release of all electrolytes (S_2). Leaf electrolyte leakage was determined according to Dionisio-Sese and Tobita (1998) [28] and calculated as follows:

$$REC = (S_1/S_2) \times 100\% \tag{2}$$

2.8. Soluble Sugars, Soluble Proteins, and Proline

Leaf soluble sugars content was measured by the method of Magné et al. (2006) [29]. First, 0.1 g of leaves was incubated in hermetic tubes containing 10 mL of deionized water, placed in boiling water for 30 min, cooled and filtered into a 25-mL volumetric flask that was then filled with distilled water. The reaction mixtures contained 0.5 mL of extract, 1.5 mL of distilled water, 0.5 mL of anthrone ethyl acetate and 5 mL of 98% sulfuric acid. Each tube was thoroughly shaken with the oscillator and immediately put into boiling water for 1 min. The absorbance of the supernatant was measured at 630 nm after cooling. Then, the value was put into glucose standard curve to find the appropriate content.

The content of soluble proteins was measured by the method of Coomassie brilliant blue G-250 staining. The reaction mixtures contained 0.1 mL of enzyme solution and 5 mL of Coomassie brilliant blue G-250 solution. The absorbance of the supernatant was measured at 595 nm after standing for 5 min. The leaf soluble proteins content was determined in the supernatant according to the method of Bradford (1976) [30] using bovine serum albumin (BSA) as the standard.

Proline was extracted and determined using the method of Steinert et al. (1990) [31]: Samples were homogenized with 3% sulfosalicylic acid and then centrifuged at 4000 rpm at 4 °C for 20 min. The supernatant was treated with acetic acid and acid ninhydrin and boiled in water for 1 h; then, the absorbance was measured at 520 nm. Proline was used to generate a standard curve.

2.9. Measurement of Antioxidant Enzyme Activities

Frozen leaves weighing 0.3 g were ground at 4 °C in a mortar in 6 mL of a pH 7.8 phosphate buffer solution. The homogenate was centrifuged at 9000 rpm at 4 °C for 20 min. The supernatant was collected as a crude enzyme extract for enzyme measurements and stored at 4 °C.

The SOD activity was analyzed according to Beyer et al. (1987) [32] with some modifications. The reaction mixture contained 1.5 mL of sodium phosphate buffer, 0.1 mL of enzyme solution, 0.3 mL of 130 mmol L^{-1} methionine, 0.3 mL of 20 μmol L^{-1} riboflavin, 0.3 mL of 100 μmol L^{-1} EDTA, 0.3 mL of 750 μmol L^{-1} nitroblue tetrazolium (NBT), and 0.5 mL of distilled water. The reaction was started by exposing the mixture to white fluorescent light for 15 min, and reduced NBT (blue color) was measured at 560 nm such that one unit of SOD activity caused 50% inhibition of NBT reduction per min.

The POD activity was determined according to Civello et al. (1995) [33]. The reaction mixtures contained 1 mL of enzyme solution, 3.8 mL of 0.3% guaiacol reaction solution, and 0.1 mL of 3% H_2O_2. The increase in absorbance (tetraguaiacol formation) was recorded at 470 nm every 1 min 3 times.

2.10. Measurements of the Sodium (Na), Potassium (K), Calcium (Ca), and Magnesium (Mg) Content

The roots, stems and leaves of seedlings were separately dried at 85 °C for 48 h. The dried samples were later digested with concentrated HNO_3 acid and $HClO_4$ acid (5:1 *v:v*). The Na, K, Ca, and Mg ion concentrations in the digested samples were measured using a plasma emission spectrometer (OPTIMA PE-4300DV, Waltham, MA, USA). Each test was repeated three times, taking the average.

2.11. Statistical Analysis

All data were analyzed by calculating the means and standard deviation (SD), using the one-way ANOVA, and the means were separated with Duncan's multiple range test at the 5% probability level using the SPSS statistical package version 22.0 (IBM Corp, Amonk, NY, USA).

3. Results

3.1. Changes in Growth under Increasing Levels of Soil Salinity

Growth inhibition and biomass reduction are the most sensitive physiological responses of plants under salt stress. The results showed that (Table 1) the reductions in growth became more pronounced with the increasing levels of salt in the soil; however, the root length, leaf area, dry weight of leaves were slightly higher under low salinity conditions (S1) than those in the control plants, which indicated that moderate (S3) and high (S4, S5) salt concentrations had obvious inhibitory effects on *C. betulus* but that the low concentration of salt treatment (S1) may promote the growth of the roots and leaves. The root/shoot ratio continued increasing with soil salinity, and the ratio of the treatment group was significantly higher than that of the control group.

Table 1. Effect of increasing levels of soil salinity on height growth, diameter growth, root length, leaf area, leaf number increment, dry weight of roots, stems and leaves, root/shoot ratio (R/S) and total biomass of *Carpinus betulus* L. after 42 days. S0 ~ S5 represents 0 ~ 85 mM NaCl respectively. The data in the table are the mean $\pm$ standard deviation (n = 3); different lowercase letters in each column indicate significant differences between treatments ($p < 0.05$).

Treatment	Height Growth (cm)	Diameter Growth (cm)	Root Length (cm)	Leaf Area (cm^2)	Leaf Number Increment
S0	4.89 $\pm$ 0.40 a	0.07 $\pm$ 0.008 a	18.55 $\pm$ 1.90 ab	10.48 $\pm$ 0.57 a	16.33 $\pm$ 3.06 a
S1	4.10 $\pm$ 0.72 ab	0.068 $\pm$ 0.004 a	20.04 $\pm$ 1.11 a	11.00 $\pm$ 0.98 a	14.33 $\pm$ 2.08 ab
S2	3.80 $\pm$ 0.60 b	0.058 $\pm$ 0.005 b	16.79 $\pm$ 0.80 bc	9.64 $\pm$ 1.52 ab	11.00 $\pm$ 2.65 bc
S3	2.50 $\pm$ 0.44 c	0.047 $\pm$ 0.005 c	15.23 $\pm$ 1.18 c	8.74 $\pm$ 1.08 abc	9.00 $\pm$ 2.00 cd
S4	1.97 $\pm$ 0.25 c	0.043 $\pm$ 0.006 cd	12.35 $\pm$ 1.16 d	7.52 $\pm$ 1.45 bc	7.67 $\pm$ 2.08 cd
S5	1.73 $\pm$ 0.25 c	0.037 $\pm$ 0.004 d	10.29 $\pm$ 1.56 d	6.81 $\pm$ 1.55 c	6.00 $\pm$ 1.00 d

Treatment	Root (DW, g)	Stem (DW, g)	Leaf (DW, g)	Root/Shoot Ratio (R/S)	Total Biomass (DW, g)
S0	0.16 $\pm$ 0.006 a	0.23 $\pm$ 0.008 a	0.27 $\pm$ 0.008 a	0.314 $\pm$ 0.002 e	0.66 $\pm$ 0.02 a
S1	0.16 $\pm$ 0.007 a	0.18 $\pm$ 0.012 b	0.28 $\pm$ 0.007 a	0.363 $\pm$ 0.005 d	0.62 $\pm$ 0.03 b
S2	0.14 $\pm$ 0.005 b	0.16 $\pm$ 0.007 c	0.21 $\pm$ 0.006 b	0.383 $\pm$ 0.002 c	0.52 $\pm$ 0.02 c
S3	0.13 $\pm$ 0.007 c	0.15 $\pm$ 0.010 c	0.16 $\pm$ 0.011 c	0.423 $\pm$ 0.006 b	0.43 $\pm$ 0.03 d
S4	0.12 $\pm$ 0.005 cd	0.13 $\pm$ 0.006 d	0.12 $\pm$ 0.008 d	0.476 $\pm$ 0.008 a	0.37 $\pm$ 0.02 e
S5	0.11 $\pm$ 0.006 d	0.11 $\pm$ 0.007 e	0.11 $\pm$ 0.007 d	0.482 $\pm$ 0.003 a	0.33 $\pm$ 0.02 e

3.2. Stomata Density, Size, and Leaf Section Characteristics

At the low salinity concentration (S1), the leaf stomata and leaf section characteristics of the *C. betulus* leaves were not significantly different from those with the S0 treatment (Table 2). However, with the increased salinity concentrations (S2 ~ S5), the stomata density of leaves decreased gradually. At a salt concentration of 85 mM, the stomata density was 65% of that of the control. Opened stomata and leaf thickness decreased with the increasing salinity. Only 17.66% open stomata were observed with the salinity concentration of 85 mM. The stomata shape, palisade and parenchyma tissues of *C. betulus* seedlings had different degrees of change under various levels of salinity (Figure 1a,b). The opening degree of the stomata decreased with the increase in salt stress, and the stomata formed only a thin seam at a salt concentration of S3. At S4, the guard cells were deformed, and the epidermal tissue was atrophied. In the S5 treatment, the stomata were nearly closed. The palisade tissue of *C. betulus* was elongated and arranged orderly in the control (Figure 1b, S0). The difference of leaf section characteristics under S1 and S2 salinity treatments was insignificant with the control, so we didn't attach the pictures. However, at the salinity concentrations of the S3–S5 treatments, the palisade and spongy parenchyma tissues were loosely arranged, and part of these tissues had started to shrink.

Table 2. The main effects of salinity on leaf stomata and leaf section characteristics of *C. betulus*. The parameters measured include stomata density, stomata length, opened stomata, leaf thickness, palisade tissue thickness, and palisade tissue thickness/leaf thickness. S0 ~ S5 represents 0 ~ 85 mM NaCl respectively. The data in the table are the mean $\pm$ standard deviation (n = 3); different lowercase letters in each column indicate significant differences between treatments ($p < 0.05$).

Treatment	Stomata Density (Number mm^{-2})	Stomata Length (µm)	Opened Stomata (%)	Leaf Thickness (µm)	Palisade Tissue Thickness (µm)	Palisade Tissue Thickness/Leaf Thickness (%)
S0	193.33 $\pm$ 16.04 ab	14.87 $\pm$ 1.70 a	98.04 $\pm$ 2.49 a	69.33 $\pm$ 4.21 a	19.30 $\pm$ 19.30 ab	27.79 $\pm$ 1.28 ab
S1	215.00 $\pm$ 15.00 a	15.70 $\pm$ 1.75 a	90.66 $\pm$ 2.14 a	67.89 $\pm$ 6.52 ab	20.32 $\pm$ 2.93 a	29.85 $\pm$ 2.01 a
S2	184.67 $\pm$ 19.43 bc	13.50 $\pm$ 1.70 ab	80.70 $\pm$ 1.66 b	64.26 $\pm$ 6.00 ab	19.10 $\pm$ 2.07 ab	29.70 $\pm$ 0.59 a
S3	158.67 $\pm$ 13.01 cd	11.60 $\pm$ 1.28 bc	68.57 $\pm$ 5.42 c	59.47 $\pm$ 5.39 b	16.13 $\pm$ 1.59 bc	27.12 $\pm$ 0.43 bc
S4	147.33 $\pm$ 13.01 de	10.57 $\pm$ 1.66 c	42.87 $\pm$ 2.00 d	57.75 $\pm$ 4.06 b	15.63 $\pm$ 1.95 bc	27.00 $\pm$ 1.52 bc
S5	125.67 $\pm$ 15.63 e	9.17 $\pm$ 1.25 c	17.66 $\pm$ 1.45 e	56.73 $\pm$ 7.91 b	14.13 $\pm$ 1.88 c	24.92 $\pm$ 0.20 c

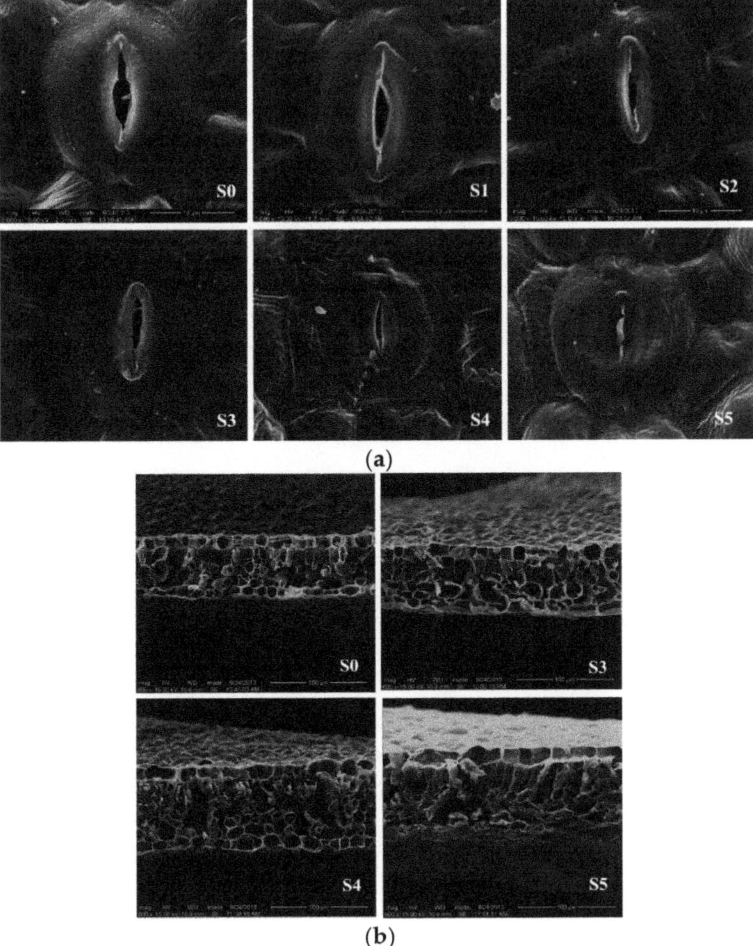

Figure 1. Stomata structure on the leaf surface (**a**), ×5000 and leaf section (**b**), ×600 of *C. betulus* under salinity treatments; S0 ~ S5 represents 0 ~ 85 mM NaCl respectively.

3.3. Leaf Gas Exchanges, and Chlorophyll Fluorescence Parameters

A significant reduction in the net photosynthetic rate was observed after 42 days of the S2 treatment (Table 3). The water-use efficiency (WUE$_i$) and stomatal conductance (*Gs*) of leaves decreased with increasing salt concentrations, with S2 ~ S5 NaCl having the most pronounced effect. The intercellular CO_2 (*Ci*) did not change significantly under S1 NaCl but did increase under the S2 ~ S5 salinity treatments. However, a significant increase in transpiration rates (*Tr*) occurred in the plants treated with S1 and S2 NaCl. Leaf fluorescence parameters, such as Φ_{PSII}, *Fv/Fm*, qP and ETR were not negatively influenced at less than S3 treatment, while only NPQ exhibited the most pronounced effect under all salinity treatments, which increased with the salt concentration.

Table 3. The main effects of salinity on gas exchange and chlorophyll fluorescence parameters of *C. betulus*. The parameters measured include the net photosynthetic rate (*Pn*), transpiration rate (*Tr*), water use efficiency (WUE$_i$), stomatal conductance (*Gs*), intercellular CO_2 concentration (*Ci*), photochemical quantum efficiency (Φ_{PSII}), maximum photochemical efficiency of photosystem II (*Fv/Fm*), photochemical quenching parameter (qP), non-photochemical quenching parameter (NPQ), and the electron transfer rate (ETR). S0 ~ S5 represents 0 ~ 85 mM NaCl respectively. The data in the table are the mean $\pm$ standard deviation ($n = 3$); different lowercase letters in each column indicate significant differences between treatments ($p < 0.05$).

Treatment	*Pn* (μmol m^{-2} s^{-1})	*Tr* (mmol m^{-2} s^{-1})	WUE$_i$	*Gs* (μmol m^{-2} s^{-1})	*Ci* (μmol m^{-2} s^{-1})
S0	6.47 $\pm$ 0.81 a	1.13 $\pm$ 0.09 bc	5.77 $\pm$ 1.02 a	76.67 $\pm$ 9.87 a	243.67 $\pm$ 9.45 e
S1	6.97 $\pm$ 1.12 a	1.38 $\pm$ 0.08 a	5.08 $\pm$ 1.07 ab	70.33 $\pm$ 8.02 ab	234.00 $\pm$ 20.66 e
S2	5.50 $\pm$ 0.70 b	1.28 $\pm$ 0.07 a	4.31 $\pm$ 0.52 bc	60.00 $\pm$ 5.00 bc	268.00 $\pm$ 11.79 d
S3	3.50 $\pm$ 0.30 c	1.11 $\pm$ 0.11 bc	3.16 $\pm$ 0.27 c	52.33 $\pm$ 8.39 cd	296.33 $\pm$ 4.04 c
S4	1.53 $\pm$ 0.50 d	1.00 $\pm$ 0.11 cd	1.52 $\pm$ 0.43 d	51.33 $\pm$ 7.57 cd	355.00 $\pm$ 14.53 b
S5	0.50 $\pm$ 0.20 d	0.88 $\pm$ 0.13 d	0.57 $\pm$ 0.25 d	44.67 $\pm$ 5.13 d	380.00 $\pm$ 13.53 a

Treatment	Φ_{PSII}	*Fv/Fm*	qP	NPQ	ETR (μmol m^{-2} s^{-1})
S0	0.49 $\pm$ 0.02 a	0.63 $\pm$ 0.03 a	1.03 $\pm$ 0.13 a	0.67 $\pm$ 0.04 f	255.48 $\pm$ 17.60 ab
S1	0.53 $\pm$ 0.07 a	0.59 $\pm$ 0.06 ab	0.97 $\pm$ 0.04 ab	0.77 $\pm$ 0.03 e	260.78 $\pm$ 25.76 a
S2	0.51 $\pm$ 0.03 a	0.54 $\pm$ 0.04 abc	0.90 $\pm$ 0.07 abc	0.88 $\pm$ 03 d	254.48 $\pm$ 12.81 ab
S3	0.47 $\pm$ 0.04 a	0.49 $\pm$ 0.09 abc	0.85 $\pm$ 0.10 bc	1.12 $\pm$ 0.04 c	226.42 $\pm$ 13.22 bc
S4	0.39 $\pm$ 0.05 b	0.45 $\pm$ 0.11 c	0.75 $\pm$ 0.10 cd	1.21 $\pm$ 0.03 b	206.02 $\pm$ 15.70 c
S5	0.36 $\pm$ 0.04 b	0.41 $\pm$ 0.06 c	0.60 $\pm$ 0.08 d	1.33 $\pm$ 0.03 a	200.28 $\pm$ 17.71 c

3.4. Chlorophyll Content, and Relative Water Content (RWC)

Chlorophyll is an essential molecule for the capture and transfer of energy in the photosynthetic machinery. The results showed that there were no significant differences between the control and S1 ~ S4 salinity after 14 and 28 days, but after 42 days, a significant reduction in chlorophyll concentrations was found under moderate (S3) and high (S4, S5) salinity treatments (Figure 2a). The water status of a plant is important for maintaining cellular and metabolic functions under salt stress. Under low salt stress (S1), the leaves maintained a higher RWC than those of the control, while a strong decrease was observed with S4 and S5 treatments after 14 days. However, it was significantly reduced after 28 and 42 days under different salt stress (Figure 2b).

3.5. Malondialdehyde Concentration (MDA) and Cell Membrane Stability

With the increased salinity treatments, the MDA content continued increasing, and the low salinity (S1, S2) treatments did not significantly affect the MDA content in leaves after 14 and 28 days (Figure 2c), while after 42 days, a significant increase in the MDA content occurred under S2~S5 salt concentrations. Cell membrane stability remained unchanged after 14 days under NaCl treatments less than S4 in leaves, but a significant increase was observed after 42 days under different treatments (Figure 2d).

3.6. Soluble Sugars, Soluble Proteins, and Proline

The soluble sugars content did not change significantly in leaves under S1 ~ S4 salinity treatments after 14 days of growth but continued increasing after 28 days under various salt stress, while after 42 days of growth, the soluble sugars content in high (S4, S5) salinity treatments was not significantly different than the control as well as plants treated with S1, while it was higher in S2 and S3 treatments (Figure 2e). The soluble proteins content was not significantly affected by salinity treatments after 14 and 28 days, but after 42 days, the content in S1was higher than the control, and a great reduction in protein content was found in S5 treatment (Figure 2f). We observed that the proline content in leaves under S1 ~ S3 salt stress was not obviously increased after 14 days, however, after 28 and 42 days of growth, the proline content was significantly higher than that of the control group under S2 ~ S5 treatments (Figure 2g).

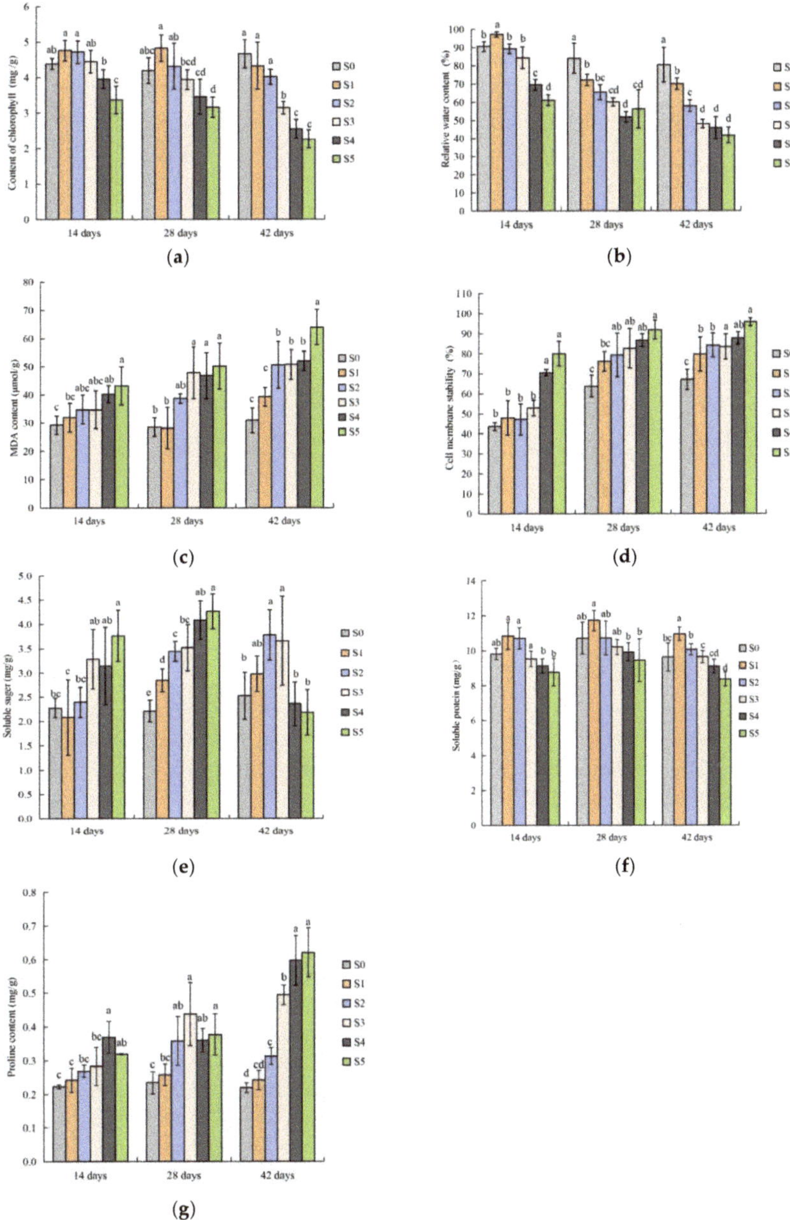

Figure 2. Chlorophyll content (**a**), relative water content (**b**), MDA content (**c**), cell membrane stability (**d**), soluble sugars (**e**), soluble proteins (**f**), and proline content (**g**) in leaves of *C. betulus* plants grown for 14, 28 and 42 days with increasing levels of soil salinity. Different lowercase letters indicate significant differences between treatments ($p < 0.05$), $n = 3$, S0 ~ S5 represents 0 ~ 85 mM NaCl respectively.

3.7. Antioxidant Enzyme Activities

The SOD activity increased at all concentrations of NaCl compared to that of the control plants after 14 and 28 days, as shown in Figure 3a. After 28 days, the SOD activity increased with increasing salt concentrations, with a peak value at S3. However, after 42 days, the SOD activity was higher at low salinity (S1) than the control and was not changed too much under S2 ~ S5 salt stress. Salt stress increased the POD activities in all treatments after 14 and 28 days (Figure 3b), while it was higher in S1 ~ S3, and POD was not obviously affected under high salinity (S4, S5) treatments than the control group after 42 days of stress.

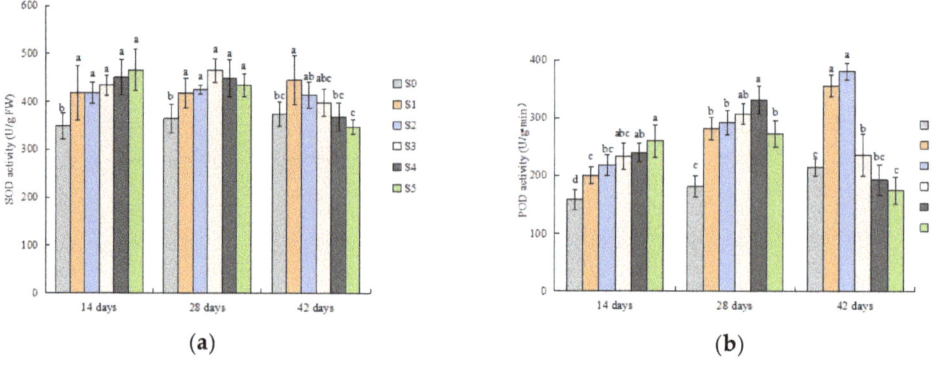

Figure 3. Effect of soil salinity on SOD activity (**a**), and POD activity (**b**) in leaves of *C. betulus* grown for 14, 28 and 42 days with increasing levels of soil salinity. Different lowercase letters indicate significant differences between treatments ($p < 0.05$), $n = 3$, S0 ~ S5 represents 0 ~ 85 mM NaCl respectively.

3.8. Sodium (Na), Potassium (K), Calcium (Ca), and Magnesium (Mg) Content of Various Organs

The Na concentration of the whole plant of *C. betulus* seedlings was increasingly pronounced with the increase in salinity concentration (Table 4). The Na content was significantly increased in stems and leaves under S2 ~ S5 salt treatments, while it was higher than the control in all salinity stress levels in the roots. In particular, the content of Na in the stem was higher than that in the roots and leaves and was significantly different from that of the control, reaching a maximum value at the concentration of S5, which was 20 times that of S0. The K concentration in leaves was significantly higher than that of the control under the salinity treatments, but the K content decreased in the roots. The concentrations of K ions in the roots and stems were lower than those in the leaves under all salt treatments. The Ca content increased in response to the salt treatment in leaves but remained unchanged in the roots and stems with NaCl concentrations less than S3. The Mg content increased significantly in stems, leaves and the whole plants in the S3 ~ S5 treatments, but it decreased in the roots. This result showed that the Ca content remained high in the stems, while the Mg content was the opposite and was higher in roots and leaves. The K/Na, Mg/Na, and Ca/Na ratios decreased significantly with salinity levels, which indicated that with the increased concentration of salt stress, the absorption of Na increased, while the absorption of other nutrients was relatively reduced. We also found that the ratio of each nutrient ion to Na in the leaves was greater than that of the roots and stems, indicating that the relative absorption of the leaves was higher than that of the roots and stems.

Table 4. Effect of different levels of soil salinity on Na, K, Ca and Mg content in roots, stems, and leaves of *C. betulus* seedlings. S0 ~ S5 represents 0 ~ 85 mM NaCl respectively. The data in the table are the mean ± standard deviation ($n = 3$); different lowercase letters in each column indicate significant differences between treatments ($p < 0.05$).

	Treatment	Na (mg g^{-1})	K (mg g^{-1})	Ca (mg g^{-1})	Mg (mg g^{-1})	K/Na	Ca/Na	Mg/Na
Root	S0	1.61 ± 0.05 e	7.92 ± 0.23 a	12.78 ± 0.36 b	4.18 ± 0.14 a	4.92 ± 0.04 a	7.59 ± 0.24 a	2.60 ± 0.11 a
	S1	3.61 ± 0.20 d	6.15 ± 0.28 b	12.43 ± 0.54 b	3.24 ± 0.24 bc	1.71 ± 0.15 b	3.45 ± 0.09 b	0.90 ± 0.02 b
	S2	4.75 ± 0.19 c	4.09 ± 0.16 cd	12.88 ± 0.40 b	2.89 ± 0.13 bc	0.86 ± 0.07 c	2.71 ± 0.07 d	0.61 ± 0.03 c
	S3	6.48 ± 0.27 a	3.85 ± 0.12 d	13.28 ± 0.60 b	2.77 ± 0.31 c	0.59 ± 0.01 d	2.05 ± 0.03 e	0.43 ± 0.05 d
	S4	5.44 ± 0.22 b	4.23 ± 0.08 c	16.29 ± 0.28 a	3.25 ± 0.38 bc	0.78 ± 0.05 c	3.00 ± 0.08 cd	0.60 ± 0.10 c
	S5	4.93 ± 0.25 c	4.39 ± 0.12 c	15.59 ± 1.11 a	3.32 ± 0.30 b	0.89 ± 0.06 c	3.17 ± 0.34 ab	0.67 ± 0.05 c
Stem	S0	0.85 ± 0.06 e	6.14 ± 0.24 b	30.22 ± 0.29 bc	2.18 ± 0.11 b	7.26 ± 0.71 a	35.68 ± 2.65 a	2.57 ± 0.22 a
	S1	1.75 ± 0.04 e	6.22 ± 0.17 b	25.17 ± 1.56 de	2.03 ± 0.09 b	3.56 ± 0.16 b	14.40 ± 1.22 b	1.16 ± 0.03 b
	S2	6.83 ± 0.12 d	3.83 ± 0.04 c	26.97 ± 0.35 de	2.01 ± 0.18 b	0.56 ± 0.01 c	3.95 ± 0.03 c	0.30 ± 0.03 c
	S3	9.83 ± 0.35 c	5.65 ± 0.17 b	33.30 ± 1.04 a	2.80 ± 0.09 a	0.58 ± 0.01 c	3.39 ± 0.04 dc	0.29 ± 0.01 c
	S4	15.87 ± 1.15 b	4.68 ± 1.24 c	31.16 ± 2.03 ab	3.01 ± 0.29 a	0.29 ± 0.06 c	1.96 ± 0.04 dc	0.19 ± 0.02 c
	S5	17.21 ± 0.40 a	7.30 ± 0.24 a	28.15 ± 1.43 cd	3.08 ± 0.18 a	0.42 ± 0.01 c	1.64 ± 0.07 d	0.18 ± 0.01 c
Leaf	S0	0.27 ± 0.03 e	7.25 ± 0.41 c	19.55 ± 0.50 d	3.47 ± 0.09 d	26.79 ± 2.52 a	72.30 ± 7.43 a	12.85 ± 1.35 a
	S1	0.45 ± 0.07 e	11.36 ± 0.22 a	22.01 ± 0.52 de	3.75 ± 0.11 bcd	25.49 ± 3.63 b	49.31 ± 6.25 b	8.38 ± 0.98 b
	S2	3.31 ± 0.64 d	9.22 ± 0.56 b	20.56 ± 1.94 cd	3.68 ± 0.30 cd	2.86 ± 0.57 c	6.36 ± 1.25 c	1.14 ± 0.23 c
	S3	9.08 ± 0.33 a	9.02 ± 0.30 b	24.29 ± 1.01 a	4.63 ± 0.20 a	0.99 ± 0.01 c	2.67 ± 0.02 c	0.51 ± 0.01 c
	S4	6.33 ± 0.07 b	8.60 ± 0.27 b	23.86 ± 0.94 ab	4.01 ± 0.23 bc	1.36 ± 0.04 c	3.77 ± 0.17 c	0.63 ± 0.04 c
	S5	5.92 ± 0.18 b	9.06 ± 0.21 b	24.32 ± 1.37 a	4.12 ± 0.26 b	1.53 ± 0.01 c	4.10 ± 0.11 c	0.70 ± 0.02 c
Whole plant	S0	2.73 ± 0.09 e	21.31 ± 0.55 b	62.54 ± 0.84 b	9.83 ± 0.25 ab	38.97 ± 2.81 a	115.93 ± 3.06 a	18.02 ± 1.19 a
	S1	5.81 ± 0.29 d	23.73 ± 0.54 a	59.61 ± 2.22 b	9.02 ± 0.42 bc	30.76 ± 3.93 b	67.16 ± 4.08 b	10.44 ± 0.94 b
	S2	14.89 ± 0.68 c	17.14 ± 0.70 d	60.41 ± 1.69 b	8.58 ± 0.26 c	4.28 ± 0.58 c	13.02 ± 2.04 d	2.04 ± 0.24 c
	S3	25.39 ± 0.25 b	18.52 ± 0.23 c	70.87 ± 0.58 a	10.21 ± 0.37 a	2.16 ± 0.01 c	8.11 ± 0.87 c	1.22 ± 0.05 c
	S4	27.64 ± 1.17 a	17.51 ± 1.13 cd	71.30 ± 2.72 a	10.27 ± 0.85 a	2.43 ± 0.10 c	8.73 ± 2.89 c	1.42 ± 0.15 c
	S5	28.07 ± 0.53 a	20.75 ± 0.52 b	68.06 ± 3.48 a	10.52 ± 0.53 a	2.85 ± 0.04 c	8.91 ± 3.96 c	1.55 ± 0.08 c

4. Discussion

The salinity stress can produce two kinds of stresses on plants: osmotic stress and ion toxicity [34], which cause the decline or injury of plant growth, photosynthetic ability, cytoplasmic membrane permeability and so on. Plants must be adapted to these two stresses to grow properly in saline conditions. Plants can change their morphological structure or enhance their own resistance by physiological and biochemical transformation to adapt to the adverse environment: for example, the protective effect of the accumulation of osmotic regulating substances and scavenging ROS by antioxidant enzymes. Different salinity stress levels may have various effects on plants, and the physiological and biochemical reactions in response to different salt stress levels of plants also differ.

In this study, we found that the salinity stress caused a prominent inhibition in the growth of *C. betulus* seedlings by the changes of height growth, diameter growth, the increment in leaf number, and dry weight of total biomass. A reduction in plant growth is a common effect of salt stress in other tree species, such as *Morus alba* [35] and *Melia azedarach* [36], and morphological changes and growth reductions provide visual evidence of the degree of injury caused by salt [37]. Salinity stress had a significantly higher inhibitory effect on the aboveground parts of *C. betulus* seedlings than on the belowground parts, which caused a significant increase in the root/shoot ratio. It is noteworthy that the growth parameters were significantly changed under S3 ~ S5 (6.2 ~ 10.5 dS m^{-1}) salt concentrations, which indicates that *C. betulus* cannot endure high salinity. According to Richards (1954) [38], sensitive crops do not thrive as the electrical conductivity of the soil is over 4 dS m^{-1}, and hence we can know that *C. betulus* is sensitive to salt stress.

Salinity stress will change the osmotic potential of roots, causing physiological drought in plants. Stomatae provide a channel for controlling the exchange of water and gas in plants, which has a direct effect on transpiration in plants. The characteristics of stomatae on the surfaces of the leaves, as well as their distribution, density, and open stomata, are greatly influenced by environmental moisture conditions. In this study, the stomata density, size, and leaf section characteristics were unchanged until the S2 treatment, while a significant decrease in those properties was found under the S3 ~ S5 treatments. The open stomatae and the degree of opening in *C. betulus* seedlings decreased with increasing salinity, and the guard cells became deformed and shrank at high salt concentrations; these changes may produce negative effects on the photosynthetic efficiency of plants. Salt stress can lead to the closing of stomatae on leaf surfaces to different degrees and can reduce the size of the stomatal aperture [39]; a similar phenomenon has also been observed in *Fragaria ananassa* [40]. This change may be a physiological adaptation of plants to environmental stress, and our study also confirmed the observation. The leaf thickness, palisade tissue thickness and palisade tissue/leaf thickness ratio of *C. betulus* decreased significantly, the palisade and spongy parenchyma tissues became loosely arranged, and part of the group started to shrink when the salt concentration was over 51 mM. Some research found that the main reason for the reduction of leaf thickness under salt stress was that the length of palisade cells and the thickness of spongy parenchyma tissues decreased, and these changes caused an inhibition of photosynthesis and stunted the growth of seedlings [41]. The degree of the stomatal opening directly determines the transpiration and photosynthetic rates of plants [42]. The closing of stomatae can slow the rate of gas exchange, resulting in a decline in the photosynthetic and transpiration rates. These reasons may explain the reduction in the gas exchange parameters of *C. betulus* under salinity stress.

Soil salinity stress makes it difficult for plants to absorb nutrient elements (N, P), inhibits the synthesis of chlorophyll, accelerates the decomposition of chlorophyll, and reduces the absorption of light energy by chloroplasts. The chlorophyll content of *C. betulus* increased first and then decreased with the degree of salt stress. Some studies have postulated that low salt stimulation can increase the chlorophyll content in plants; for example, NaCl stress (8.5 ~ 34 mM) caused an increase in chlorophyll content in strawberry plants [43], and our study supported this idea. Stomatal regulation was adversely affected in plants under salinity stress [44]. In this study, the reduction of Pn, Tr, and WUE_i under moderate (S3) and high salinity (S4, S5) conditions may have been due to the negative regulation of stomatal conductance, while Ci was increased with the salt stress and negatively correlated with the other photosynthetic indices. It is generally believed that with the reduction of the photosynthetic rate in adversity, including porosity and non-stomatal factors, we can judge the main factor through the change trend of Ci [45], and the two factors can both exist under salt stress. In our study, the reduction of the photosynthetic rate of *C. betulus* contained the two factors. Of the different chlorophyll fluorescence parameters measured in *C. betulus* leaves in this study, we found that the NPQ was increased with salinity concentrations, and this effect may be a mechanism of *C. betulus* to protect the photosystems from damage caused by photoinhibition under salt stress [46]. The leaf chlorophyll fluorescence parameters were not significantly affected when subjected to S1 and S2 NaCl treatments, indicating that the electron transport and photosystems were not impaired, but a NaCl concentration over S3 caused severe damage to the plants, reducing the conversion and utilization efficiency of light energy by chloroplasts and inhibiting photosynthesis.

With the increase in the salt concentration after 28 and 42 days, the RWC of *C. betulus* leaves decreased significantly, which may be closely related to the accumulation of sodium ions in the leaves. Soil salinity can affect soil osmotic potential, which makes it difficult for the plant to absorb water and inhibits the growth of plants. Membrane damage is one of the main negative impacts of salinity stress on plants. The accumulation of MDA content in cells is the result of lipid peroxidation, which is often used to evaluate the membrane damage [47]. Cell membrane permeability can reflect the extent of lipid peroxidation caused by ROS and is also used as an important criterion for salt tolerance in the plants [48]. In this study, the content of MDA and cell membrane stability in the leaves of *C. betulus* seedlings increased as the period of salinity stress continued and reached a maximum at last, indicating that salinity stress caused damage to cell membranes and that the membrane protection system was destroyed under high salt concentrations. To relieve the oxidative stress, most plants can synthesize antioxidant enzymes to counteract the ROS in response to salt stress. SOD and POD can effectively eliminate the ROS and prevent the oxidative damage to the cell membrane [49]. In this study, the activities of the two enzymes increased after 14 and 28 days of different treatments, but a reduction in antioxidant enzyme activities was found at S3 ~ S5 NaCl stress after 42 days. This result was similar to the result of *Quercus variabilis* seedlings [50]. Under low salt concentrations, the activity of protective enzymes in *C. betulus* was increased to scavenge the free radicals, but this ability was limited under moderate and high salt stress due to the decrease in protein synthesis, thus affecting the survival and growth of seedlings.

The ability of osmotic regulation is a key factor for plants to adapt to environmental stress. Osmotic regulation substances mainly include inorganic ions and some soluble organic substances, such as soluble sugars, soluble proteins, and proline [51]. Certain halophyte plants are resistant to salinity up to 200 mM NaCl or more, and can endure salinity by controlling the synthesis of organic 'compatible' solutes [52]. Soluble sugars are the product of photosynthesis, and the accumulation of soluble sugars can regulate the osmotic potential of plant tissue and help improve the resistance of plants to environmental stress. The results showed that the amount of soluble sugars in the *C. betulus* seedlings continuously increased before 28 days but changed insignificantly from that in the control as the salt concentration increased to S4 after 42 days; this result may be because of the decline of photosynthesis under high salt stress. Soluble proteins are materials involved in plant metabolism. The soluble proteins of *C. betulus* were negatively affected by high salinity treatments after 42 days, which was probably caused by the reduction in the assimilation of nitrogen substances or protein proteolysis induced by high salt levels. Proline plays an important role to maintain the osmotic pressure in salt-stressed plants. Reports have verified the function of proline in osmotic adjustment and cell structure protection, and a positive correlation has been observed between proline and tissue sodium content under salinity stress, as in where proline in rice (*Oryza sativa* L.) increased with salinity levels ($4 \sim 12$ dS m^{-1}) [53]. In *C. betulus* seedlings, the proline content continuously increased as the degree of salinity increased after 42 days' stress, which was of positive significance for the plant to adapt to the salt environment.

The separation distribution of ions is one of the important characteristics of plant salt tolerance. The accumulation of Na ions in photosynthetic tissues is extremely deleterious because it will interfere with K functions, including deactivating enzyme activity [54]. The normal activities of many cytosolic enzymes and the maintenance of membrane integrity are mainly decided by Na and K homeostasis in cells. The Na concentration of *C. betulus* significantly increased with the salinity concentration and mainly accumulated in the roots at low salt stress, thus reducing the damage caused by salt ions in the leaves. The K concentration in leaves was significantly higher than that of the control under salinity treatments. The Ca and Mg content changed minimally during the salt treatments. However, the K/Na, Mg/Na and Ca/Na ratios decreased significantly with salinity levels, which was mainly attributed to the increase in Na ions. These results indicated that *C. betulus* plants may be able to maintain ion homeostasis in the plants under low salt stress, while this capability is limited under moderate and high salt stress, causing the decline of biomass and photosynthesis.

5. Conclusions

This study showed that the growth and physicochemical parameters of *C. betulus* seedlings were inhibited under high salinity stress, including reduced gas exchange and increased MDA content, membrane permeability, and sodium accumulation, suggesting a limited degree of salt tolerance. We found that *C. betulus* seedlings can adapt to low salinity stress (S1, S2) by increasing the accumulation or synthesis of key osmotic adjustment substances and improving the activity of antioxidant enzymes and ionic homeostasis. However, the salt tolerance of *C. betulus* is limited under moderate (S3) and high salinity (S4, S5); the plant develops an insufficient resistance to the soil environments and has a difficult time growing. The inhibition of photosynthetic gas exchange, the restriction of the ability to synthesize osmotic adjustment substances and antioxidant enzymes, and the failure to restrict sodium exclusion may be the main reasons for the reduction of growth under high salinity levels for *C. betulus* seedlings. Therefore, *C. betulus* can be planted and popularized in low salinization areas, and its adaptability should be further observed.

Author Contributions: Q.Z. and Z.Z. conceived and designed the experiments; Q.Z., M.S. and L.X. performed the experiments and analyzed the data; and Q.Z. wrote the paper. All authors read and approved the final paper.

Funding: This research was financially supported by the National Natural Science Foundation of China (No. 31770752), the Jiangsu Science and Technology Support Program (BM2013478, BY2015006-01), Jiangsu Province Six Big Talent Peak Project (NY-029), the Fifth Stage Funded Research Projects of 333 in Jiangsu Province and the Excellent Doctoral Dissertation of Nanjing Forestry University.

Acknowledgments: We would like to thank Jing Yang, a laboratory specialist at Advanced Analysis Testing Center (AATC), Nanjing Forestry University, China for the assistance on stomata measurements. Our gratitude to Feibing Wang, research assistant at College of Landscape Architecture, Nanjing Forestry University, China for her help on the management of the seedlings.

Conflicts of Interest: The authors declare no conflict of interest.

References

1. Zhang, J.F.; Zhang, X.D.; Zhou, J.X. World resources of saline soil and main amelioration measures. *Res. Soil Water Conserv.* **2005**, *12*, 32–34. (In Chinese)

2. Bhatnagarmathur, P.; Vadez, V.; Sharma, K.K. Transgenic approaches for abiotic stress tolerance in plants: Retrospect and prospects. *Plant Cell Rep.* **2007**, *27*, 411–424. [CrossRef] [PubMed]

3. Wang, B.S.; Zhao, K.F.; Zou, Q. Advances in mechanism of crop salt tolerance and strategies for raising crop salt tolerance. *Chin. Bull Bot.* **1997**, *14*, 26–31. (In Chinese)

4. Beckerman, J.; Lerner, B. *Salt Damage in Landscape Plants*; Purdue Extension: West Lafayette, IN, USA, 2009; pp. 1–11.

5. Cristiano, G.; Camposeo, S.; Fracchiolla, M.; Vivaldi, G.A.; De Lucia, B.; Cazzato, E. Salinity differentially affects growth and ecophysiology of two mastic tree (*Pistacia lentiscus* L.) accessions. *Forests* **2016**, *7*, 156. [CrossRef]

6. Devitt, D.A.; Morris, R.L.; Fenstermaker, L.K. Foliar damage, spectral reflectance, and tissue ion concentrations of trees sprinkle irrigated with waters of similar salinity but different chemical composition. *HortScience* **2005**, *40*, 819–826.

7. Gupta, B.; Huang, B. Mechanism of salinity tolerance in plants: Physiological, biochemical, and molecular characterization. *Int. J. Genom.* **2014**, 1–18. [CrossRef] [PubMed]

8. Hakim, M.A.; Juraimi, A.S.; Hanafi, M.M.; Mohd, R.I.; Ahmad, S.; Rafii, M.Y.; Latif, M.A. Biochemical and anatomical changes and yield reduction in rice (*Oryza sativa* L.) under varied salinity regimes. *Biomed. Res. Int.* **2014**, 1–11. [CrossRef]

9. Giuseppe, C.; Giuseppe, D.M.; Mariano, F.; Cesare, L.; Vincenzo, T.; Barbara, D.L.; Eugenio, C. Morphological characteristics of different Mastic tree (*Pistacia lentiscus* L.) accessions in response to salt stress under nursery conditions. *J. Plant Sci.* **2016**, *11*, 75–80. [CrossRef]

10. Cassaniti, C.; Romano, D.; Flowers, T.J. The response of ornamental plants to saline irrigation water. In *Irrigation-Water Management, Pollution and Alternative Strategies*; InTech: London, UK, 2012; pp. 131–150. [CrossRef]

11. Hajiboland, R.; Norouzi, F.; Poschenrieder, C. Growth, physiological, biochemical and ionic responses of pistachio seedlings to mild and high salinity. *Trees* **2014**, *28*, 1065–1078. [CrossRef]

12. Shi, M.; Zhou, Q.; Zhu, Z.L. Nutrition physiology of *Carpinus betulus* in different regions. *J. Northwest For. Univ.* **2017**, *32*, 41–45. (In Chinese) [CrossRef]

13. Dirr, M.A. Hornbeams choice plants for American gardens. *Am. Nurserym.* **1978**, *148*, 10–11.

14. Sikkema, R.; Caudullo, G.; Rigo, D.D. *Carpinus betulus* in Europe: Distribution, habitat, usage and threats. In *European Atlas of Forest Tree Species*; Publications Office of the European Union: Luxembourg, 2016; pp. 73–75.

15. Prinz, K.; Finkeldey, R. Characterization and transferability of microsatellite markers developed for *Carpinus betulus* (Betulaceae). *Appl. Plant Sci.* **2015**, *3*, 1500053. [CrossRef] [PubMed]

16. Rocchi, F.; Quaroni, S.; Sardi, P.; Saracchi, M. Studies on anthostoma decipiens involved in *Carpinus betulus* decline. *J. Plant Pathol.* **2010**, *92*, 637–644.

17. Hofmann, T.; Nebehaj, E.; Albert, L. Antioxidant properties and detailed polyphenol profiling of European hornbeam (*Carpinus betulus* L.) leaves by multiple antioxidant capacity assays and high-performance liquid chromatography/multistage electrospray mass spectrometry. *Ind. Crops Prod.* **2016**, *87*, 340–349. [CrossRef]

18. Chalupa, V. Micropropagation of hornbeam (*Carpinus betulus* L.) and ash (*Fraxinus excelsior* L.). *Biol. Plant.* **1990**, *32*, 332–338. [CrossRef]

19. Zhu, Z.L.; Xu, Y.Y.; Wang, S. Structure changes of *Carpinus betulus* seed under fluctuating temperature stratification. *J. Northeast For. Univ.* **2013**, *41*, 1–5. (In Chinese) [CrossRef]

20. Santamour, S.F. Interspecific hybridization in Carpinus. In *Proceedings of the First Conference of the Metropolitan Tree Improvement Alliance (METRIA), Lanham, MD, USA, 27 July 1976 (1978)*; US Forest Service, Northeastern Area, State & Private Forestry: Newtown Square, PA, USA, 1978; pp. 73–79.

21. Shi, M.; Zhu, Z.L. Heat resistance evaluation of different cultivars of *Carpinus betulus* over summering. *J. Northwest For. Univ.* **2015**, *30*, 59–64. (In Chinese) [CrossRef]

22. Stojnić, S.; Pekeč, S.; Kebert, M.; Pilipović, A.; Stojanović, D.; Stojanović, M.; Orlović, S. Drought effects on physiology and biochemistry of Pedunculate Oak (*Quercus robur* L.) and Hornbeam (*Carpinus betulus* L.) saplings grown in urban area of Novi Sad, Serbia. *Southeast Eur. For.* **2016**, *7*, 57–63. [CrossRef]

23. Wang, S.; Zhou, Q.; Zhu, Z.L. Physiological and biochemical characteristics of *Carpinus betulus* seedlings under drought stress. *Acta Bot. Boreal. Occident. Sin.* **2013**, *33*, 2459–2466. (In Chinese)

24. Zhu, Z.L.; Shi, M.; Zhou, Q.; Wu, Y.F.; Yu, W.W.; Zhong, X.L.; Xu, H.Q. Regional adaptability trials for *Carpinus betulus* in China. *J. Zhejiang A F Univ.* **2016**, *33*, 834–840. (In Chinese) [CrossRef]

25. Porr, R.J.; Thompson, W.A.; Kriedemann, P.E. Determination of accurate extinction coefficients and simultaneous equations for assaying chlorophylls a and b extracted with four different solvents: Verification of the concentration of chlorophyll standards by atomic absorption spectroscopy. *Biochim. Biophys. Acta* **1989**, *975*, 384–394. [CrossRef]

26. Vivekanandan, M. Drought-induced responses of photosynthesis and antioxidant metabolism in higher plants. *J. Plant Physiol.* **2004**, *161*, 1189–1202. [CrossRef]

27. Hodges, D.M.; Delong, J.M.; Forney, C.F.; Prange, P.K. Improving the thiobarbituric acid-reactive-substances assay for estimating lipid peroxidation in plant tissues containing anthocyanin and other interfering compounds. *Planta* **1999**, *207*, 604–611. [CrossRef]

28. Dionisio-Sese, M.L.; Tobita, S. Antioxidant responses of rice seedlings to salinity stress. *Plant Sci.* **1998**, *135*, 1–9. [CrossRef]

29. Magné, C.; Saladin, G.; Clément, C. Transient effect of the herbicide flazasulfuron on carbohydrate physiology in *Vitis vinifera* L. *Chemosphere* **2006**, *62*, 650–657. [CrossRef] [PubMed]

30. Bradford, M.M. A rapid and sensitive method for the quantitation of microgram quantities of protein utilizing the principle of protein-dye binding. *Anal. Biochem.* **1976**, *72*, 248–254. [CrossRef]

31. Steinert, A.; Bielka, S. Determination of free proline in stressed plants. *Arch. Züchtungsforsch.* **1990**, *20*, 199–204.

32. Beyer, W.F.; Fridovich, I. Assaying for superoxide dismutase activity: Some large consequences of minor changes in conditions. *Anal. Biochem.* **1987**, *161*, 559–566. [CrossRef]

33. Civello, P.M.; Martinez, G.A.; Chaves, A.R.; Anon, M.C. Peroxidase from strawberry fruit (*Fragaria ananassa* Duch.): Partial purification and determination of some properties. *J. Agric. Food Chem.* **1995**, *43*, 2596–2601. [CrossRef]

34. Hosseini, T.; Shekari, F.; Ghorbanli, M. Effect of salt stress on ion content, proline and antioxidative enzymes of two safflower cultivars (*Carthamus tinctorius* L.). *Lang. Teach. Res.* **2013**, *17*, 282–302.

35. Lu, N.; Luo, Z.; Ke, Y.; Dai, L.; Duan, H.D.; Hou, R.X.; Cui, B.B.; Dou, S.H.; Zhang, Y.D.; Sun, Y.H.; et al. Growth, physiological, biochemical, and ionic responses of *Morus alba* L. seedlings to various salinity levels. *Forests* **2017**, *8*, 488. [CrossRef]

36. Rao, G.G.; Rose, B.V.; Rao, G.R. Salt induced anatomical changes in the leaves of pigeon pea (*Cajanus indicus* Spreng.) and cluster bean (*Cyamopsis tetragonoloba* (L.) Taub.). *Proc. Indian Acad. Sci.* **1979**, *88*, 293–301.

37. Ren, A.X.; Wang, Y.M. Effects of salt stress on stomatal differentiation and movements of amaranth (*Amaranthus tricolor* L.) leaves. *Acta Hortic. Sin.* **2010**, *37*, 479–484. (In Chinese)

38. Richards, L.A. Diagnosis and improvement of saline and alkaline soils. *Soil Sci.* **1954**, *78*, 154. [CrossRef]

39. Wang, B.X.; Zeng, Y.H.; Wang, D.Y.; Zhao, R.; Xu, X. Responses of leaf stomata to environmental stresses in distribution and physiological characteristics. *Agric. Res. Arid Areas* **2010**, *28*, 112–122. (In Chinese)

40. Turhan. Growth and stomatal behaviour of two strawberry cultivars under long-term salinity stress. *Turk. J. Agric. For.* **2007**, *31*, 55–61.

41. Parida, A.K.; Das, A.B.; Mittra, B. Effects of salt on growth, ion accumulation, photosynthesis and leaf anatomy of the mangrove, *Bruguiera parviflora*. *Trees* **2004**, *18*, 167–174. [CrossRef]

42. Hetherington, A.M.; Woodward, F.I. The role of stomata in sensing and driving environmental change. *Nature* **2003**, *424*, 901–908. [CrossRef] [PubMed]

43. Turhan, E.; Eris, A. Changes of growth, amino acids, and ionic composition in strawberry plants under salt stress conditions. *Commun. Soil Sci. Plant Anal.* **2009**, *40*, 3308–3322. [CrossRef]

44. Akram, M.S.; Ashraf, M. Exogenous application of potassium dihydrogen phosphate can alleviate the adverse effects of salt stress on sunflower. *J. Plant Nutr.* **2011**, *34*, 1041–1057. [CrossRef]

45. Ning, D.; Guo, W.H.; Zhang, X.R.; Wang, R.Q. Morphological and physiological responses of *Vitex negundo* L. var. *heterophylla* (Franch.) Rehd. to drought stress. *Acta Physiol. Plant.* **2010**, *32*, 839–848.

46. Müller, P.; Li, X.P.; Niyogi, K.K. Non-photochemical quenching. A response to excess light energy. *Plant Physiol.* **2001**, *125*, 1558–1566. [CrossRef] [PubMed]

47. Ellouzi, H.; Ben, H.K.; Cela, J.; Munné-Bosch, S.; Abdelly, C. Early effects of salt stress on the physiological and oxidative status of *Cakile maritima* (halophyte) and *Arabidopsis thaliana* (glycophyte). *Physiol. Plant.* **2011**, *142*, 128–143. [CrossRef] [PubMed]

48. Ashraf, M.; Ali, Q. Relative membrane permeability and activities of some antioxidant enzymes as the key determinants of salt tolerance in canola (*Brassica napus* L.). *Environ. Exp. Bot.* **2008**, *63*, 266–273. [CrossRef]

49. Sreenivasulu, N.; Grimm, B.; Wobus, U.; Weschke, W. Differential response of antioxidant compounds to salinity stress in salt-tolerant and salt-sensitive seedlings of foxtail millet (*Setaria italica*). *Physiol. Plant.* **2000**, *109*, 435–442. [CrossRef]

50. Li, Z.P.; Zhang, W.H. Growth and physiological response of *Quercus variabilis* seedlings under NaCl stress. *Acta Bot. Boreal. Occident. Sin.* **2013**, *33*, 1630–1637. (In Chinese)

51. Hamouda, I.; Badri, M.; Mejri, M.; Cruz, C.; Siddique, K.H.M.; Hessini, K. Salt tolerance of *Beta macrocarpa* is associated with efficient osmotic adjustment and increased apoplastic water content. *Plant Biol.* **2015**, *18*, 369–375. [CrossRef] [PubMed]

52. Flowers, T.J.; Colmer, T.D. Salinity tolerance in halophytes. *New Phytol.* **2008**, *179*, 945–963. [CrossRef] [PubMed]

53. Azooz, M.M.; Shaddad, M.A.; Abdel-Latef, A.A.A. The accumulation and compartmentation of proline in relation to salt tolerance of three sorghum cultivars. *Indian J. Plant Physiol.* **2004**, *9*, 1–8.

54. Flowers, T.J.; Munns, R.; Colmer, T.D. Sodium chloride toxicity and the cellular basis of salt tolerance in halophytes. *Ann. Bot.* **2015**, *115*, 419–431. [CrossRef] [PubMed]

MDPI

St. Alban-Anlage 66

4052 Basel

Switzerland

Tel. +41 61 683 77 34

Fax +41 61 302 89 18

www.mdpi.com

Forests Editorial Office

E-mail: forests@mdpi.com

www.mdpi.com/journal/forests